Unsteady Transonic Aerodynamics

Edited by
David Nixon
Nielsen Engineering & Research, Inc.
Mountain View, California

Volume 120
PROGRESS IN
ASTRONAUTICS AND AERONAUTICS

Martin Summerfield, Series Editor-in-Chief
Princeton Combustion Research Laboratories, Inc.
Monmouth Junction, New Jersey

Published by the American Institute of Aeronautics and Astronautics, Inc.
370 L'Enfant Promenade, SW, Washington, DC 20024-2518

American Institute of Aeronautics and Astronautics
Washington, D.C.

Library of Congress Cataloging in Publication Data

Unsteady transonic aerodynamics/edited by David Nixon.
p. cm.—(Progress in astronautics and aeronautics; v.120)
Includes bibliographical references.
1. Unsteady flow (Aerodynamics) 2. Aerodynamics, Transonic.
I. Nixon, David. II. American Institute of Aeronautics and
Astronautics. III. Series.
TL507.P75 vol. 120 629.1 s—dc20 89-15204
[TL574.U5] [629.132′304]
ISBN 0-930403-52-5

Table of Contents

Preface

This book is a complementary volume to *Transonic Aerodynamics* (Vol. 81 of *AIAA Progress in Astronautics and Aeronautics*) that is concerned with steady flow. The reason for the production of the present volume is that there is no other book that addresses the subject of unsteady transonic aerodynamics, a field that is different from steady aerodynamics in many ways. The most pronounced difference is the complex shock wave motions that can occur that are extremely difficult to predict. This and other physical phenomena are described in Chapter 1 by D. Mabey.

Chapter 2 derives a hierarchy of equations that can be used to represent unsteady transonic flow, ranging from massively separated flow to the attached, essentially inviscid, flow. The question of multiple solutions is also addressed. Chapter 3 is a detailed description of how to solve these equations using finite difference techniques; Chapter 4 discusses the use of such techniques in the various computer codes used to predict methods in actual design. Also, Chapter 5 describes some real aeroelastic problems.

Not all unsteady transonic flows are concerned with external airplane applications, and Chapters 6 and 7 describe other interesting problem areas that are frequently more complex than the external problems.

Most of the prediction methods described in this book use finite difference methods, but before these became available many different approximation techniques were used. Many of these are described in Chapter 8 in case some of the ideas may still prove useful, or at least interesting, to engineers today.

Finally, I would like to express my utmost gratitude to the authors, particularly Dennis Mabey and Joe Verdon, who have been extremely patient during the production of this book.

David Nixon
October 1988

Chapter 1

Physical Phenomena Associated with Unsteady Transonic Flows

D. G. Mabey*
Royal Aerospace Establishment, Bedford, United Kingdom

Nomenclature

A = empirical constant; Eq. (7)
a = local velocity of sound
b = reference semichord for flutter calculations
C = tunnel area; Eq. (16)
C_L = lift coefficient
$C_{L_{max}}$ = maximum lift coefficient
C_p = pressure coefficient
$C_{p_{crit}}$ = critical pressure coefficient
$c, \bar{c}$ = local and average chords
f = frequency (Hz)
H = tunnel height
$h, \dot{h}$ = heave displacement and velocity (Fig. 22)
L = bubble length
M = freestream Mach number
M_{crit} = critical Mach number
M_e, M_e^1 = local and effective local Mach numbers
M_1 = Mach number immediately upstream of shock
M_{2D} = equivalent two-dimensional Mach number; Eq. (4)
M_{sc} = Mach number at which shock first reaches trailing edge
M_{te} = local Mach number at trailing edge
$\sqrt{[nF(n)]}, \sqrt{[nG(n)]}$ = tunnel unsteadiness and buffet excitation parameters (defined in Refs. 2 and 49)

*Senior Principal Scientific Officer, Aeronautics Laboratory.

$n = fc/u$ = frequency parameter often preferred for buffeting measurements
p_1, p_2 = static pressures upstream and downstream of shock, respectively
Δ_{p_2} = periodic oscillatory pressure behind shock
$\tilde{p}$ = rms pressure fluctuation
q, q_e = freestream and local kinetic pressures
R = Reynolds number based on $\bar{c}$
R_{xt}, R_{xT} = Reynolds numbers for onset and completion of transition, respectively
r = relaxation factor; Eq. (14)
S = model wing area; Eq. (16)
t = time
U = freestream velocity
U_F = flutter speed
x = streamwise distance from leading edge
x_0 = shock amplitude
α = angle of incidence
α_{2D} = effective angle of incidence
$\beta = \sqrt{(1 - M^2)}$
δ_0, δ_1 = upwash and curvature factors
δ_A, δ_B = control displacements (Fig. 27)
Λ = sweep angle
μ = density ratio of (wing/freestream) for flutter calculations
$\nu = 2\pi fc/U$ = frequency parameter
τ = shear stress
ϕ = phase angle
$\omega = 2\pi f$ = circular frequency (rad/s)
ω_α = first uncoupled torsional frequency for flutter calculations

1. Introduction

This chapter presents a review of the wide range of physical phenomena associated with unsteady transonic flow.

Some of the problems (such as the prediction of attached flow about oscillating control surfaces) will be the subject of a full theoretical treatment using recent developments in computational fluid dynamics (CFD) in Chaps. 3 and 4. The discussion of these flows depends primarily on the pioneering experiments and the analysis of unsteady shock motions provided by Tijdeman.[1] However, other problems (such as the prediction of small-scale random pressure fluctuations or buffet[2] on rigid wings in separated viscous flow) are currently beyond the scope of prediction methods and may remain a challenge to the CFD community for some time.[3] These practical problems are discussed for aircraft in Chap. 5 and for helicopters in Chap. 7.

Local transonic flow can occur over a much wider range of freestream speeds than is generally appreciated (Fig. 1). For hypersonic speeds local

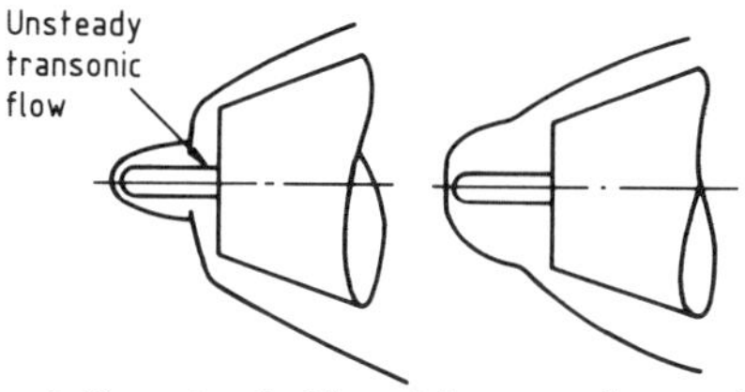

a) Re-entry bodies at hypersonic speeds.

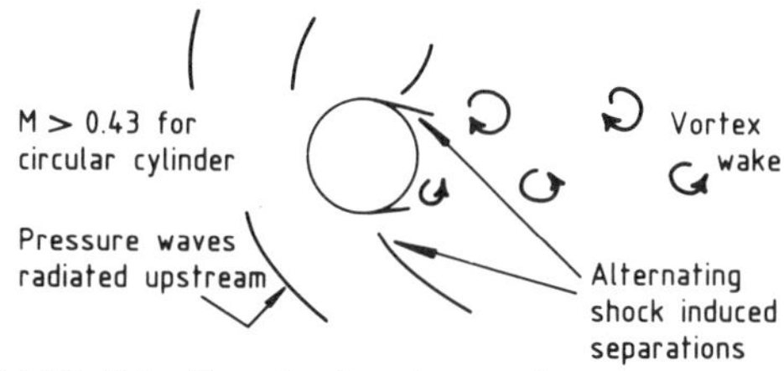

b) Bluff bodies at subsonic speeds.

c) Low-speed stall with leading- and trailing-edge flaps.

Fig. 1 Wide occurrence of unsteady transonic flow.

regions of transonic flow may be formed on the noses of bluff re-entry bodies.[4] With many configurations (e.g., the body sketched in Fig. 1a), these transonic flows will be unsteady and generate high local pressure fluctuations because of the high local kinetic pressures.

Local regions of unsteady transonic flow are found also on bluff bodies at low subsonic speeds. The critical Mach number on a two-dimensional circular cylinder is only about $M = 0.43$. Above that speed the oscillatory separations become associated with unsteady shocks that create flow patterns of great interest and complexity, as observed originally by Naumann and Pfeiffer[5] and sketched in Fig. 1b.

Small regions of transonic flow can occur on many wings close to maximum lift at low speeds, due to the high local suctions in the leading-edge region (Fig. 1c). For a freestream Mach number of $M = 0.2$ the critical pressure coefficient (Ref. 6, p. 28) is $C_{p_{\text{crit}}} = -16$, and this may be reached at quite modest lift coefficients on some sections. If as a consequence the low-speed stall is initiated either by the rapid thickening of the boundary layer far downstream or even by local shock-induced separation, the subsequent steady and unsteady flows may be sensitive to scale effects. This consideration has prevented any serious investigation of scale effects on low-speed stall in unpressurized wind tunnels where Reynolds number can be increased only by increasing freestream Mach number.

Figure 2 illustrates the effects of local regions of transonic flow by recent measurements on a model of an advanced transport aircraft with flaps

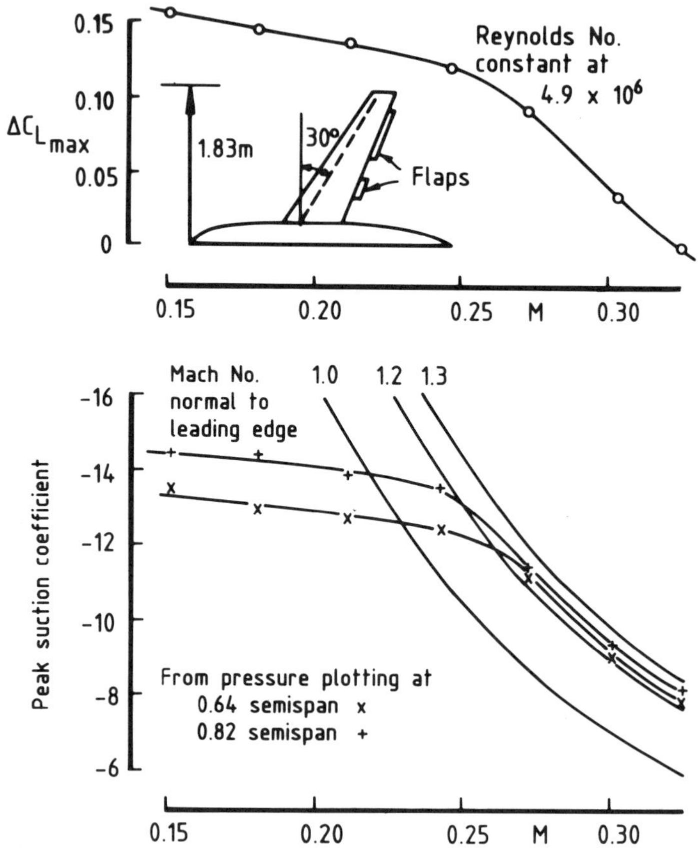

Fig. 2 Mach number effects at constant Reynolds number with flaps deflected. (British crown copyright reserved 1984.)

deflected 25 deg, made in the Royal Aircraft Establishment (RAE) 5-m pressurized low-speed tunnel.[7] For freestream Mach numbers above about $M = 0.25$, the low-speed stall is determined by locally transonic flow on the outboard wing sections. Sometimes these locally transonic flows may be separated and unsteady; then transonic buffeting may prevent the full utilization of the measured (steady) low speed $C_{L_{max}}$. This is an example of the coupling between steady and unsteady flowfields discussed later. After this important digression the remaining sections of this chapter are now outlined. Flexible structures can allow large, unbounded responses to grow from infinitesimally small initial disturbances, if self-exciting forces are generated by the motion. This process, known as flutter, is particularly important at transonic speeds and can occur with both attached and separated flows.[8]

The controversial question of the nature of unsteady tunnel interference at transonic speeds is considered briefly. This question has implications with respect to both routine, time-dependent measurements and to the design and operation of new transonic facilities.

Throughout this chapter the boundary layer at the shock will be generally assumed to be turbulent, for model tests are normally made with fixed transition in an attempt to simulate turbulent boundary layers at full scale.[9] However, interesting effects do occur in unsteady transonic flow with laminar boundary-layer–shock interactions, and these are discussed briefly as a special case of the modification of surface boundary conditions.

2. Development of Unsteady Transonic Flows

Coupling of Steady and Unsteady Flows

For attached inviscid flows that are everywhere completely subsonic or supersonic and where perturbations about the freestream velocity are small, a linear approximation to the equations of motion is adequate. Hence, solutions for steady and time-dependent flows are essentially independent or uncoupled, and combined solutions may be obtained by superposition. This simplification is particularly important for the aeroelastic analysis of aircraft structures.

In contrast, once local transonic flows occur or perturbation velocities become large (as a consequence of strong shocks or large-amplitude motions), the equations of motion become nonlinear. Then the solutions for steady and time-dependent flows are coupled and combined solutions cannot be obtained by superposition. This is an important feature of transonic flows discussed by Tijdeman.[1]

Hence, any description of the character of unsteady transonic flows must be related closely with the description of the corresponding steady flow. Similarly, any experimental investigation of unsteady transonic flows should be related with the corresponding steady flow. In this respect, the ability to measure both steady and unsteady pressures simultaneously from a single transducer[10] confers many advantages, as illustrated by Fig. 10 and Ref. 11.

Two illustrations of the coupling between steady and unsteady transonic flows will be discussed here. The first relates to the flow about a rigid airfoil at zero incidence. The second relates to the prediction of pressures on an oscillating airfoil in inviscid flow. Both of these illustrations relate to symmetric, two-dimensional flow. However, the phenomena described also appear in asymmetric flow (symmetric airfoils at an angle of incidence or supercritical airfoils) and in three-dimensional flows, often only on the upper side. Recent examples are given in Refs. 12–14.

Transonic Flow on a Rigid Airfoil

Figure 3 gives a simplified schematic view of the development of the transonic flow over an airfoil. The thick symmetric airfoil is assumed to be

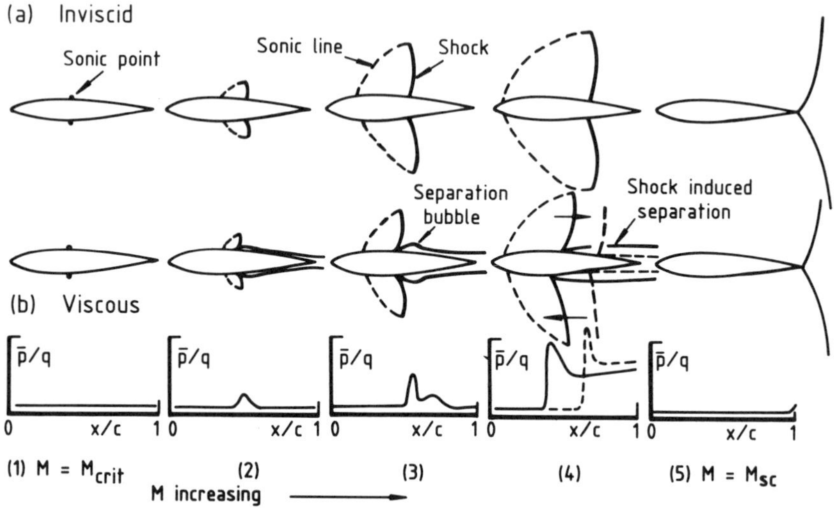

Fig. 3 Transonic flow about an airfoil ($\alpha = 0$): a) Inviscid; b) Viscous.

rigid and is set at zero incidence to simplify the discussion. Five stages of flow development are considered, with Mach number increasing.

For an inviscid flow (Fig. 3a), the local Mach number becomes sonic first at the maximum thickness at the freestream M_{crit} (stage 1). With further increases in freestream Mach number (stages 2–4), the region of supersonic flow extends downstream and away from the airfoil as the shock terminating the supersonic region moves progressively further downstream. When the terminal shock reaches the trailing edge (for $M = M_{sc}$, stage 5), the part of the transonic flow regime causing sudden changes in aerodynamic forces has been traversed, and the pressures reach the "sonic range distribution" (see discussion of Fig. 44 in Ref. 15). Strictly, Fig. 3 should include a low supersonic Mach number, say $M = 1.05$, so that a detached bow shock appears. However, this further increase in speed does not normally alter the character of the flow from that at $M = M_{sc}$. All of the flow changes shown in Fig. 3a occur in a monotonic fashion and without the production of any surface pressure fluctuations.

In marked contrast, in a real, viscous flow, changes occur in a non-monotonic fashion and produce surface pressure fluctuations, even in the absence of flow unsteadiness. The effects of shock–boundary-layer interaction initially thicken the boundary layer and then cause flow separation (Fig. 3b). Considering first the time-mean flow, the critical Mach number is reached close to the maximum thickness, as in the inviscid flow (stage 1). Then as Mach number increases the shock moves more slowly downstream than in the inviscid flow (stage 2), due to the interaction with the boundary layer. (This experimental observation is consistent with theoretical predictions, which show the shock further upstream for viscous flow than for inviscid flow for a given freestream Mach number.) The shock strengthens

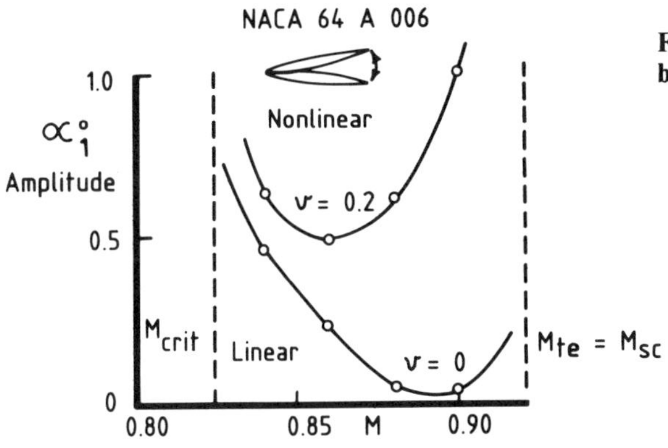

Fig. 4 Linear/nonlinear boundaries.

as it moves downstream and eventually becomes sufficiently strong to separate the boundary layer (stage 3). The M_1 required to provoke shock-induced separation varies with surface curvature and Reynolds number, but a normal shock on a flat plate gives incipient separation at about $M_1 = 1.22$. Initially the flow separation may reattach downstream of the shock for a range of speeds, but with many thick airfoils the separation extends quite rapidly to the trailing edge. Simultaneously the shock generally moves upstream a short distance to maintain compatibility of the pressure rise across the shock and the separated flow between the shock and the trailing edge, which in turn is determined by the wake. Further increases in Mach number cause the shock to move downstream again (stage 4) until it reaches the trailing edge (stage 5) at a freestream Mach number, M_{sc}, somewhat higher than that in the inviscid flow. A classic, detailed description of this process is given by Pearcey in Ref. 15.

The corresponding time-dependent flow, which produces surface pressure fluctuations, is described now. Even an attached turbulent boundary layer produces local pressure fluctuations on the airfoil. For the critical Mach number these are constant across the chord and are very small, close to the rms flat-plate value:

$$\bar{p}/\tau = 2.5 \tag{1}$$

often approximated by

$$\bar{p}/q = 0.006 \tag{2}$$

This excitation is confined to high frequencies and has small length scales relative to the airfoil thickness. When the excitation is integrated in space and time across the airfoil, there is no net integrated unsteady force.

For higher speeds (stages 2–4) the coupling between the steady and time-dependent flows modifies the surface pressure fluctuations in a complicated manner. Initally (stage 2) the thickening of the attached boundary layer and its intermittency causes a small shock oscillation, which in turn modifies the boundary-layer thickness and further increases the local pressure fluctuations. The pressure at the trailing edge provides a link between the top and bottom surfaces of the airfoil. The requirement of constant mean pressure at the trailing edge and the wake development impose the condition that these oscillations are in antiphase on opposite surfaces, as illustrated by recent measurements.[17] The excitation is appreciably higher than that under an attached flat-plate boundary layer and has larger length scales, though it is still small relative to the airfoil thickness. The excitation has a net value when integrated in space and time across the airfoil because of the differing phase of the pressures on the opposite surfaces. In practice this small net value is often masked by flow unsteadiness in transonic facilities (as discussed later) or by atmospheric turbulence in flight.

When the flow first separates (stage 3), there are maxima in the pressure fluctuations at the shock and the reattachment point. When the flow is fully separated from the shock to the trailing edge (stage 4), the pressure fluctuations usually have a well-defined maximum at the shock and may have a weakly defined maximum near the trailing edge. The frequency of the excitation becomes much lower and the length scale becomes much larger, of the order of the thickness of the separated shear layer. It is important to notice that the pressure fluctuations upstream of the mean shock position are small, because these can only propagate upstream across the top of the sonic region. For the speeds represented by stages 3 and 4, there is a rapid increase in the net, integrated, time-dependent force compared to that in stage 2. This increase should not be masked either by flow unsteadiness in transonic facilities (but see discussion of Fig. 32) or atmospheric turbulence in flight.

When the shock reaches the trailing edge (stage 5), the local pressure fluctuations revert to those of an attached boundary layer and are again of low level, high frequency, and small scale. The excitation has no significant net value when integrated in space and time across the chord.

This simplified description shows that the coupling between the steady and time-dependent viscous flows on a rigid airfoil is important in the speed range from

$$M_{\text{crit}} < M < M_{sc} \tag{3}$$

and that the conditions at the trailing edge are crucial. Pearcey's description[15] of the steady flow emphasizes the importance of the mean trailing-edge pressure, the trailing-edge angle, and the radical change in aerodynamic characteristics that occurs once the trailing edge becomes supersonic (stage 5). In addition, Pearcey had suggested earlier[18] that there would be a fair correlation between the onset of significant flow separation (and excitation) and the divergence of the static trailing-edge pressure (see lower portion of Fig. 24). When the separation bubble extends rapidly from the shock to the trailing edge, trailing-edge pressure divergence occurs suddenly. Then, for airfoils there is a fair correspondence with the onset of significant aerodynamic excitation. In contrast, when the separation bubble extends slowly from the shock to the trailing edge, the trailing-edge pressure divergence is relatively slow, and this criterion does not give a good indication of the onset of significant aerodynamic excitation.

Although Fig. 3 is restricted to symmetric nonlifting airfoils, the general concepts can be extended to three-dimensional lifting wings. Thus, an infinite yawed wing (with a Λ) has a two-dimensional equivalent in which the section is the same normal to the leading edge and for which the Mach number and incidence may be derived from the sweep analogy

$$M_{2D} = M \cos\Lambda$$
$$\alpha_{2D} = \alpha \sec\Lambda \qquad (4)$$

For general three-dimensional wings the viscous transonic flow development is much more complex than that represented in Fig. 3 (see, e.g., Fig. 18). The description of the time-mean flow development for thin wings given by Rogers and Hall[19] emphasizes the importance of sonic leading and trailing edges (corresponding approximately to stages 1 and 5, respectively, in Fig. 3b), but there is no corresponding description of the associated time-dependent flow. This deficiency is remedied by some comparisons between the steady and time-dependent pressures measured on a low-aspect-ratio planform (an AGARD SMP standard tail plane). These extensive measurements cover both attached and separated flows.[20]

It is not generally appreciated that the divergence of trailing-edge pressure does not give a consistent indication of separation and unsteady flow boundaries for three-dimensional wings.[2] Ultimately this is because of the difficulty of distinguishing the effects of increased boundary-layer thickness at the trailing edge (due to spanwise drift in this region or local separation and reattachment well upstream) from the local effects of trailing-edge separation.

3. Inviscid Transonic Flow on an Oscillating Airfoil

The description of the development of inviscid transonic flow about an oscillating airfoil given by Dowell et al. in Ref. 21 makes an interesting comparison with that given for the inviscid flow about a rigid airfoil. The airfoil oscillates about its leading edge with a mean incidence of zero and has an NACA 64A-006 section. The calculations are made with a low-frequency approximation to a transonic small perturbation theory (LTRAN 2). The criterion Dowell et al. adopted for linearity was that the shock movement for an amplitude of 0.25 deg was to be restricted to 5% of the chord. This criterion may be contrasted with that suggested by Landahl[22]:

$$\nu >> |1 - M_e|_{max} \qquad (5)$$

For freestream Mach numbers below $M_{crit} = 0.825$, the flow was subsonic everywhere (Fig. 4). Here linear solutions were obtained both for quasisteady ($\nu = 0$) and oscillatory motions. The subsonic, linearized, flat-plate theory generally gives good comparisons with experiments because the corrections due to airfoil thickness and boundary-layer thickness are of roughly equal magnitudes and opposite signs. (This is illustrated in the discussion of Fig. 13.3 in Ref. 1). For Mach numbers above $M = M_{sc} = 0.920$, the shock terminating the local region of supersonic flow reached the trailing edge, and linear solutions were obtained again for both quasisteady and oscillatory motions.

For intermediate Mach numbers

$$M_{\text{crit}} < M < M_{sc}$$

there was a small linear region for quasisteady flow and a much larger linear region for oscillatory flow. For inviscid transonic flow Nixon[23] has shown that, although large changes in shock motion and phase occur at the low-frequency parameters appropriate to rigid body modes (say $\nu = 0.02$), frequency effects stabilize for frequency parameters appropriate to structural modes (say as low as $\nu = 0.2$). This is consistent with the much larger linear region for $\nu = 0.2$ compared with $\nu = 0$ in Fig. 4.

The changes in oscillatory flow in Fig. 4 may be compared to the corresponding changes in quasisteady lift-curve slope shown in Fig. 5. This includes predictions according to classical subsonic theory, the LTRAN 2 theory, the Garabedian and Korn full potential theory, and a local linearization theory using M_{te}. The Garabedian and Korn theory offers realistic predictions throughout the transonic Mach number range. In particular, it illustrates the large increase in lift-curve slope, which is a significant factor in the "transonic flutter dip" discussed later.

Note on the Nonlinearities in the Chordwise Variation of Pressures

Dowell et al.[21] considered the nonlinearities on the integrated forces on the oscillating airfoil. A rather different picture of the nonlinearities would have been obtained by considering the chordwise time-dependent pressures.

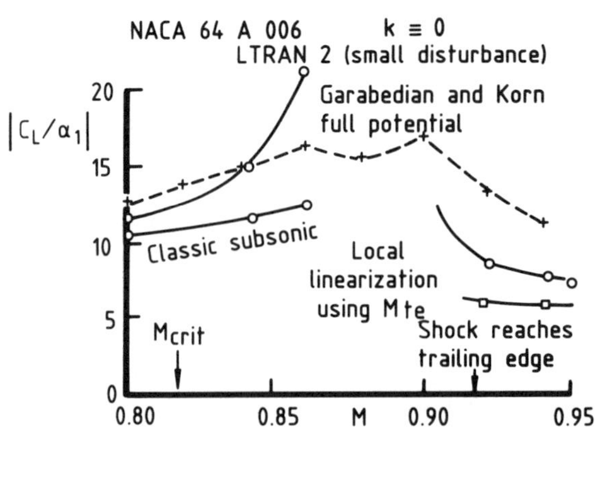

Fig. 5 Lift curve slope vs freestream Mach number.

As a first approximation for an inviscid flow, nonlinearities are concentrated in the vicinity of the sonic point close to the leading edge and in the vicinity of the shock. In contrast, elsewhere the local pressures are generally linear. This may be seen from Fig. 10 of Ref. 24, which shows the first three harmonics of the pressure developed in an NLR airfoil with an oscillating flap. The second and third harmonics are only significant in the vicinity of the shock and when integrated have a negligible effect on the lift, as discussed in Ref. 1, p. 89. McOwat et al.[25] give an interesting illustration of the radical chordwise changes in the character of the time-dependent pressures.

Figure 6 shows typical time-dependent pressure measurements on one section of RAE wing A with an oscillating trailing-edge flap. The freestream Mach numbers are $M = 0.85$, 0.90, and 0.95, and the steady flows were judged subcritical, critical, and supercritical, respectively. Figures 6a–6c show the chordwise variations in the pressure spectra. The flap motion spectrum is dominated by the fundamental component of the drive signal, with very small components at its first harmonic and at the first two wing eigenfrequencies. This spectrum is relatively insensitive to the change in flow, and only the component at the first wing eigenfrequency increases with freestream Mach number. In subcritical flow the pressures are dominated by the drive fundamental component at all positions across the chord, with no significant harmonic or model response components (Fig.

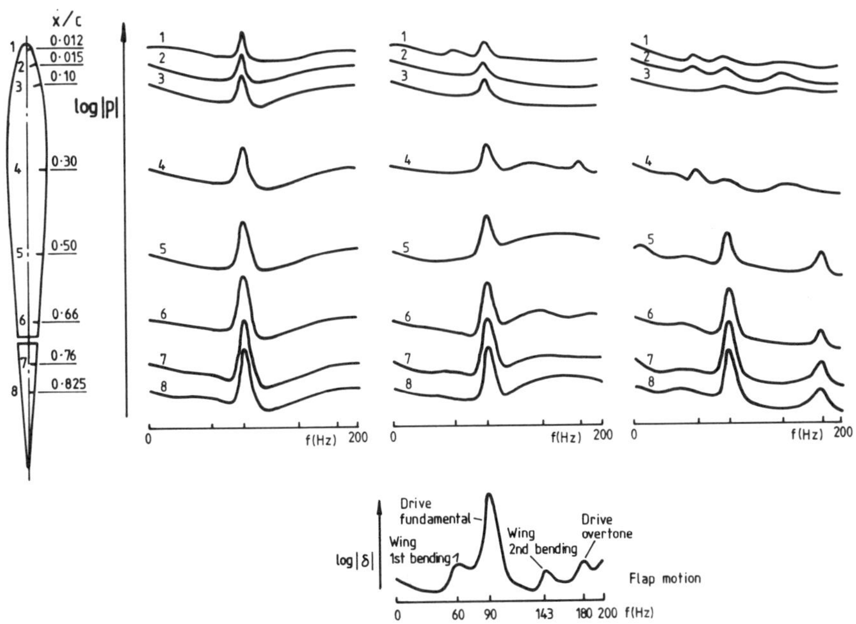

a) Subcritical ($M = 0.85$) b) Critical ($M = 0.90$) c) Supercritical ($M = 0.95$).

Fig. 6 Typical spectra: $\alpha = 0$; $\eta = 0.60$. (British crown copyright reserved 1981.)

6a). In critical flow the spectra show characteristics of both sub- and supercritical flows (Fig. 6b). The drive fundamental component dominates downstream of the intermittent shock at $x/c = 0.3$, where the flow is subsonic. The drive harmonic component is significant at the shock, indicative of local transonic flow nonlinearities. Upstream of the shock the drive fundamental component is attenuated sufficiently to indicate the leading-edge singularity produced by the first wing bending mode. In supercritical flow upstream of the shock the drive fundamental component is attenuated to very low levels, comparable with the wing response of its bending modes (see Fig. 6c, $x/c = 0.50$). The spectra at and downstream of the shock are dominated by the drive fundamental and its first harmonic. The increase of the first harmonic component relative to the fundamental is indicative of the increased nonlinearity in this large region of transonic flow.

4. Unsteady Shock Motions

The formation and movement of shocks is a fundamental problem of time-dependent transonic aerodynamics addressed by Lambourne[26] and by Tijdeman.[1] The whole of Ref. 1 deserves careful study, and the section on shock motions is summarized here. In keeping with the coupling of steady and unsteady flows, the formation of steady shocks is considered briefly.

In an inviscid one-dimensional flow the strength of a normal shock (the pressure ratio p_2/p_1) is determined by the Rankine-Hugoniot relation for the static pressure ratio:

$$\frac{p_2}{p_1} = 1 + \frac{2\gamma}{\gamma + 1}(M_1^2 - 1) \tag{6}$$

where the subscripts 1 and 2 relate to conditions immediately upstream and downstream of the shock, respectively. This pressure ratio is never observed on airfoils because of the effects of shock–boundary-layer interaction. The foot of the normal shock causes a sudden local increase in the rate of change of boundary-layer displacement thickness, which in turn produces a small oblique shock forward of the normal shock. The combination of oblique and normal shocks and the thickened boundary layer causes a smaller rise in pressure than that given by Eq. (6) and may be written

$$\frac{p_2}{p_1} = 1 + \frac{2\gamma}{\gamma + 1}A(M_1^2 - 1) \tag{7}$$

where $0.55 < A < 0.75$ as illustrated in Fig. 7. This equation is valid only up to Mach numbers of about $M_1 = 1.3$, because above that speed shock-induced separation occurs. The subsequent discussion of unsteady shock movements is restricted to attached boundary layers and assumes that Eq. (6) is valid for steady flow in the freestream.

Oscillatory Motion of Normal Shocks

Tijdeman[1] derived an approximate model for the oscillatory motion ($x_0\, e^{i\omega t}$) of a normal shock (Ref. 1, Appendix C). The relation between the sinusoidal pressure p_2 behind the shock and the shock movement is

$$\frac{\Delta p_2}{p_1} = \frac{2\gamma M_1}{\gamma + 1}\left[\left\{\frac{3 + \gamma - 2M_1^2}{2 + (\gamma - 1)M_1^2}\right\}\frac{\delta M_1}{\delta x} - \frac{2i\omega}{a_1}\right] x_0 e^{i\omega t} \tag{8}$$

An important assumption implicit in Eq. (8) is that the "sonic range pressure distribution,"[15] commonly known as the "Mach number freeze" is preserved throughout the shock motion. Subject to this assumption, the Rankine-Hugoniot relations are satisfied instantaneously. All of the main features of shock motion for attached flows can be derived from Eq. (8).

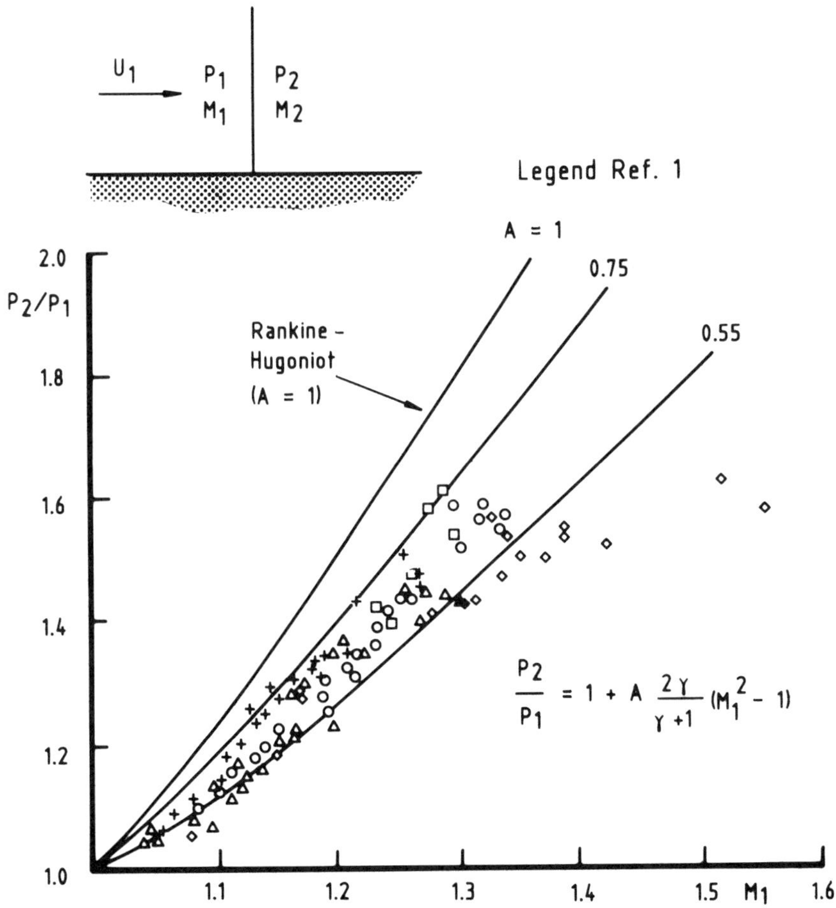

Fig. 7 Pressure ratio across the shock wave.

For quasisteady flows $\omega \to 0$ and Eq. (8) becomes

$$\frac{\Delta p_2}{p_1} = \frac{2\gamma M_1}{\gamma + 1}\left[\frac{(3+\gamma-2M_1^2)}{(2+(\gamma-1)M_1^2)}\frac{\delta M_1}{\delta x}\right] x_0 \tag{9}$$

The sign of Δp_2 is determined by the term $3+\gamma-2M_1^2$. This becomes zero for $M_1 = \sqrt{[(\gamma+3)/2]} = 1.483$, which corresponds to a maximum pressure behind the shock of $p_2/p_t = 0.671$. For $M_1 < 1.483$, the term is positive. Then the pressure ratio across the shock increases according to Eq. (6) and the loss of total pressure is small, the absolute level of p_2/p_t increasing with M_1. In contrast, for $M_1 > 1.483$, the term is negative. The pressure ratio across the shock still increases, but now the loss of total pressure is appreciable, so that the absolute level of p_2/p_t decreases with M_1. In practice, values of M_1 greater than about 1.5 have never been observed in experiments on two-dimensional airfoils. Laitone confirms this observation.[27]

The same criterion for stability was recently derived again for normal shocks in diffusers.[28] Despite the unanimity of Refs. 1, 27, and 28, another report[29] suggests that there may be no such restriction. This anomaly deserves further consideration because it could have important implications for CFD.

For the Mach number range observed in experiments, $M_1 < 1.483$, Δp_2 is in phase with the shock movement for quasisteady frequencies. According to the sonic range pressure distribution, an increase in p_2 corresponds to a decrease in p_1. Hence, in inviscid, quasisteady flow there is a change of phase angle of 180 deg across an oscillating shock. For weak shocks, where viscous effects are small, this change in phase angle is observed in experiments.

For higher frequencies changes in phase angle according to Eq. (8) will occur between Δp_2 and the shock movement. For the special case $\delta M_1/\delta x = 0$ the phase change is always 90 deg because Eq. (8) becomes

$$\frac{\Delta p_2}{p_1} = -\frac{i4\gamma M_1 \omega}{(\gamma+1)a_1} x_0\, e^{i\omega t} \tag{10}$$

This expression corresponds with earlier predictions for a flow with a constant Mach number ahead of the shock.

The phase change according to the general Eq. (8) is determined by the frequency, the shock upstream Mach number, and the Mach number gradient just ahead of the shock in the sonic range pressure distribution. The modulus of the complex function on the right-hand side of Eq. (8) increases with frequency, so that for a fixed level of disturbance, Δp_2, the shock amplitude must decrease steadily with frequency. This important inference is consistent with Nixon's prediction.[23] For moderate values of frequency parameter ν the amplitude of the shock motion is proportional to $(\nu)^{-1}$ (see Ref. 24, p. 212).

Although Eq. (8) does not include the effects of viscosity represented roughly by the factor A in Eq. (7), it can be used to derive the three different types of shock motion observed in experiments, with both attached and separated boundary layers, which are described below.

Types of Oscillatory Shock Motion

With the help of Eq. (8), Tijdeman was able to explain the different types of shock motion observed in his classic experiments with attached flow on an NACA 64A006 airfoil at zero incidence with an oscillating trailing-edge flap (Fig. 8).

For the type A motion observed at $M = 0.90$ with $M_1 = 1.18$, the steady shock is at $x_1/c = 0.67$, well down the sonic range pressure distribution. A small oscillatory shock motion of amplitude $x_0 = 0.05c$ causes an oscillatory variation of shock strength, both for a quasisteady motion and for motion at $f = 120$ Hz, although here the shock strength is amplified and

a) Type A $x_1/c = 0.67$. b) Type B $x_1/c = 0.57$. c) Type C $x_1/c = 0.45$.

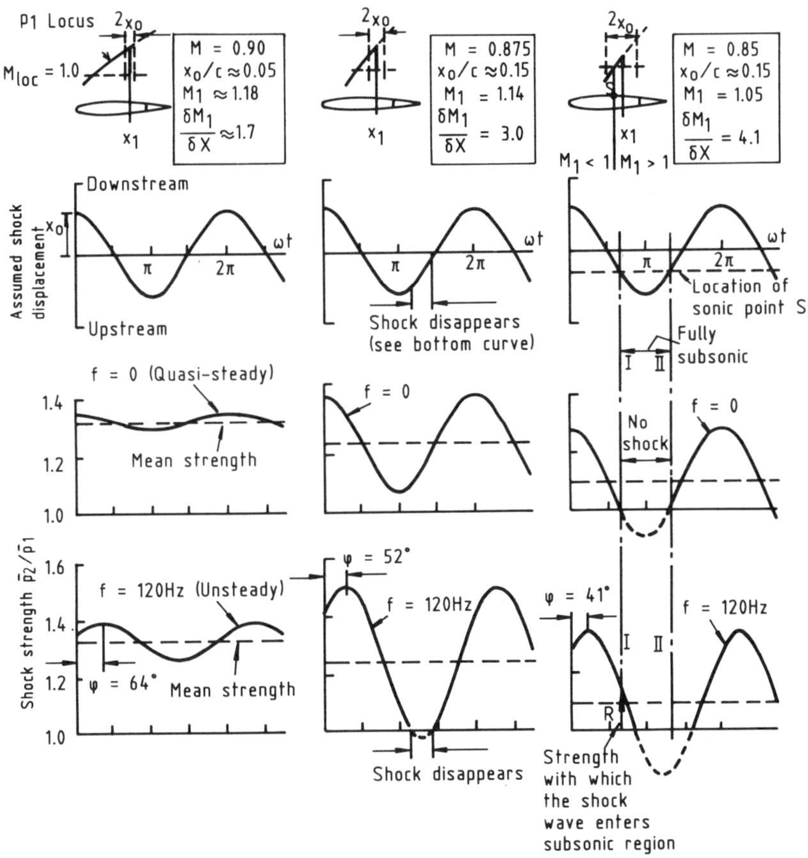

Fig. 8 Explanation of difficult types of shock-wave motion.

there is a phase change of 64 deg. For type A motion the nonlinear features of the flow are comparatively small and are confined to the vicinity of the quasisteady shock motion, i.e., in this example within a length a little larger than $2x_0 = 0.1c$.

For the type B motion observed at a lower Mach number, $M = 0.875$ with $M_1 = 1.14$, the steady shock is at $x_1/c = 0.57$, farther upstream on the sonic range pressure distribution than at $M = 0.90$. A much larger oscillatory shock amplitude of $x_0 = 0.15$ still causes an oscillatory variation of shock strength for the quasisteady flow. However, for the motion at $f = 120$ Hz the shock actually disappears as it moves downstream, and there is a phase change of 52 deg. For type B motion the nonlinear features of the flow are large but are confined to a region somewhat larger than that of quasisteady shock motion, i.e., $2x_0 = 0.3c$.

For the type C shock motion observed at the lowest Mach number, $M = 0.85$ with M_1 only 1.05, the steady shock is at $x_1/c = 0.45$ and is close to the extreme forward position of the shock in the sonic range pressure distribution. (With the maximum thickness at $x/c = 0.38$, this is where sonic flow first occurs.) With a shock amplitude of $x_0 = 0.15$, the quasisteady shock disappears over part of the cycle. In contrast, for the motion at 120 Hz, because of the 42-deg phase change, the shock actually enters the subsonic region with a finite shock strength and then propagates upstream from that point past the leading edge of the airfoil. Thus, for type C motion the nonlinear features of the flow are pronounced and extend from the steady shock position (at $x_1/c = 0.45$) upstream to the leading edge and downstream to about $x/c = x_1/c + x_0/c = 0.60$. It is important to recall that the nonlinear pressure distributions at the shock location do not necessarily lead to noticeable nonlinear forces and moments, as discussed in Ref. 1, p. 89. Type C shock motions may be of importance in the generation of aerodynamic noise.

All three types of motion can occur at a single speed. Thus, for the intermediate Mach number of $M = 0.875$, a type A motion could be obtained either by reducing the frequency or by reducing the shock amplitude to $x_0/c = 0.01$. Alternatively, a type C motion could be obtained either by increasing the frequency or by increasing the shock amplitude. (An increase in shock amplitude to $x_0/c = 0.20$ would take the quasisteady shock forward of the sonic point, a necessary but not insufficient condition for type C flow.)

The radical changes in shock motion that can occur with both amplitude and frequency at a given Mach number suggest that for transonic speeds several amplitudes (say 0.25, 0.50, and 1.00 deg) should be used in both experiments and calculations for a particular frequency. This would ensure that any changes in the type of shock motion within the range of interest are established. According to Dowell's criterion for nonlinearity[21] ($x_0 > 0.025C$), all three types of shock motion illustrated in Fig. 8 would be considered nonlinear. Manifestly a necessary test of any transonic theory is its ability to predict all three different types of shock motion.

Tijdeman also observed type A and B shock motions on a supercritical airfoil oscillating in pitch and suggested that type C motions would occur at higher frequencies. Although for airfoil pitching oscillations there is no

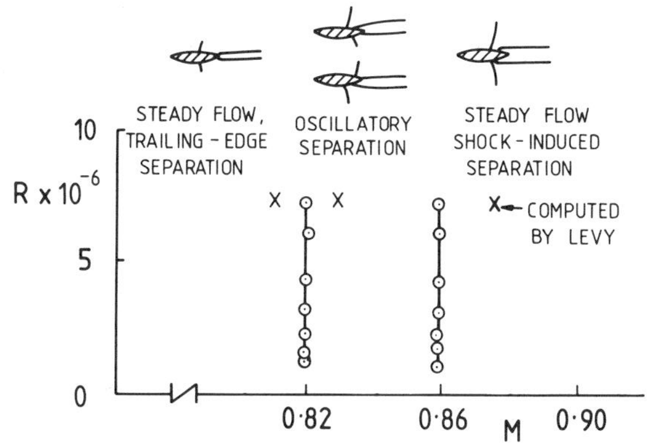

Fig. 9 Flow domains for a 14% thick biconvex airfoil; $\alpha = 0$ deg (fixed transition). (British crown copyright reserved 1981.)

unique sonic range pressure distribution. Tijdeman[1] succeeded in deriving an expression equivalent to Eq. (8), which predicted shock motions of the three different types. It was omitted from Ref. 1 because of its complexity.

Periodic Shock Oscillations on Rigid Airfoils

Shock motions of types A and B have been observed on thick biconvex airfoils at transonic speeds (see Refs. 30–32). For these rigid airfoils the oscillatory pressure behind the shock, Δp_2, is provided by the periodic oscillation of a separated shear layer and the wake, instead of the oscillation of a trailing-edge flap, as in Tijdeman's experiments in attached flow. Levy has succeeded in predicting the separated periodic flows observed on both the 18 and 14% thick biconvex airfoils using thin-layer Navier-Stokes computations (Refs. 33 and 34, respectively). The prediction of these periodic flows offers a unique challenge to the CFD community; thus, the main results of the experimental investigation on the 14% thick biconvex airfoil[32] are enumerated here.

Figure 9 shows the flow domains for $\alpha = 0$ deg with fixed transition as a function of Reynolds number (invariant) and Mach number. For Mach numbers

$$0.74 < M < 0.82$$

a shock forms downstream of the maximum thickness. The flow is almost steady, (cf. discussion of Fig. 3), with a small, symmetric separation well downstream of the shock, close to the trailing edge.

For Mach numbers from

$$0.82 < M < 0.86$$

the flow is periodic, with shocks moving upstream on alternative surfaces. The chordwise pressures are correlated strongly in both space and time (by virtue of the shock motion), so that the fluctuating time-dependent lift coefficient, $C_L(t)$, has an amplitude of about $C_L = 0.2$ at a frequency parameter $\nu = 1.0$. For Mach numbers such that

$$M > 0.86$$

the flow is almost steady, with a separation extending from the shock to the trailing edge.

Figure 10 shows the corresponding pressure distributions. The steady pressure distributions (Fig. 10a) are expressed as the local external Mach number M_e. For the sonic condition ($M = 0.74$), the pressure distribution is close to that predicted by Spreiter. For higher Mach numbers a shock forms and moves downstream progressively, with the last steady flow at $M = 0.81$, for which the trailing-edge pressure has not diverged. For speeds a little higher the flow becomes periodic, the mean shock position moves upstream, and the trailing-edge pressure diverges, e.g., as for $M = 0.85$. For higher speeds the flow becomes almost steady again and the shock moves downstream with the trailing-edge pressure remaining diverged, as for $M = 0.88$. These features of the steady flow are classic and are exactly as described with respect to Fig. 3 or by Pearcey.[15] In particular, the Mach

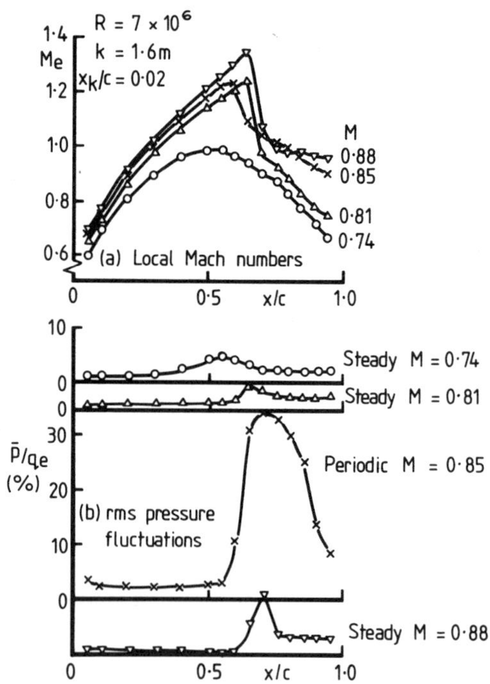

Fig. 10 Typical flow development; $\alpha = 0$ deg. (British crown copyright reserved.)

number freeze upstream of the shock is preserved for

$$0.81 \leqslant M \leqslant 0.88$$

In marked contrast to the steady pressure distributions, the unsteady rms pressure distributions (Fig. 10b) are quite different in character from those normally observed, as described in Fig. 3b. For the critical Mach number, $M = 0.74$ in the absence of a shock, there is an anomalous peak of $\bar{p}/q_e = 0.045$ in the pressure fluctuations. This peak (at the point where the mean flow is sonic) is caused by the amplification of disturbances propagating upstream from the tunnel diffuser, as illustrated in the predictions of Ref. 35. For $M = 0.81$ the pressure fluctuations upstream of the shock are small (being shielded from the diffuser noise by the shock), and there is a small peak of $\bar{p}/q_e = 0.04$ at the shock and a constant, somewhat lower level downstream to the trailing edge. For $M = 0.85$, with the periodic flow, there are extremely large pressure fluctuations (typically $\bar{p}/q_e > > 0.20$) for $0.60 < x/c < 0.80$. These pressure fluctuations are at a single frequency, $\nu = 1$. For $M = 0.88$ the pressure fluctuations upstream of the shock are small, and there is a peak of $\bar{p}/q_e = 0.10$ at the shock and a lower level downstream to the trailing edge. For this condition the pressure fluctuations are random and cover a wide range of frequencies.[2]

Periodic flows of the type observed at $M = 0.85$ have been observed also at off-design conditions on advanced supercritical airfoils and wings. These periodic flows also produce large pitching moment fluctuations, $C_m(t)$, often $\nu = 1$, close to that of wing torsional modes. Hence, there is a need to predict these localized regions of type B shock motions due to shock-induced separation.

These periodic flows due to shock-induced separation on rigid airfoils can be amplified by oscillation of the airfoil at the same frequency (e.g., see discussion of Figs. 12–14 of Paper 4 of Ref. 12). Hence, the periodic frequency ($\nu = 0.5$ in that experiment) may be considered as an "aerodynamic natural frequency" of some autonomous mechanism brought into "resonance" by the airfoil oscillations. The dependence of this aerodynamic

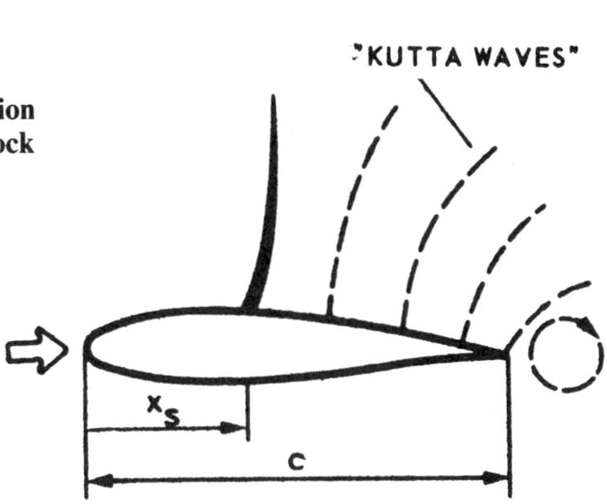

Fig. 11 Propagation of information from the trailing edge to the shock wave.

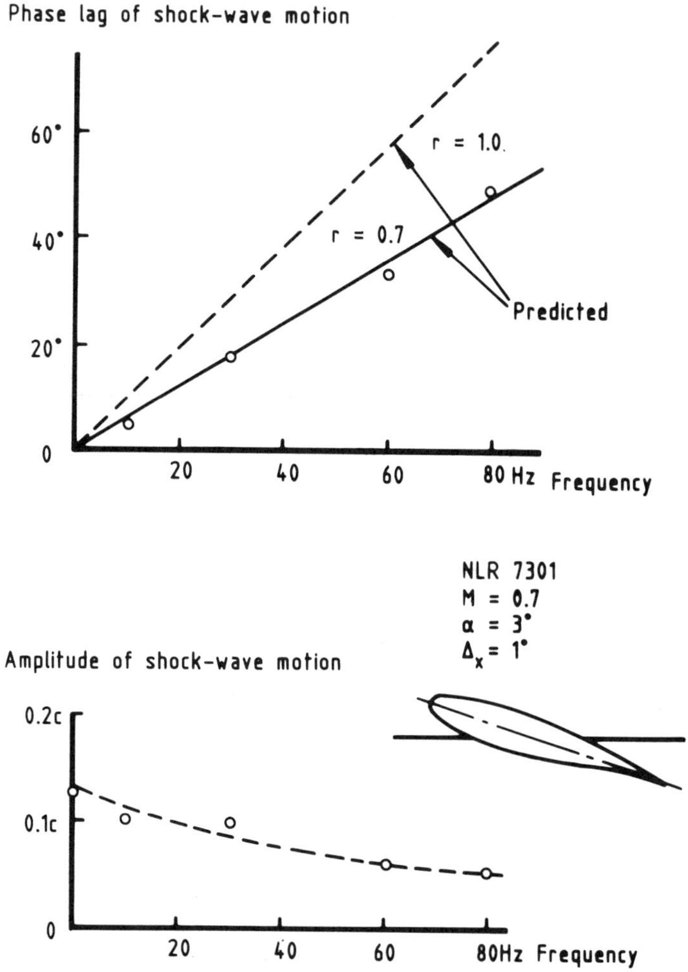

Fig. 12 Effect of frequency on the amplitude and phase lag of the periodic shock-wave motion.

natural frequency on Mach number and angle of incidence is not well understood.

Type C shock motions are responsible for the upstream waves observed in previous experiments on rigid circular cylinders (as sketched in Fig. 1b); again the oscillatory pressure behind the shock is provided by the movement of the wake.

Influence of Trailing-Edge Conditions

Pearcey showed how the trailing-edge static pressure controls the development of the steady transonic flow about a rigid airfoil[15] and how the divergence of steady trailing-edge pressure may sometimes indicate the onset of unsteady separated flow.[18] Tijdeman's experiments[1] confirmed that the time-dependent trailing-edge pressure applies a similar control to the Δp_2. Hence, through Eq. (8) the trailing-edge pressure influences both the type of shock motion and the overall forces.

The steady and unsteady Kutta-Joukowski conditions at the trailing edge are essentially the same for attached flow; namely, the pressures on both surfaces are identical.[36] In an inviscid subsonic flow with a finite trailing-edge angle the steady trailing-edge pressure is the same as the total pressure, and the local velocity is zero. For an attached viscous flow the trailing-edge pressure is close to freestream pressure, and the local velocity is close to the freestream velocity because of the presence of the boundary layer. For small-amplitude motions of the trailing edge (due either to a flap displacement or airfoil motion), these pressure and velocity conditions must be preserved. Hence, impulsive pressure waves are generated at the trailing edge, and these waves propagate upstream toward the shock. Magnus and Yoshihara[37] referred to these pressure signals as the "Kutta waves" (Fig. 11).

The time taken for a wave to travel upstream from the trailing edge to the shock is

$$t = -\int_{x=c}^{x=x_s} \frac{\mathrm{d}x}{(1 - M_e^1)a} \tag{11}$$

For a flat plate there are no velocity gradients normal to the surface outside the boundary layer; thus,

$$M_e^1 = M_e \tag{12}$$

In contrast, for an airfoil the Mach number gradient normal to the surface makes the Kutta waves propagate away from the airfoil and lowers the effective Mach number, so that

$$M < M_e^1 < M_e \tag{13}$$

Tijdeman[1] suggested that it was possible to write

$$M_e^1 = rM_e + (1 - r)M \tag{14}$$

where r is the empirical relaxation factor to be found from each particular experiment. A value of $r = 0.75$ was appropriate for experiments with attached flow on an NLR 7301 airfoil at 3 deg incidence at $M = 0.70$ oscillating with an amplitude of 1 deg with a type A shock motion.

The time delay according to Eq. (11) is equivalent to a phase lag at the shock of

$$\phi^{(0)} = (\Delta t) \cdot f \cdot 360 \tag{15}$$

Figure 12 suggests that, for attached, oscillatory flows, Kutta waves travel upstream from the trailing edge to the shock according to Eqs. (11), (14), and (15). Figure 12 also shows the characteristic reduction of shock amplitude as frequency increases, as discussed previously.

Unsteady measurements by Roos[14] on rigid airfoils in wind tunnels confirm (Fig. 13a) that, with attached flow, random disturbances due to flow unsteadiness also propagate upstream from the trailing edge to the shock at about the speeds required by Eqs. (11) and (14). (Roos also showed that such upstream disturbances do not exist in flight tests with attached flow.) However, for separated flows in both wind-tunnel (Fig. 13b) and flight tests, strong disturbances propagate downstream from the shock toward the trailing edge, and the speed of progation varies strongly with frequency (Fig. 13c). This change in the direction of propagation is important for the description of the buffet excitation. It also raises the question of whether a different expression should replace Eq. (11) on an airfoil oscillating with separated flows at transonic speeds. Resolution of this question by careful experiments would be of great interest. Modern calculation methods combining unsteady transonic flow theory with boundary-layer theory[38,39] should be able to compute the convection of pressure waves between the shock and the trailing edge.

5. Transonic Buffet and Buffeting

This introductory discussion of the nature and importance of the aerodynamic excitation or buffet at transonic speeds is based primarily on Ref. 2. A full discussion of these problems is included in Chap. 5.

Figure 14 shows a fighter aircraft maneuvering at transonic speeds. There are two large areas of separated flow on the aircraft at the root of the canard and the tip of the wing. These provide excitation, which at a given point may be characterized by the rms level, the frequency spectrum (the spectrum is often fairly flat at low frequencies), the degree of correlation in space and time, and the length scale. The pressure fluctuations excite a response of the structural modes, called buffeting. The aircraft structure acts as a selective filter for the excitation so that spectra of buffeting always contain pronounced peaks at structural frequencies. In the example sketched both the wing and the canard are excited. Rigid body modes may also be excited, such as "wing rocking," "wind dropping," or "nose slicing," but these are at much lower frequencies and can be regarded as aircraft handling problems. Buffet onset is defined as the first apppearance of a significant area of separated flows, although aerodynamicists often argue about how large the area must become before it is significant.

Classification of Wing Flows and Buffeting

A broad classification of wings with separated flows that excite buffeting is useful as a framework, even if the classification suggested is incomplete (Fig. 15).

Wings with low angles of sweep are characterized beyond the buffet boundary at subsonic speeds by leading-edge or trailing-edge separations.

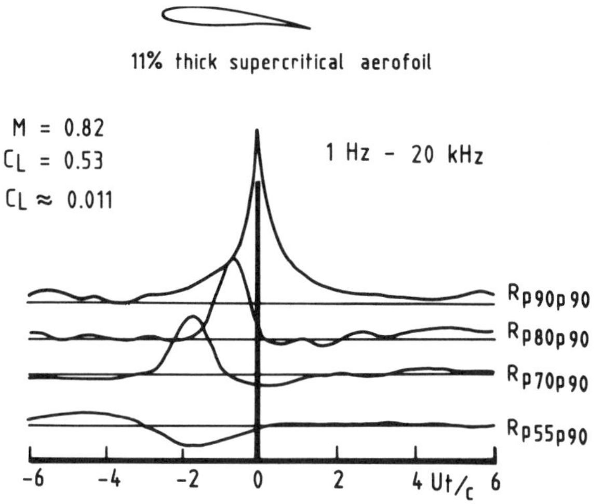

a) Attached flow.

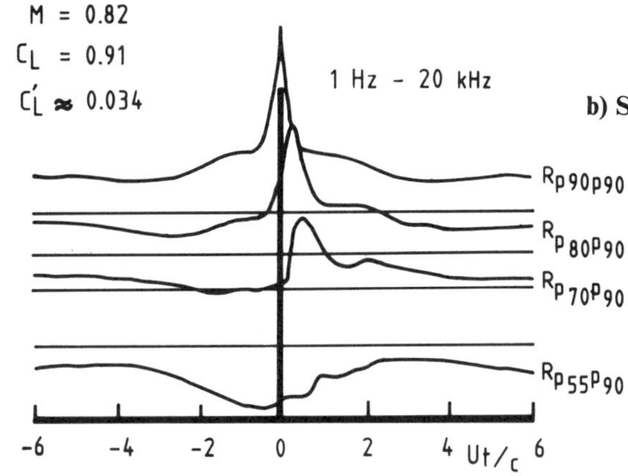

b) Separated flow, heavy buffet.

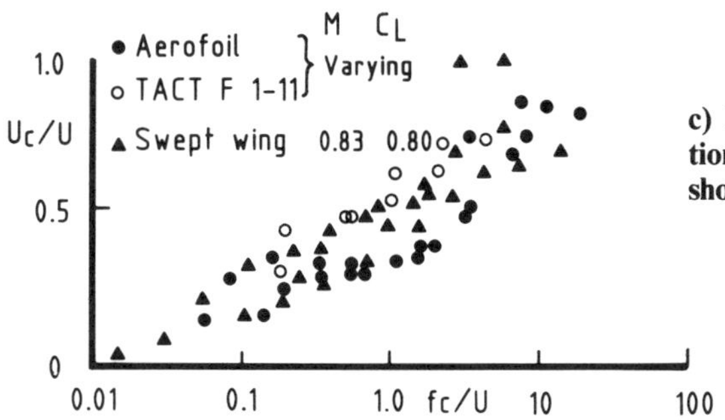

c) Pressure fluctuation convection velocity downstream of shock-induced separation.

Fig. 13 Disturbance propagation on a supercritical airfoil.

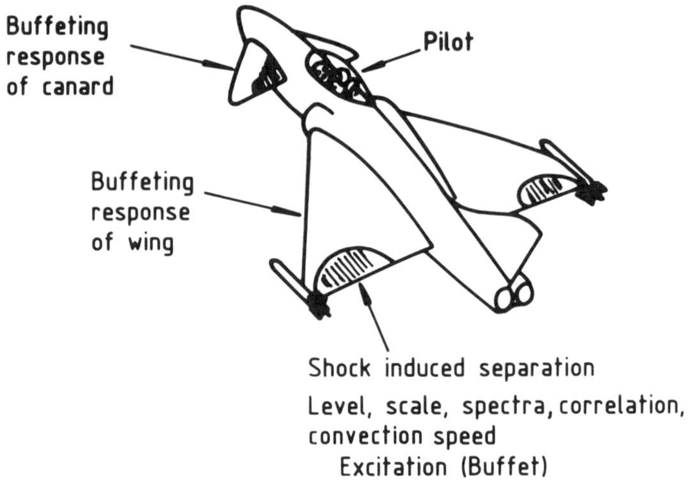

Fig. 14 Buffet and buffeting at transonic speeds.

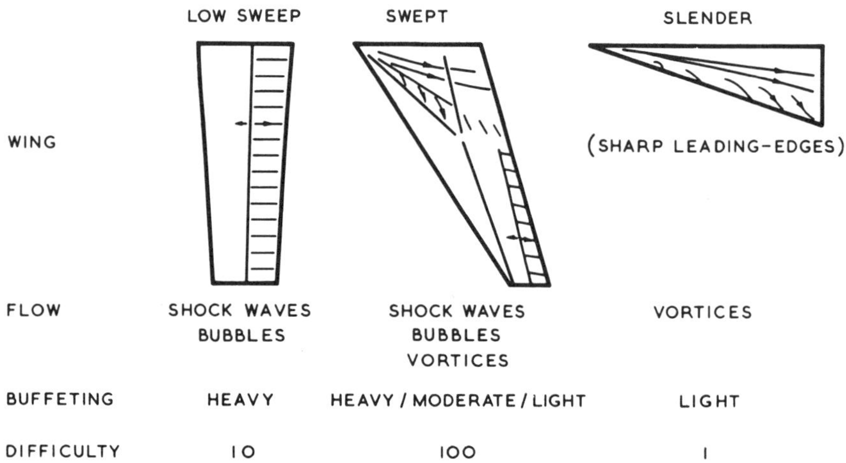

Fig. 15 Classification of flows and associated buffeting. (British crown copyright reserved 1973.)

These separations form bubbles on the wing which usually excite heavy buffeting. At transonic speeds the presence of strong shock waves nearly parallel with the leading edge add to the difficulties of predicting the flow; thus, this flow is given a prediction rating of 10. (These prediction ratings are arbitrary and are not used in any calculations; an increase in prediction rating represents increased difficulty of prediction.) Swept wings are characterized by a combination of mixed flows (Ref. 19) that is difficult to predict. The separated flows on a swept wing at transonic speeds may include shock waves (which vary in intensity across the span), bubbles (from the leading edge or the trailing edge), and vortices. Thus, a small increase in Mach number may dramatically alter the position of a shock wave or the reattachment point of a bubble. Similarly, an increase in unit Reynolds number or a change of the roughness band used to fix transition on the model in wind-tunnel tests may completely alter the character of the shock-wave–boundary-layer interaction.[40,41] These difficulties justify a prediction rating for swept wings of 100.

Slender wings with sharp leading edges are characterized by a simple vortex type of flow, which prevails over the complete speed range from subsonic to supersonic speeds. This unified type of flow is a powerful argument for the application to supersonic aircraft.[42] The vortices, which produce significant nonlinear lift, have a well-defined, small-scale structure and do not produce severe buffeting unless vortex breakdown occurs at high angles of attack outside the normal flight envelope. In addition, the vortices on slender wings with sharp leading edges are relatively insensitive to wide variations in Reynolds number. Hence, slender wings are given a prediction rating of 1.

Unswept Wings

Bubbles found for a wide range of flows at low speeds are characterized[2] by a level of excitation increasing steadily from separation and reaching a maximum of about $\bar{p}/q = 0.04$–0.10 just upstream of reattachment. The bubble excitation is random in character and covers a wide range of frequencies, with an ill-defined peak at a frequency parameter based on bubble length of about 0.5–0.8. A similar pattern is preserved at transonic speeds, with the addition of large, low-frequency pressure fluctuations in the vicinity of the shock and at reattachment, as discussed previously with respect to Fig. 3.

Coe's investigation of the loads on launch vehicles[43] included some measurements at subsonic and transonic speeds of the mean and fluctuating static pressures caused by a separation bubble downstream of a step in a body of revolution (Fig. 16). Both the mean and fluctuating pressure distributions aft of the step correspond very well in general character with the low-speed pressure distributions over the complete speed range of $M = 0.60$–1.19, although the maximum fluctuating pressure falls steadily from

$$\bar{p}/q = 0.06 \qquad \text{at} \qquad M = 0.80$$

to

$$\bar{p}/q = 0.03 \qquad \text{at} \qquad M = 1.19$$

This fall in the pressure fluctuations is probably due to the improved stability of the mean bubble flow because of the reduction of upstream influence from the reattachment region as the region of supersonic flow expands. The base-pressure fluctuations on a body of revolution also fall

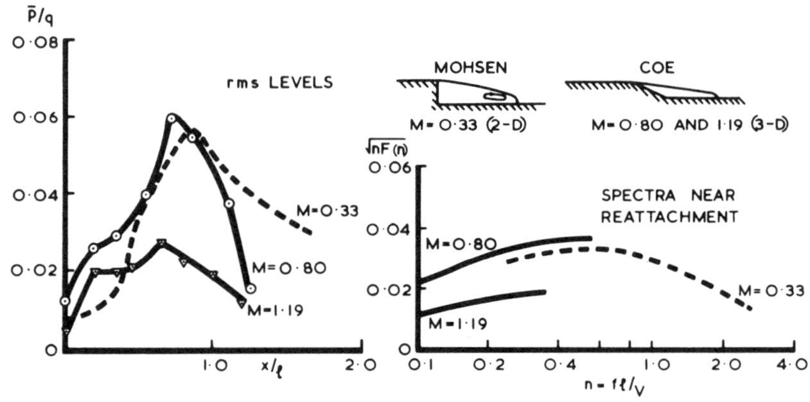

Fig. 16 Comparison of excitation caused by the flow down a step at $M = 0.33$, 0.80, and 1.19. (British crown copyright reserved 1973.)

from subsonic to supersonic speeds, and a similar explanation may be applicable.[44] The mean pressure distribution suggests that the length of the bubble does not change significantly from $M = 0.60$ to 1.19; hence, within this speed range, there is probably no major change in the internal structure of the bubble.

The apparent universality of the pressure fluctuations caused by bubble flows at subsonic speeds is also well illustrated in Fig. 16, which includes the pressure fluctuations measured[45] behind a two-dimensional step at $M = 0.33$ as well as those measured behind a step on a body of revolution at $M = 0.80$. The similarity at subsonic speeds between both the rms pressure fluctuations and the spectra for the two different experimental configurations and Reynolds numbers is good.

Some interesting measurements of the excitation on a two-dimensional lifting airfoil at transonic speeds were made by Moss and Mundell and reported in Ref. 46 (Fig. 17). The condition selected for this ($M = 0.82$, $\alpha = 6.7$ deg) is just beyond buffet onset. Although the trailing-edge pressure has not yet diverged, there is a short separation bubble on the airfoil (with a length of about $1 = 0.1c$) immediately downstream of the shock wave. The position of the bubble was inferred from the shape of the mean pressure distribution, because it could not be seen in oil flow tests.

The excitation measurements along the chord are presented for two-frequency parameters:

$$n = fc/U = 0.08 \qquad \text{and} \qquad 0.8$$

At a frequency parameter of $n = 0.08$, a typical value for wing structural modes, there is a large local increase in excitation in the vicinity of the shock wave. This local excitation decreases rapidly downstream of the shock wave but then shows a small local maximum in the vicinity of the reattachment region before decreasing again. This variation in low-frequency excitation

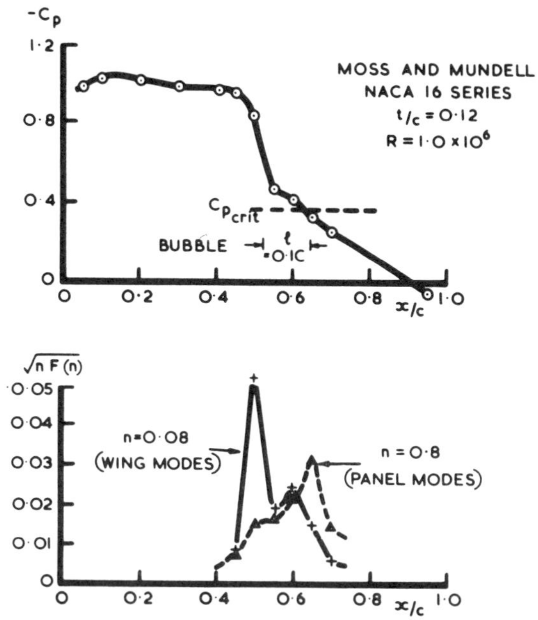

Fig. 17 Excitation on an airfoil near buffet onset; $M = 0.82$, $\alpha = 6.7$ deg. (British crown copyright reserved 1973.)

must be caused by the coupling of the shock-wave motion (at separation) with the development of the bubble and with conditions at reattachment. In contrast, at a frequency parameter of $n = 0.08$, a typical value for wing panel modes, the excitation increases progressively downstream from the shock position, reaches a maximum close to the reattachment line, and then falls rapidly as in the other bubble flows discussed in this section.

To find how the excitation develops, the angle of attack may be increased at constant Mach number. The separation bubble then extends rapidly toward the trailing edge and the trailing-edge pressure diverges, while the shock wave starts to move upstream slowly. Thus, the area of the airfoil influenced by both the low-frequency and high-frequency excitation increases, and a progressive increase in buffeting would be expected. It should be noted that, as the bubble extends rapidly in length from about $1 = 0.1c$ to $1 = 0.5c$, the predominant bubble frequency parameter will fall from $n = 8$ to $n = 1.6$; thus, there should be a large increase in excitation in a frequency range centered on this lower value.

Recent measurements of the excitation of the BGK no. 1 shockless airfoil[47] also conform to the general pattern described in Fig. 3, except for a narrow range of conditions (M,α) when the excitation is much larger and periodic. This periodic excitation has been predicted by Navier-Stokes computations[48] and is similar to that observed on biconvex airfoils,[31–33] as discussed earlier. It is important to remember that, when the flow is separated on the top surface of an airfoil, excitation will also be found on the lower surface. This is illustrated in Ref. 17, Fig. 24.

Swept Wings

Figure 18 shows the complex separated flow on a typical model with swept wings at a Mach number of 0.80 at low Reynolds number. These sketches illustrate some of the features that make buffet prediction difficult for swept wings and justify the prediction rating of 100 allotted in Fig. 15. At buffet onset there are at least three shock waves on the wing and a small shock-induced separation bubble (with a length of perhaps $0.05c$) immediately downstream of the strongest shock wave, which runs roughly parallel with the leading edge. At moderate buffeting there are complex shock patterns on the wing, areas of separated flow, and areas with attached flow having a strong spanwise velocity component. (Great care is always necessary to achieve an optimum transition fix for the boundary layer under high lift conditions like this at transonic speeds if large-scale effects are to be avoided. The essential aim in the model test is to achieve a similar shock-wave–boundary-layer interaction as at full scale. This aim can often be achieved by comparing the results of modern boundary-layer calculation methods at model and full-scale Reynolds numbers, say by achieving equality of momentum as thickness just ahead of the shock, or length of bubble downstream of it, as Fulker and Ashill[82] suggested.)

Pressure fluctuations are presented for a single point P on the wing. Figure 19 shows the variation of rms pressure fluctuations with the angle of attack. An attempt is made to explain this variation, but it is speculative because of the difficulty of discriminating between local events at P and what is happening simultaneously elsewhere on the wing. A local Mach number of 1.0 is reached at a small angle of attack at a point on the wing near the leading edge and close to the tip. A local region of supersonic flow then develops as the angle of attack increases. This supersonic region is terminated by a shock wave, which oscillates upstream and downstream.

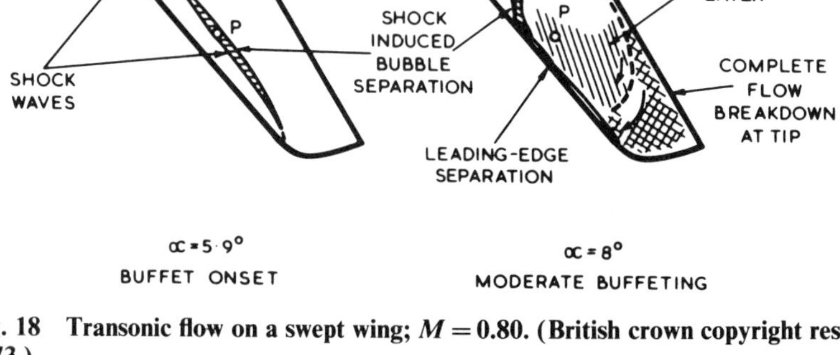

Fig. 18 Transonic flow on a swept wing; $M = 0.80$. (British crown copyright reserved 1973.)

The pressure fluctuations at P first increase slowly with the angle of attack (point A, $\alpha = 1.8$ deg) because of the combined effect of the shock oscillation and the pressure fluctuations caused by an attached boundary layer growing under an increasingly adverse pressure gradient. As the terminal shock wave approaches and passes the transducer position (point B, $\alpha = 5.0$ deg), the pressure fluctuations increase rapidly; the major part of this increase must come from the shock-wave oscillation. Shortly after point B the boundary layer separates at the terminal shock, and the terminal shock then starts to move forward and buffeting is detected by the wing-root strain gages. (Thus, the wing-root strain gages give a measure of the integrated excitation on the wing.) When the terminal shock moves upstream of the pressure transducer, the local pressure fluctuations fall rapidly to a minimum at about $\alpha = 6$ deg because point P is no longer influenced directly by the shock oscillation. The pressure fluctuations then increase to a maximum (point C, $\alpha = 7.2$ deg), when the reattachment line crosses the pressure transducer, just as on the airfoil (Fig. 17). The local pressure fluctuations then decrease as the bubble extends downstream, although the wing buffeting, which is the response to the total excitation on the wing, continues to increase steadily from moderate to heavy.

The spectra of the pressure fluctuations provide additional information. Figure 20 shows that, at point A, well below the buffet onset, the increase in pressure fluctuations is above the frequency range of wing modes. At point B, with the terminal shock oscillating across the transducer, the peak pressure fluctuations become larger and move to lower frequencies, within the range of wing modes. At point C, this peak is lower, because the shock

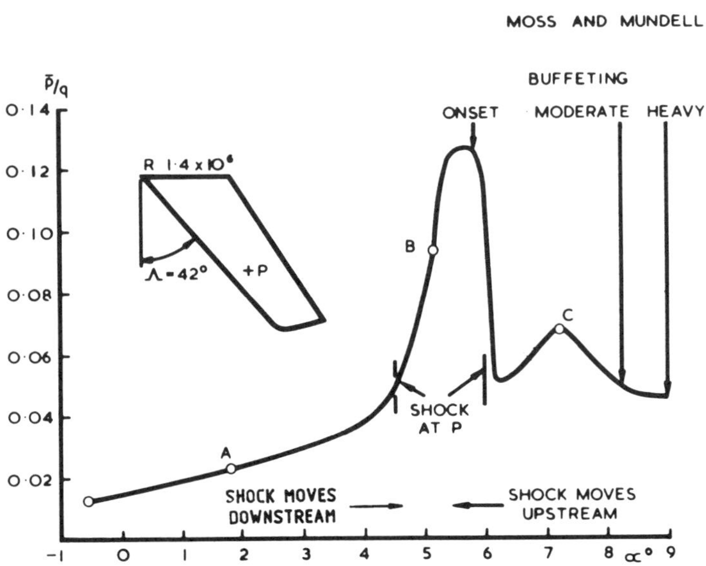

Fig. 19 Excitation of a swept wing; $M = 0.80$. (British crown copyright reserved 1973.)

is upstream of the transducer. However, at point C there is an increase in pressure fluctuations at high frequencies ($n > 0.5$). This may be associated with the separation bubble downstream of the shock.

This is obviously only a simplified account of the development of the excitation, using data from a single point on a wing. A complete description of the excitation all over a wing (including rms levels, spectra, and correlations) is difficult to achieve. Extensive computing facilities are also needed to utilize this data to predict buffeting, and there are currently few measurements of excitation available for swept wings at transonic speeds.

Slender Wings

The well-ordered vortex flows observed on slender wings at subsonic and supersonic speeds are preserved at transonic speeds, as long as the angle of incidence for vortex breakdown is not exceeded. The angle of incidence for vortex breakdown decreases slowly as Mach number increases, probably because of the development of local transonic flow around the vortices. Thus, for a 65-deg delta wing, the angle of incidence for vortex breakdown falls from about 22 deg at $M = 0.35$ to about 20 deg at $M = 0.70$.[49] This decrease in the usable range of incidence is compensated partially by a small increase in lift-curve slope.

Heavy Buffet Limit for Unswept, Swept, and Slender Wings

Some careful comparisons between measurements of wing buffeting in wind tunnels and in flight are now available.[50] These comparisons relate to the wing first bending mode and reveal an interesting result of great importance.

The buffet excitation parameter, $\sqrt{[nG(n)]}$, has a maximum value of about 0.0030 for all wings in both wind tunnels and in flight. This maximum is found at both subsonic and transonic speeds near the point of maximum buffet penetration (i.e., maximum lift coefficient or angle of incidence) in flight. A physical explanation of this common limit for the widely differing separated flows about unswept, swept, and slender wings has yet to be provided. The common limit for both subsonic and transonic speeds may imply that for well-separated flows the shocks contribute a relatively small additional excitation to that provided by the shear layers.

6. Transonic Flutter Problems

The transonic speed range introduces some novel flutter problems that do not succumb to a simple explanation. These flutter problems are listed briefly here and are discussed fully in Chap. 7.

Transonic Dip in the Flutter Boundary

Classic coupled torsion/bending flutter occurs when, with increase in kinetic pressure, the bending frequency increases and the torsional frequency falls, until the frequencies become coincident.[8] Then a single divergent mode (comprising combined bending and torsion) can be excited

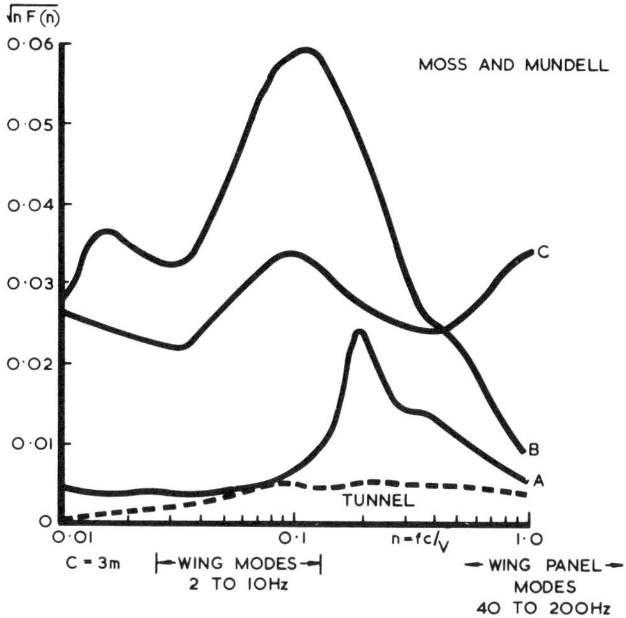

Fig. 20 Spectra of excitation on a swept wing; $M = 0.80$. (British crown copyright reserved 1973.)

even by an infinitesimally small disturbance, because energy can be extracted from the freestream.

In contrast, for unswept and swept wings with attached flow at transonic speeds, the flutter mechanism is somewhat different and frequently causes a pronounced dip in the flutter boundary. Although this "transonic dip" is associated with the increase in quasisteady lift-curve slope (illustrated for an NACA 64A006 airfoil in Fig. 5) and the corresponding downstream movement of the center of pressure (for an airfoil from $x/c = 0.25$ at subsonic speeds to $x/c = 0.50$ at supersonic speeds), no adequate general explanation can be given in terms of quasisteady derivatives. Instead, calculations must be made with realistic structural properties (e. g., bending and torsional stiffnesses and center of gravity positions) and aerodynamic coefficients (e.g., aerodynamic stiffness and damping derivatives).

Zwaan[51] gives results of such calculations for an unswept flat-plate wing with freedoms of motion in heave and pitch that are reproduced in Fig. 21. These calculations show the characteristic transonic dip in the flutter speed about $M = 1$. In addition, the nondimensional flutter speed

$$U_F/(b\omega_\alpha\sqrt{\mu})$$

decreases as the mass density ratio μ increases (Fig. 21a). The calculations also show that in the vicinity of the transonic dip a sudden change in flutter

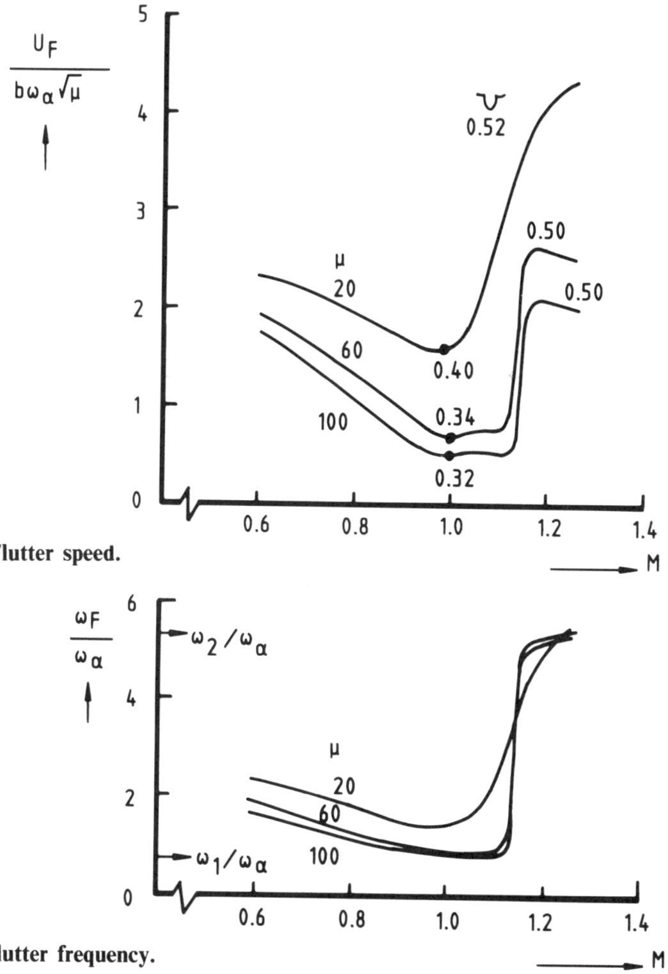

a) Flutter speed.

b) Flutter frequency.

Fig. 21 Flutter speed and frequency of a flat plate for various mass ratios, μ.

mode occurs. Here, at the bottom of the dip, near $M = 1$, the flutter mode is virtually pure heave at a frequency almost identical with the wind-off bending frequency (in this particular example, $\omega_F/\omega_\alpha = \omega_1/\omega_\alpha = 0.71$). In marked constrast, at a slightly higher Mach number, where the flutter speed is much higher, the flutter mode is virtually pure torsion at a frequency almost identical with the wind-off torsional frequency ($\omega_F/\omega_\alpha = \omega_2/\omega_\alpha = 5.34$), as shown in Fig. 21b. These predictions are in accord with many experiments and thus describe the nature of the phenomenon adequately.

Zwaan[51] then relates this change in the predicted flutter characteristics with changes in the unsteady aerodynamic coefficients. The unsteady aerodynamic forces due to heaving always have a damping effect and therefore cannot cause the dip. However, the unsteady aerodynamic forces due to pitching may be able to drive the system. This mechanism for flutter is illustrated in Fig. 22. In Fig. 22a the wing is oscillating in its first mode with a nodal line far ahead of the leading edge. Both heaving and pitching contribute to this mode. The principle aerodynamic forces are the damping force, L_h, due to heaving and the lifting force, L_α, due to pitching, which has a phase lag ϕ_α. In the vector diagram, Fig. 22b, all forces are indicated, aerodynamics as well as structural. (For simplicity the moments are omitted.) The component $L_\alpha \sin\phi_\alpha$ works in the same direction as $\dot{h}$ during the whole cycle and is able to feed energy into the system. When this component exceeds the energy dissipation provided by L_h, flutter instability occurs. According to this explanation the transonic dip is determined primarily by the development of the ϕ_α with Mach number, rather than by variations in the quasisteady coefficients.

The predicted values of modulus, L_α, and phase, ϕ_α, are given in Fig. 23 for frequency parameters of 0.10, 0.30, and 0.60 (derived from Fig. 10 of Ref. 51). For a frequency parameter of $\nu = 0.30$, phase lag increases with Mach number up to about $M = 1.10$ and then decreases, corresponding to the rise and fall in the flutter boundary shown in Fig. 21a. For higher mass ratios the frequency parameter is lowered. This decrease in frequency parameter further increases the phase lag according to Fig. 23b. Hence, the transonic dip is deepened as the mass ratio increases, as shown clearly in Fig. 21a.

Zwaan's review[51] suggests that in general the ϕ_α is larger at transonic speeds for supercritical wings than for conventional wings. Hence, supercritical wings generally will have a more pronounced transonic dip than conventional wings.

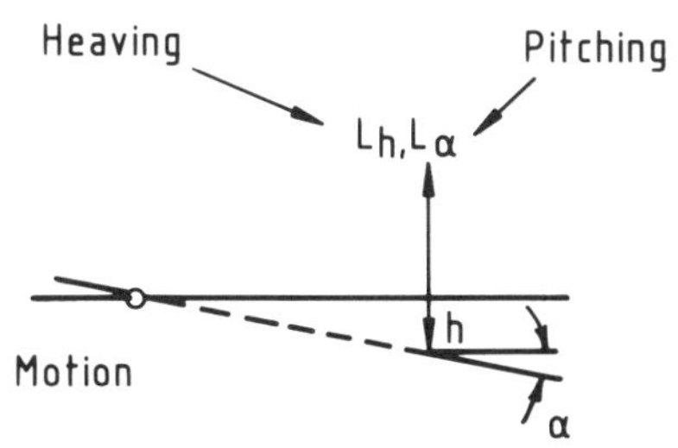

a) First mode.

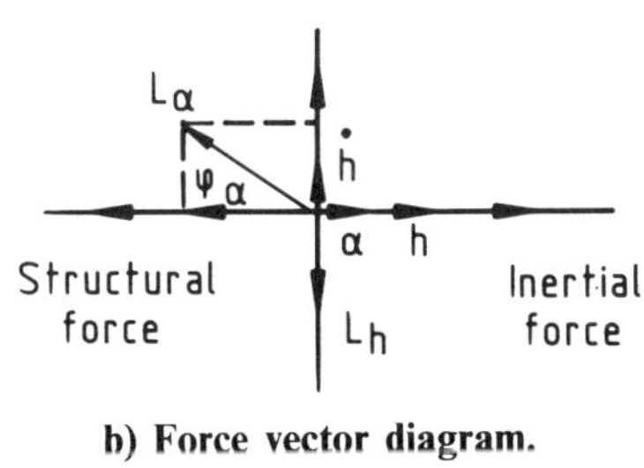

b) Force vector diagram.

Fig. 22 Fundamental flutter mechanism for oscillating flat plate.

Sometimes the moving shock makes the largest contribution to the increase in phase lag. This point was raised by Ashley[52]; a more recent example was given by Isogai in Ref. 12, Paper 17, Sec. 3.2.4. However, a moving shock is not the only cause of a flutter dip (cf. the flat-plate computations discussed earlier and the computations and experiments for a supercritical wing described in Ref. 53).

Zwaan's review[51] also stresses the importance of making transonic wind-tunnel flutter tests with fully turbulent boundary layers to avoid unrealistic single-degree-of-freedom flutter caused by transitional boundary layers at low Reynolds number. However, it is now known that single-degree-of-freedom flutter can be excited by transitional boundary layers over a wide range of speeds (see the experiments cited in Ref. 54) and is therefore not peculiar to the transonic speed range.

Single-Degree-of-Freedom Flutter due to Shock-Induced Separation

Many different types of single-degree-of-freedom flutter have occurred due to shock-induced separation. Zwaan[51] suggests that most of these can

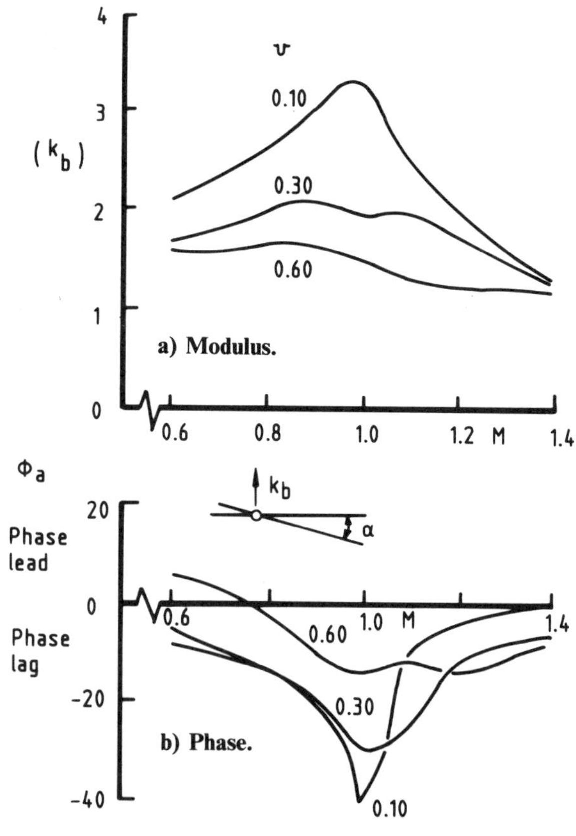

Fig. 23 Unsteady lift coefficient for flat-plate pitching about 0.25*c*.

be explained in terms of phase angles of measured aerodynamic coefficients for separated flow being completely different (even in sign) from those of calculated coefficients for attached flow. An example is given in Fig. 24, which shows a flutter boundary with two dips. The first corresponds with mainly wing bending in mainly attached flow, and the second corresponds with almost pure torsional motion in well-separated flow (cf. Refs. 53 and 55), as shown by the divergence of the trailing-edge pressure at 70% semispan.

Moss and Pierce[55] studied the responses in fundamental bending and torsion on a series of swept wing models at transonic speeds. The illustrative results, for a Mach number of 0.75 and a leading-edge sweep angle of 27 deg, relate to a specially constructed flexible model that represented the full-scale static bending and torsional stiffness of an aircraft design. On the model no attempt was made to represent the aircraft mass distribution; thus, the frequency parameters were incorrect. The spectra of the wing accelerometer signals were used to estimate the total damping in the primary bending and torsional modes, and these were plotted against the mean angle of incidence in Fig. 25.

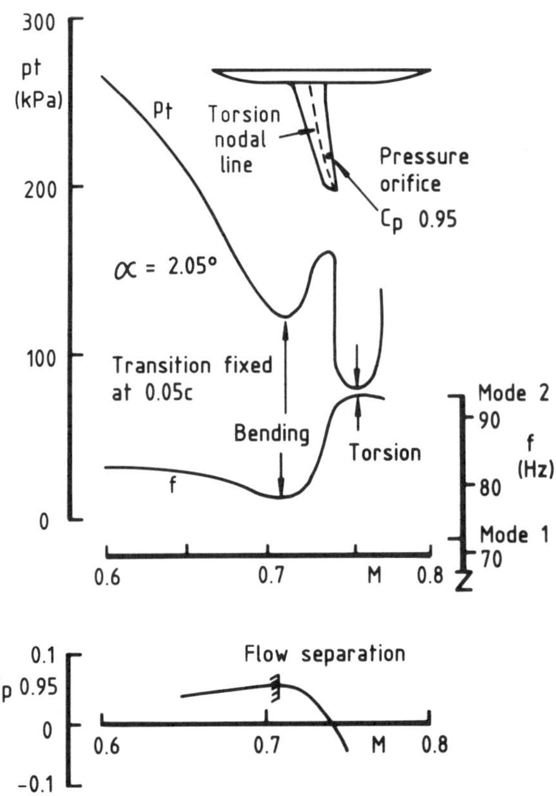

Fig. 24 Two different types of flutter.

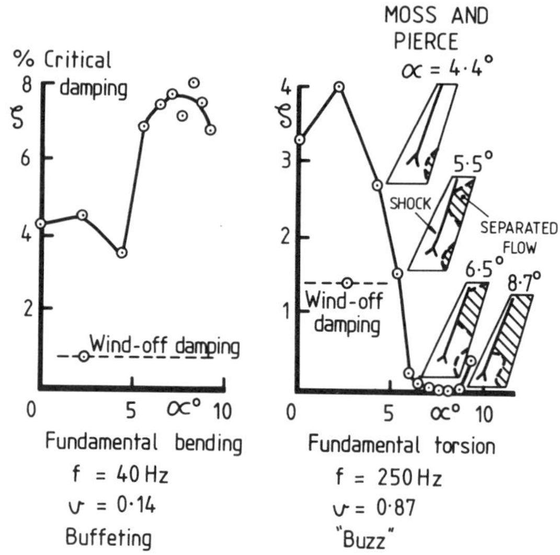

Fig. 25 Aeroelastic responses on a flexible wing; $M = 0.75$. (British crown copyright reserved 1977.)

For the primary bending mode, at a frequency parameter of $\nu = 0.14$, the total damping increases rapidly with angle of incidence, after buffet onset at about $\alpha = 4.4$ deg. This increase is comparable with the increase measured recently on a flutter model of a typical strike aircraft[50] at a higher-frequency parameter, $\nu = 0.50$. In marked contrast, for the primary torsion mode, at a frequency parameter $\nu = 0.87$, the total damping falls rapidly as the angle of incidence increases after buffet onset. For $\alpha = 6.5$ and 8.7 deg, the total damping is almost zero. This condition is indicative of a limit cycle torsional oscillation or "buzz." Here the amplitude achieved is determined by the equality of the net work done by the air throughout the cycle and the energy dissipated by the aerodynamic and structural damping, integrated throughout the cycle. Both positive and negative aerodynamic damping can occur at different parts of the cycle (Fig. 25).

Figure 25 includes sketches of the shock-induced flow separations that create this loss of damping in the torsional mode. The strong shock lies upstream of and approximately parallel to the torsional node line (which is at about $x/c = 0.35$). The quasisteady movement of this shock with angle of incidence is such that the slope of the static characteristic of the local pitching moment about the nodal line with respect to incidence is negative (see Fig. 26b). These criteria are probably necessary but insufficient conditions for the establishment of wing buzz. The extent to which aerodynamic damping at a particular frequency parameter becomes negative and varies with the amplitude has been predicted by Beddoes[56] from the response measurements made in this experiment.

Beddoes assumed that the character of torsional buzz would be determined by conditions at 90% semispan on the wing, for which static lift

(Fig. 26a) and pitching moments (Fig. 26b) were available from pressure measurements on a geometrically similar but effectively rigid model. Thus, steady-induced effects were included and pitch angle and angle of incidence were used synonymously. Then the aerodynamic damping was calculated for a range of reduced frequencies corresponding to the measured values of frequencies and tip angular amplitudes, using a two-dimensional method developed for helicopter blades.[57] A major feature of this method is the assumption of a time delay for the unsteady force development, which produces nonlinear hysteresis loops for both lift and pitching moment. The results indicate, via the regions of net negative damping (Fig. 26c), that the range over which sustained or growing oscillations may occur by deriving energy from the airstream. These predictions are in excellent general agreement with the measurements (Fig. 25). The success of this semiempirical method of predicting the wing torsional response is of considerable interest, because subsequent to the wind-tunnel tests, wing torsional oscillations (generally of small amplitudes) have been identified in flight on several military aircraft. In principle these oscillations could be greatly reduced by an appropriate active control system. For the present model a trailing-edge flap located where the flow is still attached (at roughly 30–50% semispan even at $\alpha = 6.5$ deg) should be particularly effective.

Aileron buzz was one of the earliest single-degree-of-freedom flutter problems encountered at subsonic speeds.[58] It occurs when the aerodynamic damping provided by the control hinge moment becomes sufficiently negative to overcome any mechanical damping in the system. Lambourne[59] remarked that for separated flows at transonic speeds the hinge moment

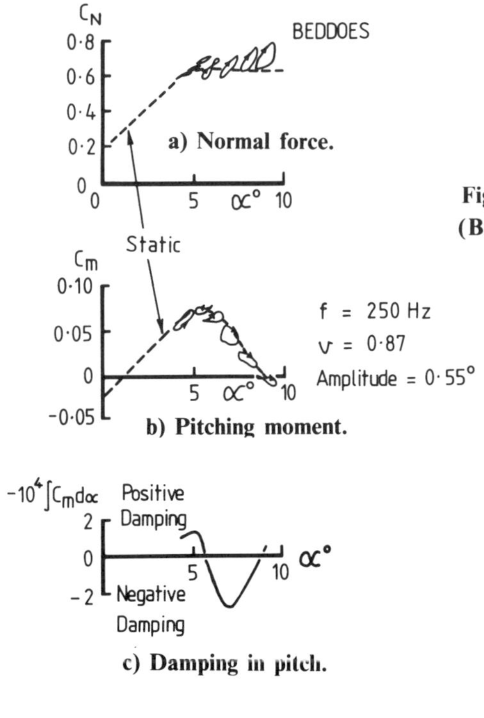

Fig. 26 Prediction of wing buzz. (British crown copyright reserved 1984.)

damping is generally a nonlinear function of the oscillation amplitude. Thus, the possible existence of a limit cycle depends on the initial value of the displacement, as illustrated in Fig. 27. With zero friction damping for free system A the aileron instability starts spontaneously and reaches a limit cycle amplitude $\bar{\delta}_A$. In contrast, for free system B the aileron becomes unstable only after an initial displacement δ_B and then reaches a final amplitude $\bar{\delta}_B$. Lambourne showed how this concept relates to the recent predictions of aileron buzz using the Navier-Stokes thin-layer approximation made by Steger and Bailey.[60] For the viscous calculations (Fig. 28a), there was a stable limit cycle at $M = 0.83$ with an amplitude of about 10 deg and another at $M = 0.85$ with an amplitude of about 5 deg. In addition, the aileron was stable at $M = 0.82$ for an initial flap angle of zero, but unstable for an amplitude of 4 deg. Lambourne shows how these predictions are consistent with the stable (continuous curve) and unstable (dashed curve) Poincare limit cycles. These viscous calculations were in good agreement with the original measurements. It will be interesting to see whether the alternative inviscid/viscous strong interaction methods now being developed can also give such good agreement. In marked contrast, for the inviscid calculations (Fig. 28b), the aileron was stable at $M = 0.88$ but had a divergent oscillation at $M = 0.84$. Although the inviscid calculation does give the correct frequency of the aileron buzz, it gives an incomplete overall description of the phenomenon.

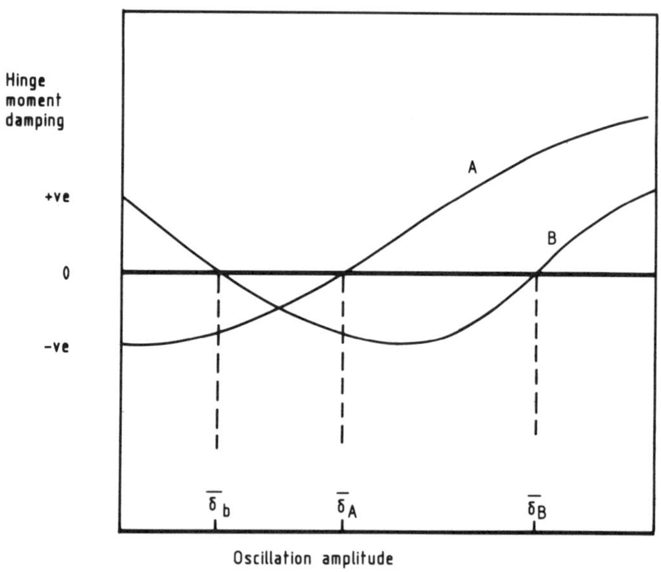

Fig. 27 Aerodynamic damping of flap. Variation with amplitude.
Free system A; instability starts spontaneously then reaches limit cycle amplitude $\bar{\delta}_A$.
Free system B; Unstable only after initial excitation $\bar{\delta}_b$ then reaches amplitude δ_B.

7. Some Experimental Problems

This section gives an introduction to some of the experimental problems in time-dependent aerodynamics that may be important at transonic speeds. It should help the reader to have confidence in the comparisons between wind-tunnel measurements and predictions given in subsequent chapters.

Steady and Unsteady Wall Interference

Most readers are probably familar with the different types of steady wall interference (e.g., that due to upwash δ_0 and curvature δ_1) and the correction methods described in Ref. 61]. Thus, the incidence correction may be written as

$$\Delta\alpha = \frac{SC_L}{C}\left[\delta_0 + \frac{\bar{c}}{2\beta H}\delta_1\right] \tag{16}$$

Ideally the curvature term should always be zero for transonic tunnels, because this occurs in Eq. (16) in conjunction with the term $\bar{c}/2\beta H$, which becomes large as $\beta \to 0$ at transonic speeds. If closed boundaries are used, the corrections are large but known exactly. If ventilated tunnels are used, the corrections can be much smaller but are rarely known with precision. If some form of adaptive wall is used, in principle the need for steady wall corrections can be eliminated completely.

The situation with respect to unsteady wall interference is less satisfactory, particularly at transonic speeds. Closed boundaries again offer a well-defined boundary condition but provide strong reflections of unsteady pressure waves. Ventilated boundaries offer a quite different boundary condition and a diffused reflection of unsteady pressure waves. There seems to be no possibility of making an active controlled adaptive wall capable of coping with simultaneous steady and time-dependent flow perturbations. However, it has been suggested[62] that the boundaries of adaptive wall tunnels could be made from sound-absorbing material, preferably giving "passive" attenuation over a wide frequency range. The fact that this concept offers both attenuation of reflected waves from oscillating models and a reduction in flow unsteadiness at transonic speeds over a useful range of frequencies has already been demonstrated in the RAE 3-ft tunnel.[63]

The pioneering work of Garner et al.[64] established a direct relationship between steady and unsteady interference characteristics of small models. In particular, the interference on an oscillating model has important contributions due to the curvature (represented by δ_1) and the upwash (represented by δ_0). It has been shown[65] that, for a wide range of symmetric wind tunnels with differing closed or open boundaries, there is a unique linear relation between δ_1 and δ_0:

$$\delta_1 = 1.68\delta_0 + 0.045 \tag{17}$$

It follows that, if a wall condition is selected with $\delta_1 = 0$ [the "ideal" condition for steady interference at transonic speeds according to Eq. (17)]:

$$\delta_0 = -0.027$$

Garner[66] has also demonstrated that it may be possible to achieve

$$\delta_1 = \delta_0 = 0$$

by the introduction of a small asymmetry between the roof and floor of the tunnel. This desirable characteristic with respect of lift and pitching moment would be at the expense of a small uncertainty in the blockage

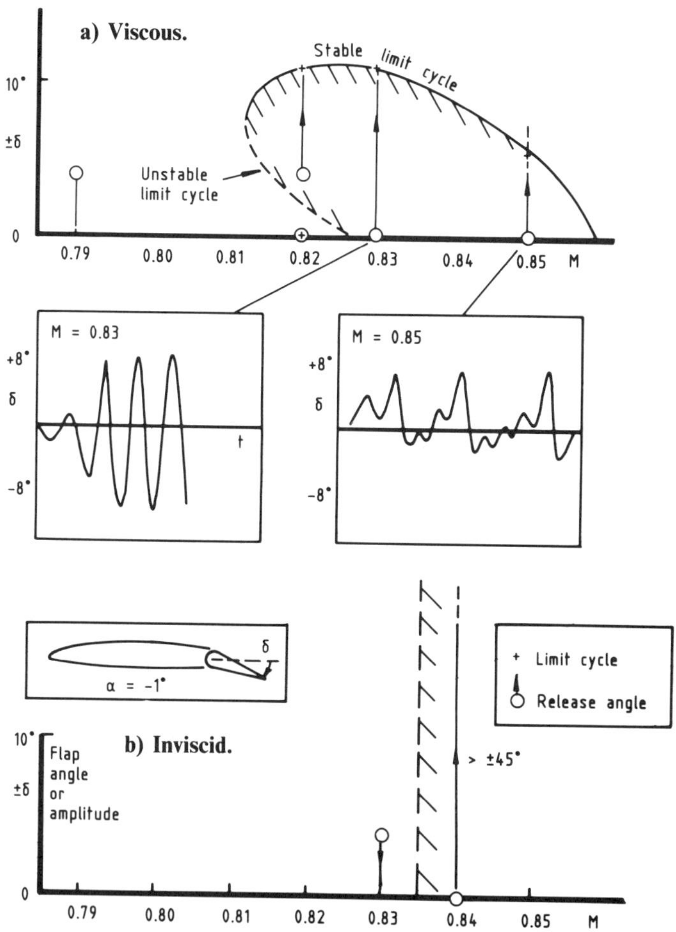

Fig. 28 Prediction of aileon buzz.

correction, which assumes symmetry between the roof and floor and between the side walls. If $\delta_1 = \delta_0 = 0$ could be achieved, it would confer unique advantages on both static and dynamic wind-tunnel tests at high subsonic and transonic speeds.

This discussion of attempts to achieve particular values of δ_1 and δ_0 is important. Some years ago, guided by Garner's theory, Moore and Wight[67] developed a hybrid slotted tunnel that gave low dynamic interference on models of widely varying planforms at Mach numbers as high as 0.95. This research should be more widely exploited.

A particularly severe form of dynamic wall interference can occur when pressure waves from an oscillating model are reflected from the walls onto the model in such a way as to reinforce or cancel the pressure on the model and create a resonance condition. The resonance frequency for a two-dimensional closed tunnel is given by the relation

$$f = n \frac{a}{2H} \sqrt{1 - M^2} \tag{18}$$

where $n = 1$ for the lowest mode.

As Mach number increases, the resonance frequency falls and the time taken for the return journey of a wave between the model and the wall increases. This may allow some attenuation. For two-dimensional perforated or slotted tunnels, the resonance frequencies are given by Eq. (18), but n now has a value dependent on the wall geometry and the depth of the plenum chamber. References 68 and 69 give the correct equations for the resonance frequencies in perforated and slotted tunnels, respectively. Figure 29 shows typical values for a tunnel of height $H = 1$ m.

Wind-tunnel resonances have been excited by two-dimensional models at transonic speeds in both closed and slotted[31,63] working sections. Although there is no evidence currently available of resonances excited by three-dimensional models, such a possibility should always be considered when anomalous aerodynamic measurements are found over a narrow range of frequencies coinciding with a tunnel resonance according to Eq. (18).

Thus far the types of interference discussed have been essentially subsonic in character, although locally the flow around the model is transonic. A completely different type of interference occurs when the shocks on the model strengthen and extend to the tunnel walls. This may happen when the Mach number increases at constant incidence (as in Fig. 3) or when the angle of incidence increases at constant Mach number (as in Figs. 18 and 19). When unsteady shocks from the model reach the tunnel walls, even a fully adaptive wall (controlled by steady flow equations) will produce significant time-dependent interference on the model. Hence, it is likely that this condition will represent an important upper limit for oscillatory time-dependent measurements. Once shock-induced separation occurs, the shock becomes unsteady (see discussion on Fig. 3). Hence, if the unsteady shock reaches the tunnel walls, this condition also represents an important upper limit for measurements of buffet excitation or buffeting.[46] It follows that for two-dimensional unsteady tests it would be advantageous to ensure that the schlieren window extends to the roof and floor of the tunnel.

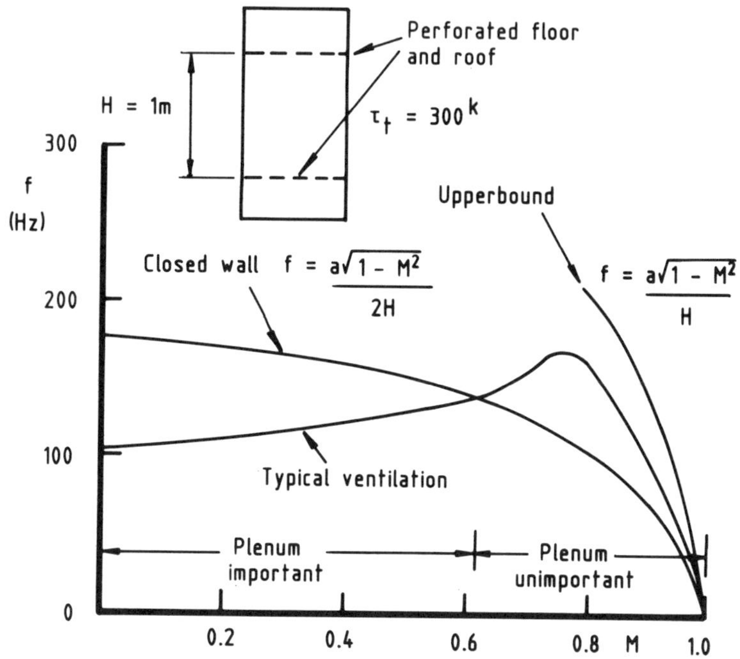

Fig. 29 First resonance frequency of transonic tunnel.

Alternatively, pressure transducers could be mounted (as a standard control for every experiment) on the roof of the tunnel in the vicinity of the model to detect the first characteristic signal indicative of time-dependent shock motion.

Flow Unsteadiness

Flow unsteadiness has three components: the pressure fluctuations, the velocity fluctuations or turbulence, and the temperature fluctuations, as discussed in Ref. 70. Most transonic tunnels have a high level of flow unsteadiness (particularly pressure fluctuations), and this has significant adverse effects on both time-dependent and steady measurements.

The pressure fluctuations are generated primarily in the extraction region at the end of the diffuser (for ventilated tunnels with diffuser suction) or in the first diffuser immediately downstream of the working section[71] (Fig. 30). The diffuser pressure fluctuations can be eliminated completely by the installation of a second throat, but because this requires additional power this feature is rarely incorporated in large transonic tunnels. The pressure fluctuations generated in the extraction region may be reduced significantly by covering the slots with perforated screens.[71] (This same modification is required to convert a conventional slotted tunnel to a hybrid slotted tunnel[65]).

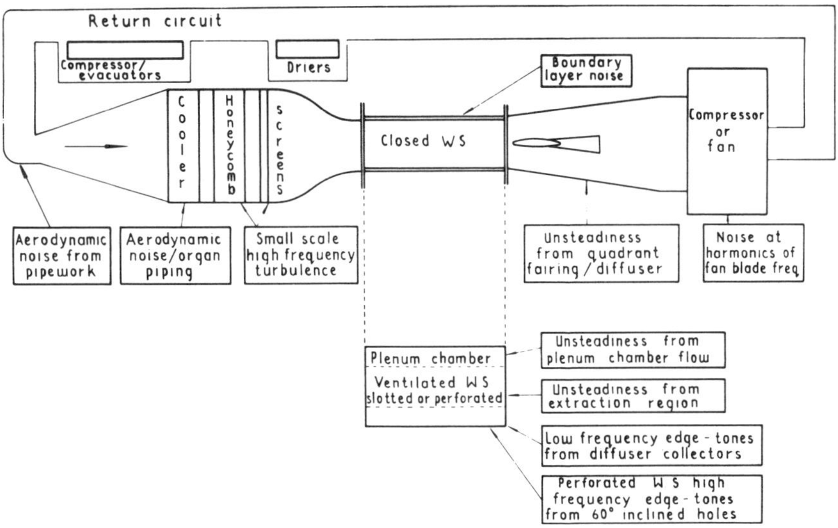

Fig. 30 Sources of unsteadiness in transonic tunnels. (British crown copyright reserved 1971.)

The turbulence level in the working section is determined primarily by the turbulence level in the settling chamber of the wind tunnel. This can be controlled by all the measures appropriate to the design of good low-speed wind tunnels, e.g., by honeycombs, screens, and carefully chosen contraction designs.[72] Temperature fluctuations are normally important only for blow-down wind tunnels. However, temperature fluctuations could be more important in cryogenic tunnels.

Effects of Flow Unsteadiness

For measurements of oscillatory time-dependent pressures at discrete frequencies the effect of high levels of pressure fluctuations is to increase the measurement time to achieve a given coherence level between the oscillatory pressure and the oscillatory motion by the model. Figure 31 illustrates this by some recent oscillatory pressure measurements on the AGARD SMP tail plane.[20,73] For $M = 0.65$, where the pressure fluctuations are generally high, many more periods are required at 3 Hz to obtain an acceptable coherence level than at 70 Hz. This is consistent with the much higher level of pressure fluctuations in the tunnel at 3 Hz than at 70 Hz. In marked contrast, for $M = 1.20$, where the pressure fluctuations are generally low, quite a small number of periods suffice to achieve acceptable coherence levels at both frequencies. Reference 73 also shows how the scale of flow unsteadiness can be gleaned from comparisons of oscillatory pressures measured at discrete frequencies (the traditional method) or with the frequency sweep technique (a new method currently being evaluated).

For measurements of random time-dependent pressures (say buffet excitation due to shock-induced separation as in Fig. 20), the main effect of high levels of pressure fluctuations is to increase the shock amplitude. Typically the shock amplitude due to shock–turbulent boundary-layer interaction is about $\pm 0.005c$, but this can be easily increased to $\pm 0.05c$ (due to a level of pressure fluctuations of $\sqrt{[nF(n)]} = 0.02$). Similarly, for measurements of random response (say buffeting derived from wing-root strain gages or wing-tip accelerometers), the main effect of flow unsteadiness is to mask the onset of light buffeting. The well-established criteria for the levels of pressure fluctuations acceptable for buffeting tests are reproduced from Ref. 71 in Fig. 32. Many transonic tunnels still exceed these levels and need improvement to make them suitable for buffeting measurements.

For steady measurements, the main effect of high levels of flow unsteadiness at transonic speeds is to alter the mean position of the shock, since this is a nonlinear region, even for attached flow (see discussion of Fig. 6). In addition, high levels of flow unsteadiness, particularly at low frequencies, make it difficult to measure this mean position. A good example of this problem is provided by Fig. 23 of Ref. 74, reproduced here as Fig. 33. High levels of flow unsteadiness can adversely effect both the mean signal level and the fatigue life of wind-tunnel balances. Experience shows that axial force balances are most sensitive to flow unsteadiness at transonic speeds,[71] probably because there is little aerodynamic damping in this mode. This is unfortunate, because the accurate measurement of drag after the onset of shock induced separation is often a prime requirement at transonic speeds.

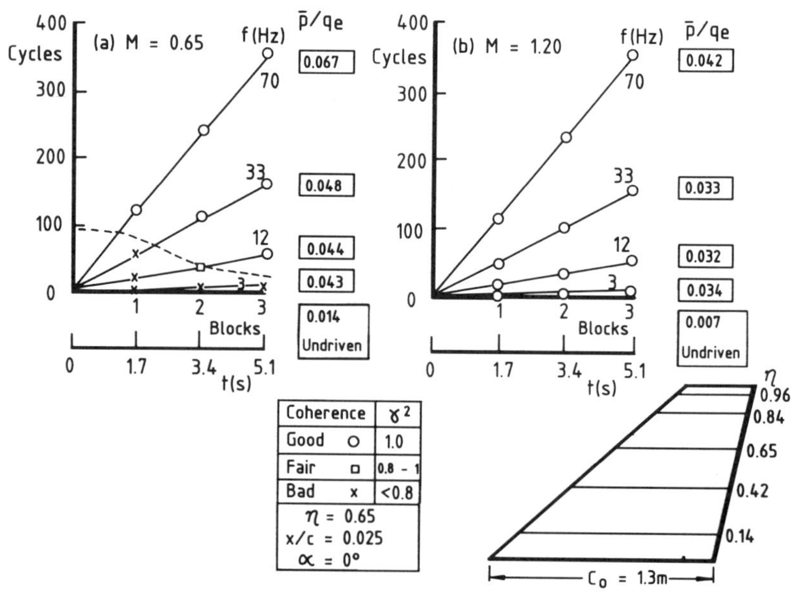

Fig. 31 Transonic section: coherence function close to leading edge at two speeds for discrete frequencies. (British crown copyright reserved 1984.)

Reference 75 gives a concise overview of the effects of flow unsteadiness on measurements (both steady and time-dependent) in intermittent wind tunnels.

Note on the Difficulty of Measuring Unsteady Pressures

There are two different methods of measuring time-dependent pressures, each of which has its own advantages and disadvantages.

The method developed originally at NLR is the distribution over the model of a large number of pressure tubes, connected in groups to a relatively small number of scanning valves.[76] These scanning valves measure each of the time-dependent pressures in sequence. The method has the advantage of cheapness but has the serious disadvantage that it is impossible to measure transient pressures or to perform cross correlations between successive pressures when only one scanning valve is used. (The measurement of cross correlations between pressures is admittedly not a frequent requirement but is illustrated in Fig. 13.)

The method developed by RAE is the use of a large number of in situ pressure transducers, each having its own amplifier.[10] This method is expensive but allows the simultaneous measurement of the mean component, the time-organized component (i.e., the component coherent with the model motion), and the random component of every pressure. It also allows transient and cross-correlation measurements.

Careful comparative tests sponsored by AGARD suggest that, for the restricted objective of the measurement of harmonic oscillatory pressures for small amplitude motions (say 1 deg) in attached flow, either method gives satisfactory results if applied carefully.[77] However, there is currently no absolute dynamic method available for the calibration of either system.

A combined system has recently been introduced at the National Aerospace Laboratory [(NLR) Amsterdam]. It consists of many tubes

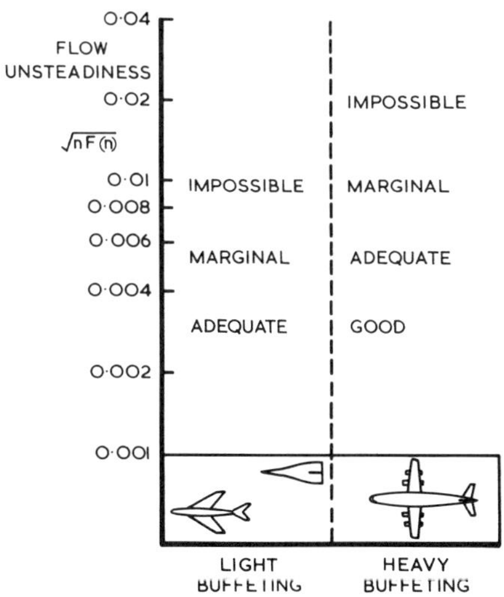

Fig. 32 Flow unsteadiness criteria for tunnel buffeting tests.
(British crown copyright reserved 1971.)

connected to scanning values supplemented by a few in situ pressure transducers. This combined system allows direct comparisons of the two methods for particular test conditions.[78]

Evaluation of NLR Unsteady Measurements

The main experimental problems thus identified are tunnel interference, flow unsteadiness, and the difficulty of measuring unsteady pressures. Hence, some evaluation may be attempted of the extensive collection of unsteady measurements made by NLR. (Typical examples are data sets 1 and 2 of Ref. 13, which include information about the accuracy of the unsteady pressure measurements.)

Most of these measurements were made on airfoils in the NLR pilot tunnel. This has closed side walls and slotted roof and floor designed to alleviate tunnel interference at transonic speeds. The ratio of model chord-to-tunnel height is 0.33; hence, wall interference is likely. There is evidence (Ref. 1, Sec. 13.2.3) that the quasisteady transonic correction due to wall interference might be about 20%, and no reliable correction methods are available. Figure 11 of Ref. 78 also suggests that even at subsonic speeds the dynamic interference is largest (about 20%) at a nonzero, though small, value of v.

Model drive frequencies are generally much less than 120 Hz. For the tunnel height ($H = 550$ mm), the predicted resonance frequency only falls as low as 120 Hz at a Mach number of about $M = 0.96$ (scaling from the curve with ventilation from Fig. 29). Hence, no transonic resonances should have been excited within the tunnel for tests below about $M = 0.94$.

The level of flow unsteadiness in the NLR pilot tunnel is low with respect to most other transonic tunnels (Ref. 79, Fig. 13). It has an rms level of only $\bar{p}/q = 0.004$, almost as good as the best closed working section combined with a second throat. This extremely low level of flow unsteadiness would facilitate oscillatory pressure measurements (as long as the flow on the model remains attached) and would permit the detection of light buffeting (Fig. 32).

Most of the oscillatory pressure measurements are derived from pipes as described above. However, these measurements are generally restricted to oscillatory motion of small amplitudes (as required for aeroelastic calculations) with attached flows and hence should be reliable. However, there could well be doubts about the reliability of oscillatory measurements for separated flows.

Although the test Reynolds number was only about 2×10^6, transition was fixed carefully for most experiments, so that shock–boundary-layer interactions were turbulent. In general no investigation of different transition fixes was attempted (as suggested in Ref. 9); thus, there is no internal evidence within the experiments that scale effects are small. A few comparative measurements with free transition are presented: these will be totally unrepresentative of full-scale conditions because the shock–boundary-layer interactions were laminar.

Thus, the main restrictions on the NLR measurements are wall interference, the low Reynolds number of the measurements, and attached flows.

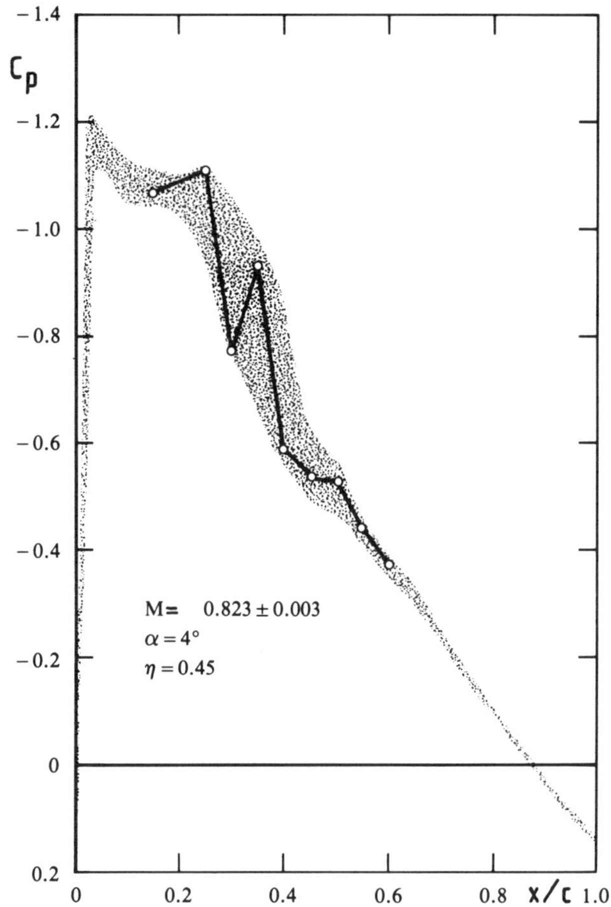

Fig. 33 Envelope of 12 successive measurements of the upper-surface static pressure distribution for one test condition.

These measurements explored the physical phenomena of unsteady transonic flow when computational methods for such flows did not exist.

8. Modification of Time-Dependent Boundary Conditions

There is good evidence that the modification of time-dependent boundary conditions can have a strong influence on the development of unsteady transonic flows. The two examples discussed here suggest that further research in this area would be profitable.

The first example relates to the suppression and reestablishment of periodic flow due to shock-induced separation on a 14%-thick biconvex wing, when tests are made with free transition over a wide range of

Reynolds numbers. The measurements described in Ref. 32 show that, at the lowest Reynolds number, $R = 1 \times 10^6$, the periodic flow extends from $M = 0.80$ to 0.88. As the Reynolds number increases, the Mach number range for periodic flow becomes progressively smaller, until it disappears very suddenly at a Reynolds number of 3.4×10^6 (Fig. 34a).

Earlier tests[31] indicate that an increase in boundary-layer thickness will reduce the region of periodic flow with both laminar or turbulent boundary layers. Hence, in the present tests the reduction in the region of periodic flow with the increases in Reynolds number may be reasonably be attributed to an increase in the thickness of the laminar boundary layer at the shock. The increase in boundary-layer thickness is caused by the forward

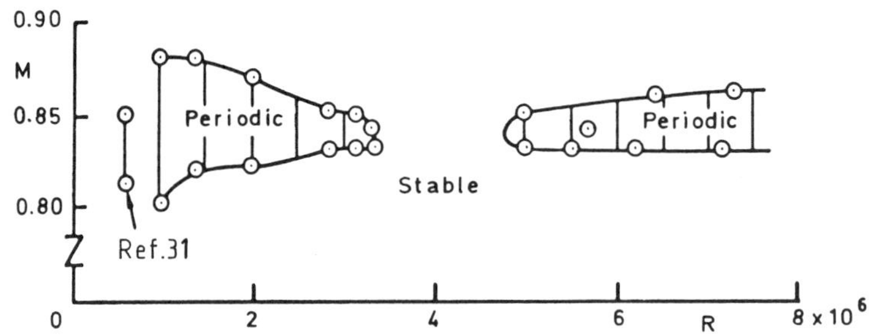

a) Free transition.

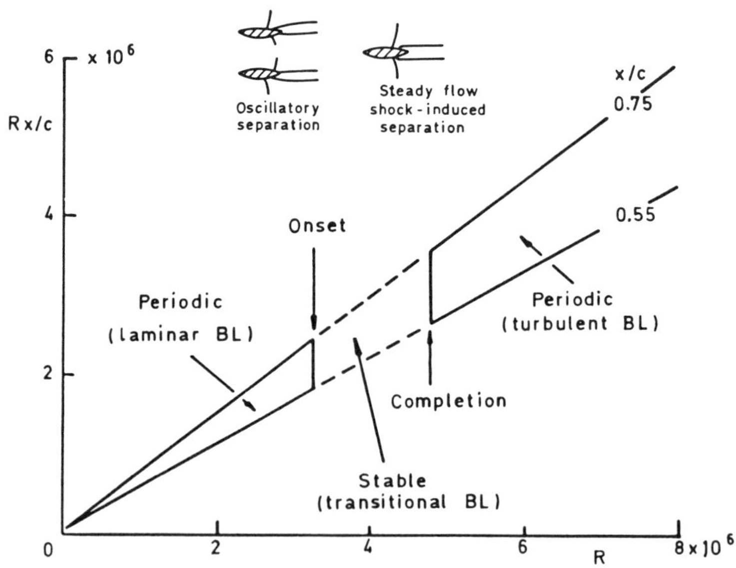

b) Influence of transition on periodic flow.

Fig. 34 Boundaries for period flow ($\alpha = 0$ deg), $M = 0.85$. (British crown copyright reserved.)

movement of transition, either in an attached laminar boundary layer or in a separated laminar shear layer.

The disappearance of the periodic flow over the Reynolds number range from 3.6×10^6 to 4.8×10^6 was unexpected. However, an explanation can be offered. The periodic flow involves large chordwise movements of the shock, from about $x/c = 0.75$ to 0.55, on opposite surfaces of the airfoil (Fig. 34b). Within this region of the airfoil, the hypothesis is made that the boundary layer is in the same state (i.e., either fully laminar or fully turbulent), for otherwise the intermittency characteristic of a transition region would surely seriously inhibit the development of periodic shear layers and shock movements. The intermittency of the transition region is determined by the rate of growth and decay of turbulent spots within the laminar boundary layer. This process varies both in time and in space. Hence, the intermittency of the transition region could interrupt the large chordwise movements of the shock and destroy the two-dimensionality of the motion inferred from shadowgraph and oil flow pictures.

According to this hypothesis, the sudden collapse of the periodic flow (at $R = 3.3 \times 10^6$) corresponds to R_{xt} at $x/c = 0.75$, where

$$R_{xt} = 3.3 \times 10^6 \times 0.75 = 2.7 \times 10^6 \tag{19}$$

The re-establishment of the periodic flow (at $R = 4.8 \times 10^6$) corresponds to R_{xt}, at $x/c = 0.55$:

$$R_{xt} = 4.8 \times 10^6 \times 0.55 = 2.7 \times 10^6 \tag{20}$$

The agreement between Eqs. (19) and (20) confirms the hypothesis and indicates that the boundary-layer transition is sudden, as the direct result of the shock-wave–boundary-layer interaction, and is probably associated with separation. This is, of course, in complete contrast to the more gradual transition process observed on flat plates or slender cones, where there is no shock and where the pressure gradient is small. On a slender cone the transition region at transonic speeds typically extends from onset at about $R_{xt} = 2 \times 10^6$ to completion at about $R_{xt} = 3 \times 10^6$, in a fairly noisy ventilated wind tunnel, such as the RAE 3-ft tunnel. Hence, the Reynolds number given by Eqs. (19) and (20) is of the correct order of magnitude.

Once the periodic flow is re-established (at $R = 4.8 \times 10^6$), further increases in Reynolds number slightly decrease the thickness of the turbulent boundary layer and hence slowly increase the Mach number range for periodic flow, until at $R = 7 \times 10^6$ periodic flow extends from $M = 0.83$ to 0.86. Hence, this type of periodic flow could certainly occur at full-scale Reynolds numbers as high as 10^8. The complete and misleading disappearance of the periodic flow over the Reynolds number range from $R = 3.3 \times 10^6$ to 4.8×10^6 (typical of model tests in many transonic tunnels) is a caution against attempting to simulate full-scale conditions more closely by tests with free transition at moderate Reynolds numbers at transonic speeds. It justifies the decision made in the Introduction to this chapter to restrict all of the subsequent discussions to fully turbulent boundary layers.

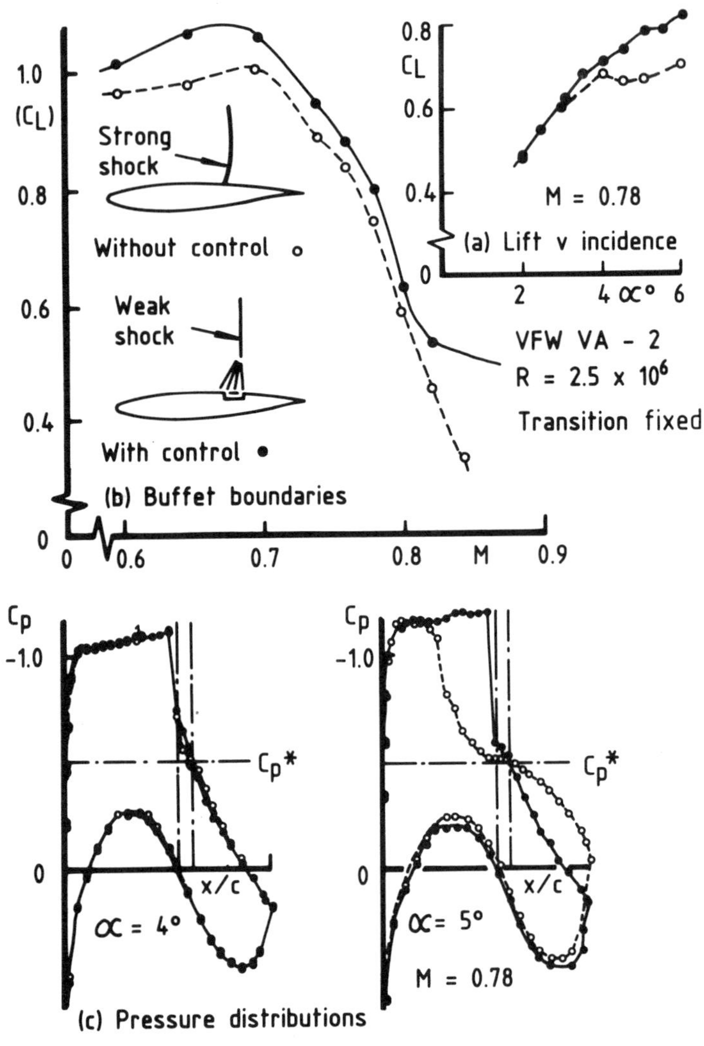

Fig. 35 Effect of modification of boundary condition.

The second example relates to the reduction of buffeting on a supercritical airfoil at transonic speeds by passive ventilation through a plenum of the flow upstream and downstream of the shock. Full details of the experiments are given in Ref. 80: tests were made with fixed transition at a Reynolds number of 2.5×10^6 with several different types of ventilation control.

The most effective form of control was the double-slot configuration, which produced a favorable effect on the steady off-design characteristics of all speeds. This is illustrated by the C_L vs α curve for $M = 0.78$ (Fig. 35a), which has a progressive increase in lift coefficient with angle of incidence. The effect of the control on the buffet boundaries was considerable (Fig. 35b). The authors of Ref. 80 made the following observation: "For the double slot in the Mach number range investigated, contrary to the other configurations, no shock oscillations could be detected in schlieren observations above the buffet boundary, nor was severe buffeting indicated by the rms signals. Moreover, buffeting apparently ceased completely for $M > 0.82$, although the flow was separated downstream of the shock. It seems that the strong coupling in the flow between the shock and the trailing edge is almost suppressed by double slot ventilation." The final remark emphasizes the importance of the connection between the shock and the trailing edge discussed earlier.

Currently no measurements of local unsteady pressures are available. However, Fig. 35c shows the effect of the control on the mean pressure distributions at $M = 0.78$. For $\alpha = 4$ deg the pressures with and without the control are virtually identical. Without the control, for $\alpha = 5$ deg the flow is separated at the shock at about $x/c = 0.30$. The flow is unsteady and the mean trailing-edge pressure has diverged.[18] In marked contrast, with the control the flow is attached and the shock is displaced downstream almost to the double slot. The flow is steady and the mean trailing-edge pressure has not diverged. Changes in static pressure distribution like those shown in Fig. 35c, accompanied by increases in lift like those in Fig. 35a, have been predicted for an NACA 0012 airfoil with a porous area in the vicinity of the shock.[81] The main effect on the steady forces appears due to the generation by the control of a series of weak oblique shocks upstream of the final terminal shock. Thus, the terminal shock is weaker than that without the control (cf. sketches in Fig. 35b).

There seems no a priori reason why similar improvements should not be found at off-design conditions at higher Reynolds numbers on swept wings, and such tests are planned.

References

[1]Tijdeman, H., "Investigation of the Transonic Flow Around Oscillating Aerofoils," National Aerospace Lab., Amsterdam, The Netherlands, TR-77-090U, 1977.

[2]Mabey, D. G., "Beyond the Buffet Boundary," *Aeronautical Journal*, Vol. 77, No. 748, April 1973, pp. 201–215.

[3]Chapman, D. R., "Computational Aerodynamics Development and Outlook," *AIAA Journal*, Vol. 17, Dec. 1979, pp. 1293–1313.

[4]Calarese, W. and Hankey, W. L., "Modes of Shock-Wave Oscillations on Spike Tipped Bodies," *AIAA Journal*, Vol. 23, Feb. 1985, pp. 185–192.

[5]Naumann, A. and Pfeiffer, H., "Über die Grenzschichtablösung am Zylinder bei Hohen Geshwindigkeiten," *Proceedings of the 2nd International Council of Aeronautical Sciences Meeting*, Pergamon, New York, 1960.

[6]Liepmann, H. W. and Puckett, A. E., *Aerodynamics of a Compressible Fluid*, Wiley, New York, 1954.

[7]Fiddes, S. P., Kirby, D. A., Woodward, D. S., and Peckham, D. H., "Investigations into the Effects of Scale and Compressibility on Lift and Drag in the RAE 5m Pressurised Low-Speed Wind Tunnel," AGARD CP-365, Paper 2, 1984.

[8]Jones, W. P., (ed.), *AGARD Manual on Aeroelasticity*, AGARD, 1961.

[9]Haines, A. B., "Possibilities for Scale Effect on Swept Wings at High Subsonic Speeds," AGARD CP-83-71, Paper 14, Aug. 1971.

[10]Welsh, B. L., "Recent Developments in the Measurement of Time-Dependent Pressure," AGARD CP-348, Paper 35, Sept. 1983.

[11]Mabey, D. G., "A Review of Some Recent Research on Time-Dependent Aerodynamics," *Aeronautical Journal*, Vol. 88, No. 872, Feb. 1984, pp. 23–37.

[12]Anonymous, "Transonic Unsteady Aerodynamics and its Aeroelastic Applications," AGARD CP-374, 1984.

[13]Lambourne, N. C., "Compendium of Unsteady Aerodynamic Measurements," AGARD Rept. 702, 1982.

[14]Roos, F. W., "Some Features of the Unsteady Pressure Field in Transonic Airfoil Buffeting," *Journal of Aircraft*, Vol. 17, No. 11, pp. 781–788.

[15]Pearcey, H. H., "Shock Induced Separation and Its Prevention by Design and Boundary Layer Control," *Boundary Layer and Flow Control*, Vol. 2, Pergamon, Oxford, UK, 1961, pp. 1166–1344.

[16]Laganelli, A. L., Martellucci, A., and Shaw, L. L., "Wall Pressure Fluctuations in Attached Boundary-Layer Flow," *AIAA Journal*, Vol. 21, April 1983, pp. 495–501.

[17]McDevitt, J. B. and Okuno, A. F., "Static and Dynamic Pressure Measurements on NACA 0012 Airfoil in the AMES High Reynolds Number Facility," NASA TP-2485, 1985.

[18]Pearcey, H. H. and Holder, D. W., "Simple Methods for the Prediction of Wing Buffeting Resulting from Bubble Type Separations," Aeronautical Research Council, ARC 23884, and National Physical Lab., NPL Aero. Rept. 1024, 1962.

[19]Rogers, E. W. E. and Hall, I. M., "An Introduction to the Flow About Plane Swept Back Wings at Transonic Speeds," *Aeronautical Journal*, Vol. 64, No. 596, Aug. 1980, pp. 449–464.

[20]Mabey, D. G., Welsh, B. L., and Cripps, B. E., "Measurements of Steady and Oscillatory Pressures on a Low Aspect Model at Subsonic and Supersonic Speeds," Royal Aircraft Establishment, TR-84-095, 1984; also, "Measurements and Calculation of Steady and Oscillatory Pressures on a Low Aspect Ratio Model at Subsonic and Transonic Speeds," *Journal of Fluids and Structures*, Vol. 1, 1987, pp. 445–468.

[21]Dowell, E. H., Williams, M. H., and Chi, M. R., "An Assessment of Theoretical Models for Viscous and Transonic Flow," AGARD CP-296, Paper 1, 1980.

[22]Landahl, M. T., *Unsteady Transonic Flow*, Pergamon, Oxford, UK, 1961.

[23]Nixon, D., "On Unsteady Transonic Shock Motions," *AIAA Journal*, Vol. 17, Oct. 1979, pp. 1143–1145.

[24]Tijdeman, H. and Seebass, R., "Transonic Flow Past Oscillating Aerofoils," *Annual Review of Fluid Mechanics*, 1980, Vol. 12, pp. 181–222.

[25]McOwat, D. M., Welsh, B. L., and Cripps, B. E., "Time-Dependent Pressure Measurements on a Swept Back Wing with an Oscillating Flap at Subsonic and Transonic Speeds," Royal Aircraft Establishment, TR-81-033, 1981.

[26]Lambourne, N. C., "Some Instabilities Arising from the Interaction Between Shock Waves and Boundary Layers," AGARD Rept. 182, 1958.

[27]Laitone, E. V., "Local Supersonic Region on a Body Moving at Subsonic Speeds," *Symposium Transsonicum (Aachen 1962)*, edited by K. Oswatitsh, Springer-Verlag, Berlin, FRG, 1964, pp. 57–70.

[28]Culick, F. E. and Rogers, T., "The Response of Normal Shocks in Diffuser," *AIAA Journal*, Vol. 21, Oct. 1984, pp. 1382–1390.

[29]Salas, M. D., "Local Stability Analysis for a Planar Shock Wave," NASA TP-2387, 1984.

[30]McDevitt, J. B., "Supercritical Flow About a Thick Circular Arc Airfoil," NASA TM-78549, 1979.

[31]Mabey, D. G., "Oscillatory Flows from Shock-Induced Separations in Biconvex Airfoils in Ventiled Wind Tunnels," AGARD CP-296, Paper 11, 1980.

[32]Mabey, D. G., Welsh, B. L., and Cripps, B. E., "Periodic Flows on a Rigid 14% Thick Biconvex Wing at Transonic Speeds," Royal Aircraft Establishment, TR-81-059, 1981.

[33]Levy, L. L., Jr., "Experimental and Computational Steady and Unsteady Transonic Flows about a Thick Airfoil," *AIAA Journal*, Vol. 16, June 1978, pp. 564–572.

[34]Levy, L. L., "Predicted and Experimental Steady and Unsteady Flows About a Biconvex Airfoil," NASA TM-81262, 1981.

[35]Hariharan, S. I. and Lester, H. C., "A Finite Difference Solution of the Propagation of Sound in Near Sonic Flows," NASA TM-84663, 1983.

[36]Radwan, S. F., Rockwell, D. C., and Johnson, S. H., "Critical Review of the Trailing Edge Condition in Steady and Unsteady Flow," NASA CR-169705, 1982.

[37]Magnus, R. and Yoshihara, H., "The Transonic Oscillating Flap," AIAA Paper 76-327, 1976.

[38]Houwink, R., "Computations of Separated and Transonic Flow About Airfoils in Unsteady Motion," National Aerospace Lab., Amsterdam, The Netherlands, NLR MP-84-094U, 1984.

[39]Le Balleur, J. C., " A Viscous-Inviscid Interaction Method for Computing Unsteady Transonic Separation," Office National d'Etudes et de Recherches Aéronautiques (ONERA) TP-1985-5, Jan. 1985.

[40]Green, J. E., "Some Aspects of Viscous-Inviscid Interactions at Transonic Speeds, and their Dependence on Reynolds Numbers," AGARD CP-83, 1971.

[41]Blackwell, J. A., "Preliminary Study of the Effects of Reynolds Number and Boundary Layer Transition Location on Shock Induced Separation," NASA TN D-5003, 1969.

[42]Kuchemann, D. and Weber, J., "An Analysis of Some Performance Aspects of Various Types of Aircraft Designed to Fly Over Different Ranges at Different Speeds," Royal Aircraft Establishment, TR-66-188, 1966.

[43]Coe, C. F., "The Effects of Some Variations in Launch Vehicle Nose Shape on Steady and Fluctuating Pressures at Transonic Speeds," NASA TM-X646, March 1962.

[44]Mabey, D. G., "Some Measurements of Base Pressure Fluctuations at Subsonic and Supersonic Speeds," Royal Aircraft Establishment, Aeronautical Research Council, TR-70 148, Aug. 1970; also published as ARC CP-1204, 1972.

[45]Mohsen, A. M., "Experimental Investigation of the Wall Pressure Fluctuations in Subsonic Separated Flows," Boeing Co. Rept. D6-17094, AD669214, Jan. 1967.

[46]Mundell, A. R. G. and Mabey, D. G., "Pressure Fluctuations Caused by Transonic Shock-Boundary Layer Interactions," Royal Aircraft Establishment, TM-2047, 1985.

[47]Lee, B. H. K. and Ohman, L. H., "Unsteady Pressure and Force Measurements Associated with Transonic Buffeting of a Two-Dimensional Supercritical Airfoil," National Aeronautical Establishment, NAE AN 14, 1983.

[48]Levy, L. L. and Bailey, H. E., "Computation of Airfoil Buffet Boundaries," *AIAA Journal*, Vol. 19, Dec. 1981, pp. 1488–1490.

[49]Mabey, D. G. and Butler, G. F., "Measurements of Buffeting on Two 65° Delta Wings of Different Materials," AGARD CP-226, Paper 6, 1977.

[50]Mabey, D. G. and Cripps, B. E., "Some Measurements on Buffeting on a Flutter Model of a Typical Strike Aircraft," AGARD CP-339, Paper 13, 1982.

[51]Zwaan, R. J., "Aeroelastic Problems of Wings in Transonic Flow," NLR, MP 81-005U, National Aerospace Lab., Amsterdam, The Netherlands, 1981; also "Unsteady Airloads and Aeroelastic Problems in Separated and Transonic Flow," VKI Lecture Series, von Kármán Inst., Belgium.

[52]Ashley, H., "On the Role of Shocks in the 'Sub-Transonic' Flutter Problem," AIAA Paper 79-0765, 1979.

[53]Zwaan, R. J., "Verification of Calculation Methods for Unsteady Airloads in the Prediction of Transonic Flutter," *Journal of Aircraft*, Vol. 22, Oct. 1985, pp. 833–839.

[54]Mabey, D. G., Ashill, P. R., and Welsh, B. L., "Aeroelastic Oscillations Caused by Transitional Boundary Layers and Their Attenuation," *Journal of Aircraft*, Vol. 24, July 1987, pp. 463–469.

[55]Moss, G. F. and Pierce, D., "The Dynamic Response of Wings in Torsion at High Subsonic Speeds," AGARD CP-226, Paper 4, April 1977.

[56]Beddoes, T. S., "Stall Flutter of a Finite Wing Westland RP 584," Aeronautical Research Council, ARC 37909, Oct. 1978.

[57]Beddoes, T. S., "A Synthesis of Unsteady Aerodynamic Effects Including Stall Hysteresis," *Vertica*, Vol. 1, No. 2, 1976, pp. 113–123.

[58]Erickson, A. L. and Stephenson, U. D., "A Suggested Method of Analyzing for Transonic Flutter of Control Surfaces Based on Available Experimental Evidence," NACA Rept. TIB 1505, 1947.

[59]Lambourne, N. C., "Unsteady Aerodynamic and Aeroelastic Properties of Control Surfaces in Transonic Flow," *Unsteady Airloads and Aeroelastic Problems in Separated Flow*, VKI Lecture Series, von Kármán Inst., Belgium, VKI LS 1981-4, 1981.

[60]Steger, J. L. and Bailey, H. E., "Calculation of Transonic Aileron Buzz," AIAA Paper 79-0134, Jan. 1979.

[61]Garner, H. C., Rogers, E. W. E., Acum, W. E., and Maskell, E. C., "Subsonic Wind Tunnel Wall Corrections," AGARDograph 109, 1966.

[62]Mabey, D. G., "The Use of Sound-Absorbing Walls to Reduce Dynamic Interference in Wind Tunnels," Aeronautical Research Council, R&M 383, 1976.

[63]Mabey, D. G., " The Reduction of Dynamic Interference by Sound-Absorbing Walls in the RAE 3 ft Wind Tunnel," Aeronautical Research Council, R&M 3837, 1977.

[64]Garner, H. C., Moore, A. W., and Wight, K. C., "The Theory of Interference Effects on Dynamic Measurements in Slotted-Wall Wind Tunnels at Subsonic Speeds and Comparison with Experiment," Aeronautical Research Council, R&M 3550, 1966.

[65]Mabey, D. G. and Steinle, F. W., "Computer Studies of Hybrid Slotted Working Sections with Minimum Steady Interference at Subsonic Speeds," *Aeronautical Journal*, Vol. 89, No. 884, April 1985, pp. 135–148.

[66]Garner, H. C., "Wall Interference on Small Lifting Wings in Tunnels with Differing Roof and Floor," Royal Aeronautical Establishment, TM-Aero-2066, April 1986.

[67]Moore, A. W. and Wight, K. C., "On Achieving Interference-Free Results from Dynamic Tests on Half Models in Transonic Wind Tunnels," Aeronautical Research Council, R&M 3626, 1969.

[68]Mabey, D. G., "The Resonance Frequencies of Ventilated Wind Tunnels," R&M 3841, 1978.

[69]Mabey, D. G., "Homogeneous Wall Boundary Condition for Oscillatory Flow in Slotted Tunnels," Royal Aircraft Establishment, TR-85007, 1985.

[70]Mabey, D. G., "Some Remarks on the Design of Transonic Tunnels with Low Levels of Flow Unsteadiness," NASA CR-2722, 1976.

[71]Mabey, D. G., "Flow Unsteadiness and Model Vibration in Wind Tunnels at Subsonic and Transonic Speeds," Aeronautical Research Council, ARC CP-1155, 1971.

[72]Bradshaw, P. and Pankhurst, R. C., "The Design of Low-Speed Wind Tunnels," *Progress in Aeronautical Sciences*, Vol. 5, edited by D. Küchemann and L. H. G. Sterne, (eds.), Pergamon, New York, 1964, pp. 1–69.

[73]Mabey, D. G., Welsh, B. L., and Cripps, B. E., "Effects of Flow Unsteadiness on the Measurement of Oscillatory Pressures on a Wind Tunnel Model with Discrete and Sweep Excitation," Royal Aircraft Establishment, TM Aero. 2016, 1984; also, "The Measurement of Oscillatory Pressures with Discrete and Sweep Excitation," *Journal of Fluids and Structures*," Vol. 1, 1987, pp. 95–105.

[74]Spaid, F. W. and Roos, F. W., "An Experimental Study of the Turbulent Boundary Layer on a Transport Wing in Transonic Flow," AIAA Paper 83-1687, 1983.

[75]Küchemann, D., (ed.), "On the Flow Quality Necessary for the Large European High-Reynolds Number Transonic Wind Tunnel LEHRT," AGARD Rept. 644, 1976.

[76]Bergh, H., "A New Method for Measuring the Pressure Distribution on Harmonically Oscillating Wings," *Proceedings of the 4th International Council of Aeronautical Sciences Conference*, Paris, ICAS Paper 64-28, 1964; also NLR MP-224, 1964.

[77]Destuynder, R. and Tijdeman, H., "An Investigation on Different Techniques for Unsteady Pressure Measurements in Compressible Flow and Comparison with Lifting Surface Theory," AGARD Rept. 617, 1974.

[78]Horsten, J. J., "Recent Developments in the Unsteady Pressure Measuring Techniquc," National Aerospace Lab., Amsterdam, The Netherlands, NLR 81-055U, 1981.

[79]Anonymous, *Users Guide to the* $1.6 \times 2.0\,m^2$ *High Speed Wind Tunnel of NLR*, National Aerospace Lab., Amsterdam, The Netherlands, 1976.

[80]Thiede, P., Krogmann, P., and Stanewsky, E., "Active and Passive Shock Boundary Layer Interaction Control on Supercritical Airfoils," AGARD CP-365, Paper 23, 1984.

[81]Chen, C. L., Chow, C. Y., Holst, T. L., and Van Dalsem, W. R., "Numerical Study of Porous Airfoils in Transonic Flow," NASA TM-86713, 1985.

[82]Fulker, J. L. and Ashill, P. R., "A Model of the Flow Over Swept Wings with Shock Induced Separation," *IUTAM Symposium on Turbulent Shear-Layer/Shock Wave Interactions*, Springer-Verlag, New York, 1985, pp. 233–245.

Chapter 2

Basic Equations for Unsteady Transonic Flow

David Nixon*
Nielsen Engineering & Research Inc., Mountain View, California

1. Flow Phenomena to be Modeled

The most complex flow that is encountered at transonic speeds occurs when the airplane undergoes an unsteady motion of large amplitude, due either to aeroelastic forces or maneuver. In such a motion shock waves on the airplane will move location considerably, may vanish, or be generated. The airflow may separate, due to angle-of-attack effects or to shock strength, and may separate and reattach in a time-dependent fashion. The patterns of shock waves and separation lines can be quite complicated. At its best any prediction method must be capable of modeling all of these types of flow and must be based on the nonlinear Navier-Stokes equations. Because these equations are very difficult to solve, it is desirable to look for simplications.

If the flow is always attached, the dominant nonlinearity is due to the shock waves and the viscous terms can be neglected. A second simplification is possible if the shock-wave geometry is relatively simple across a wing, since a two-dimensional prediction method may give adequate results if used in a "strip theory" approach. It is not possible to simplify the problem further without removing the dominant transonic phenomena of shock waves; thus, the simplest physical problem that must be modeled is that of two-dimensional, attached flow with shock waves.

2. Basic Equations for Transonic Flow

Navier-Stokes Equations

The Navier-Stokes equations[1] are generally accepted as the basic equations governing most fluid dynamic phenomena of interest to aerodynamicists. The equations are capable of representing mathematically the physical phenomena encountered in transonic flows, including mixed subsonic-

---

*President.

supersonic flow, shock waves, boundary layers, and separation. They also apply to turbulence, a random, dissipative, three-dimensional phenomenon that involves many characteristic scales.

The unsteady, compressible Navier-Stokes equations for a perfect gas are a set of equations representing the conservation of the physical quantities of mass, momentum, and energy. In addition, there are two equations of state. The complete set of equations are as follows:

Mass:

$$\frac{\partial \rho}{\partial t} + \frac{\partial(\rho u_j)}{\partial x_j} = 0 \tag{1}$$

Momentum:

$$\frac{\partial(\rho \boldsymbol{u}_i)}{\partial t} + \frac{\partial(\rho \boldsymbol{u}_i u_j)}{\partial x_j} = -\frac{\partial p}{\partial x_i} + \frac{\partial \tau_{ij}}{\partial x_j} \tag{2}$$

Energy:

$$\frac{\partial(\rho h)}{\partial t} + \frac{\partial(\rho h u_j)}{\partial x_j} = \frac{\partial p}{\partial t} + u_j \frac{\partial p}{\partial x_j} + \tau_{ij} \frac{\partial \boldsymbol{u}_i}{\partial x_j} - \frac{\partial \boldsymbol{q}_i}{\partial x_j} \tag{3}$$

where ρ is the density, $\boldsymbol{u}_i$ the velocity vector, p the pressure, h the specific enthalpy, τ_{ij} the stress tensor, and $\boldsymbol{q}_i$ the heat flux vector. The τ_{ij} and $\boldsymbol{q}_i$ are given by

$$\tau_{ij} = \lambda \delta_{ij} \frac{\partial u_m}{\partial x_m} + \mu\left(\frac{\partial \boldsymbol{u}_i}{\partial x_j} + \frac{\partial \boldsymbol{u}_j}{\partial x_i}\right) \tag{4}$$

$$\boldsymbol{q}_i = -k \frac{\partial T}{\partial x_i} \tag{5}$$

where δ_{ij} is the Kronecker delta, $\lambda = -\frac{2}{3}\mu$ the bulk viscosity, μ the dynamic viscosity, T the temperature, and k the thermal conductivity. The equation set is completed by using equations of state for a perfect gas, namely,

$$p = \rho R T \tag{6a}$$

and

$$h = C_p T \tag{6b}$$

where R is the gas constant and C_p is the specific heat at constant pressure.

Numerical solutions of the Navier-Stokes equations are available,[2] but only for Reynolds numbers of the order of 10^3. The reason is that all of the turbulence length scales must be resolved, and computer limitations severely restrict the range of length scales that can be represented. The computer time required for even a simple simulation of turbulence at a Reynolds number of about 5000 is approximately[3] 100 CPU hours on a

Cray-2 computer; the necessary grid is 256^3. Such computer times are totally out of the question for general computations; thus, some form of simplification of the Navier-Stokes equations is necessary.

Reynolds-Averaged Navier-Stokes Equations

Although there are several different forms of averaging, the most commonly used form for aerodynamics is Reynolds averaging. This is a time average over some characteristic time Δt, which is large relative to the characteristic time of the turbulence but which is small relative to the characteristic time of any motion of the body. In this form of averaging the general variable, $f_i(x_i,t)$, is written as

$$f_i(x_i,t) = \bar{f}(x_i) + f'(x_i,t) \tag{7}$$

where the bar denotes a time average of $f_i(x_i,t)$ and is defined at a time t_0 by

$$\bar{f}_i(x_i) = \lim_{\Delta t \to \infty} \frac{1}{\Delta t} \int_{t_0}^{t_0 + \Delta t} f(x_i,t)\,\mathrm{d}t \tag{8}$$

where the time Δt is large relative to the time scale of turbulent fluctuations and $f'(x_i,t)$ is the fluctuating remainder. Introducing this concept into the conservation equations gives, for example, the modified momentum equation

$$\frac{\partial}{\partial t}(\bar{\rho}\bar{\boldsymbol{u}}_i + \overline{\rho'\boldsymbol{u}'_i}) + \frac{\partial}{\partial x_j}(\bar{\rho}\bar{\boldsymbol{u}}_i\bar{u}_j + \bar{\boldsymbol{u}}_i\overline{\rho' u'_j} + \bar{u}_j\overline{\rho'\boldsymbol{u}'_i} + \bar{\rho}\,\overline{\boldsymbol{u}'_i u'_j} + \overline{\rho'\boldsymbol{u}'_i u'_j}$$
$$= -\frac{\partial \bar{p}}{\partial x_i} + \frac{\partial}{\partial x_j}\bar{\tau}_{ij} \tag{9}$$

The disadvantage of some sort of averaging is now apparent, since the process has introduced additional correlations such as the $\overline{\rho'\boldsymbol{u}'_i u'_j}$ terms involving the Reynolds stresses $\overline{\boldsymbol{u}'_i u'_j}$, which must be found as part of the solution. In short, therefore, there are more unknowns than equations; hence, additional equations must be derived by some form of turbulence modeling. Since the science, or art, of turbulence modeling has been studied for many years, there are numerous modeling schemes, ranging from simple algebraic methods to those involving one or more additional differential equations. A discussion on turbulence modeling is beyond the scope of this book. The interested reader is referred to other sources of information, for example, Refs. 4 and 5. However, some remarks on the validity of turbulence modeling are appropriate.

For attached flows such as the flow over wings, a simple mixing length turbulence model, such as that of Baldwin and Lomax,[6] may be quite adequate. If the flow is more complicated, such as separated flow, then a more complex model may be required, such as a two-equation k-ε representation of turbulence. For certain flows in which the turbulence may be highly anisotropic, it is necessary to use a Reynolds stress transport model in which a differential equation for each of the Reynolds stresses and for the dissipation is required. Because of the lack of experience in computing unsteady separated flows, it is not yet clear what form of turbulence model should be used. A further area of ignorance about turbulent flow behavior occurs at a shock wave; the topic of shock-turbulence interaction is one for which little knowledge is available, leading to uncertainty about the accuracy of a given turbulence model.

An important point to remember at this junction is that, because of limitations on present computer size, it is not really practical to use anything other than a simple algebraic turbulence model, since the addition of, say, a two-equation model, would reduce the maximum number of grid points available for a solution and hence lead to poorer resolution of the flow variables.

The Navier-Stokes equations will describe any type of flow over an airplane in any speed range; however, the problem of turbulence modeling can constrain the applications to certain physical problems, depending on the modeling scheme used. At present, the necessary use of the algebraic model restricts the applicable flows to those with little or no separation, although separated flows have been investigated qualitatively.[7]

A slight simplification of the Navier-Stokes equations is the thin-layer approximation.[6] In this approximation the viscous terms in the streamwise or near streamwise direction are neglected. The rationale behind this approximation is that the streamwise diffusion terms are small compared to the normal diffusion terms. A second important reason is that because of limitations on computer storage it is not possible to provide sufficient grid points to adequately resolve the viscous terms in all directions, and only those in a thin layer close to the surface are resolved. The thin-layer equations are similar to the more classical boundary-layer equations with the main exception that the normal momentum equation is retained. This prevents the occurrence of the usual boundary-layer singularity at a separation point.

Euler Equations

It is sometimes the case that the viscous effects may be small compared to the inviscid effects in a flow, for example, in a well-behaved attached flow, and it is then possible to neglect the viscous terms entirely. This approximation leads to the following Euler equations:

Mass conservation:

$$\frac{\partial \rho}{\partial t} + \frac{\partial (\rho u_j)}{\partial x_j} = 0 \tag{10}$$

Momentum conservation:

$$\frac{\partial (\rho \boldsymbol{u}_i)}{\partial t} + \frac{\partial (\rho \boldsymbol{u}_i u_j)}{\partial x_j} = -\frac{\partial p}{\partial x_i} \tag{11}$$

Energy conservation:

$$\frac{\partial(\rho h)}{\partial t}+\frac{\partial(\rho h u_j)}{\partial x_j}=\frac{\partial p}{\partial t}+u_j\frac{\partial p}{\partial x_j} \qquad (12)$$

It is assumed that the gas is nonconducting. The Euler equations will represent inviscid rotational flows in all speed ranges. However, the set of Eqs. (10–12) is not yet complete, since the neglect of the viscous terms in the Navier-Stokes equations leads to some nonuniqueness of the solutions. First, from the second law of thermodynamics, entropy cannot decrease; this results in the impossibility of expansion shock waves. In the Navier-Stokes equations this condition is enforced by the viscous terms. However, in the Euler equations it is possible to have expansion shock waves; therefore, the Euler equations must be solved subject to an entropy condition that eliminates these unrealistic solutions.

The second point concerns the trailing edge of an airfoil. At a sharp trailing edge the local flow is controlled by the action of the viscous stresses; this determines the circulation around the airfoil and hence the lift. In an inviscid flow the behavior at the trailing edge is no longer controlled, and so a multiplicity of solutions can be found. Therefore, to model the real viscous flow accurately, an additional condition, the Kutta-Joukowski condition of zero pressure jump at the trailing edge, is imposed to give a physically realistic flow. It is possible that numerical dissipation can represent the effect of real viscosity, but this is not a satisfactory substitute because of unknowns in the nature of numerically generated viscosity.

Potential Equation

The Gibbs relation is given by

$$T\,\mathrm{d}S=\mathrm{d}h-(1/\rho)\,\mathrm{d}p \qquad (13)$$

where S is the specific entropy. If the Gibbs relation is used to replace $\partial p/\partial x_i$ in the momentum equation [Eq. (11)], then, on using the mass conservation equation [Eq. (10)],

$$\frac{\partial \boldsymbol{u}_i}{\partial t}+u_j\frac{\partial \boldsymbol{u}_i}{\partial x_j}=T\frac{\partial S}{\partial x_i}-\frac{\partial h}{\partial x_i} \qquad (14)$$

Equation (14) can be rewritten as

$$\frac{\partial \boldsymbol{u}_i}{\partial t}+u_j\left(\frac{\partial \boldsymbol{u}_i}{\partial x_j}-\frac{\partial \boldsymbol{u}_j}{\partial x_i}\right)=T\frac{\partial S}{\partial x_i}-\frac{\partial h_0}{\partial x_i} \qquad (15)$$

where

$$h_0=h+\tfrac{1}{2}u_ju_j \qquad (16)$$

The term $-(\partial \boldsymbol{u}_i/\partial x_j) - (\partial \boldsymbol{u}_j/\partial x_i)$ is the vorticity $\boldsymbol{\Omega}_{ij}$ and Eq. (15) can be written in vector form as

$$T\,\mathrm{grad}(S) + \boldsymbol{u} \times \boldsymbol{\Omega} = \mathrm{grad}h_0 + \frac{\partial \boldsymbol{u}}{\partial t} \tag{17}$$

where $\boldsymbol{u}$ is the velocity vector and $\boldsymbol{\Omega}$ is the vorticity vector. This is the unsteady version of Crocco's equation.

If the entropy production is zero, then Eq. (17) implies that a function exists such that

$$\frac{\partial \Phi}{\partial x_i} = \boldsymbol{u}_i \tag{18a}$$

and

$$\frac{\partial \Phi}{\partial t} = -h_0 = -(h + \tfrac{1}{2}u_j u_j) \tag{18b}$$

The function Φ is the velocity potential, and Eq. (18a) ensures that the vorticity $\boldsymbol{\Omega}$ is zero; Eq. (18) is valid only for irrotational flow.

If the flow is isentropic, then a combination of Eq. (6) and Eq. (13) gives

$$p/\rho^{\gamma} = \mathrm{const} \tag{19}$$

If the subscript ∞ denotes freestream values, then combining Eqs. (6), (18b), and (19) gives

$$\rho/\rho_{\infty} = \{1 + [(\gamma - 1)/2]M_{\infty}^2[1 - \Phi_t - \tfrac{1}{2}u_j u_j]\}^{1/(\gamma - 1)} \tag{20}$$

where M_{∞} is the freestream Mach number. In Eq. (20) u_j is normalized with respect to the freestream velocity U_{∞}, and Φ_t is normalized with respect to U_{∞}^2. Equation (20) is valid only if the freestream is stationary, in other words,

$$(\Phi_t)_{\infty} = 0 \tag{21}$$

If Eq. (20) is used in conjunction with Eqs. (10) and (18a), then an equation for Φ, called the full potential equation, is obtained.

In an inviscid fluid the only internal sources of entropy production are shock waves, and the entropy production through a normal shock is given by

$$\frac{\Delta S}{R} = \ell n\left\{\left[1 + \frac{2\gamma}{(\gamma + 1)}(M_1^2 - 1)\right]^{-1/(\gamma - 1)}\left[\frac{(\gamma + 1)M_1^2}{(\gamma - 1)M_1^2 + 2}\right]^{-\gamma/(\gamma - 1)}\right\} \tag{22}$$

where M_1 is the Mach number just ahead of the shock. For values of M_1 close to unity, Eq. (22) can be approximated by

$$\frac{\Delta S}{R} = \frac{2\gamma}{(\gamma + 1)^2}\frac{(M_1^2 - 1)^3}{3} \tag{23}$$

In transonic flow shock waves are weak, and it can be assumed that M_1 is sufficiently close to unity to make $\Delta S/R$ negligible. Thus, in transonic flow the isentropic assumption may be made even if shocks are present. Hence, it is frequently assumed that as a consequence of Eqs. (15) and (23) a potential formulation can be used to represent the flow. However, in a potential calculation of transonic flow with shock waves, momentum is not conserved. Since the derivation of Eq. (15) and the potential equation require that momentum is conserved, it follows that the potential formulation is not correct for transonic flows with shock waves. A more complete explanation of this point for steady flow is given by Klopfer and Nixon.[8] However, in spite of the inconsistencies in its derivation, the transonic potential equation has been shown to be an adequate model for transonic flows with shock waves. As for the Euler equations, an entropy condition and the Kutta-Joukowski condition are necessary to close the problem.

It is assumed sometimes that the deviations of the flow from the freestream values are small compared to the freestream and shock waves are nearly normal to the freestream direction. Thus, if u, v, and w are the normalized velocity components in the coordinate directions x,y,z, respectively, and x is the freestream, then

$$\Phi = x + \phi \tag{24}$$

$$U = 1 + u = 1 + \phi_x \tag{25a}$$

$$V = v = \phi_y \tag{25b}$$

$$W = w = \phi_z \tag{25c}$$

If these equations are combined with Eq. (20) and if second-order terms in ϕ_y, ϕ_z, and ϕ_t and third-order terms in ϕ_x are neglected, then

$$\frac{\rho}{\rho_\infty} \approx 1 - M_\infty^2 \phi_t - M_\infty^2 \phi_x - \frac{\phi_x^2}{2} M_\infty^2 [1 - (2-\gamma) M_\infty^2] \tag{26}$$

The reason for retaining the second-order terms in ϕ_x is that these terms are the minimum necessary to represent nearly normal shock waves. If Eqs. (24) and (25) are combined with the conservation of mass equation

$$\frac{\partial \rho}{\partial t} + \frac{\partial(\rho V)}{\partial x} + \frac{\partial(\rho V)}{\partial y} + \frac{\partial(\rho W)}{\partial z} = 0$$

and if all second-order terms in the perturbation velocity potential ϕ are neglected except those in the x derivatives, where third-order terms are neglected, then the resulting equation is

$$\{1 - M_\infty^2 - M_\infty^2[3 + M^2(\gamma - 2)]\phi_x\}\phi_{xx} + \phi_{yy} + \phi_{zz} - \frac{2M_\infty^2}{U_\infty}\phi_{xt} - \frac{M_\infty^2}{U_\infty^2}\phi_{tt} = 0 \tag{27}$$

Equation (27) is the unsteady transonic small-disturbance (TSD) equation. It can be derived by a more rigorous approach, but it is felt that the present treatment adequately illustrates the relation between the small-disturbance and mass conservation equations. A great advantage of the TSD equation is that to the order of accuracy of the equations the boundary conditions can be applied on a mean surface. This avoids the need to treat moving boundaries in a computation.

The TSD equation is formally valid as $M_\infty \to 1$. Thus, in Eq. (27) the term in the brackets may be replaced by $(\gamma + 1)$, and Eq. (27) becomes

$$\{1 - M_\infty^2 - (\gamma + 1)M_\infty^2 \phi_x\}\phi_{xx} + \phi_{yy} + \phi_{zz} - \frac{2M_\infty^2}{U_\infty}\phi_{xt} - \frac{M_\infty^2}{U_\infty^2}\phi_{tt} = 0 \tag{28}$$

Finally, if the frequency of motion is low, the small-disturbance solution can be further simplified[9] by neglecting the ϕ_{tt} term in Eq. (27) or (28). Thus, Eq. (28) becomes

$$\{1 - M_\infty^2 - (\gamma + 1)M_\infty^2 \phi_x\}\phi_{xx} + \phi_{yy} + \phi_{zz} - \frac{2M_\infty^2}{U_\infty}\phi_{xt} = 0 \tag{29}$$

This is the classical low-frequency small-disturbance equation. A further modification is sometimes made in which the exponent on M_∞ in the $(\gamma + 1)M_\infty^2$ term is replaced by an arbitrary q. The various choices of $\boldsymbol{q}$, varying from 1.69 to 2.0, are referred to as the transonic scaling. A fuller description of this scaling is given in Ref. 9.

It should be noted that, if the nonlinear term in Eq. (27) or (28) is neglected, then the resulting equation is the common linear subsonic equation, which can be solved using vortex lattice theory,[10] or by using integral equations.[11] In the author's opinion, the minimum complexity of equation that should be used for transonic flow prediction is Eq. (29), since equations lower in the hierarchy, e.g., the linear subsonic equation, cannot represent shock waves.

Summary of Equations Sets

In summary, then, the systems of equations that may be used for a transonic flow prediction are as follows:

Reynolds-Averaged Navier-Stokes Equations

These equations should be used for the most complex aerodynamic phenomena, including flow separation. A good turbulence model is a necessity if useful results are to be obtained.

Thin-Layer Navier-Stokes Equations

This simplified system of equations can be used for flows when viscous effects are important and when separation, if it occurs, is mild. A relatively simple turbulence model may be adequate.

Euler Equations

The Euler equations should be used when the viscous effects are small and can be neglected but when the inviscid flow is highly rotational, for example, for flows with strong shock waves.

Full Potential Equation

The full potential equation may be used when any shock waves in the flow are weak and when accuracy requirements demand the implementation of the boundary conditions on the moving wing.

Small-Disturbance Equation

This equation is a subset of the full potential equation and should be used when the accuracy requirement for the boundary conditions for the potential equation can be relaxed. The low-frequency approximation can be used when the reduced frequency is less than 0.2. [9]

3. Boundary Conditions

A correct choice of boundary conditions for the equations described in the previous section is essential if an adequate solution procedure is to be developed. It is sometimes difficult to determine the correct set of boundary conditions because of the complexity of the equations. However, a description of boundary conditions that have been used for computations is given later, although some doubts remain as to their validity. It should be noted that no existence or uniqueness proof for transonic flow with shock waves has been derived, and, consequently, some degree of faith is required. The following conditions are for flows with a subsonic freestream.

Navier-Stokes Equations

For the incompressible Navier-Stokes equations in two dimensions, it is necessary and sufficient to specify all the velocity components on the boundary.[12] It is expected, but not proven, that these are also correct boundary conditions for incompressible three-dimensional flow. No rigorous set of boundary conditions for the compressible Navier-Stokes equations has been derived, although several attempts to derive computationally adequate conditions have been made (for example, Refs. 13 and 14). A discussion of computational boundary conditions is given in Chapter 4 by Steger and Van Dalsem. It should be noted that, for the Reynolds-averaged Navier-Stokes equations, boundary conditions for the equations modeling turbulence are also required.

Euler Equations

It may be shown[15] that the correct boundary conditions for two dimensions for the Euler equations are to specify the normal or tangential velocity on the boundary. It may be expected that these boundary conditions are also applicable in three dimensions. In addition to the boundary conditions, the Kutta-Joukowski condition and an entropy condition are necessary to close the problem.

Potential Equation

The boundary conditions for the potential equation are relatively well understood and require that either the velocity potential or its normal derivative be specified on a boundary. The Kutta-Joukowski condition must be satisfied for a lifting flow, and, if shock waves are present, an entropy condition is necessary.

4. Some Aspects of Transonic Potential Theory

The essential nonlinearity of the governing equations for transonic flow is due to the need to model shock waves. Although little is known about the nature of nonlinear equations, in general, it is possible to perform some analysis on the TSD equations to gain some insight into transonic flow phenomena. The first topic discussed in this section is the linearization of the TSD equation for unsteady flow. The second topic concerns the uniqueness of the solutions of the TSD equation. For convenience the discussion is concerned only with two-dimensional flows.

Indicial Theory

Consider the change in the pressure coefficient $C_p(x,y,t)$ due to an infinitesimal change in some parameter $\varepsilon(\tau)$ at some time τ. If the pressure distribution varies continuously with ε, then a Taylor series expansion gives

$$\Delta C_p(x,y,t) = C_{p_\varepsilon}(x,y,t,\tau)\,\frac{d\varepsilon(\tau)}{d\tau}\,\Delta\tau + \begin{pmatrix}\text{higher-order}\\ \text{terms}\end{pmatrix} \tag{30}$$

where t is the time, $C_{p_\varepsilon}(x,y,t,\tau)$ the rate of change of $C_p(x,y,t)$ with ε at some time τ, and $\Delta C_p(x,y,t)$ the change in pressure distribution. Neglecting the higher-order terms, the total effect of such steps up until t is then

$$\Delta C_p(x,y,t) = \varepsilon(0)C_{p_\varepsilon}(x,y,t,0) + \int_0^t C_{p_\varepsilon}(x,y,t,\tau)\,\frac{d\varepsilon(\tau)}{d\tau}\,d\tau \tag{31}$$

If it can be assumed that the behavior of C_p with ε is linear, then $C_{p_\varepsilon}(x,y,t,\tau)$ can be represented by its value at $\tau = 0$, provided that t is taken relative to τ. Thus,

$$C_{p_\varepsilon}(x,y,t,\tau) = C_{p_\varepsilon}(x,y,t-\tau,0) \tag{32}$$

Equation (31) then becomes

$$\Delta C_p(x,y,t) = \varepsilon(0)C_{p_\varepsilon}(x,y,t,0) + \int_0^t C_{p_\varepsilon}(x,y,t-\tau,0)\,\frac{d\varepsilon(\tau)}{d\tau}\,d\tau \tag{33}$$

By a simple change of variable, Eq. (33) can then be written as

$$\Delta C_p(x,y,t) = \varepsilon(0) C_{p_\varepsilon}(x,y,t) + \int_0^t C_{p_\varepsilon}(x,y,\tau) \frac{\mathrm{d}\varepsilon(t-\tau)}{\mathrm{d}\tau} \mathrm{d}\tau \tag{34}$$

where the functional form of $C_{p_\varepsilon}(x,y,t,0)$ has been contracted to $C_{p_\varepsilon}(x,y,t)$ for convenience in presentation. Equation (34) is the form of a typical relation in the indicial method. It is assumed that $C_{p_\varepsilon}(x,y,t)$ can be obtained by some means. A more detailed description of the indicial method is given by Tobak.[16]

Linearization of the TSD Equation

The principal objection to linearizing the TSD equation [Eq. (27)] about some value of ϕ_0 is that a linearized equation cannot model discontinuities other than those that are in the coefficients of the linearized equation.

Consider, for example, the two-dimensional version of Eq. (28):

$$\{1 - M_\infty^2 - (\gamma+1) M_\infty^2 \phi_x\} \phi_{xx} + \phi_{yy} - \frac{2M_\infty^2}{U_\infty} \phi_{xt} - \frac{M_\infty^2}{U_\infty^2} \phi_{tt} = 0 \tag{35}$$

This equation can be written in a more compact form as

$$(\beta^2 - k\phi_x)\phi_{xx} + \phi_{yy} - A\phi_{xt} - B\phi_{tt} = 0 \tag{36}$$

where

$$\beta^2 = 1 - M_\infty^2 \tag{37a}$$

$$k = (\gamma + 1) M_\infty^2 \tag{37b}$$

$$A = (2M_\infty^2)/U_\infty \tag{37c}$$

$$B = M_\infty^2 / U_\infty^2 \tag{37d}$$

If the velocity potential is expanded as a perturbation series

$$\phi(x,y,t) = \phi_0(x,y) + \varepsilon\phi_1(x,y,t) + \cdots \tag{38}$$

where ε is some small parameter, then Eq. (36) reduces to the following equations:

$$(\beta^2 - k\phi_{0_x})\phi_{0_{xx}} + \phi_{0_{yy}} = 0 \tag{39}$$

$$\beta^2 \phi_{1_{xx}} - k(\phi_{0_x}\phi_{1_x})_x + \phi_{1_{yy}} = 0 \tag{40}$$

Equation (39) is capable of representing a shock wave at some location x_s because of the nonlinear term

$$k\phi_{0_x}\phi_{0_{xx}}$$

Equation (40) is linear in ϕ_1 and cannot represent any discontinuity other than that due to ϕ_0. Consequently, the shock wave cannot change location due to the change in ϕ of ϕ_1. The ability to let the shock wave move when a boundary change is made is essential to any adequate model of transonic flow. It is shown by Nixon[17] that this aspect of modeling transonic flow can be attained in a linear equation set if a strained coordinate system is used. In the following discussion the basic idea is outlined; a fuller discussion may be found in Refs. 17 and 18. It is assumed in the discussion that the shock waves are parallel to the y axis.

Let the velocity potential $\phi(x,y,t)$ be expanded as the series

$$\phi(x,y,t) = \phi_0(x',y) + \varepsilon\phi_1(x',y,t)\ldots \tag{41}$$

where the coordinates x and x' are related by

$$x \sim x' + \varepsilon x_1(x't) + \cdots \tag{42}$$

where $x_1(x')$ is a specified straining function and is related to the shock movement $\delta x_s(t)$.

If Eqs. (41) and (42) are used in Eq. (36), then the following set of equations results:

$$(\beta^2 - k\phi_{0_x})\phi_{0_{x'x'}} + \phi_{0_{yy}} = 0 \tag{43}$$

$$\beta^2\phi_{1_{x'x'}} - k(\phi_{0_{x'}}\phi_{1_{x'}})_{x'} + \phi_{1_{yy}} - A\phi_{1_{x't}} - B\phi_{1_{tt}} = [x_1(\phi_{0_{x'}} - \phi^2_{0_{x'}})]_{x'}$$
$$+ x_{1_{x'}}\left(\phi_{0_{x'}} - \frac{\phi^2_{0_{x'}}}{2}\right)_{x'} + Ax_{1_{x't}}\phi_{0_{x'}} \tag{44}$$

Equation (44) is a linear equation for $\phi_{1_{x'}}$ and has a discontinuity at the value of x' where $\phi_0(x',y)$ has a discontinuity. However, since x' can have a different representation in physical space for the zeroth and first approximations because of the $x_1(x',t)$ term in Eq. (42), the shock location can change in the perturbation expansion.

The straining function $x_1(x,t)$ is fairly arbitrary, but it can be shown[17,18] that the straining must be such that expansion shocks are excluded. In other words, the shock must move such that there are no expansion shocks in the solution of Eq. (44). The equation representing this condition is linear in $x_1(x',t)$; hence, the complete perturbation problem is linear. This allows the use of superposition and this fact is used in the following analyses.

Effect of Frequency on Shock Motion

The indicial theory, outlined in the section *Indicial Theory* and the linearization of the TSD equation described in the section *Linearization of the TSD Equation* can be used to provide insight into the behavior of shock motions in oscillating transonic flow. The description given here is a shortened version of the analysis given in Ref. 19.

The equation for the shock movement, denoted by $\delta x_s(t)$,

$$\delta x_s(t) = \delta x_{s_\varepsilon}(t)\varepsilon(0) + \int_0^1 \delta x_{s_\varepsilon}(\tau)\,\frac{d\varepsilon(t-\tau)}{d\tau}\,d\tau \tag{45}$$

where $\varepsilon(t)$ is a time-dependent parameter typifying the unsteady motion. The expression $\delta x_{s_\varepsilon}(t)$ is the indicial response defined as the response of the shock location due to the instantaneous unit change in the parameter $\varepsilon(t)$.

Let the unsteady motion be simple harmonic motion, with the parameter $\varepsilon(t)$ given by the equation

$$\varepsilon(t) = \varepsilon_0 + \varepsilon_1\, e^{i\omega t} \tag{46}$$

where ω is the frequency of the motion.

Define the term $\Delta x_{s_\varepsilon}(t)$ by the following equation:

$$\Delta x_{s_\varepsilon}(t) = \delta x_{s_\varepsilon}(\infty) - \delta x_{s_\varepsilon}(t) \tag{47}$$

where $\delta x_{s_\varepsilon}(\infty)$ is the steady-state change in shock location due to a unit change in ε. Only the periodic motion is of interest, that is, the limit as $t \to \infty$. Then substitution of Eqs. (46) and (47) into Eq. (45) gives

$$\delta x_s(t) = \delta x_{s_\varepsilon}(\infty)\varepsilon_0 + \varepsilon_1\, e^{i\omega t}\left[\delta x_{s_\varepsilon}(\infty) - i\omega \int_0^\infty \Delta x_{s_\varepsilon}(\tau)\, e^{-i\omega\tau}\, d\tau\right] \tag{48}$$

To investigate the shock motion further, some representation of the indicial response, $\delta x_{s_\varepsilon}(t)$, is required. A suitable equation for the indicial response is

$$\delta x_{s_\varepsilon}(t) = \frac{\delta x_s(\infty)}{\varepsilon_1}\,(1 - e^{-at})\left(1 + \sum_{n=1}^{N} b_n\, e^{-ant}\right) \tag{49}$$

where $\delta x_s(\infty)$ is the steady-state motion due to a change in ε of ε_1, a is some constant, and b_n are constant coefficients. The number of terms in the summation can vary from 1 to ∞. Incidentally, the constant a is given by

$$a = \varepsilon_1 \left[\frac{\partial \Delta x_{s_\varepsilon}(t)}{\partial t}\right]_{t=0} \left[\left(1 + \sum_{n=1}^{N} b_n\right)\delta x_s(\infty)\right]^{-1} \tag{50}$$

Substitution of Eq. (49) into Eqs. (47) and (48) leads to the expression

$$\delta x_s(t) = \delta x_{s_\varepsilon}(t)\varepsilon_0 + \frac{\delta x_s(\infty)}{\varepsilon_1}\cdot \varepsilon_1\, e^{i\omega t}\left\{\frac{a(a - i\omega)}{a^2 + \omega^2} + i\omega a \sum_{n=1} b_n \frac{[a^2 n(n+1) - \omega^2] - i\omega a(2n+1)}{(a^2 n^2 + \omega^2)[a^2(n+1)^2 + \omega^2]}\right\} \tag{51}$$

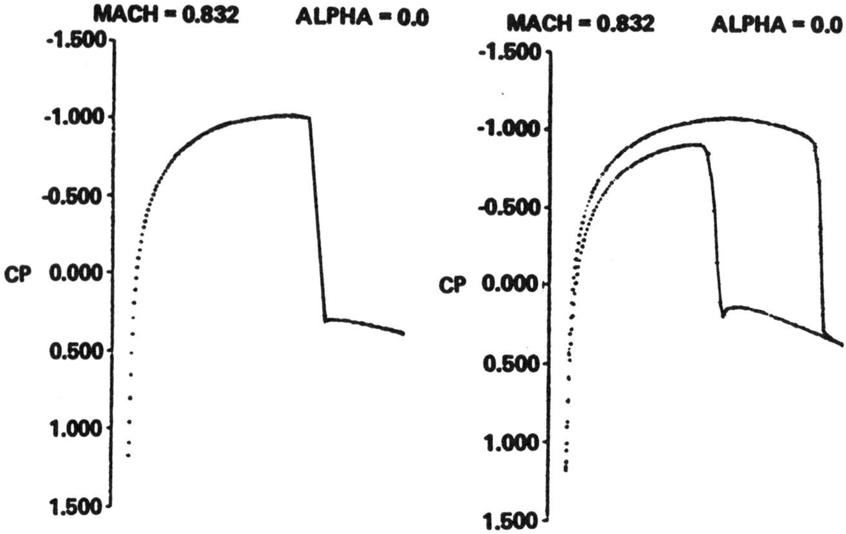

Fig. 1 Pressure distribution over a Joukowski airfoil $M_\infty = 0.832$, $\alpha = 0$ deg: a) symmetric solution; b) asymmetric solution.

For simplicity, consider only purely simple harmonic motions, in which case, $\varepsilon_0 = 0$. Equation (51) can then be written in the form

$$\delta x_s(t) = \frac{\delta x_s(\infty)}{\varepsilon_1}\varepsilon_1\left\{e^{i\omega t}A_0\,e^{i\theta_0} + \sum_{n=1}^{N} b_n A_n\,e^{i\theta_n}\right\} \tag{52a}$$

where

$$A_0 = \frac{\bar{a}}{(\bar{a}^2 + k^2)^{\frac{1}{2}}}, \qquad \theta_0 = \tan^{-1}\left(-\frac{k}{a}\right) \tag{52b}$$

and

$$A_n = \frac{k\bar{a}[\bar{a}^2 n(n+1) + k^2]}{(\bar{a}^2 n^2 + k^2)[\bar{a}^2(n+1)^2 + k^2]}$$

$$\theta_n = \tan^{-1}\left\{\frac{\bar{a}n(n+1) - k^2}{\bar{a}k(2n+1)}\right\} \tag{52c}$$

where k is the reduced frequency $\omega c/U_\infty$, c is the airfoil chord, and

$$\bar{a} = ac/U_\infty \tag{53}$$

Equation (52) is completely general, apart from the assumption of small disturbances. Consider now the behavior in the limiting cases of $k = 0$,

$k \to \infty$. When $k = 0$,

$$A_0 = 1, \qquad \theta_0 = 0, \qquad A_n = 0, \qquad \theta_n = \pi/2$$

Hence,

$$\lim_{k \to 0} \delta x_s(t) = \varepsilon_1 \delta x_s(\infty)/\varepsilon_1 \tag{54}$$

which is the steady-state change due to a change in ε of ε_1. When $k \to \infty$,

$$A_0 \to \bar{c}/k, \qquad \theta_0 \to -\pi/2, \qquad A_n \to \bar{a}/k, \qquad \theta_n \to -\pi/2$$

Hence,

$$\lim_{k \to \infty} \delta x_s(t) \to \frac{\varepsilon_1 \delta x_s(\infty)}{\varepsilon_1} \left(1 + \sum_{N+1}^{N} b_n\right) \frac{\bar{a}}{k} \exp[i(\omega t - \pi/2)] \tag{55}$$

It easily can be seen that the asymptotic value of the phase angle is a phase lag of $\pi/2$. Furthermore, the magnitude of the shock motion, $|\delta x_s(t)|$, approaches 0 like $1/k$.

For high frequencies, then, the shock wave does not move; thus, x and x' are identical. The unsteady linearized transonic equation [Eq. (44)] thus reduces to

$$\beta^2 \phi_{1_{xx}} - k(\phi_{0_x} \phi_{1_x})_x + \phi_{1_{yy}} - A\phi_{1_{xt}} - B\phi_{1_{tt}} = 0 \tag{56}$$

If the frequency is high, the term $\phi_{1_{xt}}$ will dominate the other x' derivatives, and a high-frequency form of Eq. (44) is then

$$\phi_{1_{yy}} - A\phi_{1_{xt}} - B\phi_{1_{tt}} = 0 \tag{57}$$

This is Landahl's[20] high-frequency transonic equation.

Multiple Solutions of the Potential Equation

In recent years multiple solutions to the numerical approximation to the full potential equation have appeared in the literature.[21,22] Initially, the phenomena appeared in computations of the flow over a symmetric airfoil at zero angle of attack where two lifting solutions were present in addition to the expected nonlifting solution. An example from Ref. 21 is shown in Fig. 1. In Ref. 22 some results for a nonsymmetric airfoil, a RAE 2822 section, are also presented. Steinhoff and Jameson[21] suggested that the change from one of the solutions to another is discontinuous and noted a hysteresis effect indicating that the lift coefficient C_L depended on whether the angle of attack was increasing or decreasing. Salas et al.[22] extended the computations of the flows considered by Steinhoff and Jameson to show that it is possible to construct a smooth C_L-α curve connecting the three solutions for a symmetric airfoil.

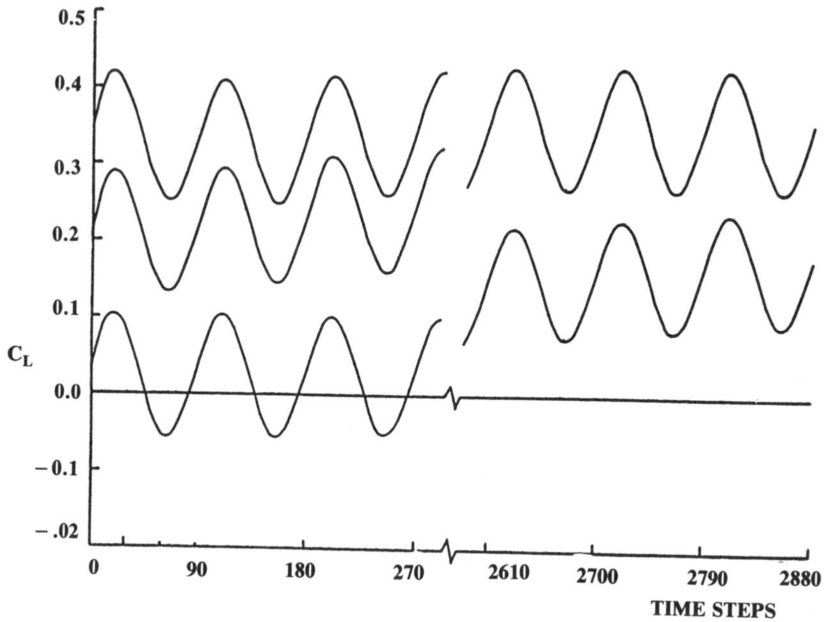

Fig. 2 Unsteady nonunique behavior.

More recent work by Williams et al.[23] has shown that solutions can appear for unsteady flow. An example of the behavior at unsteady flow is shown in Fig. 2. Nixon et al.[24] showed that the multiple solutions can exist in steady and unsteady flow through cascades and that the cascade effect seems to reduce the possibility of such solutions occurring.

The possible appearance of these solutions in routine calculations degrades the usefulness of potential calculations considerably, since often it is not clear whether a nonphysical solution is present. In recent work[25] it is suggested that the steady potential equation can give multiple solutions if the flow has a supersonic pocket; the actual appearance of a nonunique solution may be dependent on the relative stability of a particular solution and the algorithm used in the computation.

References

[1]Mehta, U. and Lomax, H., "Reynolds Averaged Navier-Stokes Computations of Transonic Flows—the State of the Art," Progress in Astronautics and Aeronautics, *Transonic Aerodynamics*, Vol. 81, edited by D. Nixon, AIAA, New York, 1982.

[2]Rogallo, R. and Moin, P., "Numerical Simulation of Turbulent Flows," *Annual Review of Fluid Mechanics*, Vol. 16, 1984, pp. 99–137.

[3]Nixon, D., "A Report on a Workshop on Turbulence," Nielsen Engineering & Research, Inc., Mountain View, CA, TR-361, 1986.

[4]Reynolds, W. C. and Cebeci, T., "Calculation of Turbulent Flows," *Turbulence*, edited by P. Bradshaw, Springer-Verlag, Berlin, FRG, 1978.

[5]Bradshaw, P., Cebeci, T., and Whitlaw, J. H., *Engineering Calculation Methods for Turbulent Flow*, Academic, New York, 1981.

[6]Baldwin, V. W. and Lomax, H., "Thin Layer Approximation and Algebraic Model for Separated Turbulent Flows," AIAA Paper 78-257, Jan. 1978.

[7]Rumsey, C. L., Thomas, J. L., Warren, G. P., and Liu, G. C., "Upwind Navier-Stokes Solutions for Separated Periodic Flows," AIAA Paper 86-0247, Jan. 1986.

[8]Klopfer, G. H. and Nixon, D., "Nonisentropic Potential Formulation for Transonic Flows," *AIAA Journal*, Vol. 22, June 1984, pp. 770–776.

[9]Ballhaus, W. F., "Some Recent Progress in Transonic Flow Computations," *Numerical Methods on Fluid Dynamics*, edited by H. J. Wirz and J. J. Smoldren, Hemisphere, Washington, DC, 1976.

[10]Albano, E. and Rodden, W. P., "A Doublet-Lattice Method for Calculating Lift Distributions on Oscillating Surfaces in Subsonic Flows," *AIAA Journal*, Vol. 7, Feb. 1965, pp. 279–285.

[11]Possio, C., "L'Azione Aerodinamica sul Profilo Oscillante in un Fluido Compressibile a Velocitá Iposonora," *L'Aerotecnica*, t. XVIII, Fasc. 4, April 1938.

[12]Temam, R., *Navier-Stokes Equations and Nonlinear Functional Analysis*, Society for Industrial and Applied Mathematics, Philadelphia, PA, 1983.

[13]Yee, H. C., "Numerical Approximation of Boundary Conditions with Application to Inviscid Equations of Gas Dynamics," NASA TM-83625, 1981.

[14]Chakrarthy, S., "Euler Equations—Implicit Schemes and Implicit Boundary Conditions," AIAA Paper 82-0228, 1982.

[15]Nixon, D., Tzuoo, K.-L., Caruso, S. C., Farshchi, M., Klopfer, G. H., and Ayoub, A., "Concepts for Radically Increasing the Numerical Convergence Rate of the Euler Equations," Nielsen Engineering & Research, Inc., Mountain View, CA, TR-380, 1987.

[16]Tobak, M., "On the Use of the Indicial Function Concept in the Analysis of Unsteady Motion of Wings and Wing-Tail Combinations," NACA Rept. 1188, 1954.

[17]Nixon, D., "Perturbation of a Discontinuous Transonic Flow," *AIAA Journal*, Vol. 16, Jan. 1978, pp. 47–52.

[18]Nixon, D., "Calculation of Unsteady Transonic Flows Using the Integral Equation Method," *AIAA Journal*, Vol. 16, Sept. 1978, pp. 972–983.

[19]Nixon, D., "On Unsteady Transonic Shock Motions," *AIAA Journal*, Vol. 17, Oct. 1979, pp. 1143–1145.

[20]Landahl, M. T., *Unsteady Transonic Flow*, Pergamon, Oxford, UK, 1961.

[21]Steinhoff, J. S. and Jameson, A., "Multiple Solutions of the Transonic Potential Flow Equation," *AIAA Journal*, Vol. 20, Nov. 1982, pp. 1521–1525.

[22]Salas, M. D., Gumbert, C. R., and Turkel, E., "Nonunique Solutions to the Transonic Potential Flow Equation," *AIAA Journal*, Vol. 22, Jan. 1984, pp. 145–146.

[23]Williams, M. H., Bland, S. R., and Edwards, J. W., "Flow Instability in Transonic Small-Disturbance Theory," *AIAA Journal*, Vol. 23, Oct. 1985, pp. 1491–1496.

[24]Nixon, D. and Tzuoo, K. L., "Notes on Unsteady Transonic Calcade Flows," Neilsen Engineering & Research, Inc., Mountain View, CA, TR-343, 1985.

[25]Nixon, D., "Observations on the Occurrence of Multiple Solutions in Transonic Potential Theory," *Acta Mechanica*, Vol. 68, July 1987, pp. 43–56.

Chapter 3

Practical Problems: Airplanes

Atlee M. Cunningham Jr.*
General Dynamics Corporation, Fort Worth, Texas

1. Introduction

The unsteady transonic phenomena described in the previous chapters will be discussed in this chapter regarding how they affect aircraft and the practical problems that result. Topics covered will include flutter, limit cycle oscillations, maneuver aerodynamics, control reversal, buzz, active control technology needs, gust response and alleviation, and unsteady shock-vortex interaction. Three-dimensional effects on real aircraft will be stressed. All of the known important effects of unsteady transonic flows will be presented as far as is possible regardless of how well we understand their nature.

2. Flutter

The determination of aircraft flutter characteristics is one of the most important safety aspects of the aircraft design and development process. Three approaches are normally employed in determining flutter characteristics: analytic, flutter model testing in an appropriate wind tunnel, and flight flutter testing of either a free-flight model or a full-scale aircraft. Analytic techniques that are the least expensive to use are primarily used to evaluate the entire flight envelope and isolate potential problem areas. Wind-tunnel flutter model testing is next employed to verify the analytic predictions and to explore possible nonlinearities such as those associated with transonic and/or separation effects. This approach is not used until well into the design process because it is much more costly than the analytic approach. The last and most costly technique to be applied is flight flutter testing, with which the most critical areas are again evaluated with full-scale hardware under actual flight conditions.

For flight speeds outside the nominal transonic range of about 0.8–1.1, linear theoretical methods may be applied to obtain reliable predictions of flutter characteristics for winglike surfaces. However, the presence of strong

---

*Engineering Specialist, Sr. Computational Fluid Dynamics Group, Aerospace Technology.

shock systems in the flowfields at transonic speeds prevents application of linear methods at these conditions. A further restriction is placed on the linear methods that excludes their application to conditions that are dominated by highly separated flows. These flows may be experienced at both fully subsonic and supersonic speeds as a result of high aircraft incidence or large deflections of control surfaces, for example. A further complication at transonic speeds is that the strong shock systems may also induce significant regions of separated flow. This aspect of transonic flow leads to difficulties in determining flutter characteristics that, because of Reynolds number effects, may cast doubt on results obtained with the analytic transonic methods described in other chapters of this book or even wind-tunnel flutter tests. In some cases, flight testing is the only way to obtain the final answer. Thus, it is important to understand how transonic flowfields influence all unsteady aerodynamic forces that can contribute to flutter.

The most notable feature of transonic flow that distinguishes it from purely subsonic or supersonic flows is the coupling of unsteady aerodynamic forces with the steady mean flowfield. As discussed in the previous chapter, the "spike" in an unsteady pressure loading due to shock motion changes both in character and position as the shock is altered. This can influence flutter speeds, as was first shown by Farmer and Hanson,[1] where the characteristic flutter speed dip in the transonic speed range was significantly different, as shown in Fig. 1 for two models with identical planform, stiffness, and mass but different airfoil sections. (Transition was fixed on both models in order to represent full-scale Reynolds number as much as possible.) This finding opened a Pandora's box for the flutter analyst, since it meant that any change in the steady mean flowfield at transonic speeds could change the flutter characteristics. Thus, airfoil shape and camber, wing twist, aircraft incidence, control surface deflection, and elastic deflections due to aeroelastic and inertia effects all have the potential for influencing transonic flutter speeds. This is not so bad for aircraft designed to cruise at transonic speeds such as transports, since the angle of attack is limited and inertia forces are nominally steady at 1 *g*. However, fighter aircraft designed to maneuver at transonic speeds are a severe example of this problem. Not only do they operate under wide ranges of incidence, but they are also subject to wide variations of inertia forces and aeroelastic effects due to the carriage of different wing stores and the pulling of high *g* at different altitudes. Such influences on flutter speeds are very difficult to obtain with wind-tunnel flutter testing due to 1) difficulty of obtaining static and dynamic structural similarity and 2) inability of flutter models to withstand large static loads necessary to simulate high incidence conditions.

The situation mentioned earlier is further complicated by the existence of shock-induced separation. Although it is present with almost all occurrences of wing shocks, the extent of shock-induced separation is usually very small for weak shocks, and its influence is minimal on the associated unsteady aerodynamic forces. With stronger shocks, either at higher incidence or higher Mach number, the area covered by shock-induced separation can spread significantly, even as far as the trailing edge. The larger the

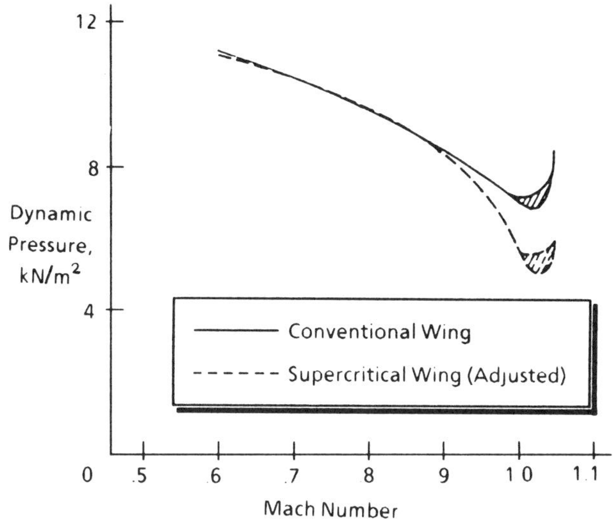

Fig. 1 Comparison of conventional wing flutter boundary with adjusted supercritical wing boundary.[1]

area covered by shock-induced separation becomes, the larger is its influence on shock position and motion due to small changes in wing incidence. As a result, significant discrepancies can exist in position for strong shocks if Reynolds number similarity is not achieved either through the use of transition fixing or high-Reynolds-number test facilities. As an example, for the C-141 airplane, shock locations measured in full-scale flight test were as much as 20% further aft than those obtained with wind-tunnel testing.[2] Thus, in addition to matching the exact geometry, some investigation of scale effects must be included in the determination of transonic flutter characteristics, particularly where strong shocks are present.

Although very little unsteady experimental data are available, it is possible to demonstrate some of the geometric effects just discussed. The experimental results of Triebstein[3] provide the means by which steady and unsteady transonic effects can be related for a variety of conditions. Figure 2 shows Mach number effects on the steady and unsteady pressure distributions for an NACA 0012 airfoil in two-dimensional flow obtained from a wind-tunnel test. (Transition was not fixed in these tests.) Since the mean angle of attack is zero and the airfoil is symmetric, the upper and lower surface pressures are similar for steady flow and simply 180 deg out of phase for unsteady flow. As can be seen, with increasing Mach number and therefore shock strength, the unsteady pressure distributions change toward the familiar shape that is dominated by a peak due to shock oscillation up to $M = 0.8$. Data for $M = 0.9$ show a significant difference, however, which is most likely due to shock-induced separation as indicated by the sign change on the shock peak.

The role of shock-induced separation becomes increasingly important as the wing incidence is increased. The separation mentioned earlier at $M = 0.9$ was induced primarily because of the increased shock strength at the higher Mach number. An increase of wing incidence at a lower Mach number will likewise produce stronger shocks and thereby promote shock-induced separation. Triebstein also investigated this effect with the NACA 0012 airfoil and obtained the results shown in Fig. 3. Although the Reynolds number was decreased at the higher incidences, the relationships between steady and unsteady pressures are still valid. The Mach number was held constant at 0.78, and incidence was increased from 0 to 7 deg for steady flow and from 0 to 5 deg for unsteady flow with the amplitude and frequency fixed at 1 deg and 30 Hz (or $k = 0.073$), respectively. The most interesting development in the steady data is the aft movement of the shock up to 3 deg and forward movement at 5 and 7 deg, as can be seen in the upper surface pressures. It should also be noticed that this reversal in shock movement with increasing incidence is accompanied by a change of the trailing-edge pressure coefficient from positive (full recovery) to negative (partial recovery). This characteristic is usually referred to as trailing-edge pressure divergence and signals the point at which significant separation has reached the wing trailing edge. This is also a point that may be used as an indicator for the onset of wing buffet, as discussed in Chapter 1.

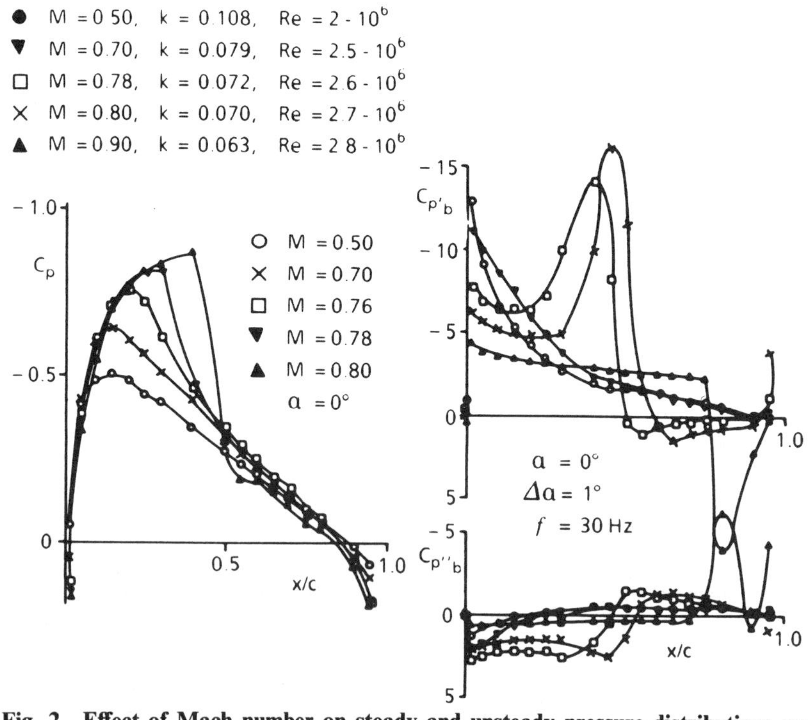

Fig. 2 Effect of Mach number on steady and unsteady pressure distributions on an NACA 0012 airfoil.[3]

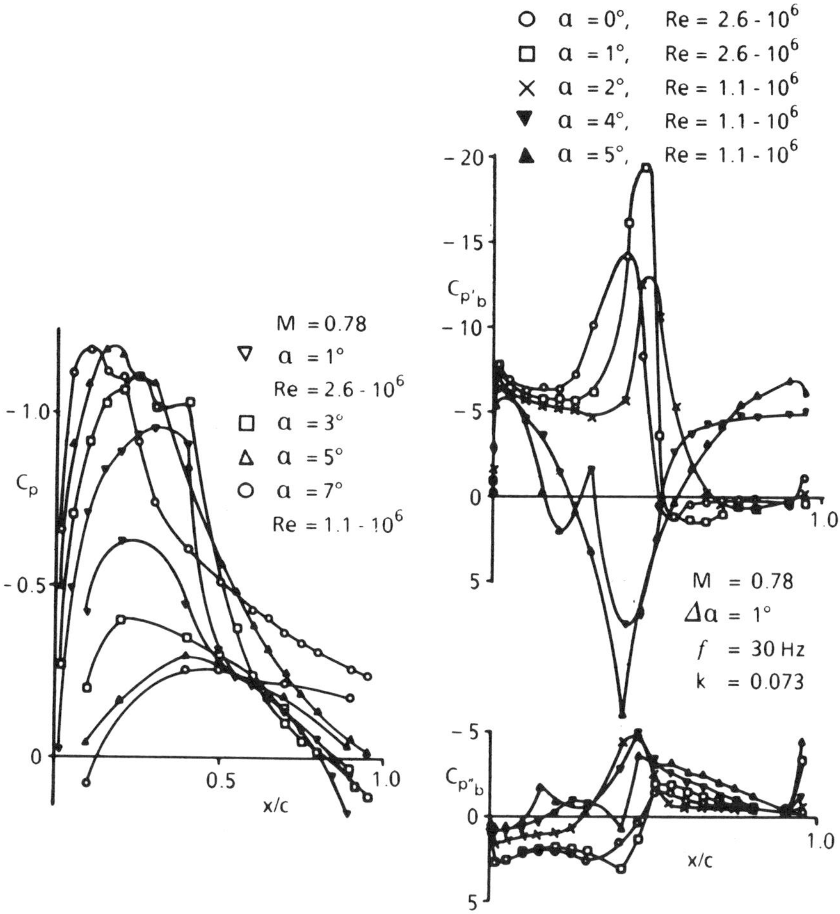

Fig. 3 Effect of angle of attack on steady and unsteady pressure distributions on an NACA 0012 airfoil.[3]

Of more interest to this discussion is the development of the unsteady pressures in Fig. 3. At a mean incidence up to 2 deg, the shock oscillation peak appears normal according to our understanding of flows with weaker shocks. At 4 and 5 deg, however, the unsteady distributions have changed drastically in nature. Although there are some differences in the imaginary results, the most notable change is in the real part. At low mean incidence, the unsteady pressures have a definite nose-up pitching moment characteristic, whereas at higher mean incidence, the characteristic has changed to either a neutral or a nose-down pitching moment. The reversal of the shock oscillation peak is a result of the reversal of shock motion with increasing incidence as noted earlier for the steady data. The high in-phase unsteady pressures aft of the shock are a result of the shock-induced separation

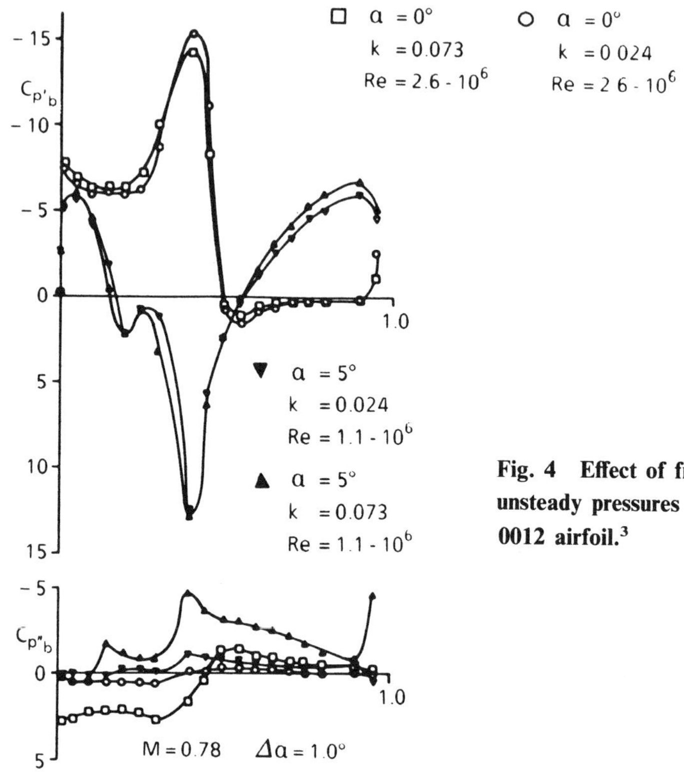

Fig. 4 Effect of frequency on unsteady pressures on an NACA 0012 airfoil.[3]

reaching the trailing edge. Figure 4 shows a better comparison of low- and high-incidence incremental pressures at different frequencies. The principal characteristics are unaffected, which might have been expected due to the low values of reduced frequency.

These characteristics are also present in three-dimensional flows as shown in Fig. 5 for steady flow. The results in Fig. 5 are the incremental pressure coefficient changes ΔC_p that occur due to a $\Delta\alpha$ change in incidence as determined from the pressure data given in Ref. 4 for the ONERA M-6 wing. The four plots show how the $\Delta C_p/\Delta\alpha$ distributions on the upper and lower surfaces at the 90% span station vary as a function of the incidence range. The aft shock is separated since it moves forward from $\alpha = 0.03$ to 3.07 deg, as indicated by the negative peaks on the upper surface. Between $\alpha = 3.07$ and 4.08 deg it merges with the forward shock, which is moving aft. The trailing edge separates due to the combined shock strengths, as indicated by the large increase in suction on the upper surface trailing edge. This distribution is very similar to that shown in Fig. 4 for C_{pb} at $\alpha = 5$ deg and either frequency and is also similar to steady distributions taken from Ref. 4 at more inboard pressure stations but at higher angles. Above $\alpha = 4.08$ deg the trailing edge is already separated; hence, the incremental pressures near the trailing edge drop significantly for α, increasing from

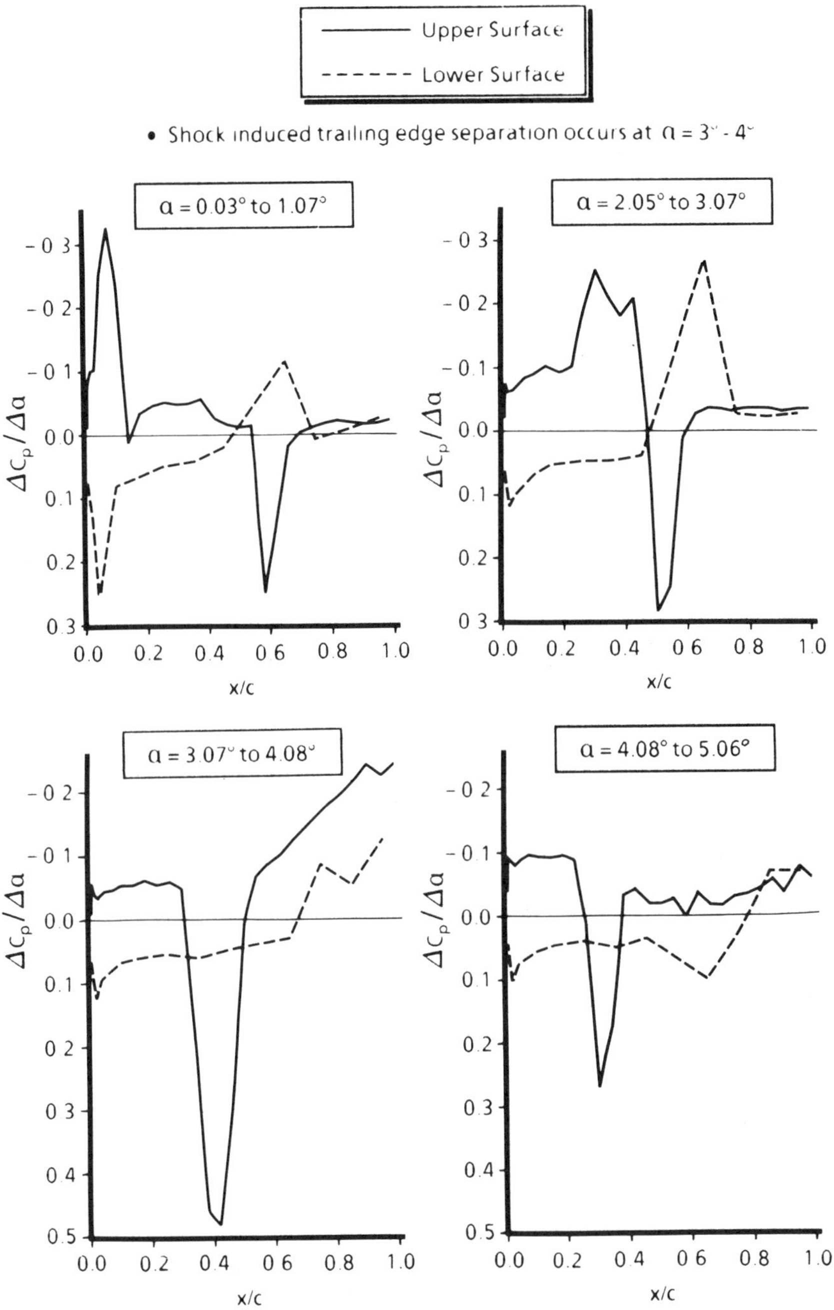

Fig. 5 Effect of angle of attack on pressure changes due to incremental α for the ONERA M-6 wing; $M = 0.92$, $\eta = 0.90$.[4]

4.08 to 5.06 deg since the large change due to the transition is no longer present. This transition to shock-induced trailing-edge separation in two-dimensional flow is similar to the Pearcy type A separation discussed by Mabey in Chapter 1. In three-dimensional flow, however, the separation grows in the spanwise direction once the Pearcy type A chordwise development has occurred at some point on the wing. The starting point and spanwise development are functions of the wing geometry and flow parameters.

It is obvious that such a change in incremental chordwise pressure distributions can have a significant impact on the generalized forces due to wing modes with significant torsional motion. This influence can have either a conservative or nonconservative effect on flutter speeds and can result in a limited amplitude oscillation that may be difficult to distinguish from incipient flutter. Since it is induced by a change in incidence, it may not be present at cruise conditions but suddenly appear if the aircraft performs a maneuver at elevated g.

An excellent summary of the dilemma faced by the flutter analyst at transonic speeds is depicted by the results shown in Fig. 6 as reproduced from Ref. 5. A comparison is made between linear theory (doublet lattice), nonlinear theory (FTRAN3), and experiment as a function of Mach number for a transonic transport. Included are results for the nonlinear theory and experiment for two angles of attack, $\alpha = -0.64$ and 1.78 deg. Thus, each experimental trace represents the Mach number trend of flutter speed for different angles of attack. Linear theory cannot distinguish between these conditions; however, it does seem to provide a baseline from which the nonlinear effects are projected. (It is interesting to note the similarity between the two experimental curves relative to the linear theory curve. The higher angle at the lower Mach number produces a dip similar in shape and magnitude to that of the lower angle but higher Mach number.) From a safety viewpoint in this case, linear theory is very unconservative depending on angle and Mach number, although the nonlinear theory is also very unconservative for the higher angle.

The more important point to be made in Fig. 6 is the influence of parameters other than Mach number and dynamic pressure on flutter speed. Consider a nominal cruise condition at $M = 0.7$ and $\alpha = -0.64$ where the flutter dynamic pressure is about 200 kPa. If we are flying at a safe dynamic pressure of 140 kPa and suddenly have to maneuver, say up to $\alpha = 1.78$ deg, then we are no longer in the flutter-free region since the flutter dynamic pressure is now only about 130 kPa—below the level at which we are flying. Thus, by simply changing the boundary conditions for the steady mean flowfield, i.e., changing angle of attack, we suddenly transition from a flutter-free condition to a flutter condition.

This is the essence of the transonic flutter dilemma and serves to illustrate the importance of other parameters as well. The experimental data in Fig. 6 were obtained from a flutter model, which most likely did not have the static aeroelastic characteristics of the full-scale aircraft. Thus, the data for $\alpha = 1.78$ deg would not be the same as that for full-scale flight test at the same angle. Also, if ailerons or spoilers were deployed during our maneuver, the flutter characteristics would probably be further changed. And, last

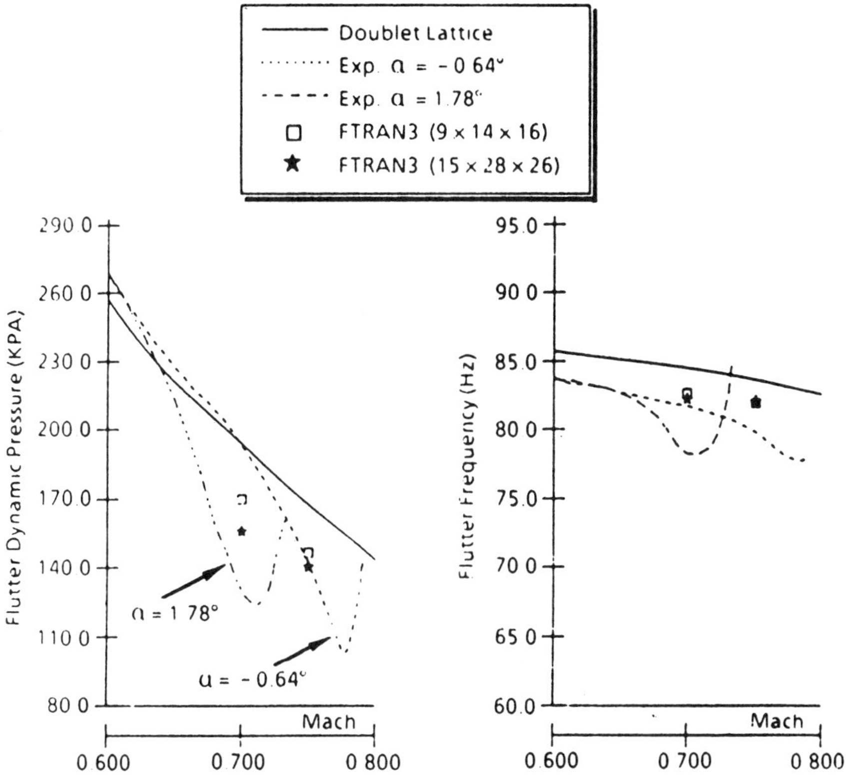

Fig. 6 Flutter results for a transonic transport wing model from doublet lattice, experiment, and FTRAN3.[5]

but not least, since strong shocks would be developed at the higher angle, the development of significant shock-induced separation would be very sensitive to Reynolds number dissimilarity between flutter model testing and flight test. One good result to come out of all of this mismatching is that full-scale tip washout for aft-swept wings (a reduction in wing tip angle as a result of aeroelastic effects) usually tends to counteract the influence of full-scale Reynolds number. This is not the case for forward-swept wings, however. Next, we will turn our attention to limited amplitude oscillations that occur, which can be confused with incipient flutter to further add to the flutter analyst's dilemma.

3. Limit Cycle Oscillations

Limit cycle oscillations (LCO) of lifting surfaces occur as a result of a nonlinear coupling between the surface structural response or similar motions and unsteady aerodynamic forces. The oscillations are usually associated with response of a single structural mode such as first wing torsion or bending and have a clean sinusoidal characteristic unlike the wide frequency band response of wing buffeting. The distinction between LCO and flutter is the limited amplitude of oscillatory motion of the former as opposed to the divergent nature of the latter, hence the name "limit cycle." Unlike flutter, LCO may not always pose a threat of catastrophic structural failure, but does introduce limitations similar to those of buffet such as severe pilot discomfort or equipment restrictions due to unacceptable dynamic environmental conditions. Fatigue damage due to LCO can be significant; however, it is usually restricted to secondary structure such as flaps and leading- or trailing-edge fairings.

The occurrence of LCO is not predictable with linear theoretical methods, since it is a result of nonlinear aerodynamic forces interacting with a structure that may also have nonlinearities such as free play. Thus, LCO is not encountered until flutter model or flight testing takes place; however, it often occurs under conditions of low but stable damping as predicted by the linear methods. Because LCO usually occurs where flutter is expected, it is particularly troublesome since the distinction between it and incipient divergent flutter is impossible to make under actual test conditions. Thus, the appearance of a sinusoidal oscillation in flutter testing is sufficient to warrant backing off from the suspected condition. The anxiety of both test pilots and engineers associated with this appearance tends to restrict investigative attempts to determine if the oscillations are LCO or flutter, since the amplitude of LCO can become quite large and alarming.

The nonlinear aerodynamics are usually associated with some form of separated flow such as leading-edge separation or that produced by strong shock/boundary-layer interaction as discussed in the previous section on flutter. In the following discussions, several examples will be presented to illustrate the diverse nature of LCO along with the description of a proposed math model for LCO.

Influence of Manufacturing Tolerances on the Transonic Vibration of a Horizontal Tail

In this example[6] mild but sustained oscillations of the horizontal tails were encountered on two fighter aircraft of the same type undergoing acceptance flight tests. The conditions were consistent at about $M = 0.92$–0.95 with little influence of altitude, which varied from 10,000 to 30,000 ft. The oscillations occurred at 1-g cruise but disappeared for attitudes beyond 2 g. The right-hand tail vibrated more than the left on one of the aircraft. Various fixes were tried, which included inspecting and changing servo-actuators, bearings, and flight control computers, but they had no effect on the vibration. It was found that, by extending the speed brakes near the horizontal tails to 10 or 20 deg or changing the tails, the vibration would disappear. Hence, it was suspected that the source of the problem was aerodynamic, since pulling more than 2 g also changed the aerodynamic flowfields through a change in angle of attack.

By chance it was noticed that one of the "bad" tails that vibrated had a flat spot on the lower surface. Surface profile measurements were then made that indicated significant deviation from the theoretical biconvex airfoil section. A "good" tail that did not vibrate was also measured, and

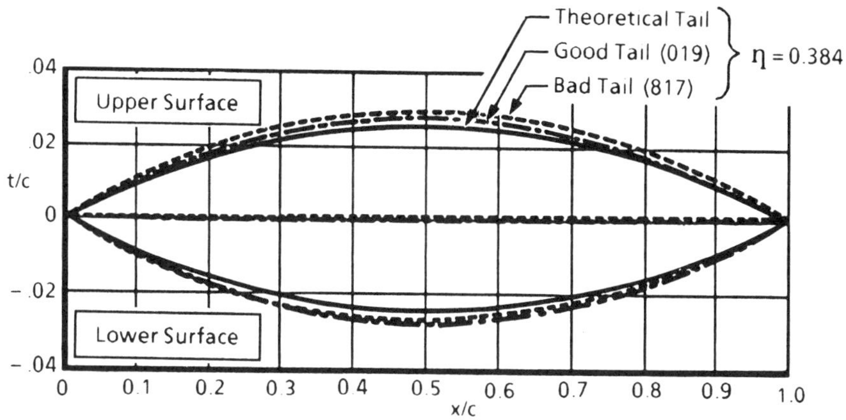

Fig. 7 Horizontal tail surface contours.[6]

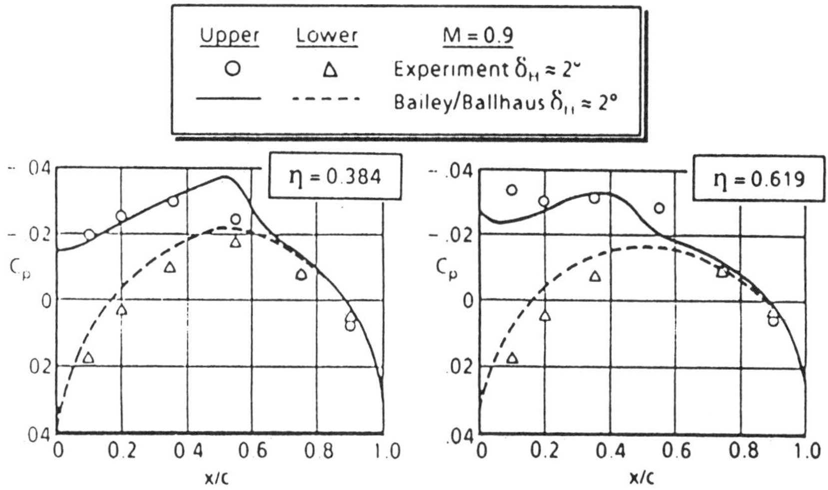

Fig. 8 Aerodynamic modeling of the horizontal tail.[6]

the three-way comparison is shown in Fig. 7 at 38.4% of the horizontal tail half-span. Although the good tail was thicker than the theoretical tail, it was symmetric. The bad tail, however, had an asymmetric section that introduced camber as well as some areas of higher curvature.

The Bailey-Ballhaus wing-body transonic code[7] was used to evaluate the influence of these contour variations on the steady transonic aerodynamic characteristics of the horizontal tails. A nonconservative solution was used with tail-body models set up for all three contours (good, bad, and theoretical). Results for the theoretical contour model were compared with experiment to verify the math model. As shown in Fig. 8, the predicted

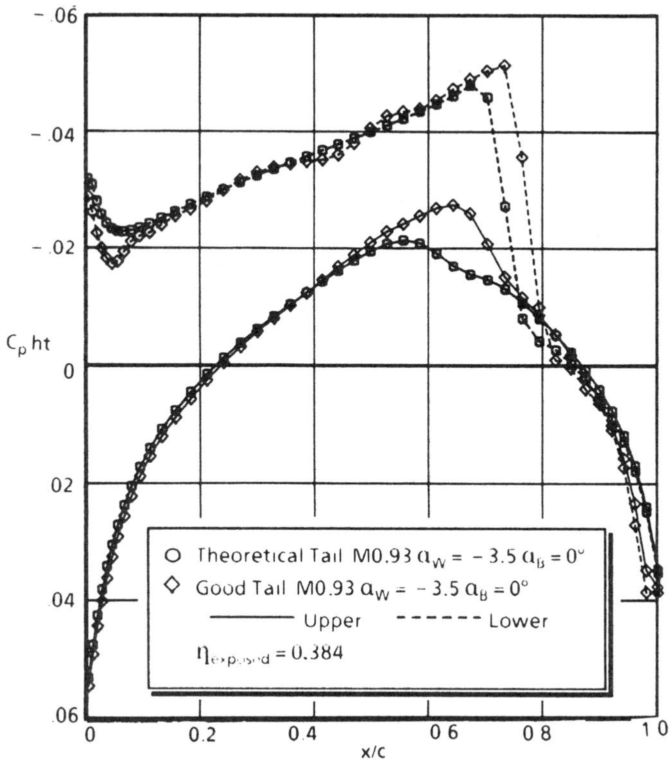

Fig. 9 Theoretical results for good and theoretical tails at −3.5 deg (nose down).[6]

results generally agree with experiment at the two span stations so that the incremental effects of contour variation should be representative of those for tails in the presence of the total configuration.

A comparison of calculated results for the theoretical and good tails is shown in Fig. 9 for a nose-down deflection of −3.5 deg at $M = 0.93$. The pressures generally agree with exception of the areas near the shock on both upper and lower surfaces, which is attributed to the thicker section of the good tail. The waviness noted in the good tail lower surface pressures is directly attributed to the use of actual contour data. Next the good and bad tails are compared as shown in Fig. 10 under about the same conditions. (In this case, the total normal force of the tails is matched; hence, the bad tail deflection is slightly higher at −3.85 deg.) The waviness noted in the good tail lower surface pressures is greatly exaggerated for the bad tail, although they show very good agreement near the shocks. Another comparison of good and bad tail pressures is shown in Fig. 11 for a nose-up deflection. Again, the waviness noted in the lower surface pressures for nose-down deflection is present but is considerably reduced for the nose-up deflection. This is a natural result of the lower speed flows on the high-pressure side being less sensitive to small contour variations.

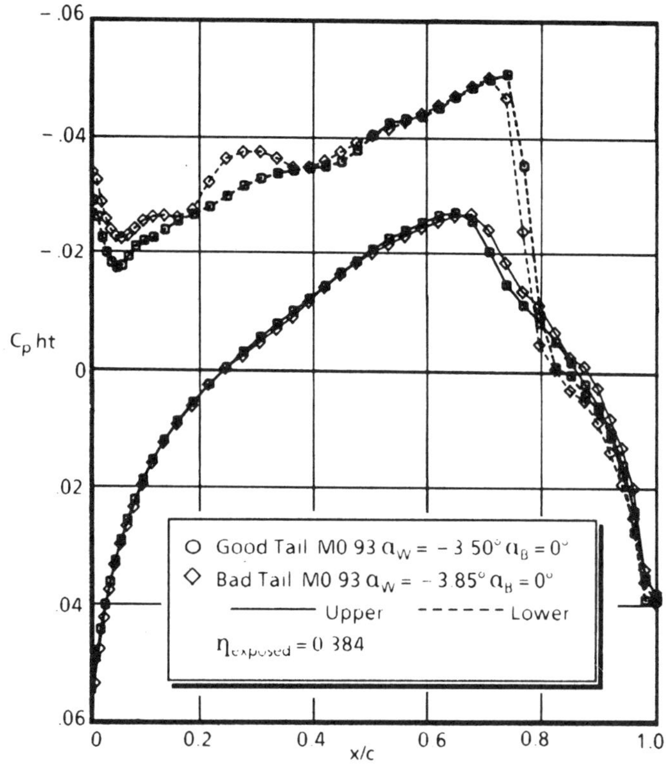

Fig. 10 Theoretical results for good and bad tails at −3.5 deg (nose down).[6]

In the comparisons of predicted pressures for the two tails in Figs. 10 and 11, nothing is evident that would suggest that the bad tail would tend to vibrate during flight operation at around $M = 0.92–0.94$. However, the variation of chordwise center of pressure for the tails, X_{cp}, with Mach number and tail deflection as shown in Fig. 12 does suggest a cause for the vibration. Relating these data with the tail hinge line indicates that both the theoretical and good tails have X_{cp} aft of the hinge line for the trimmed conditions over the Mach range as well as untrimmed conditions at $M = 0.90$ and 0.93. The characteristics for varying δ_h are also very smooth and consistently aft of the hinge line. The bad tail, on the other hand, has an X_{cp} that is forward of the hinge line for 1-g trim flight and crosses over at about $M = 0.92$ in the range of tail vibration. With the combination of X_{cp} crossing the hinge line and the change of X_{cp} with tail deflection, it was possible to construct a mechanism by which the tail would vibrate.

Considering the pitching moment on the horizontal tail at a nose-down attitude, $\delta_h < 0$, as X_{cp} crosses the hinge line, it is seen that the moment continuously varies from nose down before the crossing to nose up after the crossing. This reversal of aerodynamic moment is difficult for any servo-actuator system with position feedback to accurately respond due to

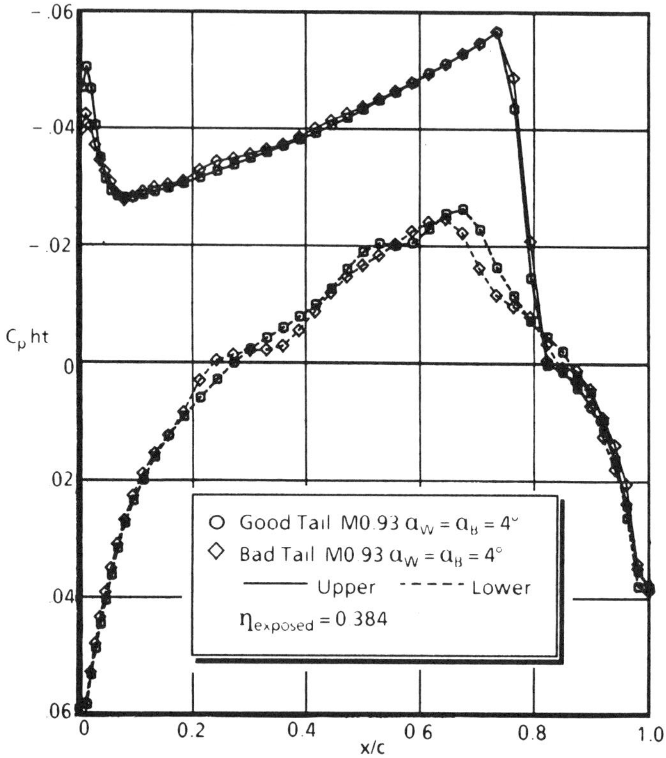

Fig. 11 Theoretical results for good and bad tails at +4 deg (nose up).[6]

unavoidable lag. This lag can provide the necessary hysteresis at the neutral point to produce an LCO. Another source of hysteresis is position free play of the horizontal tail. The latter produces hysteresis in a way that can be more easily understood from an explanatory viewpoint. However, in the actual case it is felt that the servo-actuator system was the more likely source, since free play was checked on the aircraft and found to be acceptably small.

Pursuing the idea of free-play interaction with the aerodynamic forces for purposes of illustration, it is possible to construct a simple mechanism by which the tails could vibrate. Considering the increasing nose-down tail deflection, δ_h, through the crossover point, δ_{hc}, before reaching δ_{hc} the tail is loaded nose down due to a forward X_{cp}. On reaching δ_{hc} the moment is neutral but just beyond it becomes nose up as the X_{cp} moves aft of the hinge line. Now, the tail must rotate nose up by the amount of free play, $\Delta\delta_h$, which will reduce the deflection below δ_{hc}. Because of aerodynamic lag, the moment will not reverse to nose down immediately, thus permitting development of a hysteresis effect. After some delay the moment will reverse and the tail will begin to be pitched nose down by $\Delta\delta_h$. At this point the tail had been deflected more nose up than $(\delta_{hc} - \Delta\delta_h)$, due to elastic

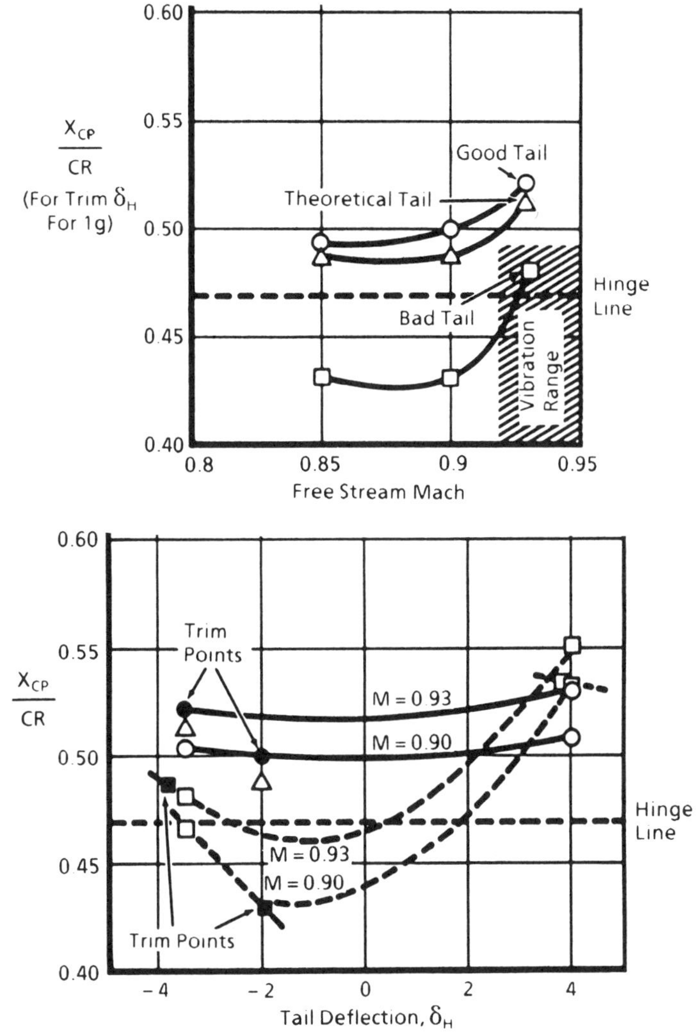

Fig. 12 Effect of Mach number and tail deflection on the chordwise center of pressure.[6]

deflection produced by work input permitted by the aerodynamic lag (or hysteresis). With the combination of strain and aerodynamic energy, the downstroke overshoots δ_{hc} by an amount that is further enhanced by aerodynamic lag completing the other leg of the hysteresis loop. At some point beyond δ_{hc}, the aerodynamic moment again reverses to nose up and the cycle begins again. The work input to the system becomes constant due to finite limits on the hysteresis loop. The motion amplitude continues to build during succeeding cycles until fixed work input of the hysteresis loop is balanced by system damping. Thus, for smaller system damping, i.e., at conditions near flutter for example, the limit cycle amplitude becomes proportionately larger.

This example serves to illustrate the necessary ingredients for LCO. In its simplest form LCO occurs when a step change in aerodynamic force is triggered by a small change in attitude of an elastic winglike surface and that change in force opposes the change in attitude. The step change in force drives attitude back to a condition at which the step change in force must disappear. Stored elastic energy is released and the wing returns to the point at which the step change in force reappears to start the cycle over again. In essence the step change in aerodynamic force acts as a nonlinear spring for which stiffness increases with increasing distance from the transition point. In the example for the vibrating horizontal tails, free-play or servo-actuator lag provided the step change characteristics, where, as in the following examples, the step change is produced by the transition of aerodynamic flowfield characteristics. The concept of a nonlinear aerodynamic spring will be explored in these examples.

Limit Cycle Oscillations on a Variable Sweep Wing

This example of LCO was extensively investigated as described in detail in Ref. 8 with both wind-tunnel and flight testing. The study was instigated by the occurrence of limited amplitude motions of the first wing-bending mode on a bomber-type aircraft flying at transonic speeds with high wing sweeps of 65–67.5 deg. The oscillations occurred over a very limited incidence range of 7–9 deg that was relatively insensitive to Mach number for a range of $M = 0.6$–1.1. For wing sweeps of 60 deg or greater, leading-edge vortex flows are prevalent even at transonic speeds; hence, the source of LCO was suspected to be connected with vortex flows because of the insensitivity to Mach number and constancy of angle-of-attack range. Although explanations were given in Ref. 8 for the possible mechanism by which the LCO was produced, the author feels that another mechanism is more appropriate as a result of his experience with other LCO occurrences. Therefore, the author will present an explanation based on his analysis of the results in Ref. 8.

The results shown in Fig. 13 are taken from Ref. 8 and are pressure data for $M = 0.80$ and leading-edge sweep, $\Lambda = 65$ deg, at various angles of attack and several span stations. The pressures are the chordwise distributions normal to the wing leading edge and hence are not streamwise. The angles shown include both those above and below the oscillation conditions at 7.44 and 8.0 deg. The signature of a vortex passing over the wing is a suction "hump" that results from the low pressure produced under the vortex. This is clearly seen at $\eta = 60.68\%$, for example, where the vortex is located at about $x/c = 0.10$ at $\alpha = 5.97$ deg and $x/c = 0.50$ at $\alpha = 9.63$ deg. At $\eta = 83.39\%$ two vortices can be seen at about $x/c = 0.05$ and 0.40 for $\alpha = 5.97$ deg, but only one is evident at $x/c = 0.70$ for $\alpha = 9.63$ deg. At $\eta = 94.90\%$ two vortices are also shown at $x/c = 0.15$ and 0.70 for $\alpha = 5.97$ deg, but again only one is indicated at $x/c = 0.75$ for $\alpha = 9.63$ deg.

The disappearance of the forward vortex at the outboard span stations at higher incidence is the key to the mechanism for LCO proposed by the

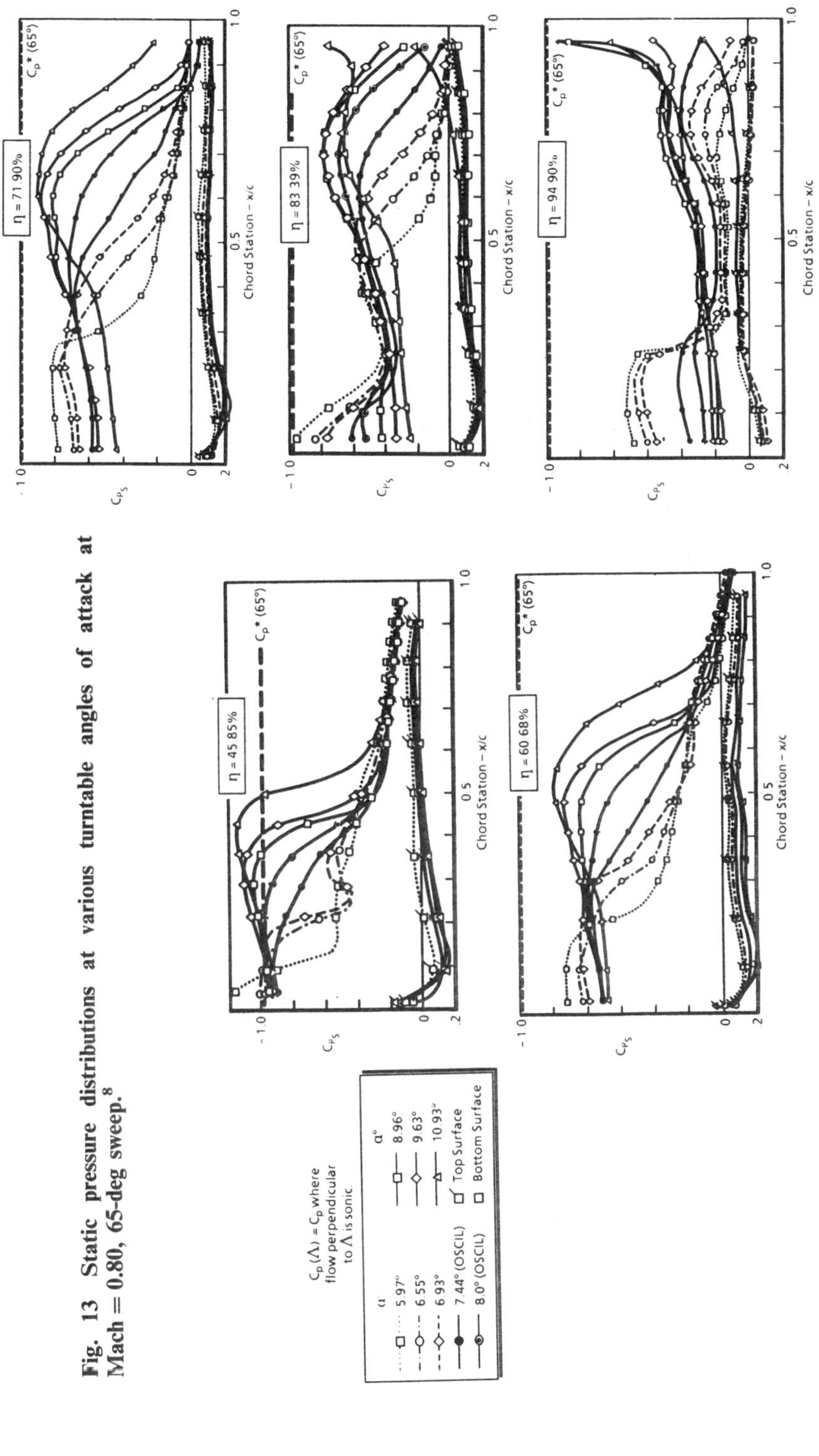

Fig. 13 Static pressure distributions at various turntable angles of attack at Mach = 0.80, 65-deg sweep.[8]

author. The concept is based on the stability of multiple leading-edge vortices on highly swept wings that also have highly swept trailing edges. Schematically, the lower incidence range multiple vortex system is shown in Fig. 14 as deduced from the data in Fig. 13. As the incidence is increased, the inboard vortex becomes stronger and dominates the outboard vortex to the point that it is pulled away from the leading edge and merged with the stronger vortex into a single vortex system that is also illustrated in Fig. 14. As a result of pulling the outboard vortex away from the leading edge, lift is lost from that area and partially reappears in the vicinity of the inboard vortex. This is seen in the pressures at $\eta = 94.90\%$ for α changing from 6.93 to 7.44 deg. In this small change in incidence, the leading-edge vortex suction fell dramatically with a subsequent shift toward the trailing edge. These results strongly suggest a rapid transition from a two-vortex system to a one-vortex system with an equally rapid transition to a nose-down pitching moment. The change in pitching moment is quite dramatic as shown by the variation of torque with incidence in Fig. 15 as reproduced from Ref. 8. Also shown in Fig. 15 is the variation of wing section lift with incidence. A more subtle change in lift at the outboard station is evident as a small loss in the vicinity of oscillations that results in a slope reversal.

The LCO mechanism proposed by the author makes use of the rapid transition from a two-vortex to a one-vortex system with its accompanying loss in wing tip lift and lift-slope reversal. A schematic relationship between wing elastic deflection and bending moment for a highly swept wing is shown in Fig. 16 for a condition at which the flexible wing incidence is near the transition from a two-vortex to a one-vortex system, α_t. It is important to understand that the relationship in Fig. 16 does not represent static equilibrium between bending moment (BM) and α_t but rather the response of BM to changes in δ. Normally, an increase of deflection for a highly swept wing at a fixed body incidence will produce a drop in aerodynamic force and consequently bending moment as a result of wing washout. This

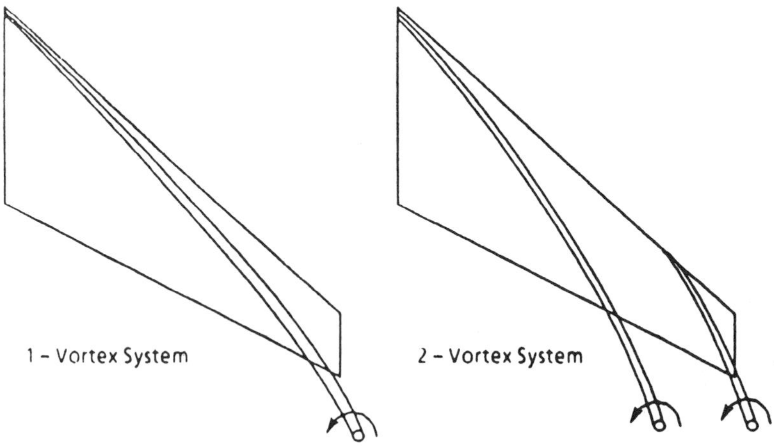

Fig. 14 One- and two-vortex systems on highly swept high-aspect-ratio wings.

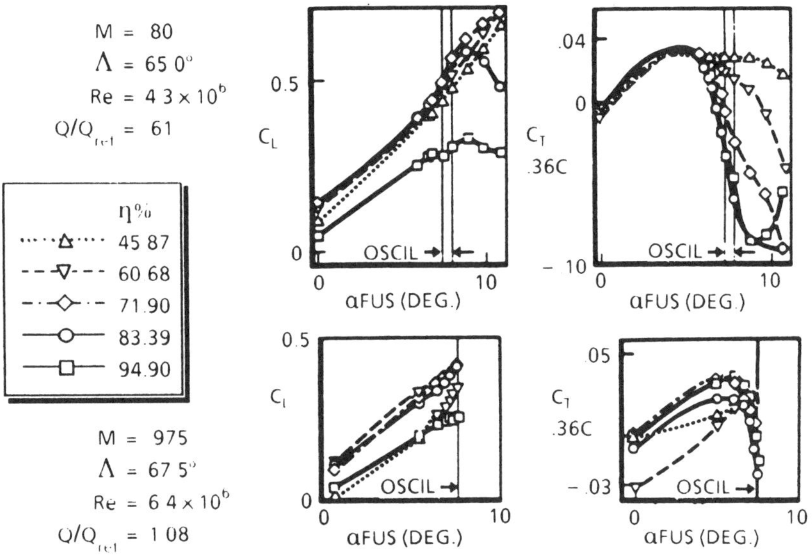

Fig. 15 Section lift and torque (about 36% chord) coefficients vs angle of attack.[8]

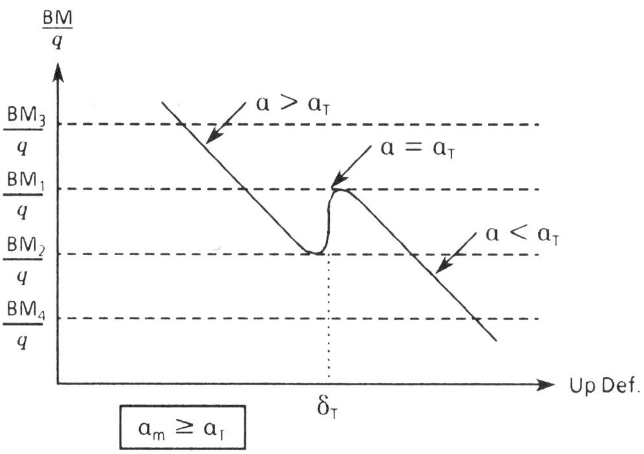

Fig. 16 Quasisteady relationship between wing bending moment per q and wing elastic deflection near the transition condition at α_T.

is illustrated in Fig. 17 (taken from Ref. 8), where the variation of wing incidence along the span is shown for zero, maximum, and minimum wing deflections during LCO. Maximum incidence occurs at minimum deflection and vice versa at maximum deflection due to washout. This is a stable situation where, if the wing deflection is disturbed either up or down, it will tend to return to the original undisturbed position. The presence of a slope reversal or a discontinuity as illustrated in Fig. 16 provides a source of local

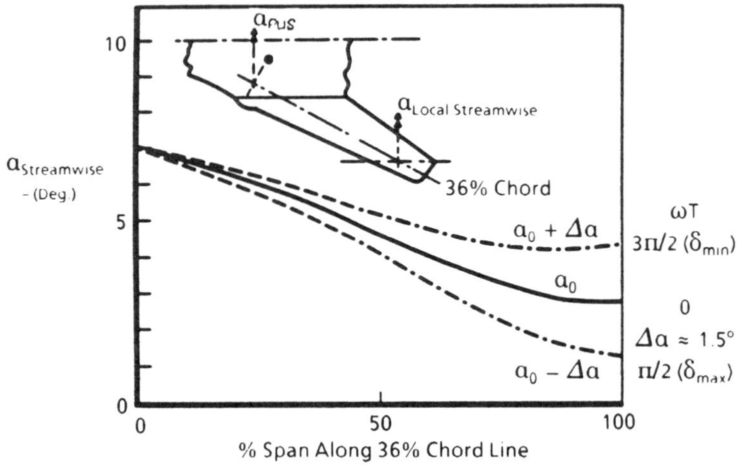

Fig. 17 Streamwise wing angle-of-attack variation during limit cycle oscillation at Mach 0.975, 67.5-deg sweep, $\alpha = 7.38$ deg.[8]

instability which can lead to a limit cycle oscillation. How this occurs can be best understood by "walking through" the physical process of LCO development.

Consider the situation where the body incidence is slowly increasing toward the point of transition α_t. Referring to Fig. 16, if α has just passed α_t, then the wing experiences a sudden drop in bending moment from BM_1 to BM_2, and as a result, the wing deflection will be reduced but the incidence will be increased as shown in Fig. 17. This reduction in deflection results from the strain energy stored in the wing, which was at static equilibrium just before passing α_t. The wing downward momentum builds up but begins to be resisted as the bending moment increases due to increasing incidence at lower deflections. At some point, the downward motion is stopped where the bending moment BM_3 is too high for the low deflection and the wing begins to move upward with decreasing incidence. Now the wing builds upward momentum until reaching BM_2 at δ_t, at which point it suddenly increases to BM_1, where the two-vortex system reforms. Although this is where it began, the wing upward momentum carries it past this initial equilibrium condition to a lower angle where the two-vortex system characteristics are in control. With increasing deflections, the wing strain energy is again building against the motion until BM_4 is reached, at which point the motion stops and the wing reverses to a downward motion. This reversal must take place because the higher deflection is not in equilibrium with BM_4. (Remember, this cycle began at δ_t and BM_1, which were in static equilibrium.) Now, because of downward momentum, BM_1 drops to BM_2, δ_t is again passed, and the cycle repeats. With each succeeding cycle, the motion amplitude increases since the work input during each cycle is about constant. The amplitude increases until the

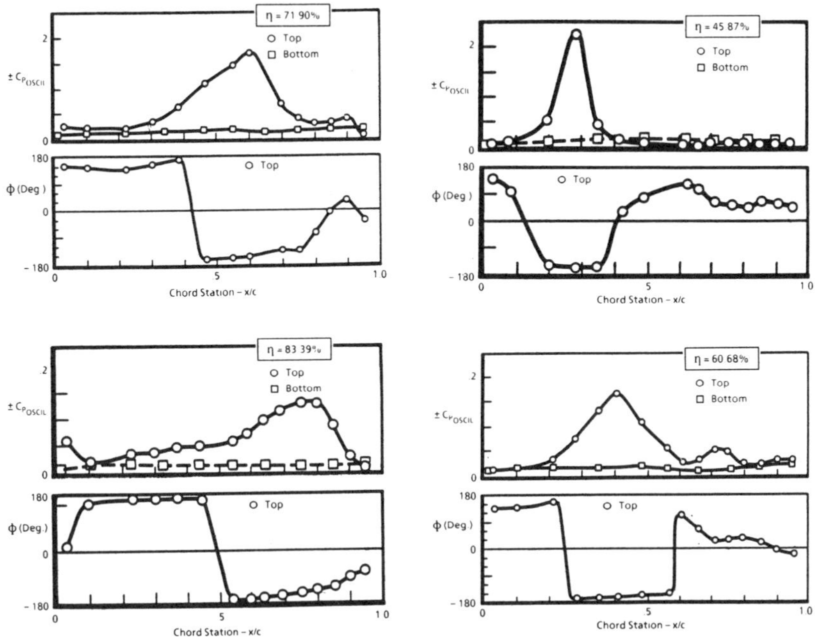

Fig. 18 Oscillating pressure distributions and phase angle between oscillating pressure and wing displacement at $\alpha = 7.44$ deg, Mach = 0.80, 65-deg sweep.[8]

work due to the nonlinear aerodynamic bending moment is balanced by that due to conventional damping, which is amplitude-dependent.

The preceding process was described in a quasisteady sense, which is also felt to be applicable to the dynamic case. Because of the finite time required for signals to propagate along a vortex, a phase lag is introduced in the vortex flow development as a result of rapid wing motions. An influence that would produce a leading phase angle would be the production of leading-edge vorticity by the plunging wing motion. On a downward motion, positive vorticity would be shed from the leading edge that would be in the same direction as the wing vortices and would therefore tend to re-enforce that system. This would likewise tend to cause an earlier transition from the two-vortex to the single-vortex system. On an upward motion, the reverse would take place, and a weakened wing vortex system would transition back to the two-vortex system at an angle higher than α_t (or δ_t). The phase lead angles for the plunging wing effect would depend on the plunging velocity and thus be higher toward the wing tip for the first wing bending mode.

The nominal phase angles of 180 deg for the pressures quoted in Ref. 8 were expressed relative to wing bending motion, which in itself has a phasing of 180 deg relative to wing incidence as a result of the swept-wing washout effect shown in Fig. 17. Thus, the phase angle variations presented in Ref. 8 in essence reflect the quasisteady characteristics just discussed.

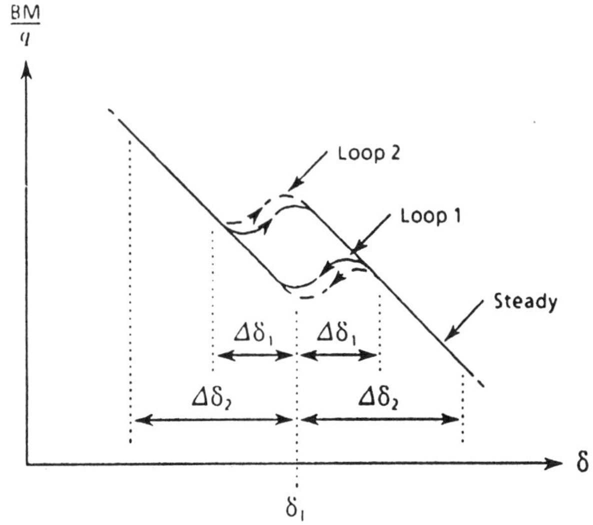

Fig. 19 Hysteresis loop in the relationship between bending moment per q and deflection near the transition point α_T.

These phase angles are shown in Fig. 18 for the condition $M = 0.8$, $\alpha = 7.44$ deg, and a wing sweep of 65 deg. Taking into consideration the wing bending moment and incidence phasing, it can be seen that the phase angles in Fig. 18 are leading in the vicinity of the unsteady pressure peaks. This results in a modification to the quasisteady characteristics in Fig. 16 to produce the dynamic characteristics shown in Fig. 19. This also implies that the lead due to wing plunging motion is the dominant unsteady effect over that due to vortex flow lag.

The direction of varying δ with time is indicated by arrows in Fig. 19. Clockwise motion around a force-displacement hysteresis loop is characteristic of work being put into the system and is therefore destabilizing. Conversely, a counterclockwise motion would be stabilizing. The clockwise motion in Fig. 19 is consistent with the concept that work is being put into the wing spring mass system. The magnitude of this work is relatively independent of wing motion beyond a certain point, as illustrated in Fig. 19, where the areas within two hysteresis loops of different maximum deflection are about the same. The damping force for each would be much different, however, since it is directly proportional to the maximum amplitude.

The mechanism proposed earlier therefore involves a nonlinear aerodynamic spring as exemplified by the variation of BM with δ near α_t. The apparent reduction described in Ref. 8 of measured damping during the oscillation is believed to be a result of the existence of sustained oscillations that occur due to a coupling of wing motion with the aerodynamic forces. (If the oscillation is sustained, then the apparent damping must be near zero.) In the next example of LCO, the idea of a nonlinear spring will be further explored and a mathematical model will be developed that demonstrates the idea that a reduction of damping is not necessary in order to have LCO in a nonlinear system. The constancy of damping, on the contrary, is necessary in order to achieve LCO with the nonlinear aerodynamic spring; otherwise, the oscillations would become quickly divergent, since work input per cycle for the aerodynamic spring is nearly constant.

Limit Cycle Oscillations on a Variable Sweep Wing with a Supercritical Airfoil

The F-111 TACT aircraft was the subject of an extensive buffet investigation that included both flight and wind-tunnel testing. The aircraft had a variable sweep wing with a supercritical airfoil section that was flight buffet tested at wing sweeps of 26 and 35 deg and incidences up to about 12 and 13 deg, respectively. The $\frac{1}{6}$-scale model instrumented with high-response pressure transducers was tested with two wings, one of aluminum and the other of steel, in order to separate Reynolds number and aeroelastic effects. Transition was not fixed on these models. The results of the buffet investigation are presented and discussed in Ref. 9.

As a major part of the effort described in Ref. 9, the wind-tunnel model fluctuating pressure data were used to make predictions of full-scale buffet characteristics. The predictions compared quite favorably with flight-test data for almost all of the conditions with the exception of the wing torsion modes at 26-deg wing sweep and $M = 0.8$. Buffet onset was well predicted; however, as shown in Fig. 20, the rms amplitudes were far underpredicted, especially when compared with the torsion results for 35-deg sweep also shown in Fig. 20 and bending mode results shown in Ref. 9. An examination of power spectra for the aircraft fluctuating pressure data revealed strong response in the vicinity of the wing torsion modes at conditions where flight-test buffet response was significantly underpredicted. It was therefore suspected that a coupling existed between the wing torsion modes and fluctuation pressures on the full-scale aircraft that did not exist on the wind-tunnel model and was not accounted for in the buffet prediction technique.

A clue to a possible means for the coupling can be deduced from the static pressures in Fig. 21. The occurrence of trailing-edge pressure divergence at about 10 deg also corresponds to a large forward movement of the upper surface main shock, as shown by comparing the pressure distributions at 9 and 10 deg in Fig. 21. It will also be noted that the forward shock movement for the aircraft is much larger than that for the model. The condition of this transition is the occurrence of shock-induced trailing-edge separation (SITES), which was discussed earlier in this chapter (see Figs. 2–5). It was shown that this transition was accompanied by a step change in pitching moment with either increasing or decreasing angle of attack. With increasing angle, the forward shock movement produced a loss in lift forward and the trailing-edge divergence produced a gain in lift aft. The net result was to provide a step change in pitching moment that was nose down for increasing angle. For decreasing angle, the opposite took place and produced a step change in pitching moment that was nose up. Such a step

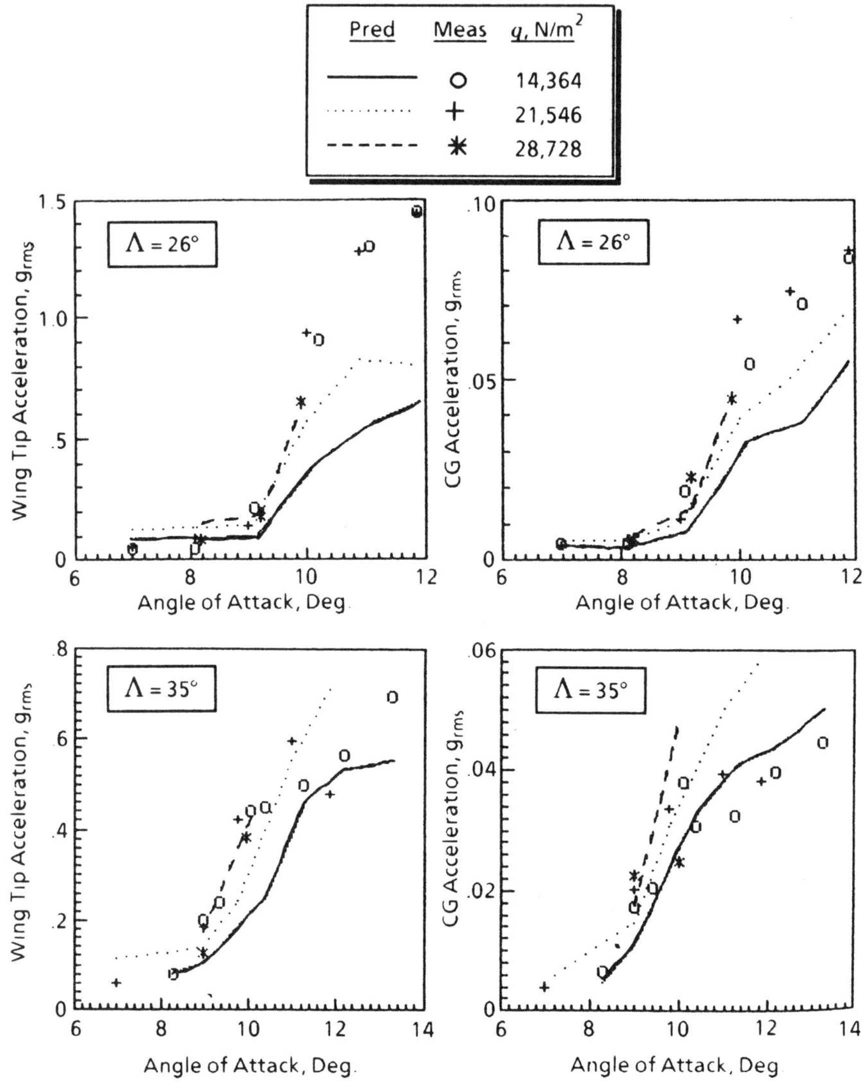

Fig. 20 Torsion mode response for the F-111 TACT airplane at various altitudes.[9]

change in effect resulted in a nonlinear spring that would provide an increased resistance to wing motion as the point of SITES was passed for either increasing or decreasing angle.

The nonlinear spring described earlier can potentially produce a limit amplitude, self-sustaining oscillation, the existence of which could explain the higher torsion mode response exhibited by the full-scale aircraft. How this is possible can be described by considering an airfoil with a torsion spring. Slowly increasing incidence at angles below that of SITES allows

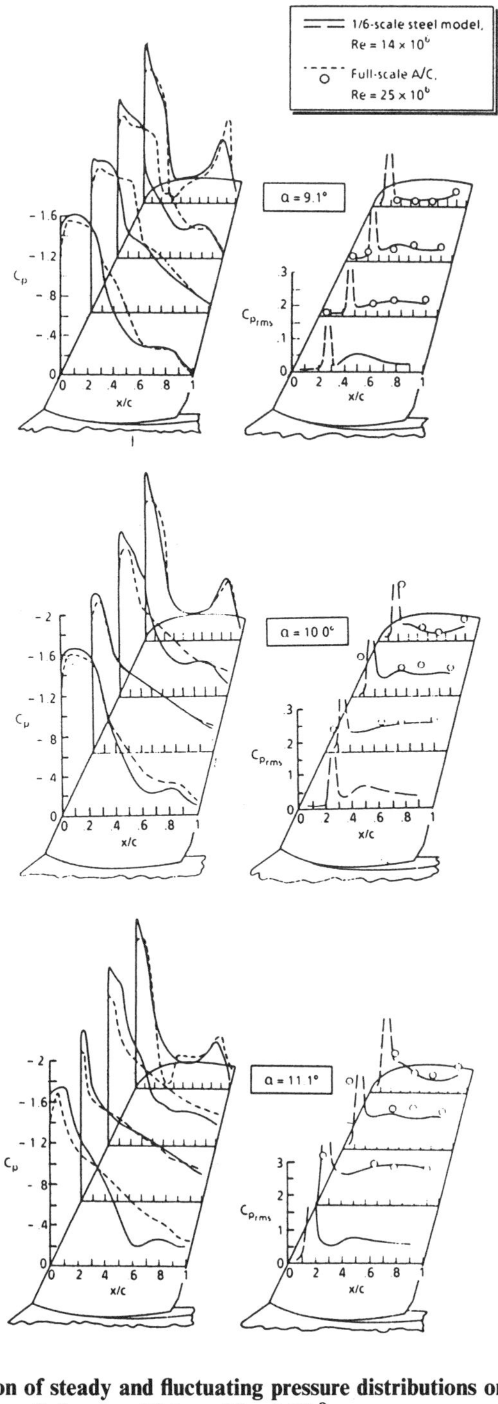

a) b) c)

Fig. 21 Correlation of steady and fluctuating pressure distributions on TACT $\frac{1}{6}$-scale steel model and aircraft for $\Lambda - 26$ deg, $M - 0.80$.[9]

the torsion spring to attain a continuous state of equilibrium with aerodynamic pitching moment. When SITES is reached, a sudden nose-down increment is imposed on the aerodynamic pitching moment that will tend to reduce wing incidence. This will be a dynamic negative pitch rate, which will delay reattachment and permit the nose-down moment to put work into the system. At some point, reattachment does take place and the nose-down moment disappears. Accelerations become negative and the wing experiences a reducing pitch rate until it reaches zero and begins nose-up motion. Positive pitch rate now takes over, which will produce a delay in SITES and allow an overshoot of the initial starting point due to stored elastic energy during the downstroke. When SITES does occur, the cycle then repeats itself.

Thus, the idea that a nonlinear aerodynamic spring can be a primary source of LCO has again been proposed as a result of the transition associated with SITES. The common characteristic of the nonlinear spring in this example and previous examples is that they all have a rapid reversal of slope in either pitching or bending moment that opposes wing motion. This reversal occurs in the vicinity of conditions at which LCO frequently occurs. Based on this observation, it should be possible to develop a mathematical model that is capable of describing the essential features of related LCO mechanisms.

Proposed Mathematical Model for LCO

Since the common feature noted earlier is a sudden stiffening of a nonlinear aerodynamic spring, a one-degree-of-freedom spring mass system should be capable of responding in an LCO manner. This assumption is justified on the basis of observations of LCO where a single mode is usually involved as indicated by response at a single frequency.

The example given in the previous subsection where SITES is the apparent source of the nonlinear spring represents a case for which the explanation is most straightforward. Restricting our attention to a single-degree-of-freedom spring mass system involving the first wing torsion mode, the appropriate equation of motion is

$$M_{\text{tor}}\ddot{q} + 2M_{\text{tor}}\omega_{\text{tor}}\delta_{\text{tor}}\dot{q} + M_{\text{tor}}\omega_{\text{tor}}^2 q = F(q)$$

where M_{tor}, ω_{tor}, δ_{tor} are the mass, frequency, and total damping, respectively, for the torsion mode; q is the response of the torsion mode; and $F(q)$ is the nonlinear generalized force due to the step change in the static pressure distribution.

The nonlinear right-hand-side term $F(q)$ is obtained by integrating the torsion mode deflections with the step change in the C_p values that occur during the transition to SITES. The function of q is expressed as

$$\begin{aligned} F(q) &= 0, & q &< q_{\text{trans}} \\ &= F, & q &\geqslant q_{\text{trans}} \end{aligned}$$

To include the hysteresis effects, pitch rate must be taken into account so that

$$q_{\text{trans}}(\dot{q}) > q_{\text{trans}}(0), \qquad \dot{q} > 0$$
$$q_{\text{trans}}(\dot{q}) < q_{\text{trans}}(0), \qquad \dot{q} < 0$$

which simply parallels the idea that positive pitch rate delays separation and negative pitch rate delays reattachment. This equivalence is possible when the mode of interest is a torsion mode or has a significant streamwise angular deflection. The relationship of the step change F and hysteresis is shown in Fig. 22 where the coordinate system has been referenced to $q = 0$ at the occurrence of SITES.

By writing the derivatives in finite-difference form:

$$q = q_n$$
$$\dot{q} = \frac{q_{n+1} - q_{n-1}}{2\Delta t}$$
$$\ddot{q} = \frac{q_{n+1} - 2q_n + q_{n-1}}{(\Delta t)^2}$$

a finite-difference solution to the preceding equation of motion can be cast in the form

$$q_{n+1} = \frac{1}{1+\lambda\delta}\left[(2-\lambda^2)q_n - (1-\lambda\delta)q_{n-1} + \lambda^2\varepsilon\right], \qquad q_n \geqslant q_{\text{trans}}(\dot{q})$$

and

$$q_{n+1} = \frac{1}{1+\lambda\delta}\left[(2-\lambda^2)q_n - (1-\lambda\delta)q_{n-1}\right], \qquad q_n < q_{\text{trans}}(\dot{q})$$

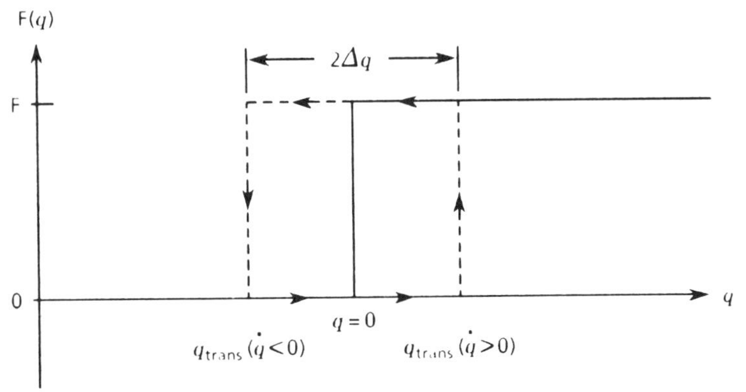

Fig. 22 Schematic diagram of variation of the step function due to SITES.

where

$$\lambda = \Delta t \omega_{\text{tor}}$$

$$\delta = \delta_{\text{tor}}$$

$$\varepsilon = \frac{F}{M_{\text{tor}} \omega_{\text{tor}}^2}$$

The initial conditions are

$$\left.\begin{aligned} q &= 0 \\ \dot{q} &= 0 \end{aligned}\right\} \qquad t = 0$$

which are satisfied in the finite-difference form as

$$q_0 = 0$$

$$q_1 = 0$$

$$q_2 = \frac{\lambda^2 \varepsilon}{1 + \lambda \delta}$$

By denoting Δq as the half-width of the hysteresis (shown in Fig. 22),

$$\Delta q = \tfrac{1}{2}[q_{\text{trans}}(\dot{q} > 0) - q_{\text{trans}}(\dot{q} < 0)]$$

the solution to the equation of motion can be presented as a function of the parameters ε and Δq for the system M_{tor}, ω_{tor}, and δ_{tor}.

The algorithm described earlier was programmed and estimates were made for the appropriate ε value to be used for the torsion mode. The estimate of ε was obtained by using the C_p distributions shown in Fig. 21 at 9 and 10 deg in the following form:

$$\varepsilon = \frac{\frac{1}{2}\rho V^2}{M_{\text{tor}} \omega_{\text{tor}}^2} \int_A h_{\text{tor}}(x,y)[C_{p_{10}}(x,y) - C_{p_9}(x,y)] \, dx \, dy$$

where

$h_{\text{tor}}(x,y)$ = deflection of the torsion mode at point x,y

$C_{p_{10}}(x,y) = C_p$ value at 10 deg and point x,y

$C_{p_9}(x,y) = C_p$ value at 9 deg and point x,y

A = area of the wing

The preceding integral was evaluated in the same manner as were the buffet excitation forces by summing the product of C_p values, mode deflections, and panel areas over the exposed wing area. The value obtained was

$$\varepsilon_{10-9} = -0.00337 \text{ m } (-0.0127 \text{ ft})$$

at the intermediate altitude of 6.1 km (20,000 ft). Other parameters were

$$M_{\text{tor}} = 809.4 \text{ lb}, \qquad f_{\text{tor}} = 14.17 \text{ Hz}, \qquad \delta_{\text{tor}} = 0.07$$

where δ_{tor} was estimated from Fig. 32c in Ref. 9.

Estimates for Δq were not possible based on the available data; hence, a parameter study was conducted by letting Δq vary as a fraction of ε. A sample plot shown in Fig. 23 for $\Delta q = \varepsilon$ illustrates how the transient solution quickly approaches a limit amplitude oscillatory motion. It is also interesting to note that the apparent frequency is slightly higher than that of the torsion mode. Since the number of time steps in the plot represents 10 cycles, it is easy to calculate the apparent frequency. The results for varying Δq are listed in Table 1 for the wing tip accelerometer response in g.

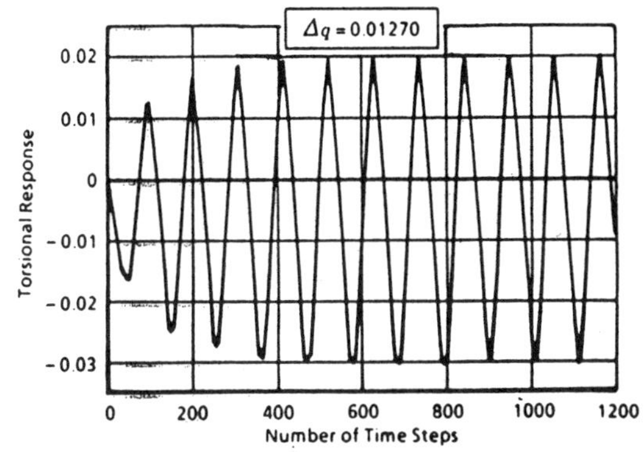

Fig. 23 Sample of limited amplitude oscillations produced by the shock-induced trailing-edge separation model.

Table 1 Effect of Δq on LCO algorithm response

		Torsion mode	Right wing tip accelerometer
$\Delta q/\varepsilon$[a]	Δq, m	Apparent f, Hz	Response, g
0.2	0.00077	18.4	0.71
0.4	0.00155	17.0	1.42
0.6	0.00232	16.4	1.79
0.8	0.00310	15.1	2.05
1.0	0.00387	14.9	2.34

[a]$\varepsilon = -0.00387$ m (0.0127 ft).

The response results presented in Table 1 are in excellent order-of-magnitude agreement with the torsion mode rms responses at 10 deg shown in Fig. 20. The slightly higher frequency also agrees with observations in the flight-test power-spectral densities (PSD) as discussed in Ref. 9. Since the estimate for ε was considered to be conservative, the results in Table 1 are especially encouraging. It is also interesting to note that the reduced pressure levels shown by Kinsey[10] with increasing dynamic pressure would also explain the peculiar tendency to maintain constant response levels nearly independent of altitude as shown in Fig. 20. Reducing or increasing the C_p values used to evaluate would tend to offset the variation in ε due to dynamic pressure. This effect would be primarily due to static aeroelasticity.

This concept also helps explain why the higher torsion mode response did not appear on the $\frac{1}{6}$-scale wind-tunnel model with which the full-scale predictions were made. An estimate for ε was made again using the pressure data shown in Fig. 21. The result was -0.000077 m (-0.000253 ft), which is far less than one-sixth of the full-scale result of -0.00387. For similar deflections between model and aircraft, the ratio should be the same as the scale factor. Hence, because amplitude of motion of the model on the basis of ε is more like 0.12 relative to the full-scale motion, this concept would predict very little influence of the coupling mechanism on the wind-tunnel model excitation forces. Another very important aspect is the mismatch of scaled torsion mode frequencies, so that, even if the coupling occurred on the model, it would not appear at the correct full-scale frequency range.

With regard to a more realistic mathematical model for the limit amplitude oscillations, it is expected that the nonlinear spring would be slightly more complicated. A calculation for the incidence change from 10 to 11 deg produced a higher ε value of -0.00651 m (-0.0232 ft), which indicates an increasing nonlinear spring constant beyond SITES. Similar results could be produced with such a spring and would explain the high responses measured in the torsion modes at angles beyond the transition to SITES (or 10 deg for this case).

In view of the promising results obtained with the simple mathematical model, further research on this approach should be very beneficial. A better definition of ε and Δq as functions of both static and dynamic motions is the most obvious first step. In addition, the mathematical model would need refinement to reflect the knowledge gained through the experimental work. The result of this effort should be a simplified method for predicting the occurrence and magnitude of limit amplitude oscillations in which wing motions couple with separated flowfields.

4. Gust Response

The gust response of aircraft flying at cruise conditions and transonic speeds does not differ significantly from that encountered at fully subsonic or supersonic speeds. The mechanism is still the same where the response is essentially proportional to the gust amplitude, lift-curve slope, and freestream dynamic pressure but inversely proportional to aircraft velocity and total damping. The influence of all of these parameters with the exception of lift-curve slope and damping is unaffected by transonic flowfields. Although structural damping is essentially constant, aerodynamic damping can be significantly changed by transonic flowfields either with or without extensive flow separation. This is illustrated in the flutter results presented in Ref. 5, where transonic effects tended to lower the flutter speeds relative to linear theory predictions for transport-type configurations with high-aspect-ratio wings. In Ref. 11, however, fighter aircraft with stiff low-aspect-ratio wings were shown to have higher flutter speeds as a result of transonic effects. This generalization is not universal, since in Ref. 1 Farmer and Hanson showed that experimental transonic flutter speeds can vary significantly for high-aspect-ratio wings with different airfoil sections.

The increase in lift-curve slope at transonic speeds is a well-known characteristic that is generally predictable with linear theory. The influence of transonic flow nonlinearities is second order on wing lift as opposed to first order on aerodynamic damping, especially for torsion-type modes. This is a result of the chordwise center of pressure being far more sensitive than total wing circulation to transonic flow effects.

From the preceding discussion, it would be expected that transonic gust response of aircraft flying at or near cruise conditions would primarily vary from that predicted by linear theory only by the amount of variation of aerodynamic damping. At elevated angles of attack during maneuvering, damping would also change and hence further modify the gust response characteristics, which would still be essentially linear. However, if significant flow separation was present either at cruise or maneuver attitudes, then the possibility would exist for the occurrence of LCO.

In the case of LCO, atmospheric turbulence merely acts as a trigger rather than the source as in the case of conventional gust response. An example is where an aircraft flying slightly below the onset of LCO encounters an upward vertical gust that drives it into LCO. Once in LCO, the aircraft attitude must drop significantly in order to get back to steady flight. If the aircraft is continuously flying in LCO, then, unless the atmospheric turbulence is severe enough, the dominant response will be that of the LCO. In this last example, if the aircraft encounters a downward gust of sufficient intensity, it can be driven out of LCO for a period of time until it returns to its original condition. Still another possibility is the case for the variable sweep wing discussed previously where LCO occurred over a very small range of incidence. For this example, it would be possible for either an up or down gust to temporarily drive it out of LCO.

In summary, there are two transonic effects that can influence aircraft gust response. The first includes a first-order modification to aerodynamic damping and a second-order modification to the lift-curve slope. This effect does not alter the basic gust response mechanism from that encountered at purely subsonic or supersonic speeds but does change the response amplitude. The second effect is when the gust field acts as a trigger for LCO. In this case, gusts can drive the aircraft into or out of LCO such that the response is composed primarily of intermittent LCO. If the aircraft is flying continuously in low turbulence at LCO conditions, then the dominant response would be LCO and not be classified as gust response. If the aircraft is flying in severe turbulence around LCO, then the gust response would be primarily that due to turbulence with bursts of LCO as the aircraft passed through LCO conditions. In this last example, the presence of LCO would probably have little or no effect on wing loads that could result in structural damage or failure to the primary structure. Secondary structure, however, is very sensitive to fatigue damage caused by LCO loadings. This is especially true for trailing-edge control surfaces and fairings that are located where the flow separation is most severe and have high-frequency panel-type modes of vibration.

5. Maneuver Aerodynamics

When an aircraft performs a maneuver, almost all aerodynamic parameters are affected. The aircraft may roll, pitch, yaw, accelerate, decelerate, or any combination of these at different rates depending on the type and abruptness of the maneuver. Response of the aerodynamic flowfields to the change of conditions is finite and therefore lags behind instantaneous aircraft attitude by amounts also dependent on the maneuver abruptness. One of the best-known examples of this effect is the dynamic maximum lift overshoot achieved by rapid pull-up maneuvers. Maneuver transient effects at transonic speeds do not differ significantly from those at either subsonic or supersonic speeds, with the exception of some Mach number effects and, more importantly, transient shock-induced separation effects. This latter effect is best illustrated by considering maneuvering aircraft buffet response at transonic speeds.

An extensive investigation of the buffet characteristics and buffet response prediction for a variable sweep configuration was reported in Refs. 12 and 13. Cunningham and Benepe discussed the influence of pitch rate on the development of buffet response in Ref. 12. The nondimensional parameter

$$\bar{\dot{\alpha}} = (\dot{\alpha} 2C)/U$$

where $\dot{\alpha}$ is the pitch rate in degrees per second, C is the wing chord, and U is the freestream velocity, was used as an indicator of unsteady influence. For $\bar{\dot{\alpha}} < 0.01$ the aircraft buffet characteristics should closely follow the steady characteristics and therefore be quasisteady in nature. As $\bar{\dot{\alpha}}$ passed 0.01 and approached 0.1 in value, unsteady effects would become quite pronounced and the development of separated flow would show considerable lag relative to the instantaneous aircraft attitude. This effect is well illustrated in the development of three maneuvers that differ primarily in pitch rate as presented in Ref. 13.

The results for wing tip accelerometer response are shown in Fig. 24 for a slow wind-up turn and two rapid pull ups as taken from Ref. 13. Also shown in Fig. 24 are the pull-up data adjusted to the dynamic pressure for the wind-up turn, as designated by the solid symbols. The adjusted pull-up results are consistently lower than the wind-up-turn results at higher angles

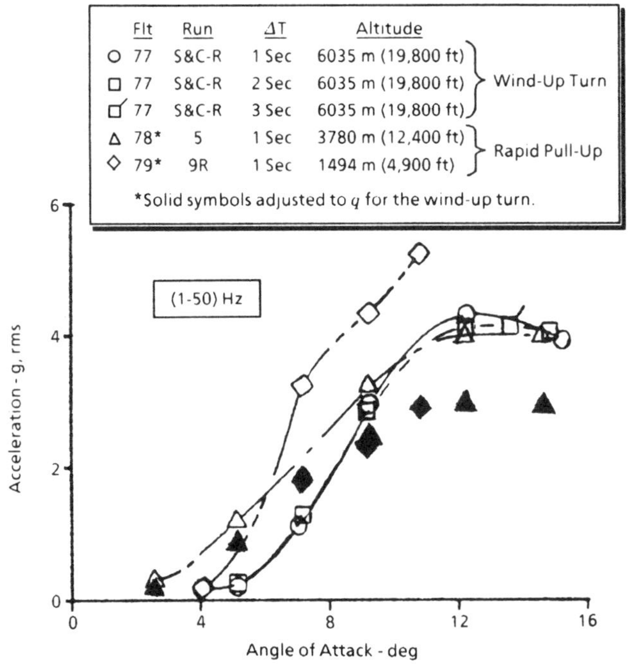

Fig. 24 Wing tip accelerometer response during buffet for a wind-up turn and two rapid pull ups.[13]

Table 2 Pitch rate parameters for maneuvers referenced in Fig. 24

Maneuver	Mach	Alt, ft	q, psf	$\dot{\alpha}$, deg/s	$\bar{\alpha}$, deg
Slow wind-up turn (flt. 77)	0.80	19,800	440	0.96	0.012
Rapid pull up (flt. 78)	0.80	12,400	600	6.64	0.080
Rapid pull up (flt. 79)	0.80	4,900	800	6.20	0.080

but show an earlier onset. Pertinent data for these maneuvers are given in Table 2.

Clearly the data in Table 2 provide a good example of the influence of pitch rate. Because buffet onset occurs at a lower angle of attack at transonic speeds than at subsonic speeds and separated flows are more sensitive to attitude rates, transonic flows are therefore more affected by maneuver transients.

Although altitude effects are present in the buffet data shown in Fig. 24, their influence does not appear to be major. The static aeroelastic effects would be different for the two rapid pull ups as indicated by the variation of q at the two altitudes. In fact, the difference between the maneuvers of

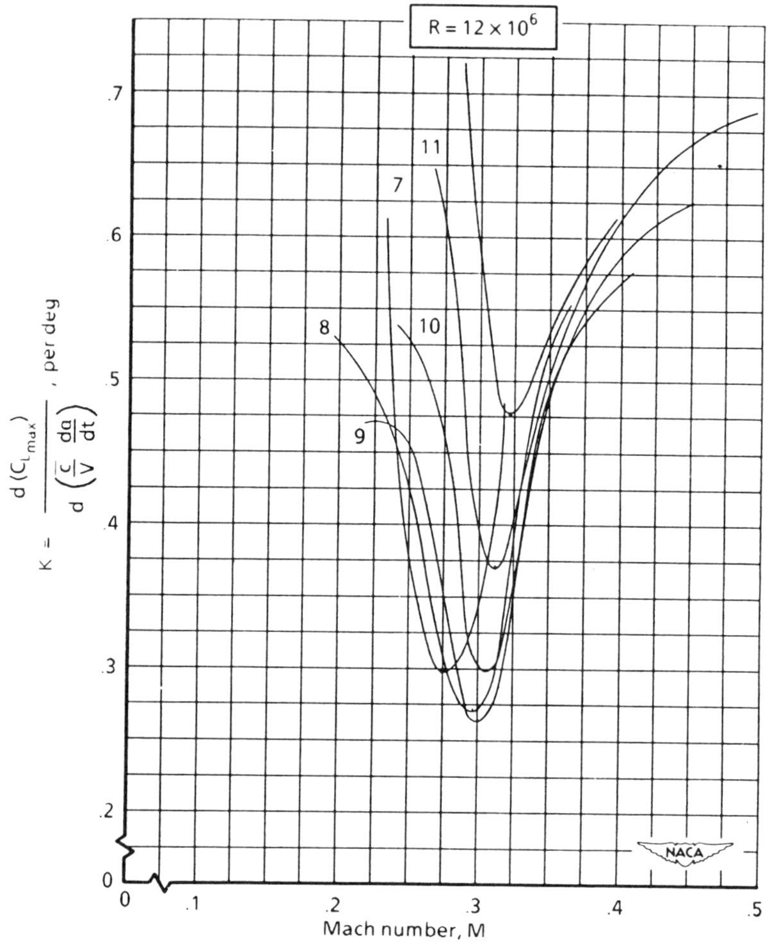

Fig. 25 Variation of *K* with Mach number for various Reynolds numbers for the P-51.[14]

flights 77 and 78 should be less than that between flights 78 and 79; thus, the slow wind-up turn at 19,800 ft would be more closely related to the rapid pull up at 12,400 ft. Since the two rapid pull ups are almost identical when adjusted for q, then the major effect shown in Fig. 24 must be pitch rate.

Evidence supporting the influence of maneuver transients on transonic aerodynamics is given in Refs. 14 and 15. These reports present data that show the influence of pitch rate on the dynamic overshoot of static maximum lift. In Ref. 14, flight-test results for the North American P-51 are given for rapid pull-up maneuvers at Mach numbers up to 0.49. The influence of Mach and Reynolds numbers on the parameter K, $C_{l_{max}}$ overshoot increment

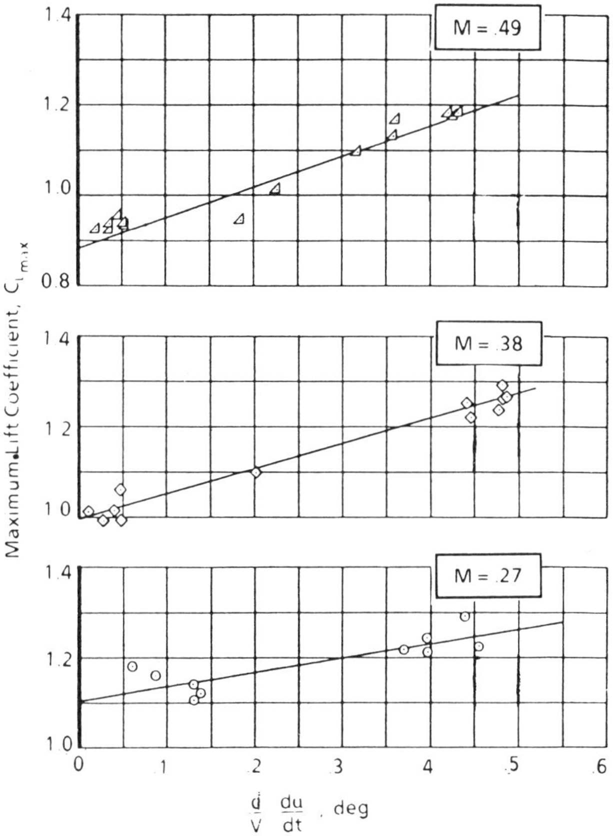

Fig. 26 Effect of rate of change of angle of attack on maximum lift coefficient at 25,000 ft altitude.[14]

divided by $\bar{\dot{\alpha}}$, is shown in Fig. 25 for the P-51 flight-test results. The dip at $M = 0.3$ corresponds to the appearance of leading-edge separation caused by formation of a nose shock as is typical on round nose airfoils. The rise in K after the dip is indicative of the increased sensitivity of dynamic effects that results from the appearance of shock-induced separation in the flowfield. Although the data shown do not go beyond about $M = 0.5$, there is an upward trend. Data for different Mach numbers at 25,000 ft altitude are shown in Fig. 26 where the change in slope as well as static $C_{l_{max}}$ with Mach number is apparent. Similar data are shown in Fig. 27 for a swept rectangular wing as reported in Ref. 15 up to $M = 0.765$. For these data $\bar{\dot{\alpha}}$ is expressed in radians rather than degrees so that $\bar{\dot{\alpha}} = 0.01$ rad in Fig. 27 corresponds to $\bar{\dot{\alpha}} = 0.57$ deg in Fig. 26.

Since $C_{l_{max}}$ tends to increase with positive pitch rate, it seems reasonable to expect separated flow effects to decrease with positive pitch rate. This is a logical assumption because of the close relationship between $C_{l_{max}}$

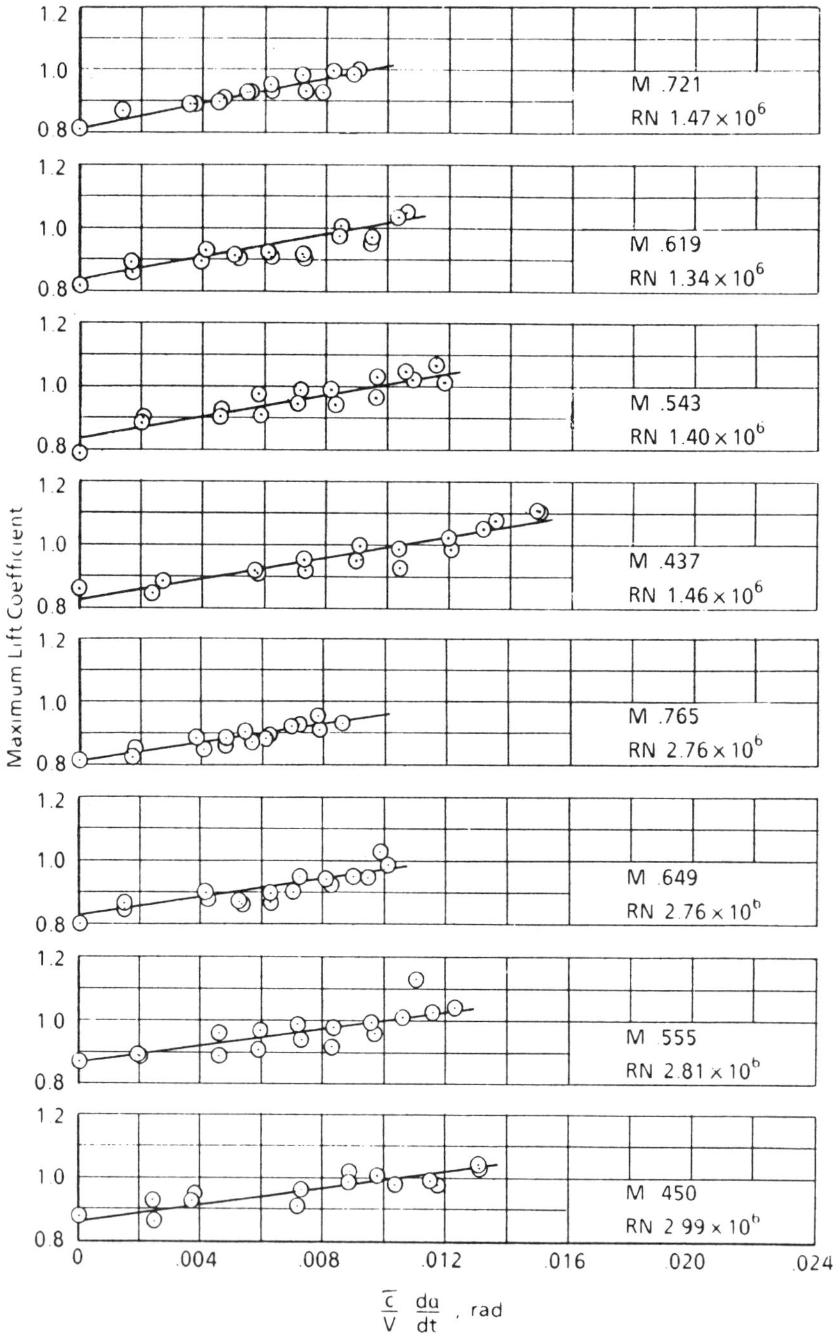

Fig. 27 Effect of angle-of-attack rate on the maximum lift coefficient of a 30-deg swept rectangular wing.[15]

and the occurrence of total flow separation. Therefore, the data shown in Figs. 26 and 27 should be applicable to the estimation of reduction in buffet response for the rapid pull ups shown in Fig. 24. Using $\bar{\dot{\alpha}} = 0.012$ deg (or $\bar{\dot{\alpha}} = 0.00021$ rad), either Fig. 26 or 27 shows very little change in $C_{l_{max}}$; thus, little change would be expected in buffet intensity for the slow wind-up turn from steady values. Using $\bar{\dot{\alpha}} = 0.08$ deg (or $\bar{\dot{\alpha}} = 0.0014$ rad) for the rapid pull ups, Fig. 26 indicates about a 6% increase in $C_{l_{max}}$, whereas Fig. 27 shows about a 2.6% increase ($Re = 2.76 \times 10^6$). The latter value as well as all other data in Fig. 27 were obtained at a low Reynolds number and are therefore suspect. Since the boundary layer at the nose shock location would most likely be laminar at such low Reynolds numbers (1.5 to 2.8×10^6), the shock/boundary-layer interaction would not be characteristic of full-scale flow phenomena. We would therefore use the value of 6% increase of $C_{l_{max}}$ to estimate the pitch rate influence for flight-test buffet data.

In Fig. 24 the lower values of normalized buffet response for the high pitch rates at angles greater than about 8 deg would be attributed directly to the reduced intensity of flow separation as indicated by an increase of $C_{l_{max}}$. The estimate of a 6% gain in $C_{l_{max}}$ obtained from Fig. 24 in the preceding discussion does not correlate well with the maximum reduction of normalized buffet intensity of about 30% shown in Fig. 24. However, if the relationship between linear lift development and separated flow effects is considered as illustrated in Fig. 28, it can be seen that a small percentage increase in $C_{l_{max}}$ due to dynamic effects could produce a larger percentage decrease of the separated flow effects. Thus, it is certainly possible that a 6% increase in $C_{l_{max}}$ could result in a 30% reduction in separated flow turbulence and hence buffet intensity.

The higher level of normalized buffet intensity below 9-deg incidence does not correlate with the previously mentioned concept of the influence of pitch rate. However, the data shown in Fig. 29 taken from Ref. 19 illustrate another effect of Mach number on dynamic airfoil lift development. It will

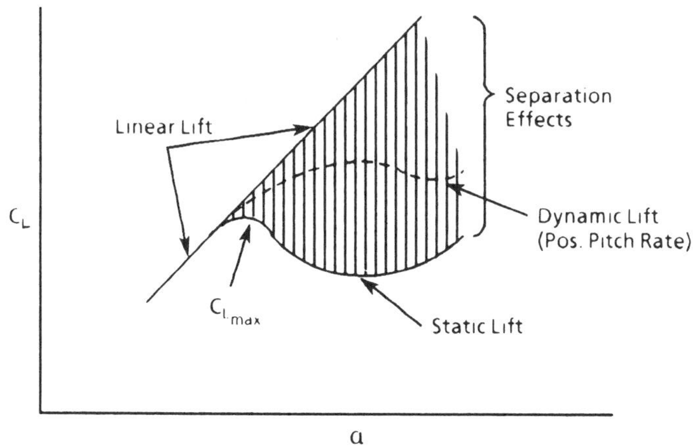

Fig. 28 Relationship between static and dynamic lift with separation effects.

be noted that the data at $M = 0.6$ show a reduction in wing lift for positive pitch rate denoted by the arrows at angles below about 10 deg, but an increase above that value results in the expected higher value of $C_{l_{max}}$. In fact, similar trends were also shown in the data presented in Ref. 15. Thus, it might be expected that where the dynamic C_l is reduced below static values, it would be accompanied by a corresponding increase in buffet intensity.

In summary, although explicit data do not exist to prove it, the differences in buffet response characteristics for the different maneuvers shown in Fig. 24 have been explained on the basis of pitch rate effects. However, it is felt that the rationale has a sound basis if for no other reason than the fact that buffet response predictions based on steady-state wind-tunnel buffet data almost always tend to be much higher than flight-test data obtained with maneuvering aircraft (except where LCO is present).

Thus, it is expected that general unsteady transonic aerodynamics are quite sensitive to aircraft maneuver transients on the basis of the preceding discussion. The connecting link is the fact that transonic buffet is mostly due to shock-induced separation, and it was just demonstrated that this buffet is very sensitive to pitch rate. The discussions presented in this chapter have repeatedly stressed the point that, although embedded shocks are certainly present in transonic flows, the most important characteristic is shock-induced separation for strong shocks. Therefore, the influence of maneuver transients on transonic flows with strong shocks would be much greater than on those with weak shocks and limited shock-induced separation.

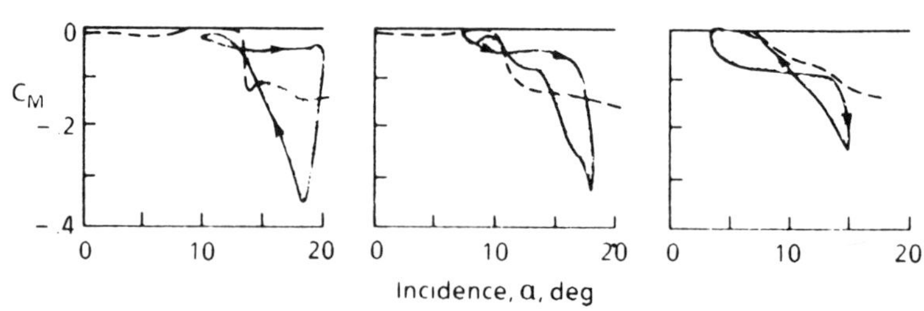

Fig. 29 Effect of Mach number on dynamic force and moment coefficients for $k = 0.25$.[16]

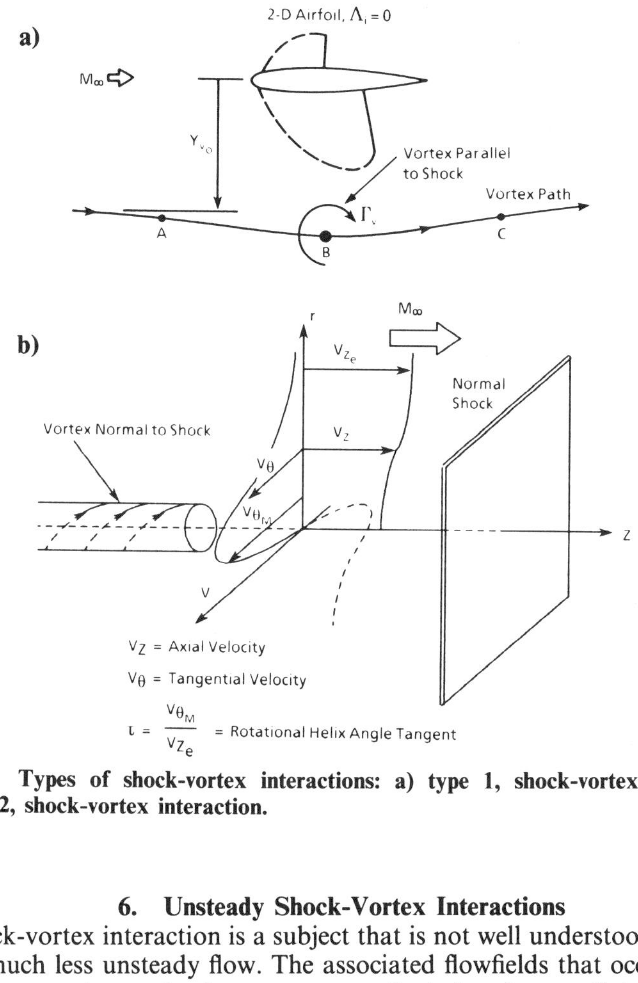

Fig. 30 Types of shock-vortex interactions: a) type 1, shock-vortex interaction b) type 2, shock-vortex interaction.

6. Unsteady Shock-Vortex Interactions

Shock-vortex interaction is a subject that is not well understood in steady flow, much less unsteady flow. The associated flowfields that occur when a vortex encounters a shock are very complicated and unpredictable as well as highly unstable. The instability leads to natural unsteadiness, which may result in wing buffeting, stability anomalies, control reversal, and/or limit cycle oscillations. Although vortex flow is usually almost independent of scale effects, the shock systems that occur at transonic speeds are very sensitive to Reynolds number as has already been discussed in this chapter. Thus, the transonic shock-vortex interaction is influenced by shock/boundary-layer coupling and may or may not be scalable between wind-tunnel and flight testing.

There are two basic types of interactions that are of importance to practical aircraft design as shown in Fig. 30. The first and simplest is the two-dimensional interaction where a vortex traverses a parallel shock system as discussed in Ref. 17. The second type is a two- or three-dimensional interaction where a vortex perpendicular to the shock plane intersects the shock as discussed in Ref. 18.

The first type of interaction, between a vortex and a parallel shock system, is more of a concern to rotary wing flows. An example of such interactions was studied in Ref. 17, where the passage of a rotor blade by its own tip vortex in forward flight was modeled by various numerical techniques. The results indicated a significant influence on the two-dimensional lift of an airfoil at zero angle of attack by a parallel vortex passing beneath the airfoil. One of the important conclusions was that recovery of airfoil circulation was quite slow after passage of the vortex, as shown in Fig. 31.

It is highly likely that the first type of interaction also occurs on fixed-wing aircraft on a downstream surface due to a rapid change of an upstream surface. An example could be the effect on a horizontal tail subjected to the starting vortex shed by rapid rotation of a trailing-edge flap on the wing. Another example is the same situation where a starting vortex from a canard passes over the wing. In either case control surface actuators on the downstream surfaces would have to respond to the induced forces caused by the vortex passage or be sufficiently damped to ignore them. The possibility of exciting LCO on these surfaces would also exist if the aircraft were flying close to an LCO boundary.

The second type of interaction between a streamwise vortex and a perpendicular shock is important for highly maneuverable aircraft that utilize controlled vortex lift (via strakes or other leading-edge extensions) and fly at transonic speeds. This interaction might resemble the flow structure illustrated in Fig. 32 where the strake vortex is burst by the wing-body normal shock. Vortex burst is an undesirable but unavoidable characteristic of vortex flows over wings at high incidence. It occurs at all speeds and is directly related to planform geometry and angle of attack. Two resulting effects that must be dealt with are 1) the loss in lift toward

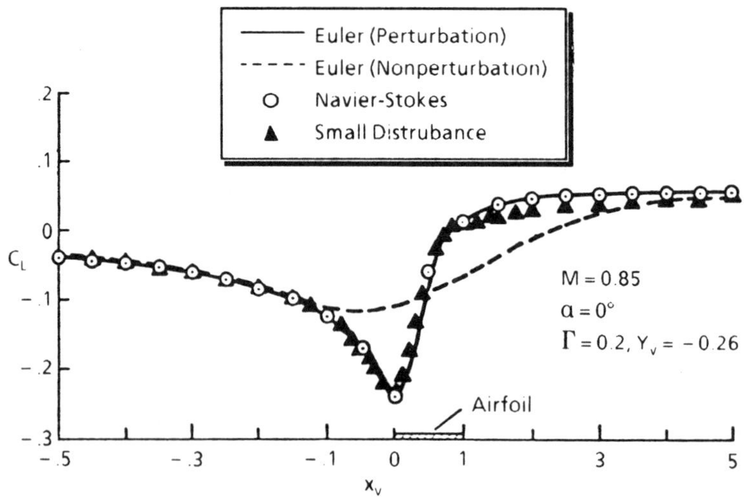

Fig. 31 Lift induced on an NACA 64A006 airfoil by passage of a parallel vortex below the wing (see Fig. 30a).[17]

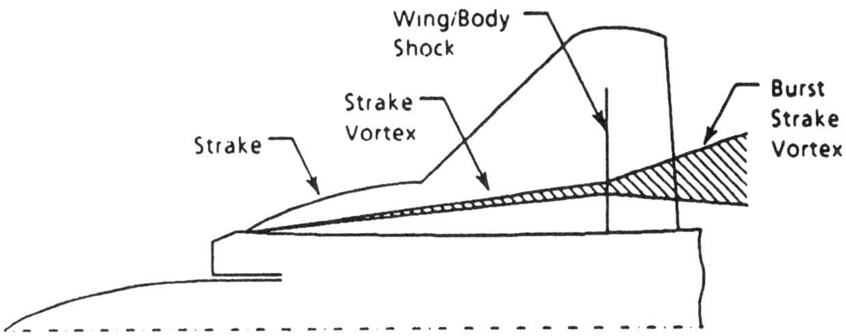

Fig. 32 Vortex bursting produced by a wing-body shock.

the wing trailing edge which produces a pitch up in pitching moment and 2) severe buffeting due to the highly unsteady flows.

Vortex bursting is usually caused by the vortex having to confront a large pressure rise such as that which occurs near the wing trailing edge at subcritical speeds or that produced by a shock embedded in the flow as illustrated in Fig. 32. Bursting or breakdown has been shown to be a function of a swirl parameter τ defined as

$$\tau = \frac{V_{\theta_{\max}}}{V_{ze}} = \frac{\text{maximum tangential velocity}}{\text{freestream velocity}}$$

where $V_{\theta_{\max}}$ and V_{ze} are shown in Fig. 30b. When τ exceeds a critical value, the vortex will burst. This property can be used to explain how a pressure rise can produce vortex burst. Since the flowfield surrounding a vortex must slow down as it encounters a pressure rise, the value of τ must increase as a result of a decrease in V_{ze}. The V_θ component is unaffected by the pressure gradient; thus, τ simply varies inversely with V_{ze}. For a leading-edge vortex produced by a highly swept delta wing, the strength of the vortex and hence $V_{\theta_{\max}}$ must increase with increasing incidence. Since the freestream pressure is always constant and the flow surrounding the vortex, i.e., V_{ze}, must return to that condition after leaving the wing trailing edge, the value of τ at the wing trailing edge for fully subsonic flow must continue to rise with $V_{\theta_{\max}}$ because V_{ze} is held constant. Thus, at some value of incidence, depending on the leading-edge sweep, τ will exceed the critical value and the vortex will burst at the wing trailing edge.

A similar situation occurs for a leading-edge vortex that encounters a normal shock. An experimental investigation of this phenomenon was summarized in Ref. 18, where a streamwise vortex in a supersonic flow was allowed to intersect a normal shock formed by a downstream constriction. The experimental setup illustrated in Fig. 33 was used to study the flowfields for burst and unburst vortices as well as to determine the influence of shock and vortex strengths on vortex burst. [The original version of Figs. 33, 34, and 36 was first published by the Advisory Group

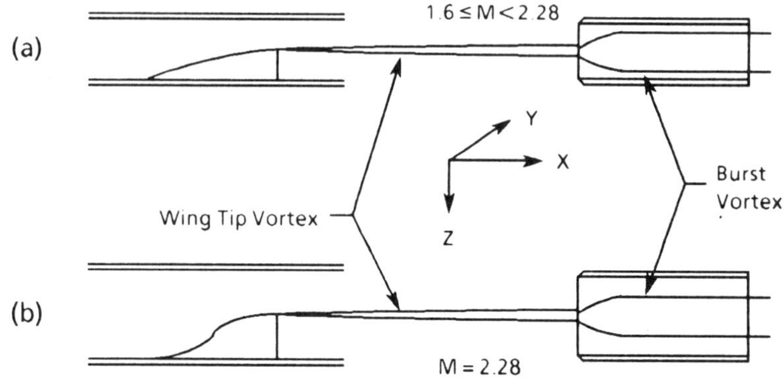

Fig. 33 Experimental setup for determining the relationship between vortex and shock strengths for vortex bursting.[18]

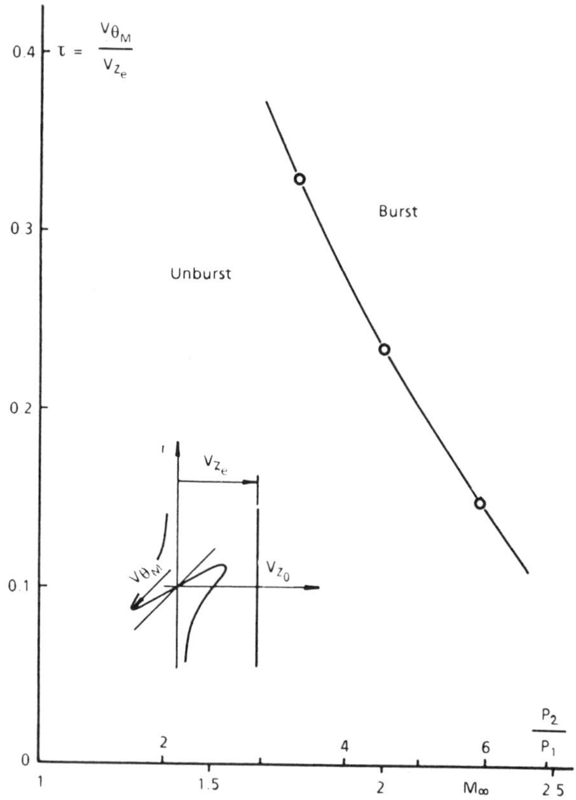

Fig. 34 Experimentally determined relationship between vortex and shock strengths for vortex bursting.[18]

for Aerospace Research and Development/North Atlantic Treaty Organization (AGARD/NATO) in Conference Proceedings 342, *Aerodynamics of Vortical Type Flows in Three Dimensions*, July 1983 (Ref. 18).] Results from this study are shown in Fig. 34, which establishes the relationship between the upstream Mach number and the critical upstream value of τ that produces vortex bursting just downstream of the shock. Other results given in Ref. 18 showed that the influence of the shock on $V_{\theta_{max}}$ was negligible but that V_{ze} dropped as expected based on normal shock relationships. Hence, the increase of τ across the shock was simply due to the drop in V_{ze}. Using the measured velocity ratio across the shocks from Ref. 18 and the critical values of upstream τ shown in Fig. 34, the τ values just downstream were obtained as plotted in Fig. 35, where

$$\tau_{\text{downs}} = \tau_{\text{ups}}/(V_{ze\,\text{downs}}/V_{ze\,\text{ups}})$$

the value shown for τ_{downs} at $M = 1.0$ is given as a reference to illustrate the effect of a normal shock relative to subsonic flow. This value was determined from 1) the results given in Ref. 19 for subsonic bursting vortex flow in a tube and 2) the vortex core size and test section dimensions given in Ref. 18. The apparent intersection of the data at higher Mach numbers with the value at $M = 1.0$ is interesting and serves to illustrate the effect of embedded shocks on vortex bursting. What is implied in Fig. 35 is that a vortex will burst at a smaller value of τ if it encounters a pressure rise due to a normal shock than it would if the same pressure rise were produced in a shock-free flow.

The unsteady rms velocity fluctuations σ_v produced by a vortex encountering a normal shock were also presented in Ref. 18 as shown in Fig. 36. The velocity components V_z and V_θ are defined in Fig. 30b, and the subscript 0 denotes vortex centerline values. These results were obtained from laser Doppler measurements for both burst ($\alpha = 22.5$ deg) and unburst ($\alpha = 10$ deg) flows. (The α values represent the incidence of the upstream model that is generating the vortex.) It is interesting to note that

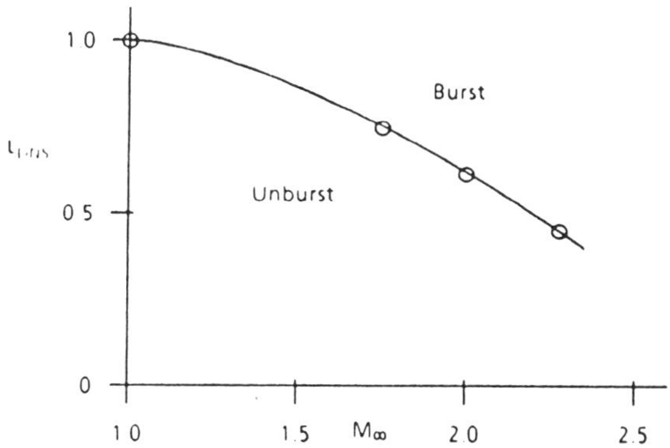

Fig. 35 Relationship between postshock values for burst vortex and upstream Mach number (deduced from Ref. 18).

the velocity fluctuations increase slightly even for the unburst vortex. However, the rise of σ_v values aft of the shock ($z > 0.0$) seen for the burst vortex demonstrates the power of vortex bursting as a source of buffet excitation. Maximum fluctuations of V_z on the order of about $0.15U_\infty$ shown in Fig. 36 can be translated into approximate fluctuations in the pressure coefficient of about 0.30.

Based on these velocity fluctuations, we may also speculate on what might happen in the case of a strake vortex encountering a wing-body shock, as suggested in Fig. 32. First, the wing-body shock would be interacting with the wing boundary layer, which would most likely be separated at the shock. Second, if the vortex burst just aft of the shock, this unsteady flow would further interact with the separated wing boundary layer. This second coupling would then tend to cause the shock to move more than it would in the case of simple shock-separated boundary-layer interaction. Increased shock motion would likewise lead to greater velocity (pressure) fluctuations aft of the shock due to shock motion and additional unsteadiness of shock-induced vortex bursting. Thus, it would appear that shock-vortex interaction could potentially produce significantly higher levels of unsteadiness on wings subject to such flows at transonic speeds than might be encountered at lower speeds.

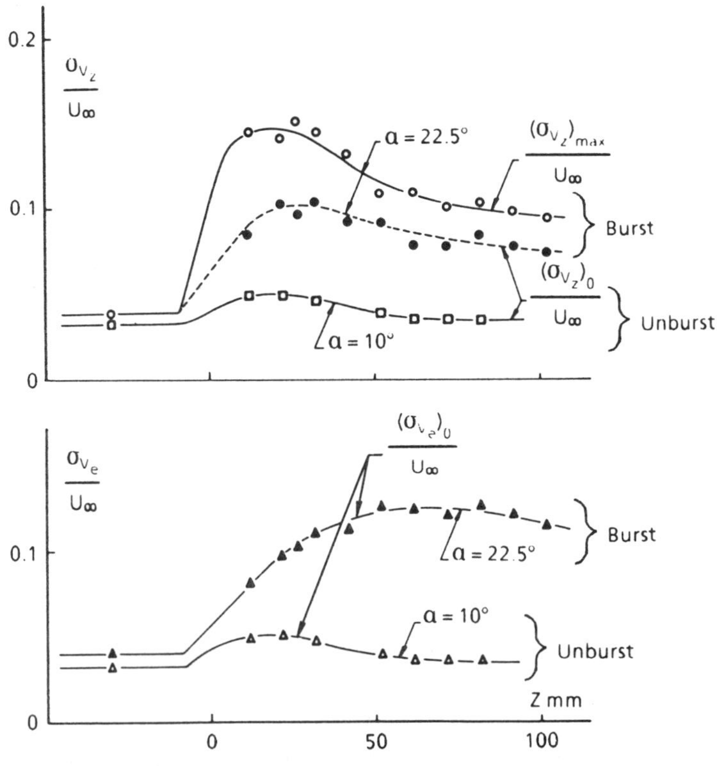

Fig. 36 Effect of shock on velocity fluctuations downstream of the shock.[18]

7. Control Reversal: Characteristics and Buzz

Control reversal is a static aeroelastic phenomenon that can occur independently of transonic effects. That is, with sufficient dynamic pressure, the force developed by a trailing-edge control surface on a wing tip, such as an aileron, can reverse. This happens when the induced load due to wing twist is greater than and opposite to that due to control rotation. As an example, rotation of a wing tip aileron trailing edge down produces an up force but a nose-down torsional moment. The wing will twist nose down in response to the moment which produces an incremental down load on the wing. It is this incremental wing twist load that counteracts the aileron up load to produce the reversal tendency. If the nose-down moment is increased for a given upward force due to control rotation, then the reversal tendency is increased. This happens as a result of transonic effects where the center of pressure for the trailing-edge control surface load is moved aft, thus increasing the nose-down pitching moment discussed in the example.

The problem of control reversal due to the static aeroelastic effects is usually most critical at low supersonic speeds. This represents the condition where the center of pressure for a trailing-edge control surface loading is the farthest aft. In fact, aileron roll effectiveness requirements at, say $M = 1.2$, frequently drive wing stiffness designs for fighter aircraft.

The transonic influence on trailing-edge control surface loads is well illustrated by the pressure distributions shown in Fig. 37 as taken from Ref. 20. These data show the steady chordwise pressure changes produced by rotation of a 25% chord trailing-edge control on an NACA 64A006 airfoil in two-dimensional flow at zero mean incidence. The distributions are regular at $M = 0.5$ and 0.75 but begin to show some distortion at $M = 0.8$ and 0.825. The "hump" shown at $M = 0.825$ becomes a peak at $M = 0.85$ and 0.875, which is the dominant feature at $M = 0.90$ and 0.92. Beyond $M = 0.92$, the shock is located on the control surface, and at $M = 0.96$, a distribution is seen on the control that is similar to that shown in Fig. 4 for the case of shock-induced trailing-edge separation on the NACA 0012 airfoil. The negative peak due to shock movement at $M = 0.96$ and 0.98 is indicative of this occurrence, as was discussed extensively for the NACA 0012 airfoil.

Considering the case for $M = 0.96$ in unsteady flow where the control surface is oscillating, a significant change takes place in the unsteady pressure distribution as the frequency is increased. This is shown in Fig. 38 for three frequencies, 30, 90, and 120 Hz, for $k = 0.054$, 0.163, and 0.216, respectively. Besides changing in character, the pressures show a significant change in phase angle toward increasing lag, as might be expected. It is also apparent that the actual distributions are more complicated than those that can be detected with the pressure instrumentation.

The integrated aerodynamic derivatives for the previously mentioned configuration are shown in Figs. 39 and 40 as a function of Mach number for steady flow and unsteady flow at 120 Hz, respectively. Also shown with the experimentally determined values are those predicted with linear theory. It is interesting to note that, where it is shown, theory agrees better with experiment for the unsteady case compared with the steady case. The

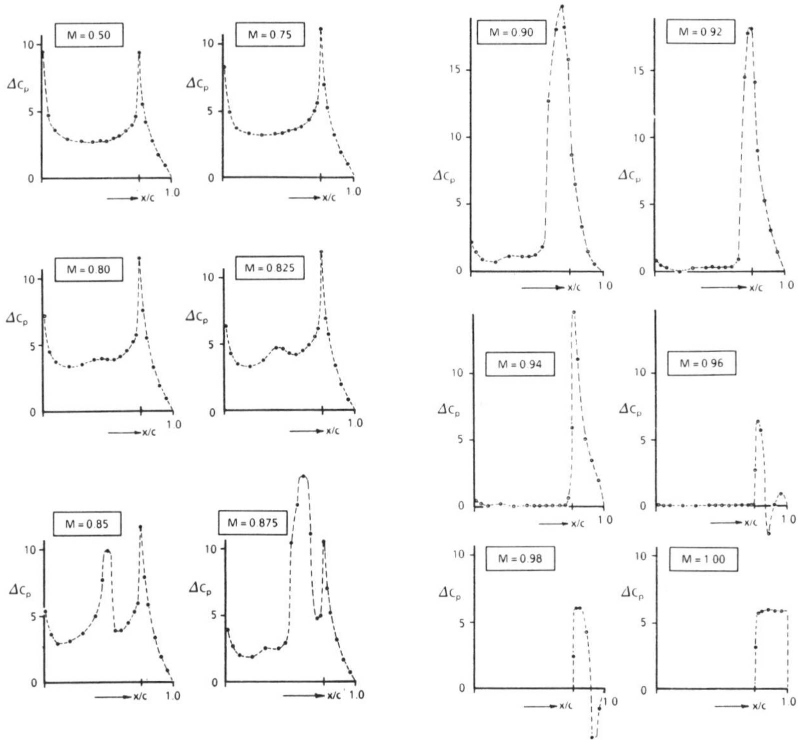

Fig. 37 Influence of Mach number on steady control surface pressure distributions on a 2D airfoil.[20]

complicated nature of the derivatives, which is illustrated for $M > 0.9$ for steady flow and $M > 0.83$ for unsteady flow, is very difficult to predict because of the dominant influence of flow separation. Progress is being made toward solving this problem, however, as shown by some calculations presented in Refs. 21 and 22 for cases that include two-dimensional, transonic, separated flowfields. Although these theoretical results are impressive, the techniques are still far from being practical design tools. The unavailability of practical three-dimensional flow design tools will probably exist into the 1990's or even beyond the year 2000.

Thus far our discussion has been concerned with two-dimensional flow over a symmetric airfoil at zero incidence. An idea of how trailing-edge controls respond in three-dimensional transonic flow can be developed by considering the data shown in Fig. 41 as extracted from Ref. 23. These data are the integrated forces measured on a part-span inboard aileron on a 45-deg swept-wing model. The unflagged symbols represent strain-gage measurements and are therefore believed to be more accurate since the separated pressure distributions are very complicated. Three sets of aileron normal force data are shown at $M = 0.80$, 0.94, and 1.03 vs angle of attack α for three aileron deflections of 0, 7.5 (trailing edge down), and -7.5 deg (trailing edge up). The data up to $\alpha = 8$ deg are reasonably symmetric for

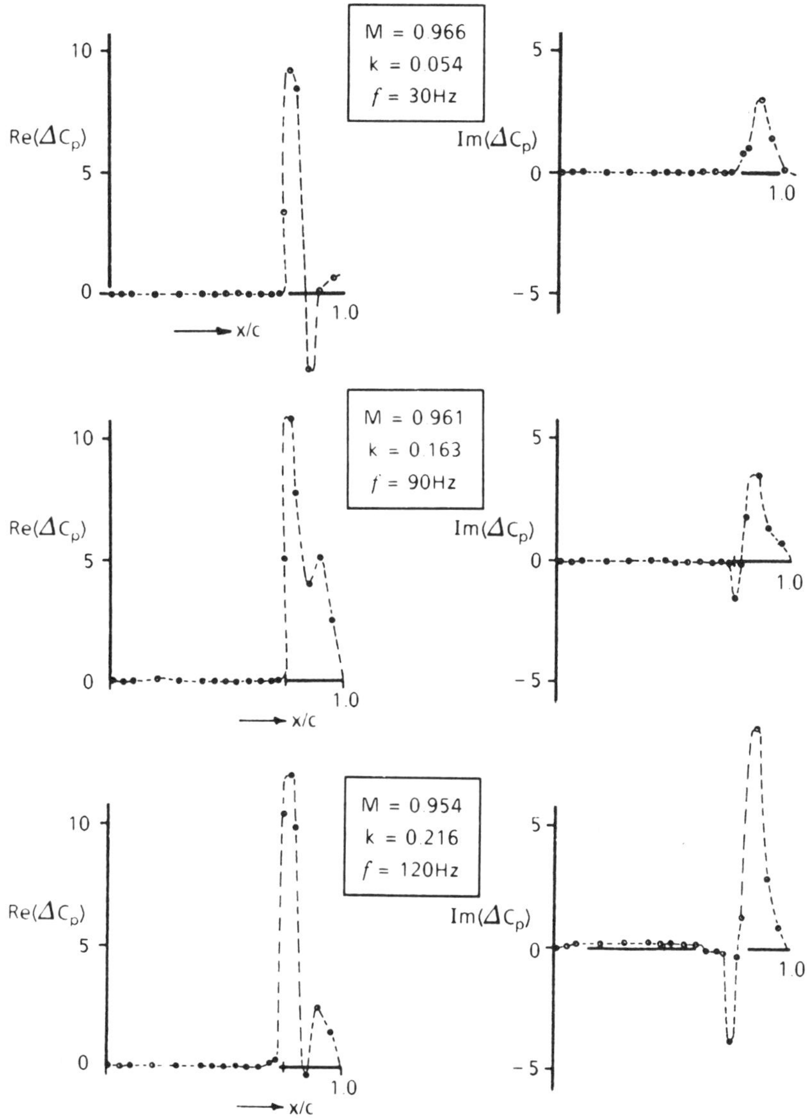

Fig. 38 Influence of frequency on unsteady control surface pressure distributions on a 2D airfoil at $M = 0.96$.[20]

the ± 7.5-deg deflections. Beyond 8 deg, however, flow separation from the wing begins to influence the forces for both 0 and $+7.5$ deg (trailing edge down) but not as much, if any, for the -7.5 deg (trailing edge up) case. This is a result of the trailing-edge-down deflection tending to aggravate wing trailing-edge separation that already exists for zero deflection at $\alpha = 8$ deg. With the aileron deflected trailing edge up at -7.5 deg, the wing

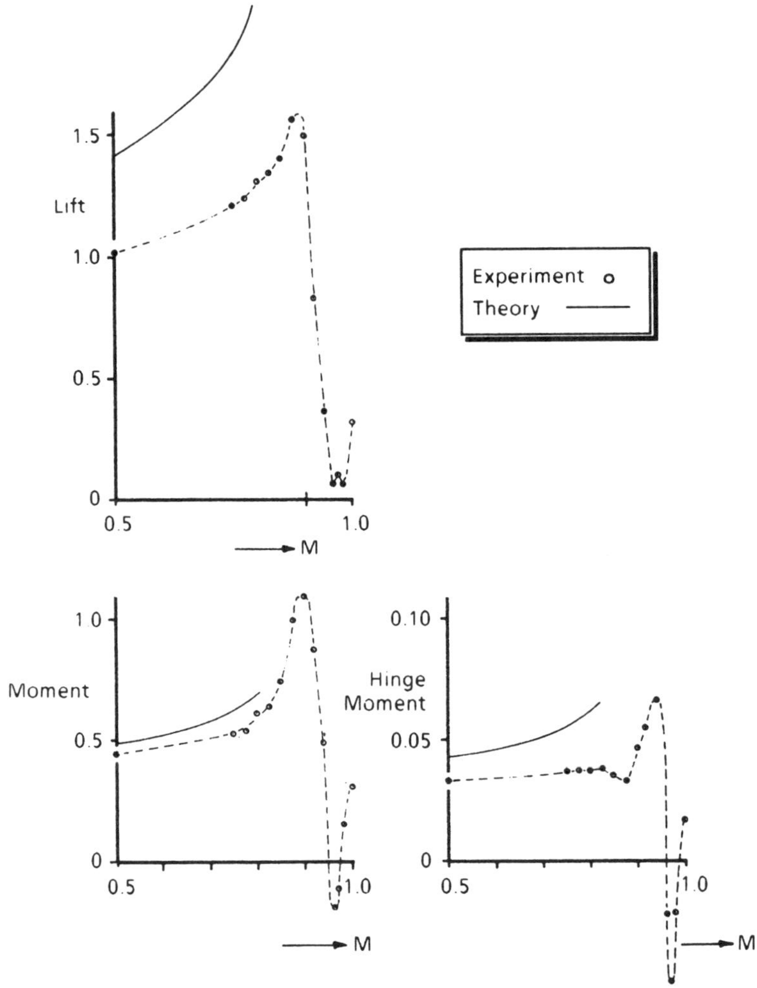

Fig. 39 Aerodynamic derivatives vs Mach number ($f = 0$) on a 2D airfoil with a trailing-edge flap.[20]

separation is either retarded or eliminated. This example demonstrates both the effect of changing the mean flow as well as the asymmetry of derivatives where flow separation is encountered. Thus, an active control system designed to operate at low angles of attack may not be appropriate for higher angles.

From the data shown in Fig. 41, if we examine the condition at $\alpha = 16$ deg, $M = 0.94$ with zero aileron deflection, we see that deflecting it downward 7.5 deg would produce no noticeable force change. Deflecting it up -7.5 deg, however, would produce a significant change in force. Since deflecting the aileron up would at some point allow the flow to become reattached, the role of separated flow becomes dominant at this condition.

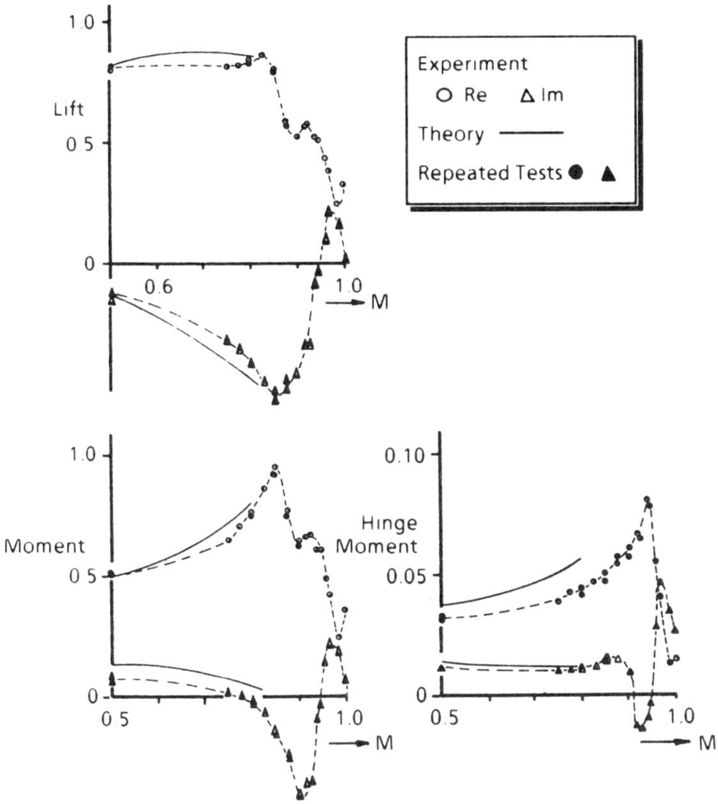

Fig. 40 Aerodynamic derivatives vs Mach number ($f = 120$ Hz) on a 2D airfoil with a trailing-edge flap.[20]

Thus, the influence of unsteady control surface motion could be very significant. This was demonstrated with the two-dimensional trailing-edge flap case just discussed. The data in Figs. 37 and 38 illustrated how the case for shock-induced trailing-edge separation at $M = 0.96$ was considerably modified as frequency was increased from 0 to 120 Hz. The results in Figs. 39 and 40 further demonstrated how the unsteady derivatives can differ from the steady values where shock-induced separation plays a dominant role.

Knowing that these problems exist does not help the designer of an active control system get better answers, but it does make him aware of the potential problem areas. Ideally, he needs prediction or extrapolation/interpolation schemes to develop a data base with which to design his system. In response to this requirement, a study was conducted for NASA to develop correction techniques for improving aerodynamic prediction methods and is described in Ref. 24. However, the techniques developed were not successful, but the conclusions did point out the shortcomings of such methods. One of the main conclusions was that, if the correction

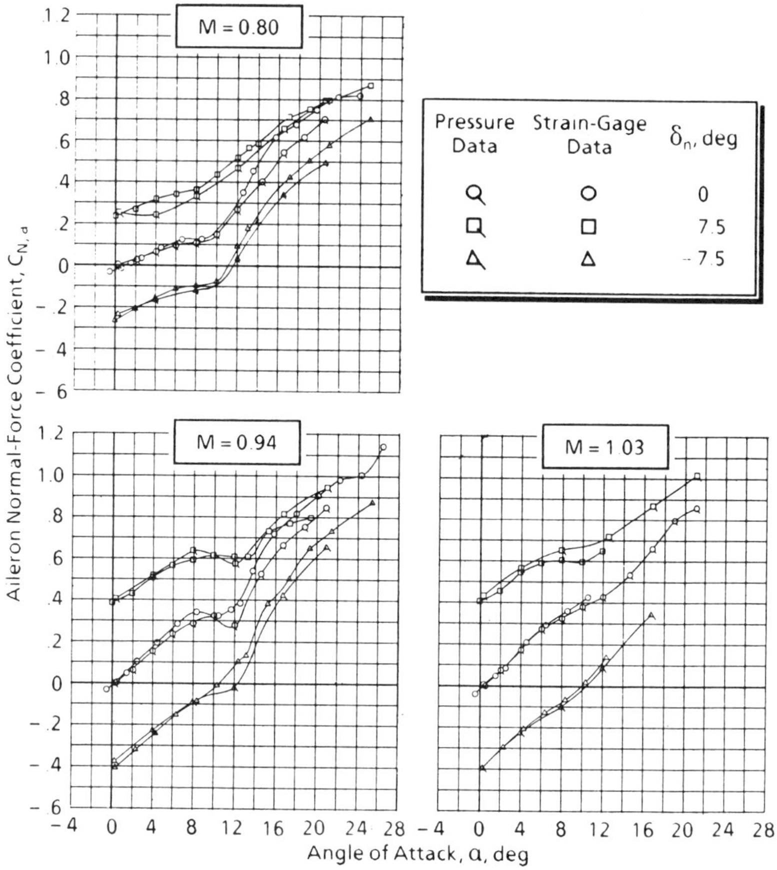

Fig. 41 Aileron normal force coefficients for an inboard aileron on a swept tapered wing.[23]

technique did not contain at least some of the physics involved in the phenomena being modeled, its generality would be very limited. For this reason, these techniques could not treat the problem discussed earlier for the aileron asymmetry with respect to deflection at $\alpha = 16$ deg. With the current advances being made in computational fluid dynamics, as shown by Refs. 21 and 22, the ultimate solution to the designer's dilemma will be computational methods that will probably be coupled with well-designed experimental programs. Until then, however, he must depend on experimental data bases that are expanded with either empirical or semiempirical techniques to suit his needs. This aspect will be further discussed in the next section of this chapter.

Returning to the subject of control surface aerodynamics in transonic flow, control surface buzz is a direct result of shock-induced trailing-edge separation. In this case, buzz may occur either with or without structural/

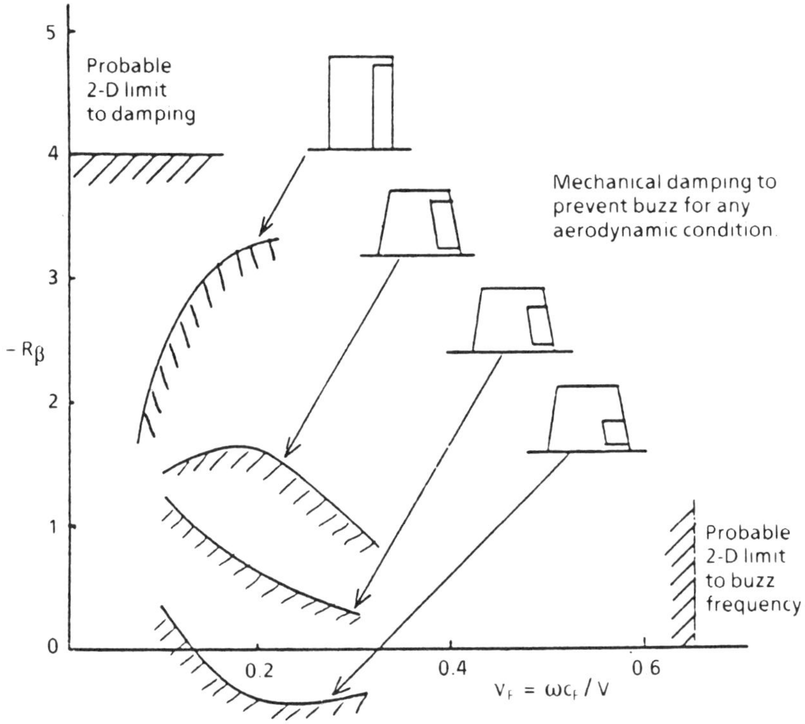

Fig. 42 Probable damping and frequency limits for control surface buzz.[25]

inertial coupling with the aerodynamic forces, since the shock-induced separation provides the hysteresis and lag necessary to drive the oscillations. Because separation plays a key role, these oscillations rarely tend to be catastrophic and are limit cycle in nature for current aircraft designs. In fact, most recent aircraft have sufficient actuator stiffness to avoid encountering control surface buzz. For this reason, buzz is not considered to be a current problem; however, with developments being made in active control technology, it is necessary to understand buzz in order to avoid its encounter with an active control system.

There are two distinct possibilities for control surface buzz to be driven by shock-induced separation. The first requires control surface motion to produce the exciting force and hence is a fully coupled mechanism. The second does not require surface motion and is more related to phenomena such as the von Kármán vortex shedding. In this case, shocks on the upper and lower surfaces oscillate naturally out of phase with each other.

An excellent discussion of the first possibility for control surface buzz is given by Lambourne in Ref. 25. His discussions of the mechanisms involved with buzz are also in agreement with those given earlier in this chapter on LCO, thus placing buzz in this latter, broader category. Recommendations based on many references for avoiding control surface buzz are given in Ref. 25 as follows: 1) sufficient mechanical damping to counter the negative damping due to aerodynamics, 2) high hinge-line (or actuator) stiffness, 3) low hinge-line moment of inertia, 4) hinge-line sweepback, and 5) low-aspect-ratio control surfaces.

These recommendations were in part based on the data given in Fig. 42, which shows the approximate hinge-line damping needed to prevent buzz for several control surface configurations. They are also based on another rule of thumb, which states that buzz will be avoided if

$$f_{cs}c_{cs} > 35$$

where f_{cs} is the flap rotation natural frequency in hertz, and c_{cs} is the flap chord in meters. Because these criteria are well known, current aircraft designs are usually based on these principles and hence avoid the problem of buzz. Again, it should be emphasized that such criteria must be accounted for in the design of active control systems to avoid adding a problem in the solution to another.

The second possibility for control surface buzz where natural unsteadiness exists usually occurs with symmetric airfoils that may be flexible or rigid. This type of buzz can also occur with all-movable surfaces such as horizontal tails and is therefore not restricted to trailing-edge controls. Some very enlightening results from an experimental investigation of this phenomenon are presented in Ref. 26. Figure 43 shows the domains for shock oscillation occurrences over a rigid 18% thick biconvex circular arc airfoil as a function of both Reynolds and Mach numbers. With the exception of the range below $Re = 6 \times 10^6$, the characteristics seem to be independent of Reynolds number for increasing Mach number. Typical of the hysteresis associated with shock-induced separation, the boundary for decreasing Mach number extends significantly farther than that for increasing Mach number. In this case, the effect of Reynolds number extends somewhat farther to $Re = 10 \times 10^6$. The minimum width at about $Re = 3 \times 10^6$ for both increasing and decreasing Mach numbers is characteristic of natural shock oscillations on rigid biconvex airfoils, where, for thinner airfoils, the oscillations disappear entirely but are present at both lower and higher Re values.

The measured nondimensional frequency of oscillation for the data in Fig. 43 was constant at about $k = 0.49$, where

$$k = (2\pi fc)/(2V)$$

f is the frequency of oscillation (Hz), c is the airfoil chord (m), and V is the freestream velocity (m/s).

This value seems to be fairly universal, since data from another reference cited in Ref. 26 indicated that the value was about 0.51 for several different rigid models that were geometrically similar with a 20% thick circular arc airfoil section.

In summary, control surface buzz, which is not considered to be a serious problem with current aircraft designs, may be a problem for active control

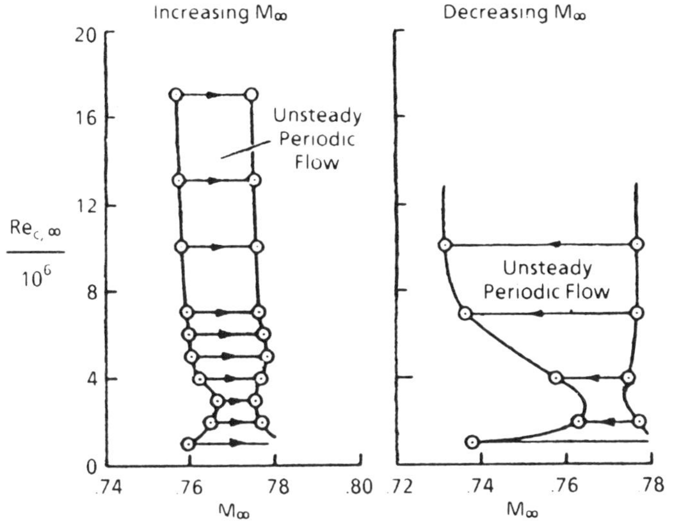

Fig. 43 Domains for airfoil shock oscillation on an 18% thick biconvex airfoil at $\alpha = 0$ deg.[26]

systems unless it is also accounted for in the design of those systems. Two types of buzz were identified, one where control surface motion is needed to couple with the unsteady aerodynamics and the other where natural unsteadiness exists that can drive the surface. Both types were shown to be due to shock-induced separation and related to the broader classification of LCO.

8. Active Controls Technology Needs for Gust Alleviation and Flutter Suppression

Active controls technology, originally developed to alleviate gust loads and improve ride quality during the 1950's and 1960's, was extended to include flutter suppression in the 1970's. These earlier systems were designed to suppress model motions for limited conditions in which the parameters involved were well defined. An example might be to improve ride quality of a transport aircraft. This could be achieved by correctly exciting wing tip control surfaces to minimize the first wing bending mode response due to atmospheric turbulence at design cruise conditions. This example would require accurate knowledge of the wing aerodynamic forces due to the first wing bending mode, wing tip control surface rotation, and atmospheric turbulence. The gains in the active control feedback loop would be determined as a function of flight conditions only, such as Mach number, altitude, and gross weight, with relatively simple schedules made up of constants, ramps, and step functions. These gain schedules would be developed on the basis of the aerodynamic information available for the wing. Thus, effectiveness of the gust alleviation system would be directly dependent on how accurately the wing aerodynamic characteristics were known.

More recent developments in active controls technology for aircraft have been directed toward adaptive systems that essentially determine the appropriate gains of phase angles needed to suppress an undesirable motion or response. With such systems, detailed knowledge of the aerodynamic characteristics involved in the feedback loop is not necessary as long as the system is designed to handle all possibilities of parameter variation. This last requirement is very important because of the ranges of aerodynamic parameter variation that can be encountered in transonic flow. Consider, for example, the normal force data due to flap rotation in two-dimensional flow shown in Figs. 39 and 40. For steady flow in Fig. 39, the force coefficient per unit rotation drops from about 1.5 at $M = 0.9$ to about 0.05 at $M = 0.95$ with a very steep gradient between. For unsteady flow, the magnitude drop is not so severe; however, the phase angle changes about 90 deg going from $M = 0.9$ to 0.95. In both cases, the changes in pitching and hinge moment are equally severe as those in normal force.

The design of an adaptive active control system must be quite flexible to work in the kind of parameter environment previously described. Considering the drop in steady force from 1.5 at $M = 0.9$ to 0.05 at $M = 0.95$, to obtain the same amount of control force at $M = 0.95$ as available at $M = 0.9$, the commanded surface rotation would have to be 30 times higher at $M = 0.95$. This means that large deflections would be required where flows are not as well behaved; thus, nonlinear characteristics would most likely be encountered, as was pointed out in the data in Fig. 41 for the aileron on a swept wing. Adding in the additional nonlinearities associated with higher angles of attack such as those encountered by fighter aircraft, the required flexibility of adaptive control systems can become quite demanding. Demonstration of their effectiveness and safety then becomes as complicated or more so than a standard flutter clearance test.

The practical capability to analytically determine these nonlinear characteristics needed for active control system design is not currently available, nor is it expected to be available before the 1990's or even beyond 2000. However, there is a wealth of information available from stability derivatives usually obtained in wind-tunnel tests of the aircraft of interest. Some of the derivatives account for rate effects such as roll, pitch, and yaw rates. These provide some information about unsteady aerodynamics of the wings, tails, etc., but not about oscillating control surfaces. Derivatives also do not provide information about large control surface deflections; however, the data from which they are obtained usually include large deflection data. Thus, for this latter purpose, the raw force data for control surface deflections are of more interest than the reduced derivatives. It is also very important that aeroelastic effects be accounted for in the stability data. This is especially true for trailing-edge control surfaces, which, as discussed in the previous section, have a tendency to reverse at transonic speeds.

The development of unsteady control surface derivatives requires considerable ingenuity and is not an easy problem to solve, as exemplified by the study summarized in Ref. 24. Generally, relating steady and unsteady characteristics for generic configurations on which these data are available and identifying the type of flowfields involved are the first steps. These relationships are then used to extract unsteady data for the configuration of interest where flowfields are similar between that and the generic configurations. As concluded in Ref. 24, this last requirement, i.e., flowfield similarity, was felt to be the one most important feature in a dependable aerodynamic estimation technique. For highly nonlinear parameter environments, adaptive active control systems are probably the only practical means to achieve the desired results. Hence, determination of the limits of parameter variation is the major requirement of the data base development.

Some final comments are needed with regard to active control technology requirements in transonic flows. Bomber- and logistic-type aircraft have well-defined flight and maneuver envelopes that do not venture too far into the nonlinear regimes. Therefore, active control systems for such aircraft can likewise be well defined and flight safety can be assured over the operational envelope. Fighter aircraft, on the other hand, do not fit into this picture, and the environment and aerodynamic characteristics are much more difficult to define. For instance, if a flutter suppression system were designed to operate under conditions based on information available up to 20-deg angle of attack, what happens if the pilot has to suddenly maneuver beyond 20 deg in combat? The flutter system cannot turn off and say "sorry, you can't do that." The system designer must be able to account for all known possibilities in the flight envelope and extrapolate where no other course is possible.

9. Concluding Remarks

The unsteady transonic phenomena described in the previous chapters were discussed in this chapter regarding how they affect aircraft and the practical problems that result. Topics covered included flutter, limit cycle oscillations, maneuver aerodynamics, control reversal, buzz, active control technology needs, gust response and alleviation, and unsteady shock-vortex interaction. Three-dimensional effects on real aircraft were stressed. All of the known important effects of unsteady transonic flows were presented as far as was possible regardless of how well we presently understand their nature.

The distinction between flutter at transonic speeds and either purely subsonic or supersonic speeds was shown to be the coupling of unsteady aerodynamic forces with the steady mean flow. The presence of shocks and flow separation influenced the unsteady forces and hence flutter characteristics. As a result, aircraft attitude, control surface deflection, aeroelastic effects, Reynolds number, and boundary-layer transition were all identified as having potential for affecting flutter at transonic speeds. An example was given in which it was possible to cruise at flutter-free conditions and maneuver into a flutter condition without changing speed but by simply increasing angle of attack. Caution was also expressed in the use of flutter models to identify these nonlinear effects as a result of not being able to accurately simulate the mean flowfields.

The occurrence of LCO's of lifting surfaces was described as a nonlinear coupling between structural response and unsteady aerodynamic forces as well as being strongly dependent on the steady mean flowfields. These oscillations usually appear as sinusoidal motions of a single vibration mode at conditions that may be close to expected flutter. Thus, the differentiation between LCO and incipient flutter is difficult to determine because of safety restrictions with either flutter models or flight test. Three examples of LCO were discussed involving both transonic and leading-edge vortex flows in which the common phenomenon was shown to be a transition that could produce a step function in force or moment. A simple mathematical model was developed using this phenomenon and was applied to one of the three examples. The mathematical model predicted LCO to occur in flight at the correct order of magnitude and not in the wind-tunnel tests in agreement with observations. The primary effects of LCO on aircraft performance were described as 1) restrictions imposed on pilot or system performance due to structural vibrations and 2) fatigue damage to secondary structure.

Two transonic effects were discussed that can influence aircraft gust response. The first included a first-order modification to aerodynamic damping and a second-order modification to the lift-curve slope. This effect does not alter the basic gust response mechanism from that encountered at purely subsonic or supersonic speeds but does change the response amplitude. The second effect was when the gust field acted as a trigger for LCO. In this case, gusts can drive the aircraft into or out of LCO such that the response is composed primarily of intermittent LCO.

Evidence was presented which indicated that unsteady transonic aerodynamics are quite sensitive to aircraft maneuver transients. The connecting link was the fact that transonic buffet is mostly due to shock-induced separation, and it was demonstrated that this buffet is very sensitive to pitch rate. Thus, the influence of maneuver transients on transonic flows with strong shocks would be much greater than on those with weak shocks and limited shock-induced separation.

Two basic types of unsteady shock-vortex interaction were discussed. The first was the two-dimensional case where a vortex traverses a parallel shock system such as might be encountered in rotorcraft. This type of interaction is more transient in nature and could excite LCO if the affected surface was close to an LCO boundary. The second type of interaction was concerned with the intersection of a streamwise vortex with a normal shock such as a strake vortex encountering a wing-body normal shock. With sufficient shock strength, the vortex could burst prematurely and produce significant unsteady aerodynamic forces. These forces, which would result from both vortex burst unsteadiness and shock oscillations, would tend to excite aircraft buffet response as well as affect aircraft stability.

The influence of transonic flowfields on trailing-edge control surfaces was shown to be significant. Trends for control reversal are affected in complicated ways and are sensitive to wing and airfoil geometry as well as flow parameters. All of these problems make it difficult to design an active control system that operates under such conditions.

Control surface buzz was not considered to be a serious problem with current aircraft designs but may be a problem for active control systems unless it is also accounted for in the design of those systems. Two types of buzz were identified, one where control surface motion is needed to couple with the unsteady aerodynamics and another where natural unsteadiness exists that can drive the surface. Both types were shown to be due to shock-induced separation and related to the broader classification of LCO.

Finally, active controls technology needs for gust alleviation and flutter suppression at transonic speeds were discussed. Active control system designs must be able to account for all known possibilities in the flight envelope, which includes the highly nonlinear nature of control surface aerodynamic forces and moments as well as those for the complete aircraft. Bomber- and logistic-type aircraft have well-defined flight and maneuver envelopes that do not venture too far into the nonlinear regimes. Therefore, active control systems for such aircraft can likewise be well defined and flight safety can be ensured over the operational envelope. Fighter aircraft, on the other hand, do not fit into this picture, and the environment and aerodynamic characteristics are much more difficult to define. The flowfields encountered at high angles of attack in transonic flows are exceedingly complicated and highly nonlinear. Even though control systems can be designed to be adaptive so that exact knowledge of the aerodynamics is not always needed, limits must be determined for the system requirements based on these aerodynamics. In addition, transitional nonlinearities such as those discussed in relation to LCO must be identified to avoid potential problems associated with unstable coupling with the control systems.

References

[1]Farmer, M. G. and Hanson, P. W., "Comparison of Supercritical and Conventional Flutter Characteristics," AIAA Paper 76-1787, May 1976; see also NASA TM X-72837, May 1976.

[2]Cahill, J. F. and Conner, P. C., "Correlation of Data Related to Shock-Induced Trailing-Edge Separation and Extrapolation to Flight Reynolds Number," NASA CR-3178, Sept. 1979.

[3]Triebstein, H., "Steady and Unsteady Transonic Pressure Distributions on NACA 0012," *Journal of Aircraft*, Vol. 23, March 1986, pp. 213–219.

[4]AGARD, "Experimental Data Base for Computer Program Assessment," AGARD-AR-138, May 1979.

[5]Hounjet, M. H. L., van der Kolk, J. T., and Meijer, J. J., "Application of NLR's Numerical Simulation Methods to the Transonic Potential Flow About Oscillating Wings," AIAA Paper 84-1564, June 1984.

[6]Cunningham, A. M., Jr., "Influence of Manufacturing Tolerances on the Transonic Vibration of F-16 Horizontal Tails," presented at the Workshop on Transonic Unsteady Aerodynamics and Aeroelasticity, NASA Langley Research Center, June 1983.

[7]Mason, W. H., Mackenzie, D., Stern, M., Ballhaus, W. F., and Frick, J. "An Automated Procedure for Computing Three-Dimensional Transonic Flow Over Wing-Body Combinations, Including Viscous Effects, Volume I—Description of Analysis Methods and Applications," AFFDL TR-77-122, Air Force Flight Dynamics Lab., Wright-Patterson AFB, OH, Vols. I–III, Feb. 1978.

[8]Dobbs, S. K., Miller, G. D., and Stevenson, J. R., "Self Induced Oscillation Wind Tunnel Test of a Variable Sweep Wing," AIAA Paper 85-0739-CP, April 1985.

[9]Coe, C. F. and Cunningham, A. M., Jr., "Predictions of F-111 Tact Aircraft Buffet Response and Correlation of Fluctuating Pressures Measured on Aluminum and Steel Models and the Aircraft," NASA CR-4069, 1987.

[10]Kinsey, D. W., "Wind Tunnel Flight Test Correlation of Static Pressures," *Proceedings of Symposium on Transonic Aircraft Technology (TACT)*, AFFDL TR-78-100, Air Force Flight Dynamics Lab., Wright-Patterson AFB, OH, Aug. 1978.

[11]Hounjet, M. H. L. and Meijer, J. J., "Application of Time-Linearized Methods to Oscillating Wings in Transonic Flow and Flutter," NLR MP-84077U, National Aerospace Lab., Amsterdam, The Netherlands, Aug. 1984.

[12]Cunningham, A. M., Jr. and Benepe, D. B., "Prediction of Transonic Aircraft Buffet Response," AGARD CP-226, Paper No. 3, April 1977.

[13]Benepe, D. B., Cunningham, A. M., Jr., and Dunmyer, W. D., "An Investigation of Wing Buffeting Response at Subsonic and Transonic Speeds: Phase I F-111 Flight Data Analysis, Volume I—Summary of Technical Approach, Results and Conclusions," NASA CR-152109, Jan. 1975.

[14]Gadeberg, B. L., "The Effect of Rate of Change of Angle of Attack on the Maximum Lift Coefficient of a Pursuit Airplane," NACA TN-2525, Oct. 1951.

[15]Conner, F., Willey, C., and Twomey, W., "A Flight and Wind Tunnel Investigation of the Effect of Angle-of-Attack Rate on Maximum Lift Coefficient," NASA CR-321, Oct. 1965.

[16]Liiva, J., "Unsteady Aerodynamic and Stall Effects on Helicopter Rotor Blade Airfoil Sections," *Journal of Aircraft*, Vol. 6, Jan.-Feb. 1969, pp. 46–51.

[17]McCroskey, W. J., "The Effects of Gusts on the Fluctuating Airloads on Airfoils in Transonic Flow," *Journal of Aircraft*, Vol. 22, March 1985, pp. 236–243.

[18]Delery, J. and Horowitz, E., "Interaction Entre une Onde de Choc et une Structure Tourbillonnaire Enroulee," AGARD CP-342, Paper No. 5, April 1983.

[19]Escudier, M. P. and Keller, J. J., "Vortex Breakdown: A Two-Stage Transition," AGARD CP-342, Paper No. 25, April 1983.

[20]Tijdeman, H. and Schippers, P., "Results of Pressure Measurements on an Airfoil with Oscillating Flap in Two-Dimensional High Subsonic and Transonic Flow," NLR TR-73078U, National Aerospace Lab., Amsterdam, The Netherlands, Aug. 1973.

[21]Le Balleur, J. C. and Girodroux-Lavigne, P., "A Viscid-Inviscid Interaction Method for Computed Unsteady Transonic Separation," presented at the Third Symposium on Numerical and Physical Aspects of Aerodynamic Flows, California State Univ., Long Beach, CA, Jan. 1985.

[22]Houwink, R., "Computations of Separated Subsonic and Transonic Flow About Airfoils in Unsteady Motion," presented at the Third Symposium on Numerical and Physical Aspects of Aerodynamic Flows, California State Univ., Long Beach, CA, Jan. 1985.

[23]Runckel, J. F. and Hieser, G., "Normal-Force and Hinge-Moment Characteristics at Transonic Speed of Flap-Type Ailerons at Three Spanwise Locations on a 4-Percent-Thick Sweptback Wing-Body Model and Pressure Distribution Measurements on an Inboard Aileron," NASA TN D-842, May 1961.

[24]Giesing, J. P., Kalman, T. P., and Rodden, W. P., "Correction Factor Techniques for Improving Aerodynamic Prediction Methods," NASA CR-144967, May 1976.

[25]Lambourne, N. C., "Unsteady Aerodynamic and Aeroelastic Properties of Control Surfaces in Transonic Flow," *Unsteady Airloads and Aeroelastic Problems in Separated and Transonic Flow*, VKI Lecture Series, Rhode-Saint-Genése, Belgium, March 1983.

[26]McDevitt, J. B., Levy, L. L., Jr., and Deiwert, G. S., "Transonic Flow About a Thick Circular-Arc Airfoil," *AIAA Journal*, Vol. 14, May 1976, pp. 606–613.

Chapter 4

Basic Numerical Methods

Joseph L. Steger* and William R. Van Dalsem†
NASA Ames Research Center, Moffett Field, California

Nomenclature

Operators

E = shifting operator, $Eu^n = u^{n+1}$, $E^{-1}u^n = u^{n-1}$, $E^m u^n = u^{n+m}$

E_x = shift operator in x, $E_x^m u_j = u_{j+m}$

∂_x = partial derivative with respect to x, $\partial_x u = \partial u/\partial x$

$\bar{\delta}$ = midpoint difference operator, $\bar{\delta} = E^{\frac{1}{2}} - E^{-\frac{1}{2}}$

δ_x = a generic difference operator with respect to x, $\delta_x \simeq \partial_x$, often taken as the three-point central difference approximation, $\delta_x = (E_x - E_x^{-1})/(2\Delta x)$

$\bar{\delta}_x$ = midpoint difference in x, $\bar{\delta}_x u_j = (u_{j+\frac{1}{2}} - u_{j-\frac{1}{2}})/\Delta x$

$\boldsymbol{\nabla}$ = vector gradient operator

∇, Δ = backward and forward differences, $I - E^{-1}$, $E - I$

∇_x = backward difference in x, $\nabla_x u = (u_j - u_{j-1})/\Delta x$

Δ_x = forward difference in x, $\Delta_x u = (u_{j+1} - u_j)/\Delta x$

Subscripts

$x, y, z, t, \xi, \ldots$ = independent variable subscript denotes partial derivative, $u_x = \partial_x u$

j, k, l = grid point location, $u_{j,k,l} = u(x,y,z)$ at point j, k, l

$1, 2, 3, \ldots$ = elements of an array, $(Q_1,Q_2,Q_3,\ldots)$ but in some contexts $u_1 = u_{j=1}$, etc.

Superscripts

-1 = inverse or reciprocal

n = usually nth iterative or time level, but sometimes an exponent

Upper case generally denotes a vector or matrix, exceptions include U, V, W, and J.

---

*Senior Staff Scientist, Fluid Dynamics Division.
†Research Scientist, Applied Computational Fluids Branch.

1. Introduction

Historically the need to treat the difficult problems of transonic flow has given considerable impetus to the development of computational fluid dynamics (CFD). For steady flow, substantial CFD capability has been achieved either by simulating flow about relatively complex configurations by using simplified equations and boundary conditions, or by simulating simple aerodynamic shapes using more complete governing equations and boundary treatment. In contrast, although "path-finding" unsteady flow calculations have been carried out for various classes of body and shock-wave motion, buffeting, buzz, vortex shedding, and flutter (see Edwards and Thomas, Chapter 5 this volume, for a discussion), the computation of unsteady transonic flow is certainly not routine. Numerical methods for solving unsteady Euler and Navier-Stokes equations have been available for several decades, but unsteady three-dimensional calculations using such procedures have been costly. As a result, perhaps only with the transonic small disturbance equations has any kind of engineering capability been reached in the numerical simulation of unsteady three-dimensional transonic flow.

The realm of unsteady transonic flow includes some of the most intriguing flow phenomena. The understanding and control of unsteady transonic flow remains of critical engineering importance, and computational fluid dynamics, in concert with experiment and traditional analysis, will be the primary method of achieving this goal. With these thoughts in mind, the quite modest goal of this article is to set down some of the basic computational methods for unsteady transonic flow.

This primer briefly reviews some of the basic finite-difference schemes that can be used to solve the nonlinear equations that describe unsteady inviscid and viscous transonic flow. Numerical schemes for solving the unsteady Euler and Navier-Stokes, boundary-layer, and nonlinear potential equations are described. The emphasis of this article is placed on the elementary ideas used in constructing various numerical procedures, not the specific details of any one procedure. As a result, the approach taken here is to first consider the nonlinear governing equation from which a model linear problem can be extracted. Representative numerical schemes for the model linear problem are then described and evaluated. Finally, application of the basic scheme to the nonlinear governing equation(s) is indicated. Some knowledge of basic numerical methods and linear algebra is assumed throughout.

Because general purpose numerical procedures for inviscid and viscous approximate methods are in many ways much less developed than numerical procedures for the Euler and Navier-Stokes equations, this primer begins with the treatment of these more complete formulations and then proceeds to boundary-layer and transonic potential schemes. Consequently, the development starts with methods for solving the Euler equations in unsteady one-dimensional flow. Analysis reveals that a simple convection equation serves as a model problem for the Euler equations, and this model equation is used to analyze differencing schemes which are then applied to the Euler equations. Techniques to bring these procedures into multidimensions are illustrated using just a few select algorithms. The techniques developed for the Euler equations are then extended to treat the Navier-Stokes equations. Here simple convection-diffusion and diffusion-alone equations serve as model equations to evaluate numerical differencing schemes. Finally, methods for unsteady boundary-layer and transonic potential flow are discussed. For the most part, the numerical analysis of many of these schemes is much less developed than the analysis for the Euler and Navier-Stokes equations, and this is especially true for viscous boundary-layer–inviscid interaction methods. Viscous-inviscid interaction procedures, although more computationally efficient, have not been adequately generalized to highly separated flow in three dimensions.

For simplicity of presentation, the developments are carried out for the equations in Cartesian coordinates. However, the methods discussed here will for the most part carry over directly to equations transformed onto curvilinear coordinates, and the transform relations are briefly indicated at the end of this article.

This primer, although lengthy, is nevertheless very limited in scope, and numerous topics are not discussed. Important issues not treated include such pertinent ones as the treatment of far-field boundary conditions, rules of how to pick a suitable time increment so as to resolve sundry unsteady flow phenomena, discussion of modern upwind shock capturing techniques, vectorized computer data structures, and so on. These are important issues, but they are not all fully resolved, and they can be experimented with once the basic numerical procedures are in hand. Finally, we note that numerical algorithms which we have experience with tend to be used for illustrative purposes.

2. Unsteady Euler Equations

For most aerodynamic flows the Reynolds number is much greater than 1, and except for thin viscous layers, which are usually confined to the wall, the overall flow can generally be considered to be inviscid. Indeed, viscous effects can even be negligible in the highly vortical region of massively separated flow. In transonic flow the nonlinear inviscid terms produce many of the interesting flow phenomena and most of the computational difficulties.

The inviscid conservation equations of mass, momentum, and energy—here not quite correctly referred to as the Euler equations—are a coupled system of partial differential equations that can be written as

$$\partial_t Q + \partial_x F + \partial_y G + \partial_z H = 0$$

where

$$Q = \begin{bmatrix} \rho \\ \rho u \\ \rho v \\ \rho w \\ e \end{bmatrix}, \qquad F = \begin{bmatrix} \rho u \\ \rho u^2 + p \\ \rho v u \\ \rho w u \\ u(e + p) \end{bmatrix}$$

$$G = \begin{bmatrix} \rho v \\ \rho uv \\ \rho v^2 + p \\ \rho wv \\ v(e+p) \end{bmatrix}, \qquad H = \begin{bmatrix} \rho w \\ \rho uw \\ \rho vw \\ \rho w^2 + p \\ w(e+p) \end{bmatrix}$$

and where ρ is density, u, v, and w are Cartesian velocity components, and e is the total energy per unit volume. For a perfect gas the pressure p is related to the other variables by

$$p = (\gamma - 1)[e - \tfrac{1}{2}\rho(u^2 + v^2 + w^2)]$$

where γ is the ratio of specific heats.

These equations contain the Rankine-Hugoniot jump relations as weak solutions and are solved for in the preceding conservative (or divergence) form so as to numerically capture shock jump conditions. The Euler equations can transport vorticity which may be generated by shock waves, separated viscous layers, or induced by sharp edges.

Before numerical procedures for solving the unsteady Euler equations can be developed, it is necessary to classify their type and to know how the equations are coupled together. These crucial ideas are illustrated using the unsteady one-dimensional form of the equations. Numerical methods are first described for the one-dimensional form of the equations and are then extended to multidimensions.

One-Dimensional Inviscid Flow Equations

In the following the unsteady Euler equations are brought into an uncoupled form for analysis purposes. It is found that the equations are of hyperbolic type. In subsonic flow the fact that eigenvalues (characteristic speeds) are both positive and negative places constraints on what are allowable difference operators.

Useful relations between the conservative and the nonconservative form of the equations are brought out by beginning with the conservative or divergence form:

$$\frac{\partial Q}{\partial t} + \frac{\partial F}{\partial x} = 0 \tag{1}$$

where

$$Q = \begin{pmatrix} \rho \\ \rho u \\ e \end{pmatrix}, \qquad F = \begin{pmatrix} \rho u \\ \rho u^2 + p \\ u(e+p) \end{pmatrix}$$

The flux vector F is a function of Q, $F = F(Q)$, and can be written as

$$F = \begin{pmatrix} F_1 \\ F_2 \\ F_3 \end{pmatrix} = \begin{pmatrix} Q_2 \\ Q_2^2/Q_1 + p \\ (Q_2/Q_1)(Q_3 + p) \end{pmatrix} \tag{2}$$

where $p = (\gamma - 1)(Q_3 - Q_2^2/(2Q_1))$ and where $Q_1 = \rho$, $Q_2 = \rho u$, and $Q_3 = e$. Because $F = F(Q)$, Eq. (1) can be brought into the nonconservative form as

$$\frac{\partial Q}{\partial t} + \frac{\partial F}{\partial Q}\frac{\partial Q}{\partial x} = 0$$

or

$$\partial_t Q + A\,\partial_x Q = 0 \tag{3}$$

where A is the Jacobian matrix, $A = [\partial F_m/\partial Q_m]$. In the case of Eq. (3),

$$A = \begin{bmatrix} 0 & 1 & 0 \\ (\gamma - 3)u^2/2 & (3 - \gamma)u & (\gamma - 1) \\ [(\gamma - 1)u^3 - \gamma eu/\rho] & \left[\dfrac{\gamma e}{\rho} - \dfrac{3(\gamma - 1)u^2}{2}\right] & \gamma u \end{bmatrix} \tag{4}$$

Although a digression, at this point it is convenient to note that, for the inviscid conservation equations,

$$F = AQ \tag{5}$$

which can be verified by direct multiplication of A with Q. (F is homogeneous of degree one; thus the preceding result is predicted by Euler's theorem of homogeneous functions.) As a consequence of Eq. (5), we can rewrite Eq. (1) as

$$\partial_t Q + \partial_x(AQ) = 0 \tag{6}$$

although this does not suit our present purpose. The flux vectors in two and three dimensions also have the homogeneous property [Eq. (5)].

The nonconservative form, Eq. (3), is not the conventional one that would use ρ, u, and p as the dependent variables. Experience shows that the conventional form is easier to analyze; hence, let

$$\tilde{Q} = \begin{pmatrix} \rho \\ u \\ p \end{pmatrix} \tag{7}$$

and note that $Q = Q(\tilde{Q})$. Because Q is a function of $Q(\tilde{Q})$, Eq. (3) can be rewritten as

$$\frac{\partial Q}{\partial \tilde{Q}}\left(\frac{\partial \tilde{Q}}{\partial t}\right) + A\frac{\partial Q}{\partial \tilde{Q}}\frac{\partial \tilde{Q}}{\partial x} = 0$$

or

$$S\,\partial_t \tilde{Q} + AS\,\partial_x \tilde{Q} = 0 \tag{8}$$

where

$$S = \frac{\partial Q}{\partial \tilde{Q}} = \begin{bmatrix} 1 & 0 & 0 \\ u & \rho & 0 \\ \dfrac{u^2}{2} & \rho u & \left(\dfrac{1}{\gamma - 1}\right) \end{bmatrix}$$

and

$$S^{-1} = \begin{bmatrix} 1 & 0 & 0 \\ -u/\rho & 1/\rho & 0 \\ \dfrac{(\gamma-1)u^2}{2} & -(\gamma-1)u & (\gamma-1) \end{bmatrix} \tag{9}$$

Multiplying through by S^{-1} gives

$$\partial_t \tilde{Q} + \tilde{A}\,\partial_x \tilde{Q} = 0 \tag{10}$$

where

$$\tilde{A} = S^{-1}AS = \begin{bmatrix} u & \rho & 0 \\ 0 & u & \rho^{-1} \\ 0 & \gamma p & u \end{bmatrix}$$

If written out, Eq. (10) is readily recognized as the familiar nonconservative form of the equations which is usually derived in aerodynamics texts by simply combining the equations in an intuitive way. It is the momentum part of Eq. (10) that is conventionally called the Euler equation(s), but in CFD the entire set of the inviscid conservation law form of the equations is often simply referred to as the Euler equations.

The eigenvalues and eigenvectors of the matrix $\tilde{A}$ are easily determined, and for a constant $\tilde{A}$, they can be used to transform the system of equations into scalar uncoupled equations. One eigenvalue of $\tilde{A}$ is u and the other two are determined from the uncoupled subblock

$$\begin{pmatrix} u & \rho^{-1} \\ \gamma p & u \end{pmatrix}$$

and are readily found to be $u \pm \sqrt{(\gamma p/\rho)} = u \pm c$, where c is the sound speed. The eigenvalues of $\tilde{A}$ are distinct; thus, there is a complete set of eigenvectors. Let the eigenvectors of $\tilde{A}$ be denoted as Ξ. Then premultiplying Eq. (10) by Ξ^{-1} results in

$$\Xi^{-1}\,\partial_t\tilde{Q} + \Xi^{-1}\tilde{A}\Xi\Xi^{-1}\,\partial_x\tilde{Q} = 0$$

or

$$\Xi^{-1}\,\partial_t\tilde{Q} + \Lambda\Xi^{-1}\,\partial_x\tilde{Q} = 0 \tag{11}$$

where Λ is a diagonal matrix of the eigenvalues of $\tilde{A}$

$$\Lambda = \begin{pmatrix} u & & \\ & u+c & \\ & & u-c \end{pmatrix} \tag{12}$$

Each eigenvector is arbitrary to a scale factor, but one choice for Ξ (and Ξ^{-1}) is

$$\Xi = \frac{1}{\sqrt{2}}\begin{pmatrix} \sqrt{2} & \rho/c & \rho/c \\ 0 & 1 & -1 \\ 0 & \rho c & \rho c \end{pmatrix}, \qquad \Xi^{-1} = \frac{1}{\sqrt{2}}\begin{pmatrix} \sqrt{2} & 0 & -\sqrt{2}c^{-2} \\ 0 & 1 & (\rho c)^{-1} \\ 0 & -1 & (\rho c)^{-1} \end{pmatrix}$$

Assuming that $\tilde{A}$ and thus Ξ^{-1} has constant elements (frozen coefficients), then

$$\Xi^{-1}\,\partial_t\tilde{Q} = \partial_t\Xi^{-1}\tilde{Q} \equiv \partial_t\boldsymbol{q}$$

$$\Xi^{-1}\,\partial_x\tilde{Q} = \partial_x\Xi^{-1}\tilde{Q} \equiv \partial_x\boldsymbol{q}$$

Thus, Eq. (11) is rewritten as

$$\partial_t\boldsymbol{q} + \Lambda\,\partial_x\boldsymbol{q} = 0 \tag{13}$$

or

$$\partial_t q_m + \lambda_m\,\partial_x q_m = 0 \tag{14}$$

where $\boldsymbol{q} = \Xi^{-1}\tilde{Q}$ and q_m is a component of $\boldsymbol{q}$. The eigenvalues λ_m are real, and clearly Eq. (14) is equivalent to three uncoupled (in the transformed variables) scalar wave equations. The wave equation is the prototype hyperbolic equation; hence, the one-dimensional unsteady Euler equations have been found to be hyperbolic. Indeed the conditions that the Jacobian matrix A of a system of equations such as Eq. (3) has a complete set of eigenvectors, and real eigenvalues are used to define a hyperbolic system of partial differential equations (cf. Garabedian[1]). Of course, the analysis has been for $\tilde{A}$, but $\tilde{A}$ is similar to A and therefore has the same eigenvalues.

The frozen coefficient form of the nonconservative Euler equations, which are similar to the conservative form, behave as the scalar wave equations [Eq. (13)]. The importance of this is that difference schemes that are found to be numerically unstable or stable for the scalar wave Eq. (14) can be expected to be unstable or stable for the system Eq. (1). The proviso is that the numerical scheme must be stable for any coefficient λ corresponding to the range of eigenvalues of the Jacobian matrix of F. In particular, if the eigenvalues of F are both positive and negative (subsonic flow, $u < c$), then the numerical scheme devised for Eq. (14) must be stable for both positive and negative values of λ.

Differencing Schemes for the Scalar Wave Equation

One-sided Operators

Having established that the scalar wave equation is the model problem for the Euler equations in one dimension, some possible differencing schemes can be surveyed. First, note that for the equation

$$\partial_t u + \lambda\, \partial_x u = 0 \tag{15}$$

with $\lambda =$ constant, u is constant along the characteristic line

$$\frac{\mathrm{d}x}{\mathrm{d}t} = \lambda \tag{16}$$

that is, the line

$$x - \lambda t = \text{const}$$

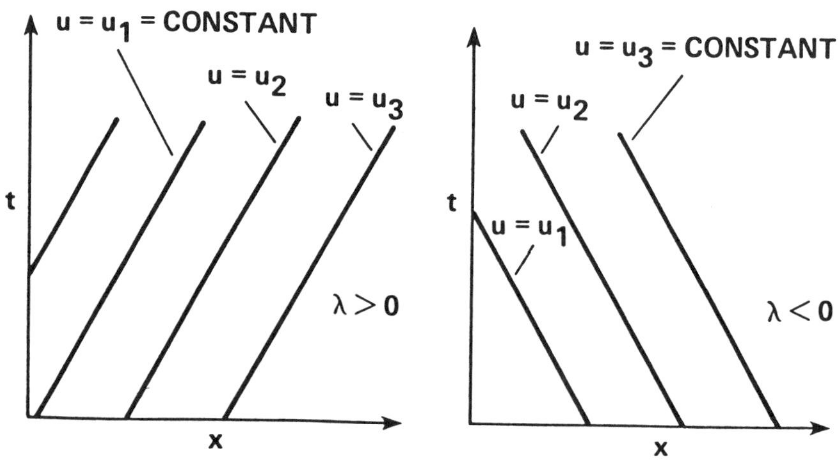

Fig. 1 Data propagation for simple wave equation, $u_t + \lambda u_x = 0$.

For $\lambda > 0$, data propagate from left to right as sketched in Fig. 1. For $\lambda < 0$, data propagate from right to left.

For positive λ it would seem that backward spatial differencing would nicely follow the characteristic direction of data propagation. The simplest such scheme is the first-order-accurate upwind scheme.

$$\Delta_t u_j^n + \lambda \nabla_x u_j^n = 0 \tag{17}$$

or

$$\frac{u_j^{n+1} - u_j^n}{\Delta t} + \frac{\lambda(u_j^n - u_{j-1}^n)}{\Delta x} = 0$$

where superscript n implies the nth time level, not an exponent. By a simple stability test employing the assumption of periodic boundary conditions in x, this scheme is numerically stable if

$$0 \leqslant \lambda \, \Delta t/\Delta x \leqslant 1$$

with $\lambda \, \Delta t/\Delta x$ the Courant or CFL number. (Examples of the stability analysis appear in later sections.)

If one chooses $\lambda \Delta t/\Delta x = 1$, $u_j^{n+1} = u_{j-1}^n$ is obtained from the previously mentioned first-order-accurate differencing scheme. This is the exact result obtained from shifting along the characteristic curve (see Fig. 2). This capability of certain explicit schemes to give an exact result at a Courant number $\lambda \Delta t/\Delta x = 1$ has been referred to as the perfect shift property.

If $\lambda < 0$, the backward scheme just described is unstable and would indeed seem to be unsatisfactory because it would propagate the data from the wrong direction. First-order backward differencing of Eq. (15) with $\lambda < 0$ and leaving the time derivative continuous leads to a set of time

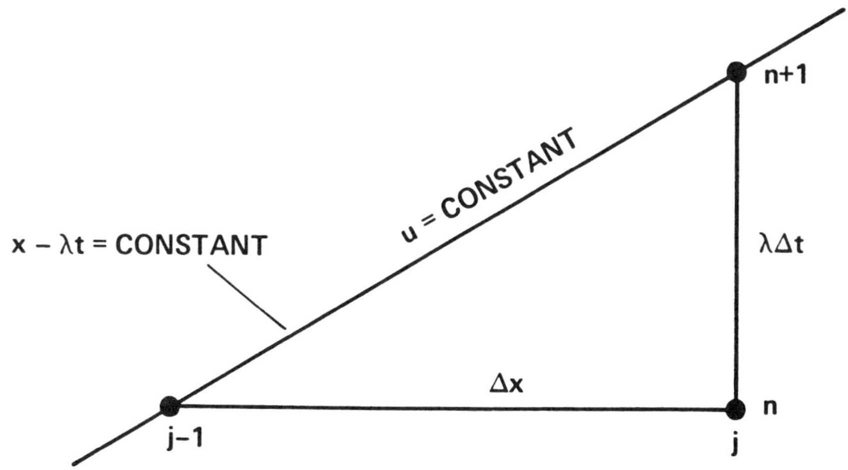

Fig. 2 Perfect shift of data along a characteristic for a Courant number equal to 1.

continuous difference equations of the form

$$\frac{\mathrm{d}\boldsymbol{u}}{\mathrm{d}t} = B\boldsymbol{u} - \boldsymbol{c}, \qquad B = \frac{1}{\Delta x}\begin{bmatrix} -\lambda & & & \\ \lambda & -\lambda & & \\ & \lambda & -\lambda & \\ & & \ddots & \ddots \end{bmatrix} = -\lambda \nabla_x \tag{18}$$

where B is lower triangular with a positive diagonal coefficient as $\lambda < 0$. Such a coefficient matrix must have positive eigenvalues, and the solution to Eq. (18) will grow as $e^{\lambda t}$ (or e^{Bt}) and will therefore become unbounded in time. Any higher-order backward spacial differencing will experience the same difficulty. However, for $\lambda < 0$ the simple upwind scheme using forward spacial differencing

$$\Delta_t u_j^n + \lambda\, \Delta_x u_j^n = 0 \tag{19}$$

or

$$\frac{u_j^{n+1} - u_j^n}{\Delta t} + \frac{\lambda(-u_j^n + u_{j+1}^n)}{\Delta x} = 0$$

is stable for $-1 \leqslant \lambda\, \Delta t/\Delta x \leqslant 0$. Data now propagate in the correct direction, and the corresponding B matrix has a negative diagonal.

From such considerations the following result can be stated for one-sided spatial difference approximations to Eq. (115) in which the time derivative is not discretized: If $\partial u/\partial x$ is approximated by a backward difference operator, then the resulting ordinary differential equation will be stable if and only if $\lambda > 0$ (i.e., the wave travels to the right). Conversely, if $\partial u/\partial x$ is approximated by a forward-difference operator, the resulting ordinary differential equation will be stable if and only if $\lambda < 0$ (i.e., the wave travels to the left). In general, no conventional backward, forward, or unsymmetric spatial operator will yield an ordinary differential equation that is simultaneously stable for both positive and negative eigenvalues.

Central Difference Operators

If one proposes the three-point central spacial difference operator for the scalar wave equation

$$\partial_t u + \lambda\, \partial_x u = 0 \tag{20}$$

then the spatial difference operator will generate a skew symmetric matrix with a zero trace

$$\delta_x = \frac{-E_x^{-1} + E_x}{2\Delta x} = \frac{1}{2\Delta x}\begin{bmatrix} 0 & 1 & & & \\ -1 & 0 & 1 & & \\ & -1 & 0 & 1 & \\ & & \cdot & \cdot & \cdot \\ & & & \cdot & \cdot & \cdot \\ & & & & \cdot & \cdot & \cdot \end{bmatrix} \tag{21}$$

where E_x is the shifting operator

$$E_x^m u_j = u_{j+m}$$

Such a matrix has pure imaginary eigenvalues. The time-continuous spatially differenced representation of Eq. (20) can be written as

$$\frac{\mathrm{d}\boldsymbol{u}}{\mathrm{d}t} = -\lambda(B\boldsymbol{u} - \boldsymbol{c})$$

where $\boldsymbol{u}$ is the vector $(u_1, u_2, \ldots, u_j, \ldots, u_{j\max})$, B is given by Eq. (21), and $\boldsymbol{c}$ contains the spatial boundary values, if any. Using the eigenvectors of B, these can be transformed into the uncoupled equations

$$\frac{\mathrm{d}\tilde{u}_j}{\mathrm{d}t} = -\lambda(i\sigma_j \tilde{u}_j - b_j) \tag{22}$$

just as the Euler equations were transformed previously. Here $i\sigma_j$ are the eigenvalues of B where i is the imaginary and σ_j are real. The exact ordinary differential equation complementary solution part of Eq. (22) is thus of the form $e^{-\lambda i\sigma_j t}$ and neither grows or decays. Consequently, the sign of λ, whether it is positive or negative, is no longer important if central spatial differencing is used.

A variety of stable time differencing schemes can be used to solve the uncoupled Eq. (22), which has a pure imaginary coefficient. The leap frog and the fourth-order Runge-Kutta schemes are conditionally stable and the Euler implicit and the trapezoidal rule are unconditionally stable for a pure imaginary coefficient (cf. Refs. 2 and 3). For example, trapezoidal-rule time differencing applied to the uncoupled problem

$$\frac{\mathrm{d}\tilde{u}}{\mathrm{d}t} = i\beta\tilde{u} \tag{23}$$

is given by

$$\frac{\tilde{u}^{n+1} - \tilde{u}^n}{\Delta t} = i\beta\left(\frac{\tilde{u}^{n+1} + \tilde{u}^n}{2}\right)$$

and is unconditionally stable. Step size is controlled by accuracy considerations, not by stability bounds. The same scheme applied to Eq. (20) is

$$\frac{u^{n+1} - u^n}{\Delta t} + \lambda\delta_x\left(\frac{u^{n+1} + u^n}{2}\right) = 0$$

where δ_x is given by Eq. (21). This scheme is unconditionally stable for either positive or negative values of λ because δ_x generates pure imaginary eigenvalues.

The more prosaic explicit time integration schemes such as Euler explicit and second-order Runge-Kutta are unstable for Eq. (23). These schemes can be stable for complex eigenvalues; thus, a potential fixup is to add small diffusive like terms to Eq. (20) that when centrally differenced provide real parts to the eigenvalues.

Lax-Wendroff and MacCormack Schemes

The Lax-Wendroff scheme[4,5] and the MacCormack scheme[6] are two-level in time second-order-accurate explicit methods that are popular in fluid mechanics. They can be interpreted as centrally spatial differenced schemes with explicitly added dissipative terms that are chosen to provide stability and to improve accuracy.

The Lax-Wendroff scheme is derived as follows for the system

$$\partial_t Q + \partial_x F = 0 \tag{24}$$

Expand Q in a Taylor series in time as

$$Q^{n+1} = Q^n + \Delta t\, \partial_t Q^n + \frac{(\Delta t)^2}{2}\partial_{tt} Q^n + \cdots \tag{25}$$

Now using Eq. (24), replace $\partial_t Q^n$ and $\partial_{tt} Q^n$ with spatial derivatives

$$\partial_t Q = -\partial_x F \tag{26a}$$

$$\partial_{tt} Q = -\partial_t\, \partial_x F = -\partial_x\, \partial_t F = -\partial_x A\, \partial_t Q = \partial_x A\, \partial_x F \tag{26b}$$

where $A = [\partial F/\partial Q]$ is given by Eq. (4) for the Euler equations. Substituting into Eq. (25) yields

$$Q^{n+1} = Q^n - \Delta t\, \partial_x F^n + [(\Delta t)^2/2]\, \partial_x A^n\, \partial_x F^n + 0(\Delta t^3) \tag{27}$$

Finally, the Lax-Wendroff algorithm is obtained by replacing ∂_x with second-order central spatial differencing that is compatible with the time accuracy

$$Q_j^{n+1} = Q_j^n - \frac{1}{2}\left(\frac{\Delta t}{\Delta x}\right)(F_{j+1}^n - F_{j-1}^n)$$
$$+\frac{1}{2}\left(\frac{\Delta t}{\Delta x}\right)^2\left[\left(\frac{A_{j+1}^n + A_j^n}{2}\right)(F_{j+1}^n - F_j^n) - \left(\frac{A_j^n + A_{j-1}^n}{2}\right)(F_j^n - F_{j-1}^n)\right] \tag{28}$$

where the midpoint operator is used in the approximation of the second derivative and the matrix A is averaged as $A_{j+\frac{1}{2}} \doteq (A_{j+1} + A_j)/2$.

The numerical stability of Eq. (28) can be analyzed by linearizing and freezing the Jacobian coefficient matrix A to obtain

$$Q_j^{n+1} = Q_j^n - \frac{1}{2}\left(\frac{\Delta t}{\Delta x}\right)A(Q_{j+1}^n - Q_{j-1}^n) + \frac{1}{2}\left(\frac{\Delta t}{\Delta x}\right)^2 A^2(Q_{j+1}^n - 2Q_n^n + Q_{j-1}^n)$$

Transforming to the diagonal or scalar form results in

$$q_j^{n+1} = q_j^n - \tfrac{1}{2}\nu(q_{j+1}^n - q_{j-1}^n) + \tfrac{1}{2}\nu^2(q_{j+1}^n - 2q_j^n + q_{j-1}^n) \tag{29}$$

where $\nu = \Delta t\lambda/\Delta x$ and λ can be one of three roots of A given by $u, u \pm c$.

Assuming periodic boundary conditions in x, the orthonormal eigenvectors of a circulant matrix can be used to diagonalize Eq. (29) to its eigenvalues as

$$\tilde{q}_j^{n+1} = \left[1 - i\nu\,\sin\theta_j - 2\nu^2\sin^2\!\left(\frac{\theta_j}{2}\right)\right]\tilde{q}_j^n, \qquad i = \sqrt{-1} \tag{30}$$

where the known eigenvalues of a circulant matrix are used (see Appendix or use the equivalent von Neumann test[2,3]). This two-level scheme has the recursive solution

$$\tilde{q}_j^{n+1} = \left[1 - i\nu\,\sin\theta_j - 2\nu^2\sin^2\!\left(\frac{\theta_j}{2}\right)\right]^n\tilde{q}_j^0$$

(with $[\ ]^n$ denoting an exponent) and becomes unbounded (unstable) unless the amplification factor

$$\left|1 - 2\nu^2\sin^2\!\left(\frac{\theta_j}{2}\right) - i\nu\,\sin\theta_j\right| \leqslant 1$$

or as $|a + ib| = \sqrt{(a^2 + b^2)}$

$$1 - 4\nu^2\sin^2\!\left(\frac{\theta_j}{2}\right) + 4\nu^4\sin^4\!\left(\frac{\theta_j}{2}\right) + \nu^2\sin^2\theta_j \leqslant 1$$

Let $\phi_j = \theta_j/2$ and use $\sin^2\theta_j = \sin^2(2\phi_j) = 4\sin^2\phi_j\cos^2\phi_j$ to obtain

$$1 + 4\nu^2\sin^2\phi_j(-1 + \nu^2\sin^2\phi_j + \cos^2\phi_j) \leqslant 1$$

or

$$1 + 4\nu^2\sin^4\phi_j(\nu^2 - 1) \leqslant 1 \tag{31}$$

Because $4\nu^2\sin^4\phi_j > 0$, the inequality (31) is only satisfied if $\nu^2 - 1 \leqslant 0$, or

$$|\nu| \leqslant 1 \tag{32}$$

The central differencing of $\partial_x F$ generates the pure imaginary term $i\nu\,\sin\theta_j$, which allows A to have real eigenvalues of either sign. The central differenced term $\partial_x A\,\partial_x F = \partial_x A^2\,\partial_x Q$ is needed in Eq. (27) to give second-order time accuracy, but it also adds a negative real part to Eq. (30) that provides stability to what would otherwise look like an unstable Euler explicit central scheme.

The Lax-Wendroff method given by Eq. (28) is a simple one-step method that is second-order accurate. It is considered computationally inefficient, however, because it is necessary to evaluate the Jacobian matrix A and perform a matrix multiplication. A way of avoiding this explicit matrix multiply is provided by the Lax-Wendroff two-step predictor-corrector scheme

$$\bar{Q}_{j+\frac{1}{2}} = \frac{1}{2}(Q^n_{j+1} + Q^n_j) - \left(\frac{\Delta t}{2\Delta x}\right)(F^n_{j+1} - F^n_j) \tag{33a}$$

$$Q^{n+1}_j = Q^n_j - \left(\frac{\Delta t}{\Delta x}\right)(\bar{F}_{j+\frac{1}{2}} - \bar{F}_{j-\frac{1}{2}}) \tag{33b}$$

where $\bar{F} = F(\bar{Q})$. Combining (33a) and (33b), and simplifying

$$\begin{aligned} Q^{n+1}_j &= Q^n_j - \frac{\Delta t}{2\Delta x}(F^n_{j+1} - F^n_{j-1}) \\ &+ \frac{1}{2}\left(\frac{\Delta t}{\Delta x}\right)^2 [\bar{A}_{j+\frac{1}{2}}(F^n_{j+1} - F^n_j) - \bar{A}_{j-\frac{1}{2}}(F^n_j - F^n_{j-1})] \end{aligned} \tag{34}$$

where the liberty of writing $\bar{F}_{j+\frac{1}{2}} - \bar{F}_{j-\frac{1}{2}}$ as $\bar{A}_{j+\frac{1}{2}}\bar{Q}_{j+\frac{1}{2}} - \bar{A}_{j-\frac{1}{2}}\bar{Q}_{j-\frac{1}{2}}$ was taken. That is, the fact that the Euler equations are homogeneous of degree one is used to show the similarity to Eq. (28).

The two-step scheme is somewhat awkward to implement in multidimensions because of the use of spacial midpoints. As a result, the MacCormack algorithm has been preferred over the two-step Lax-Wendroff scheme, at least in aerodynamic applications. The MacCormack scheme is given by the predictor-corrector:

$$\bar{Q}_j = Q^n_j - \left(\frac{\Delta t}{\Delta x}\right)(F^n_j - F^n_{j-1}) \tag{35a}$$

$$Q^{n+1}_j = \frac{1}{2}(\bar{Q}_j + Q^n_j) - \left(\frac{\Delta t}{2\Delta x}\right)(\bar{F}_{j+1} - \bar{F}_j) \tag{35b}$$

Variants of this scheme utilize the forward spatial Δ_x operator in the predictor step and the backward operator ∇_x in the corrector step. One can also alternate the use of forward and backward operators in predictor-corrector steps. Combining the predictor and corrector gives

$$\begin{aligned} Q^{n+1}_j &= Q^n_j - \left(\frac{\Delta t}{2\Delta x}\right)(F^n_{j+1} - F^n_{j-1}) \\ &+ \frac{1}{2}\left(\frac{\Delta t}{\Delta x}\right)^2 [\bar{A}_{j+1}(F^n_{j+1} - F^n_j) - \bar{A}_j(F^n_j - F^n_{j-1})] \end{aligned} \tag{36}$$

Had the forward operator been called in the predictor, the combined scheme would appear as

$$Q_j^{n+1} = Q_j^n - \left(\frac{\Delta t}{2\Delta x}\right)(F_{j+1}^n - F_{j-1}^n)$$
$$+ \frac{1}{2}\left(\frac{\Delta t}{\Delta x}\right)^2 [\bar{A}_j(F_{j+1}^n - F_j^n) - \bar{A}_{j-1}(F_j^n - F_{j-1}^n)] \tag{37}$$

so alternation of the calling sequence (35a), (35b) with one that uses Δ_x in the predictor and ∇_x in the corrector tends to remove a slight bias of $\bar{A}$ in the "dissipation-appearing term."

Although the predictor-corrector schemes are computationally more efficient than the single-step Lax-Wendroff scheme, there are some drawbacks. For one, one must call in the database twice, not just once. It is also necessary to pay more attention to proper boundary condition treatment even when the function is specified. Consider the following simple problem:

$$u_t + cu_x = 0 \tag{38}$$

in which $u(0,t)$ is given. Solving Eq. (38) using the MacCormack scheme

$$\bar{u}_j = u_j^n - \nu(u_{j+1}^n - u_j^n), \qquad \nu = (c\,\Delta t/\Delta x) \tag{39a}$$
$$u_j^{n+1} = \frac{1}{2}(\bar{u}_j + u_j^n) - \frac{\nu}{2}(\bar{u}_j - \bar{u}_{j-1}) \tag{39b}$$

For Eq. (39b), $\bar{u}$ is needed at $x = 0$ ($j = 1$ of Fig. 3). Now $\bar{u}(0,t) \neq u(0,t)$, rather from Eq. (39a) $\bar{u}_1 = u_1^n - \nu(u_2^n - u_1^n)$. Thus, it is necessary to apply the

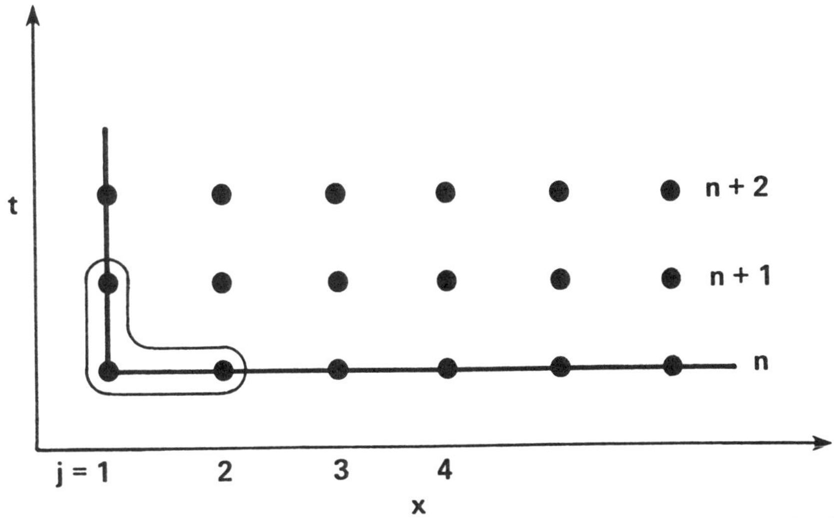

Fig. 3 The corrector step should use the predicted $n+1$ value at $j=1$, $\bar{u}_1 = u_1^n - \nu(u_2^n - u_1^n)$, not the specified value, u_1^{n+1}.

predictor on the boundary curve to supply $\bar{u}_1$, even though $u(0,t)$ is known there. Frequently, one sees the algorithm written with $n + 1$ tagged onto the $\bar{u}$, $\bar{F}$ terms, e.g., $\bar{u}^{n+1}$, $\bar{F}^{n+1}$ to indicate that the predictor is a consistent time-accurate predictor. Unfortunately, the uninitiated user sometimes takes this too seriously and replaces $\bar{u}^{n+1}$ with u^{n+1} along the boundaries. In this way he or she degrades the accuracy of the scheme at the boundary when it is not difficult to maintain.

Implicit Schemes, Local Linearization, and Numerical Dissipation

Time-implicit numerical procedures are often used to solve flowfields when it is desirable to take a larger time step than that permitted by a scheme subject to an explicit stability limit. Such a situation may occur if the dependent variables experience a more rapid variation with space than with time, if time accuracy is controlled by forcing functions (say through the boundary conditions), or if one needs to suppress a local high-frequency time variation by taking such large time steps that the phenomenon is not resolved.

A two level second-order-accurate differencing of Eq. (1) is given by

$$Q_j^{n+1} + \left(\frac{\Delta t}{4\Delta x}\right)(F_{j+1}^{n+1} - F_{j-1}^{n+1}) = Q_j^n - \left(\frac{\Delta t}{4\Delta x}\right)(F_{j+1}^n - F_{j-1}^n) \qquad (40)$$

where trapezoidal-rule time differencing is used. Central spatial differencing permits the Jacobian matrix of F to have real eigenvalues of either sign, and this implicit scheme is unconditionally but neutrally stable.

As it currently stands, Eq. (40) requires the solution of a nonlinear system of equations at each time level. Typically this necessitates an iterative solution. To avoid this complexity the nonlinear terms can be locally linearized at the previous time level (or any other convenient nearby location) as

$$F_j^{n+1} = F_j^n + \left[\frac{\partial F}{\partial Q}\right]_j^n (Q_j^{n+1} - Q_j^n) + O(\Delta t^2) \qquad (41)$$

where $\partial F/\partial Q = A$ is the Jacobian matrix. This expression is as accurate as the time differencing scheme; thus, its use eliminates the need to iterate.

Use of Eq. (41) in Eq. (40) gives

$$Q_j^{n+1} + \frac{\Delta t}{4\Delta x}(A_{j+1}^n Q_{j+1}^{n+1} - A_{j-1}^n Q_{j-1}^{n+1})$$

$$= Q^n - \frac{\Delta t}{2\Delta x}(F_{j+1}^n - F_{j-1}^n) + \frac{\Delta t}{4\Delta x}(A_{j+1}^n Q_{j+1}^n - A_{j-1}^n Q_{j-1}^n) \qquad (42)$$

Now $A^n Q^n = F^n$ for the conservative Euler equations; hence, the right-hand side of Eq. (42) can be simplified. In general, however, the complicated right-hand side can be avoided by rewriting Eq. (42) into the delta form:

$$\Delta Q_j^n + \frac{\Delta t}{4\Delta x}(A_{j+1}^n \Delta Q_{j+1}^n - A_{j-1}^n \Delta Q_{j-1}^n) = \frac{-\Delta t}{2\Delta x}(F_{j+1}^n - F_{j-1}^n) \qquad (43)$$

where $\Delta Q_j^n = Q_j^{n+1} - Q_j^n$. Define

$$\delta_x = \frac{E_x - E_x^{-1}}{2\Delta x} \tag{44}$$

and put Eq. (43) into the convenient operator form

$$\left(I + \frac{\Delta t}{2}\delta_x A_j^n\right)\Delta Q_j^n = -\Delta t \delta_x F_j^n \tag{45}$$

where I is the 3×3 identity matrix. Note that on the left-hand side of Eq. (45) the operator δ_x differences the produce $A_j^n\, \Delta Q_j^n$ to reproduce Eq. (43). The left-hand side of Eq. (45) forms a block tridiagonal matrix of the form

$$\left(\mathrm{I} + \frac{\Delta t}{2}\delta_x A_j\right) = \begin{bmatrix} I & \beta A_3 & & & & \\ -\beta A_2 & I & \beta A_4 & & & \\ & -\beta A_3 & I & \beta A_5 & & \\ & & -\beta A_4 & I & \beta A_6 & \\ & & & \ddots & \ddots & \ddots \\ & & & & -\beta A_{jm} & I \end{bmatrix}$$

where $\beta = \Delta t/(4\Delta x)$ and A_j is a 3×3 matrix. This block tridiagonal can be inverted with a standard block tridiagonal algorithm, and although not diagonally dominant, no serious rounding error tends to accumulate for this almost skew symmetric matrix.

A *constant* coefficient linear stability test of the scalar version of Eq. (45) reveals that the scheme is unconditionally but neutrally stable with an amplification factor

$$E = \left(1 - i\frac{v \sin\theta_j}{2}\right) \Big/ \left(1 + i\frac{v \sin\theta_j}{2}\right) \tag{46}$$

with $v = \lambda\, \Delta t/\Delta x$ and λ an eigenvalue of A. Analysis with *variable* coefficients indicates the possibility of very weak instability if the coefficient v changes sign between grid points. For this reason, and to damp high-frequency oscillations induced by any inconsistencies (for example, approximate boundary treatment), it is necessary to add some numerical dissipation to Eq. (45). Frequently a fourth-order differencing is subtracted in an explicit fashion as

$$\left(I + \frac{\Delta t}{2}\delta_x A_j^n\right)\Delta Q_j^n = -\Delta t \delta_x F_j^n - \varepsilon_e(\Delta\nabla)^2|_x Q_j^n \tag{47}$$

where $(\nabla\Delta)^2|_x$ is a difference (not a derivative approximation) given by

$$(\nabla\Delta)^2|_x Q_j = Q_{j-2} - 4Q_{j-1} + 6Q_j - 4Q_{j+1} + Q_{j+2} \tag{48}$$

Because the equations have been scaled by Δt, the scalar coefficient ε_e should also be scaled by Δt and is typically chosen to be $0(\Delta t/\Delta x)$. (Ideally, ε_e should be a positive matrix that varies like $\Delta t A$ [see, for example, the discussion preceding Eq. (83)].

The fourth-order term is not added implicitly because this would force the solution of a block pentadiagonal matrix even though the dissipation only adds elements to the diagonals of the blocks. However, analysis of the stability of the explicit dissipation by itself, i.e.,

$$\Delta Q_j^n = -\varepsilon_e(\nabla\Delta)^2|_x Q_j^n \tag{49}$$

shows that $\varepsilon_e \leqslant \frac{1}{8}$ in order to keep the amplification factor

$$E = \left[1 - \varepsilon_e\left(-4\sin^2\frac{\theta_j}{2}\right)^2\right] \tag{50}$$

less or equal to 1. If this bound is too stringent a second difference operator on ΔQ^n can be added to the left-hand side of Eq. (47), i.e.,

$$\left[I + \frac{\Delta t}{2}\delta_x A_j^n - \varepsilon_i(\nabla\Delta)_x\right]\Delta Q_j^n = -\Delta t\delta_x F_j^n - \varepsilon_e(\nabla\Delta)^2|_x Q_j^n \tag{51}$$

As long as $\varepsilon_i > 2\varepsilon_e$ any amount of dissipation can be added according to linear analysis of the dissipation terms alone (in practice $\varepsilon_i > 5/2\varepsilon_e$). Because $\varepsilon_i(\nabla\Delta)_x$ operates on ΔQ_j^n, not on Q_j^n, accuracy is not degraded by using a second difference in place of a fourth difference numerical dissipation. High ratios of ε_i to ε_e will enhance stability and will not inhibit accuracy in problems with slow time variation, but convergence to steady state will be degraded.

Other implicit time differencing can be used besides trapezoidal rule. For example, three-point backward time differencing and central space differencing of Eq. (1) gives

$$3Q_j^{n+1} - 4Q_j^n + Q_j^{n-1} + 2\Delta t\delta_x F_j^{n+1} = 0 \tag{52}$$

or with local linearization and in delta form

$$\left(I + \frac{2\Delta t}{3}\delta_x A_j^n\right)(Q_j^{n+1} - Q_j^n) = -\frac{2\Delta t}{3}\delta_x F_j^n + \frac{1}{3}(Q_j^n - Q_j^{n-1}) \tag{53}$$

This scheme is also unconditionally stable, and, unlike the trapezoidal-rule time differencing, it is not neutrally stable. Nevertheless, with use of three-point central space differencing, it is still necessary to add numerical dissipation to damp high frequencies. Following the previous development the resulting scheme is given by

$$\left[I + \frac{2\Delta t}{3}\delta_x A_j^n - \varepsilon_i(\nabla\Delta)_x\right]\Delta Q_j^n = \frac{1}{3}(Q_j^n - Q_j^{n-1})$$
$$- \frac{2\Delta t}{3}\delta_x F_j^n - \varepsilon_e(\nabla\Delta)^2|_x Q_j^n \tag{54}$$

For near-steady-state applications or low-frequency forced motion, first-order time accuracy can be sufficient. This scheme is given by

$$I + \Delta t\delta_x A_j^n - \varepsilon_i(\nabla\Delta)_x]\ \Delta Q_j^n = -\Delta t\delta_x F_j^n - \varepsilon_e(\nabla\Delta)^2|_x Q_j^n \tag{55}$$

It is relatively easy to go to higher-order spacial accuracy with the implicit scheme (but not as easy to go beyond second-order time accuracy because of the local linearization process). A simple procedure is to use Pade differencing in space:

$$\partial_x \doteq [\tilde{I}(E)]^{-1}\delta_x \equiv \tilde{\delta}_x \tag{56}$$

where for fourth-order

$$\tilde{I}(E) = \frac{(E_x + 4 + E_x^{-1})}{6} \equiv \tilde{I} \tag{57}$$

and δ_x is given by Eq. (44). Applying this to Eq. (51) and multiplying through by $\tilde{I}$ to clear the inverse gives

$$\left[\tilde{I} + \frac{\Delta t}{2}\delta_x A_j^n - \tilde{I}\varepsilon_i(\nabla\Delta)_x\right]\Delta Q_j^n = -\Delta t\delta_x F_j^n - \varepsilon_e\tilde{I}(\nabla\Delta)^2|_x Q_j^n \tag{58}$$

Now $\tilde{I} = I - [(\nabla\Delta)_x/6]$, and to avoid very messy dissipation terms, it is convenient to replace

$$\tilde{I}\varepsilon_i(\nabla\Delta)_x \doteq \varepsilon_i(\nabla\Delta)_x, \qquad \text{etc.}$$

In this way we arrive at

$$\left[\tilde{I} + \frac{\Delta t}{2}\delta_x A_j^n - \varepsilon_i(\nabla\Delta)_x\right]\Delta Q_j^n = -\Delta t\delta_x F_j - \varepsilon_e(\nabla\Delta)^2|_x Q_j^n \tag{59}$$

where ε_e,ε_i must be $0(\Delta t)$ to maintain fourth-order spacial accuracy. The computational work involved with Eq. (59) is only slightly increased over that of Eq. (51).

Eigensplittings

For the scalar wave equation it was previously shown that the use of properly selected one-sided or upwind spatial differencing can be numerically stable and that upwind differencing follows the characteristic flow of data. Although not previously indicated, use of upwind differencing in implicit time differencing methods can also lead to simpler inversion schemes. However, it was also shown that the hyperbolic system of equations

$$\partial_t Q + A\,\partial_x Q = 0 \tag{60}$$

must be centrally space differenced if the coefficient matrix A has both positive and negative real eigenvalues, assuming only a single space differencing operator is used throughout. If the equations are transformed to scalar decoupled form, however, one-sided spatial differencing can be used if the difference operators are chosen according to the sign of the eigenvalues. By "eigensplitting" A or F, upwind differencing can be applied to the Euler equations, and, compared to schemes using central space differencing, this can be especially advantageous when the eigenvalues of A are either largely of one sign or completely of one sign over significant parts of the field.

Backward and Forward Space Differencing

Multiplying the nonconservative Euler equations by the eigenvector inverse matrix of the coefficient matrix $\tilde{A}$ gives the uncoupled-like form Eq. (11):

$$\begin{pmatrix} \rho_t - c^{-2}p_t \\ u_t + (\rho c)^{-1}p_t \\ -u_t + (\rho c)^{-1}p_t \end{pmatrix} + \begin{bmatrix} u & & \\ & u+c & \\ & & u-c \end{bmatrix} \begin{pmatrix} \rho_x - c^{-2}p_x \\ u_x + (\rho c)^{-1}p_x \\ -u_x + (\rho c)^{-1}p_x \end{pmatrix} = \begin{pmatrix} 0 \\ 0 \\ 0 \end{pmatrix} \tag{61}$$

where some simplifying scaling by $\sqrt{(2)}$ was used.

For simplicity assume for the moment that all of the coefficients are frozen at a state (0). In particular, assume that the similarity transform coefficients are locally constant and can be brought inside the operator as

$$\partial_t \begin{pmatrix} \rho - \alpha p \\ u + \beta p \\ -u + \beta p \end{pmatrix} + \begin{bmatrix} u_0 & 0 & 0 \\ 0 & u_0 + c_0 & 0 \\ 0 & 0 & u_0 - c_0 \end{bmatrix} \partial_x \begin{pmatrix} \rho - \alpha p \\ u + \beta p \\ -u + \beta p \end{pmatrix} = \begin{pmatrix} 0 \\ 0 \\ 0 \end{pmatrix} \tag{62}$$

where $\alpha \equiv c_0^{-2}$, and $\beta = (\rho_0 c_0)^{-1}$ are constants. In the new variables, these are uncoupled-scalar equations. The simplest explicit differencing scheme for these equations is then given by

$$\Delta_t(\rho - \alpha p)_j^n + u_0\,\nabla_x(\rho - \alpha p)_j^n = 0 \tag{63a}$$

$$\Delta_t(u + \beta p)_j^n + (u_0 + c_0)\,\nabla_x(u + \beta p)_j^n = 0 \tag{63b}$$

and

$$\Delta_t(-u + \beta p)_j^n + (u_0 - c_0)\,\Delta_x(-u + \beta p)_j^n = 0 \tag{63c}$$

where it is assumed that $u_0 > 0$ and $u_0 < c_0$ so that the last equation uses forward not backward space differencing. If $u_0 > c_0$, backward differencing would be used throughout. This system is stable if the Courant number $[(u_0 + c_0)\,\Delta t]/\Delta x \leqslant 1$.

The same differencing can be applied to Eq. (61) without assuming constant coefficients by simply differencing the same collection of terms, i.e.,

$$\Delta_t \rho - c^{-2}\,\Delta_t p + u\,\nabla_x \rho - uc^{-2}\,\nabla_x p = 0 \tag{64a}$$

$$\Delta_t u + (\rho c)^{-1}\,\Delta_t p + (u + c)\,\nabla_x u + (u + c)(\rho c)^{-1}\,\nabla_x p = 0 \tag{64b}$$

and

$$-\Delta_t u + (\rho c)^{-1}\,\Delta_t p - (u - c)\,\Delta_x u + (u - c)(\rho c)^{-1}\,\Delta_x p = 0 \tag{64c}$$

where subscripts and superscripts j and n are deleted for convenience, i.e., $c^{-2}\,\Delta_t p \equiv (c_j^n)^{-2}\,\Delta_t p_j^n$, etc. Note again that the third equation uses forward, not backward differencing, assuming $u < c$ and $u > 0$. If $u > c$, backward differencing would be used throughout.

The differencing in Eqs. (64) is first order accurate, but various second-order-accurate upwind schemes can be generated. Using Eqs. (64) as a predictor, for example, a second-order-accurate corrector is given by

$$\Delta_t \rho - \bar{c}^{-2}\,\Delta_t p + \bar{u}\delta_x^b \bar{\rho} - \bar{u}\bar{c}^{-2}\delta_x^b \bar{p} = 0 \tag{65a}$$

$$\Delta_t u + \overline{(\rho c)}^{-1}\,\Delta_t p + (\bar{u} + \bar{c})\delta_x^b \bar{u} + (\bar{u} + \bar{c})\overline{(\rho c)}^{-1}\delta_x^b \bar{p} = 0 \tag{65b}$$

$$-\Delta_t u + \overline{(\rho c)}^{-1}\,\Delta_t p - (\bar{u} - \bar{c})\delta_x^f \bar{u} + (\bar{u} - \bar{c})\overline{(\rho c)}^{-1}\delta_x^f \bar{p} = 0 \tag{65c}$$

where

$$\delta_x^b = \nabla_x \left(\frac{3 - E_x^{-1}}{2} \right)$$

$$\delta_x^f = \Delta_x \left(\frac{3 - E_x}{2} \right)$$

and the overbar indicates a midpoint value is used in time such as

$$\bar{\rho} = [\rho_{\text{pred}}^{n+1} + \rho^n]/2, \qquad \text{etc.} \tag{66a}$$

and

$$\bar{c} = c(\bar{\rho}, \bar{u}, \bar{p}) \tag{66b}$$

This corrector uses the midpoint formula. This means that the ordinary differential equation

$$\frac{\mathrm{d}u}{\mathrm{d}t} + f(u) = 0 \tag{67}$$

is differenced in time as

$$\frac{u^{n+1} - u^n}{\Delta t} + f\left(\frac{u^{n+1} + u^n}{2}\right) = 0 \tag{68}$$

where for explicit treatment the u^{n+1} part of f is obtained from a predictor. Note that the trapezoidal rule would use instead

$$\frac{u^{n+1} - u^n}{\Delta t} + \frac{[f(u^{n+1}) + f(u^n)]}{2} = 0 \tag{69}$$

Boundary Conditions

The backward space difference operator ∇_x requires a boundary condition on the left of the computational domain, and the forward operator Δ_x requires a boundary condition on the right. Consequently, the differenced equations, Eqs. (64), which assume $u > 0$, $u < c$, require that two combinations of variables be specified on the left and that one combination of variables be given on the right (see Fig. 4). If $u > c$, all of the spatial terms would be backward differenced, and the differenced equations would require three left-side (upstream) boundary conditions and no right-side (downstream) conditions.

The sign of the characteristic values u, $u \pm c$ determine what kind of forward or backward space differencing can be used and what kind of boundary condition information must be specified at the ends. Even if central space differencing is used, the same kind of boundary condition information should ideally be given.

Suppose for illustrative purposes that ρu and total temperature are specified on the left boundary and static pressure on the right boundary. Other boundary variables could be specified provided that the chosen boundary variables along with the three characteristically combined partial difference equations, say Eqs. (64), allow ρ, u, and p to be determined on the boundaries. On the downstream boundary, for example, p was specified, and Eqs. (64a) and (64b), being backward differenced, can be used to solve for ρ and u on the ends. Thus, at $j = j\text{max}$,

$$\rho_j^{n+1} = \rho_j^n - \Delta t u_j^n(\nabla_x \rho_j^n - (c_j^n)^{-2}\nabla_x p_j^n) + (c_j^n)^{-2}(p_j^{n+1} - p_j^n) \tag{70a}$$

$$u_j^{n+1} = u_j^n - \Delta t(u_j^n + c_j^n)(\nabla_x u_j^n - (\rho_j^n c_j^n)^{-1}\nabla_x p_j^n) - (\rho_j^n c_j^n)^{-1}(p_j^{n+1} - p_j^n) \tag{70b}$$

where

$$p_j^{n+1} = f_3 \tag{70c}$$

is specified. Because specified data at the previous time level are known (or ultimately initial data are given for ρu and p), and $p_{j\text{max}}$ is given for all time (specified static pressure), Eqs. (70) allows determination of ρ^{n+1} and u^{n+1} at $j = j\text{max}$.

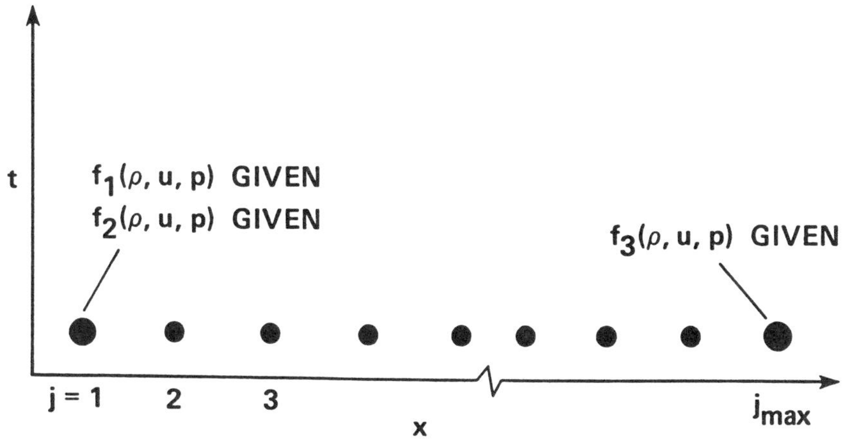

Fig. 4 For subsonic flow, $u > 0$ and $u < c$, the Euler equations need two combinations of variables specified on the left boundary and one combination of variables specified on the right boundary.

On the left boundary $j = 1$, the two conditions and Eq. (64c) result in

$$\rho_1^{n+1} u_1^{n+1} = f_1 \tag{71a}$$

$$\left(\frac{u_1^2}{2}\right)^{n+1} + \frac{\gamma}{(\gamma - 1)}\left(\frac{p_1}{\rho_1}\right)^{n+1} = f_2 \tag{71b}$$

$$\begin{aligned} -u_j^{n+1} + (\rho_j^n c_j^n)^{-1} p_j^{n+1} = {} & \Delta t (u_j^n - c_j^n)(\Delta_x u_j^n - (\rho_j^n c_j^n)^{-1} \Delta_x p_j^n) \\ & - u_j^n + (\rho_j^n c_j^n)^{-1} p_j^n \end{aligned} \tag{71c}$$

This nonlinear system of algebraic equations can be solved for ρ_1, u_1, p_1, and is most readily solved using local linearization about level n. For example,

$$\rho^{n+1} u^{n+1} \simeq u^n \rho^{n+1} + \rho^n u^{n+1} - \rho^n u^n, \qquad \text{etc.}$$

A proper number of Eqs. (64) can always be stably one-sided differenced on the boundary to supply the number of unknowns not accounted for by nondegenerate boundary conditions. Although Eqs. (64) can be used throughout the field, they are particularly useful for supplying endpoint values. A variety of boundary condition approaches are possible, but Eqs. (64) give a combination of characteristic variables that can be safely one-sided differenced without causing numerical instability.

Flux Vector and Coefficient Matrix Splitting[7,8]

Having established that noncentered space differencing can be used if the equations are properly split according to their eigenvalues, another splitting approach can be discussed. The A matrix of Eq. (3) can be diagonalized to its eigenvalues as

$$A = S\Lambda S^{-1} \tag{72}$$

and the eigenvalues can be split into positive and negative parts:

$$\Lambda = \Lambda^{+} + \Lambda^{-} \tag{73}$$

e.g., $(c > u > 0)$

$$\Lambda^{+} = \begin{pmatrix} u & & \\ & u+c & \\ & & 0 \end{pmatrix}, \qquad \Lambda^{-} = \begin{pmatrix} 0 & & \\ & 0 & \\ & & u-c \end{pmatrix}$$

Using this splitting, A can be reformed into an A^{+} and A^{-} as

$$\begin{aligned} A &= S(\Lambda^{+} + \Lambda^{-})S^{-1} \\ &= S\Lambda^{+}S^{-1} + S\Lambda^{-}S^{-1} \\ &\equiv A^{+} + A^{-} \end{aligned}$$

Equation (3) is then rewritten as

$$\partial_t Q + A^{+}\,\partial_x Q + A^{-}\,\partial_x Q = 0 \tag{74}$$

Using this splitting one can backward difference $A^{+}\,\partial_x Q$ and forward difference $A^{-}\,\partial_x Q$. The simplest explicit first-order coefficient matrix scheme is

$$Q^{n+1} - Q^{n} + \Delta t(A^{+})^{n}\,\nabla_x Q^{n} + \Delta t(A^{-})^{n}\,\Delta_x Q^{n} = 0 \tag{75}$$

Higher-order upwind schemes can also be written, for example, by using the midpoint corrector employed with Eqs. (65).

A conservative splitting that takes advantage of the homogenous property of the Euler equations is also available. That is, F can be written as

$$F = AQ$$

and this permits the splitting

$$\begin{aligned} F &= (A^{+} + A^{-})Q \\ &\equiv F^{+} + F^{-} \end{aligned} \tag{76}$$

Hence, the equations can be flux split as

$$\partial_t Q + \partial_x F^{+} + \partial_x F^{-} = 0 \tag{77}$$

and the simplest explicit first-order scheme is

$$Q^{n+1} - Q^n + \Delta t\ \nabla_x (F^+)^n + \Delta t\ \Delta_x (F^-)^n = 0 \tag{78}$$

Higher-order accuracy can be obtained in various ways such as by using the midpoint corrector given previously. Alternately, an upwind version of the MacCormack predictor corrector scheme is given by

$$\bar{Q} = Q^n - \Delta t\ \nabla_x (F^+)^n - \Delta t\ \Delta_x (F^-)^n \tag{79a}$$

$$Q^{n+1} = [(Q^n + \bar{Q}) - \Delta t(\nabla_x^2 (F^+)^n + \nabla_x \bar{F}^+) + \Delta t(\Delta_x^2 (F^-)^n - \Delta_x \bar{F}^-)]/2 \tag{79b}$$

It is interesting that the preceding matrix and flux split schemes can be represented as a central differencing and a dissipation term. This is because

$$\Lambda^{\pm} = \frac{\Lambda \pm |\Lambda|}{2}$$

and thus,

$$A^{\pm} = \frac{A \pm |A|}{2} \tag{80}$$

where

$$|A| \equiv S|\Lambda|S^{-1} \tag{81}$$

Using Eq. (80) in Eq. (75), for example, gives

$$Q^{n+1} - Q^n + \Delta t \frac{A}{2} (\nabla_x + \Delta_x) Q^n + \Delta t \frac{|A|}{2} (\nabla_x - \Delta_x) Q^n = 0 \tag{82}$$

Because

$$\frac{\Delta_x + \nabla_x}{2} Q_j = \frac{Q_{j+1} - Q_{j-1}}{2\Delta x} = \delta_x Q_j$$

and

$$\frac{\nabla_x - \Delta_x}{2} Q_j = -\left(\frac{\Delta x}{2}\right) \frac{(Q_{j+1} - 2Q_j + Q_{j-1})}{(\Delta x)^2} = -\frac{\Delta x}{2} \delta_{xx} Q_j$$

Eq. (82) is equivalent to

$$Q^{n+1} - Q^n + \Delta t A \delta_x Q^n = \frac{\Delta t\ \Delta x}{2} |A| \delta_{xx} Q^n \tag{83}$$

with δ_x and δ_{xx} central difference operators. Because $|A|$ is always positive, the right-hand side of Eq. (83) is always dissipative. In this example the upwind scheme (75) is equivalent to the central differencing scheme (83).

For stable values of the time step, the differencing scheme given with Eq. (75) or Eq. (78) is monotone for the scalar equation case. If we were to linearize these schemes and look at the matrix of the difference equations over all space and all time, then for a stable value of Δt this matrix will be diagonally dominant. That is, like a centrally differenced Laplacian

$$|a_{ii}| \geqslant \sum_{i \neq j} |a_{ij}|$$

What this means is that the point being solved for cannot obtain a value greater or less than any of the other points in its stencil. Consequently, no spurious oscillations in the difference solution are possible.

Lambda Scheme[9]

It can be argued that the equations written in terms of the primitive variables Q and $\tilde{Q}$ are not always the best choice of dependent variables for efficient numerical simulation. That is, these variables may change more rapidly or may lead to more difficult equations to solve than if the equations are formulated in other dependent variables. The Euler equations can be formulated in terms of other dependent variables, e.g.,

$$c_t + \boldsymbol{q} \cdot \nabla c + \frac{\gamma - 1}{2} c \, \nabla \cdot \boldsymbol{q} = 0 \tag{84a}$$

$$\boldsymbol{q}_t + (\boldsymbol{q} \cdot \nabla)\boldsymbol{q} + \frac{2}{\gamma - 1} c \, \nabla c - \frac{c^2}{\gamma(\gamma - 1)} \nabla_s = 0 \tag{84b}$$

$$s_t + \boldsymbol{q} \cdot \nabla s = 0 \tag{84c}$$

where c is the sound speed, $\boldsymbol{q}$ is the velocity vector, and s is the entropy. This equation set is invalid across a shock wave; thus, shock-wave discontinuities must be specially treated. Otherwise, the simple convection of entropy relation, Eq. (84c), is particularly convenient to work with and can be discarded in regions in which entropy is known to be constant.

The one-dimensional form of Eqs. (84) is given by

$$\partial_t \begin{pmatrix} c \\ u \\ s \end{pmatrix} + \begin{bmatrix} u & \dfrac{\gamma - 1}{2} c & 0 \\ \dfrac{2}{\gamma - 1} c & u & -\dfrac{c^2}{\gamma(\gamma - 1)} \\ 0 & 0 & u \end{bmatrix} \partial_x \begin{pmatrix} c \\ u \\ s \end{pmatrix} = \begin{pmatrix} 0 \\ 0 \\ 0 \end{pmatrix} \tag{85}$$

or

$$\bar{Q}_t + \bar{A}\bar{Q}_x = 0 \tag{86}$$

The eigenvalues of $\bar{A}$ are readily found to be $u, u \pm c$, and Eq. (86) can be solved with central spatial differencing or via the eigensplit forms. In particular, solution of the split form

$$\bar{Q}_t + \Xi\bar{\Lambda}\Xi^{-1}\bar{Q}_x = 0 \tag{87}$$

with spatial differencing determined according to the sign of the eigenvalues, $\bar{\Lambda}$, leads to what is called the λ scheme. For $\bar{A}$, the eigenvalue and eigenvector matrices can be written

$$\bar{\Lambda} = \begin{bmatrix} u-c & 0 & 0 \\ 0 & u & 0 \\ 0 & 0 & u+c \end{bmatrix} \tag{88a}$$

$$\Xi = \begin{bmatrix} \dfrac{\gamma-1}{4} & \dfrac{c}{2\gamma} & \dfrac{\gamma-1}{4} \\ \dfrac{-1}{2} & 0 & \dfrac{1}{2} \\ 0 & 1 & 0 \end{bmatrix} \tag{88b}$$

$$\Xi^{-1} = \begin{bmatrix} \dfrac{2}{\gamma-1} & -1 & -\dfrac{c}{\gamma(\gamma-1)} \\ 0 & 0 & 1 \\ \dfrac{2}{\gamma-1} & 1 & -\dfrac{c}{\gamma(\gamma-1)} \end{bmatrix} \tag{88c}$$

Writing Eq. (87) out gives

$$c_t + \frac{(\gamma-1)}{4}\{u-c\}\left[\frac{2c_x}{\gamma-1} - u_x - \frac{cs_x}{\gamma(\gamma-1)}\right] + \frac{c}{2\gamma}\{u\}s_x$$
$$+ \frac{(\gamma-1)}{4}\{u+c\}\left[\frac{2c_x}{\gamma-1} + u_x - \frac{cs_x}{\gamma(\gamma-1)}\right] = 0$$

$$u_t - \tfrac{1}{2}\{u-c\}\left[\frac{2c_x}{\gamma-1} - u_x - \frac{cs_x}{\gamma(\gamma-1)}\right]$$
$$+ \tfrac{1}{2}\{u+c\}\left[\frac{2c_x}{\gamma-1} + u_x - \frac{cs_x}{\gamma(\gamma-1)}\right] = 0$$

$$s_t + \{u\}s_x = 0$$

In the lambda scheme, as in most other eigensplit schemes, the collection of terms c_x, u_x, and s_x are backward differenced if the $\{\lambda_i\}$ coefficient is >0 and are forward differenced if the $\{\lambda_i\}$ coefficient is <0.

Sound Speed Splittings

Other splittings of the Jacobian matrix A are possible which do not directly use the eigenvector-similarity transforms. As noted before, the matrix $\tilde{A}$ of Eq. (10) is partially decoupled. One can difference and numerically advance the momentum and energy equations one time step independently of the continuity equation. That is, u and p can be advanced one time step from

$$\delta_t\begin{pmatrix}u\\p\end{pmatrix}+\begin{bmatrix}u & \rho^{-1}\\ \gamma p & u\end{bmatrix}\delta_x\begin{pmatrix}u\\p\end{pmatrix}=\begin{pmatrix}0\\0\end{pmatrix} \tag{89}$$

and ρ can then be advanced from

$$\delta_t\rho + u\delta_x\rho = -\rho\delta_x u$$

The 2×2 coefficient matrix of Eq. (89) has eigenvalues $u+c$ and $u-c$, and its elements are so readily identified with its eigenvalues that a backward or forward operator can be assigned to each element. To identify the elements with its eigenvalues, Eq. (89) is further split as

$$\delta_t\begin{pmatrix}u\\p\end{pmatrix}+\begin{bmatrix}u & 0\\ 0 & u\end{bmatrix}\delta_x\begin{pmatrix}u\\p\end{pmatrix}+\begin{bmatrix}0 & \rho^{-1}\\ \gamma p & 0\end{bmatrix}\delta_x\begin{pmatrix}u\\p\end{pmatrix}=\begin{pmatrix}0\\0\end{pmatrix} \tag{90}$$

Here the first coefficient matrix is seen to have eigenvalues of u and u, and the second matrix has eigenvalues of c and $-c$. Each matrix can be differenced separately. Assuming that u is positive, then $u\delta_x$ terms can be backward differenced as this part of the equation corresponds to the model scalar wave equations.

To difference the second part of the equation, consider the model problem

$$\delta_t\begin{pmatrix}u\\p\end{pmatrix}+c\begin{bmatrix}0 & 1\\ 1 & 0\end{bmatrix}\delta_x\begin{pmatrix}u\\p\end{pmatrix}=\begin{pmatrix}0\\0\end{pmatrix} \tag{91}$$

in which the coefficient matrix also has eigenvalues c and $-c$. This is a fundamental coupled wave equation that can be studied for stability just as the scalar equation. In particular, because one eigenvalue is positive while the other is negative, the simple forward-backward implicit differencing schemes can be proposed[10]:

$$\begin{pmatrix}u\\p\end{pmatrix}^{n+1}=\begin{pmatrix}u\\p\end{pmatrix}^{n}-c\,\Delta t\begin{bmatrix}0 & \nabla_x\\ \Delta_x & 0\end{bmatrix}\begin{pmatrix}u\\p\end{pmatrix}^{n+1} \tag{92}$$

or

$$\begin{pmatrix}u\\p\end{pmatrix}^{n+1}=\begin{pmatrix}u\\p\end{pmatrix}^{n}-c\,\Delta t\begin{bmatrix}0 & \Delta_x\\ \nabla_x & 0\end{bmatrix}\begin{pmatrix}u\\p\end{pmatrix}^{n+1} \tag{93}$$

Assuming periodic spacial boundary conditions and carrying out a matrix stability analysis for the scheme (92), the amplification factor E is obtained from the determinant of (with ν the Courant number)

$$\begin{bmatrix} E-1 & \nu(1-e^{-i\theta_j})E \\ \nu(e^{i\theta_j}-1)E & E-1 \end{bmatrix} = \begin{pmatrix} 0 \\ 0 \end{pmatrix}$$

or

$$\left[1+4\nu^2\sin^2\left(\frac{\theta_j}{2}\right)\right]E^2 - 2E + 1 = 0$$

from which

$$|E| = \sqrt{\frac{1}{[1+4\nu^2\sin^2(\theta_j/2)]}} \tag{94}$$

From Eq. (94), $|E| \leqslant 1$; thus, the implicit scheme Eq. (92) is unconditionally stable. The scheme Eq. (93) has the same stability properties. Both schemes, Eqs. (92) and (93), are unstable if spatial differences are treated explicitly in time, but they are conditionally stable if one spatial difference is treated implicitly, and the other is treated explicitly.

It should be remarked that the fully implicit schemes given by Eqs. (92) and (93) require 2×2 block tridiagonal inversion. Although it is interesting, this scheme is dissipative and first order accurate in space (unless p is solved at grid points centered midway to u) as well as time. Consequently, this approach does not appear to offer any advantage over using central second-order-accurate differencing for the treatment of Eq. (91) or its corresponding part in Eq. (90).

Multidimensional Explicit Schemes

For the most part, the explicit numerical schemes described in one dimension are readily applied to the Euler equations in multispace dimensions. Multispace dimensions impose more restrictive stability bounds, however, and some schemes such as the Lax-Wendroff scheme become more costly to implement. The concept of local one-dimensional space splitting can help alleviate these problems.

Denote the invisicid conservation law equations in three space dimensions as

$$\partial_t Q + \partial_x F + \partial_y G + \partial_z H = 0 \tag{95}$$

The Lax-Wendroff one-step scheme is given in two dimensions as

$$Q^{n+1} = Q^n - \Delta t(\delta_x F^n + \delta_y G^n) + \frac{(\Delta t)^2}{2}(\bar{\delta}_x A^n \bar{\delta}_x \bar{F}^n + \bar{\delta}_y B^n \bar{\delta}_y G^n + \delta_x A^n \delta_y G^n + \delta_y B^n \delta_x F^n) \tag{96}$$

where

$$\delta_x = \frac{E_x - E_x^{-1}}{2\,\Delta x}$$
$$\bar{\delta}_x = \frac{E_x^{\frac{1}{2}} - E_x^{-\frac{1}{2}}}{\Delta x}, \qquad \text{etc.}$$

As before, this algorithm is expensive because of the Jacobian matrix multiplies. It becomes even more awkward in three dimensions.

The MacCormack scheme is still easy to apply in three dimensions and is given by

$$\bar{Q} = Q^n - \Delta t(\nabla_x F^n + \nabla_y G^n + \nabla_z H^n) \tag{97a}$$
$$Q^{n+1} = \tfrac{1}{2}(Q^n + \bar{Q}) - \frac{\Delta t}{2}(\Delta_x \bar{F} + \Delta_y \bar{G} + \Delta_z \bar{H}) \tag{97b}$$

The drawback encountered with this or any other explicit formulation is that the stability bound becomes more restrictive as the spacial dimensions increase. This is difficult to show for the Lax-Wendroff class algorithms such as the MacCormack scheme, so the leap frog method will be used to illustrate the more restrictive stability bound.

Consider the scalar wave equation

$$u_t + au_x + bu_y + cu_z = 0 \tag{98}$$

with periodic boundary data. Leap frog time differencing and central space differencing of Eq. (98) gives

$$u^{n+1} - u^{n-1} + 2\Delta t(a\delta_x + b\delta_y + c\delta_z)u^n = 0 \tag{99}$$

where

$$\delta_x = \frac{E_x - E_x^{-1}}{2\Delta x}, \qquad \text{etc.}$$

Transforming into uncoupled form to deduce the stability bounds

$$\tilde{u}^{n+1} - \tilde{u}^{n-1} + 2i(v_x \sin\theta_j + v_y \sin\theta_k + v_z \sin\theta_l)\tilde{u}^n = 0 \tag{100}$$

where $v_x = a\,\Delta t/\Delta x$, etc. Let ϕ equal the terms in parentheses and rewrite Eq. (100) as

$$(E^2 + 2i\phi E - 1)\tilde{u}^{n-1} = 0$$

or

$$E = \frac{-2i\phi \pm \sqrt{-4\phi^2 + 4}}{2} = -i\phi \pm \sqrt{1 - \phi^2} \tag{101}$$

If $|\phi| > 1$, the square root term becomes imaginary and the scheme is unstable as

$$E = i(-\phi \pm \sqrt{\phi^2 - 1})$$

and

$$|E| = |\phi| + \sqrt{\phi^2 - 1} > 1$$

with $|\phi| > 1$.

If $\phi \leqslant 1$, the radicand is real and

$$|E| = (1 - \phi^2 + \phi^2)^{\frac{1}{2}} = 1$$

Thus, the scheme is neutrally stable if $\phi \leqslant 1$. For ϕ to be less than or equal to 1,

$$\Delta t\left(\frac{a}{\Delta x}\sin\theta_j + \frac{b}{\Delta y}\sin\theta_k + \frac{c\sin\theta_l}{\Delta z}\right) \leqslant 1$$

and under the worst case $\sin\theta = \pm 1$ and allowing a, b, and c to be either positive or negative

$$\Delta t \leqslant \frac{1}{\dfrac{|a|}{\Delta x} + \dfrac{|b|}{\Delta y} + \dfrac{|c|}{\Delta z}} \tag{102}$$

In one dimension the stability restriction was given as

$$\Delta t \leqslant \frac{1}{\dfrac{|a|}{\Delta x}} \tag{103}$$

or

$$\frac{|a|\,\Delta t}{\Delta x} \leqslant 1$$

If

$$\frac{|a|}{\Delta x} = \frac{|b|}{\Delta y} = \frac{|c|}{\Delta z}$$

the step size in three dimensions can be up to three times smaller.

Although leap frog was used in the preceding development, other explicit schemes show a similar fall off of step size Δt with an increase of spacial dimensions. The multidimensional Euler equations cannot be put into scalar form via similarity transform (i.e., the Jacobian matrices A, B, and C do not commute); hence, their precise stability bound (even for constant coefficients) is unknown. A bound for the MacCormack scheme when applied to the Euler equations, Eqs. (97), has been approximately determined[11] as

$$\Delta t \leqslant \frac{1}{\dfrac{|u|}{\Delta x} + \dfrac{|v|}{\Delta y} + \dfrac{|w|}{\Delta z} + c\sqrt{\dfrac{1}{(\Delta x)^2} + \dfrac{1}{(\Delta y)^2} + \dfrac{1}{(\Delta z)^2}}} \tag{104}$$

where u, v, and w are the Cartesian velocities and c is the sound speed.

One way of improving the multidimensional stability bound is to use locally one-dimensional splitting concepts—the method of fractional steps. To illustrate this ideal, consider again the Lax-Wendroff one-step scheme in two dimensions. Instead of Eq. (96), use the approximate sequence

$$\bar{Q} = Q^n - \Delta t \delta_x F^n + \frac{(\Delta t)^2}{2}(\bar{\delta}_x A^n \bar{\delta}_x F^n) \tag{105a}$$

$$Q^{n+1} = \bar{Q} - \Delta t \delta_y \bar{G} + \frac{(\Delta t)^2}{2}(\bar{\delta}_y \bar{B} \bar{\delta}_y \bar{G}) \tag{105b}$$

which when combined gives (essentially)

$$\begin{aligned} Q^{n+1} &= Q^n - \Delta t \delta_x F^n + \frac{(\Delta t)^2}{2}(\bar{\delta}_x A^n \bar{\delta}_x F^n) \\ &- \Delta t \delta_y G^n + (\Delta t)^2 \delta_y \bar{B} \delta_x F^n - \frac{(\Delta t)^3}{2}\delta_y \bar{B}(\bar{\delta}_x A^n \bar{\delta}_x F^n) \\ &+ \frac{(\Delta t)^2}{2}\bar{\delta}_y \bar{B} \bar{\delta}_y G^n - \frac{(\Delta t)^3}{2}(\bar{\delta}_y B \delta_y \bar{B} \delta_x F^n) + \frac{(\Delta t)^4}{4}(\bar{\delta}_y \bar{B} \bar{\delta}_y \bar{B} \bar{\delta}_x A^n \bar{\delta}_x F^n) \end{aligned} \tag{106}$$

Comparison with Eq. (96) shows that these schemes are equivalent except for terms $0(\Delta t)^3$ or higher provided

$$(\Delta t)^2 \delta_y \bar{B} \delta_x F^n \doteq \frac{\Delta t^2}{2}(\delta_x A^n \delta_y G^n + \delta_y B^n \delta_x F^n) \tag{107}$$

Now, if $A \equiv A_0 + 0(\Delta)$, $B = B_0 + 0(\Delta)$, then to $0(\Delta^3)$, Eq. (107) is approximately

$$(\Delta t)^2 B_0 A_0 \delta_y \delta_x Q^n \doteq \left(\frac{\Delta t^2}{2}\right)(A_0 B_0 \delta_x \delta_y Q^n + B_0 A_0 \delta_y \delta_x Q^n)$$

and the crucial difference is thus given by

$$\left(\frac{\Delta t^2}{2}\right)(B_0 A_0 - A_0 B_0)\delta_x \delta_y Q^n \tag{108}$$

If A and B commute, the approximate sequence given by Eqs. (105) is equivalent to Eq. (96) to second-order accuracy. The Jacobian matrices for the Euler equations do not commute; thus, one must either accept a less accurate method with Eqs. (105), or make a correction. The simplest correction process is to reorder the sequence on alternate time steps as

$$\bar{Q} = Q^n - \Delta t \delta_y G^n + \frac{(\Delta t)^2}{2} (\bar{\delta}_y B^n \bar{\delta}_y G^n) \tag{109a}$$

$$Q^{n+1} = \bar{Q} - \Delta t \delta_x \bar{F} + \frac{(\Delta t)^2}{2} (\bar{\delta}_x \bar{A} \bar{\delta}_x \bar{F}) \tag{109b}$$

which gives the error (108) with opposite sign on every alternative step.

The advantage of the fractional step sequence is chiefly that it has a much better stability bound. The overall time step is limited by only the one-dimensional stability bound found in the worst direction. Moreover, in the case of the one-step Lax-Wendroff scheme, the fractional step algorithm is much simpler than the standard scheme (this is not true, however, of the MacCormack scheme).

The fractional step method has a greater advantage if, say, $\Delta y \ll \Delta x$. One can implement the Δy fractional step much more frequently than the Δx step, e.g.,

$$Q^0 = Q^n - \Delta t \delta_x F^n + \frac{(\Delta t)^2}{2} (\bar{\delta}_x A^n \bar{\delta}_x F^n) \tag{110a}$$

$$Q^m = Q^{m-1} - h \delta_y G^{m-1} + \frac{(h)^2}{2} (\bar{\delta}_y B^{m-1} \bar{\delta}_y G^{m-1}), \qquad 1 \leqslant m \leqslant M \tag{110b}$$

$$Q^{n+1} = Q^M \tag{110c}$$

where $Mh = \Delta t$.

The method of fractional steps has some disadvantages. For one, the data base is called in more times as one goes through the sequence first in one direction, then in the other. The splitting also degrades accuracy somewhat and can complicate boundary condition treatment.

Multidimensional Implicit Schemes, Approximate Factorization

The extension of implicit schemes to multidimensions can be straightforward but computationally expensive unless approximations are made. As in the one-dimensional case, local linearization is used to avoid iterative solution of nonlinear equations between time steps. In multidimensions, methods of approximate factorization are generally used as well to avoid costly matrix inversions between time steps. Frequently, these approximate methods are alternating direction schemes, and consequently, they sometimes resemble the locally one-dimensional fractional step methods used for explicit schemes. Often they do not.

Consider the two-dimensional inviscid conservation law equations

$$\partial_t Q + \partial_x F + \partial_y G = 0 \tag{111}$$

and apply the trapezoidal-rule scheme (51)

$$\left[I + \frac{\Delta t}{2}(\delta_x A^n + \delta_y B^n) - \epsilon_i(\Delta\nabla)_x - \epsilon_i(\nabla\Delta)_y \right] \Delta Q^n$$
$$= -\Delta t(\delta_x F^n + \delta_y G^n) - \epsilon_e[(\nabla\Delta)_x^2 + (\nabla\Delta)_y^2]Q^n \tag{112}$$

where $A = \partial F/\partial Q$ and $B = \partial G/\partial Q$ are 4×4 Jacobian matrices that result from the time linearizations. The operators δ_x and δ_y are three-point central difference operators, e.g.,

$$\delta_y Q = \frac{Q_{k+1} - Q_{k-1}}{2\Delta y} \tag{113}$$

The dissipation operators were discussed previously, and although not indicated in Eq. (112), the ϵ_i and ϵ_e can be different in x than in y and can vary with space or time. The use of implicit dissipation is more crucial in multidimensions because the stability bound on explicitly added dissipation alone becomes more stringent with increasing dimension.

Assuming a typewriter ordering of grid points, the left-hand side of Eq. (112), which uses three-point central differencing, fills a block tridiagonal matrix, each block of which is a block tridiagonal matrix with 4×4 block elements (ordering ρ, ρu, ρv, e innermost to the jk ordering). Temporarily assume A and B are constant and set $\epsilon_i = 0$ both to simplify and to illustrate the skew symmetric property of the matrix that results from Eq. (112). This matrix has the form

$$\begin{bmatrix}
I & A & & & B & & & & & & & \\
-A & I & A & & & B & & & & & & \\
& -A & I & A & & & B & & & & & \\
& & \ddots & \ddots & \ddots & & & \ddots & & & & \\
-B & & & & I & A & & & B & & & \\
& -B & & & -A & I & A & & & B & & \\
& & -B & & & -A & I & A & & & B & \\
& & & \ddots & & & \ddots & \ddots & \ddots & & & \ddots \\
& & & & -B & & & & I & A & & \\
& & & & & -B & & & -A & I & A & \\
& & & & & & \ddots & & & \ddots & \ddots & \ddots
\end{bmatrix} \tag{114}$$

where in fact A and B are scaled by $(\Delta t/\Delta x)$ and $(\Delta t/\Delta y)$ and should appear as $[\Delta t/(4\Delta x)]A$ and $[\Delta t/(4\Delta y)]B$. The bandwidth of this matrix is $4 \times J_{max}$, where J_{max} is the number of grid points in the x direction.

Direct inversion of the block tridiagonal with block tridiagonal elements is feasible but costly because nonzero elements fill in the bandwidth. In three dimensions this can be especially costly, since the bandwidth is increased by a multiplier of the number of grid points in the z direction. To avoid this, approximate iterative solution of the matrix at each time step or approximate factorization of the left-hand-side operator of Eq. (112) [or the matrix (114)] is frequently used. For Eq. (112) an approximate factorization of the left-hand side is given by[12]

$$\left[I + \frac{\Delta t}{2}\,\delta_x A^n - \epsilon_i(\nabla\Delta)_x\right]\left[I + \frac{\Delta t}{2}\,\delta_y B^n - \epsilon_i(\nabla\Delta)_y\right]\Delta Q^n = RHS(112) \quad (115)$$

where $RHS(112)$ is the right-hand side of Eq. (112). Expanding the factors gives Eq. (112) back plus additional terms such as

$$\frac{\Delta t^2}{4}\,\delta_x A^n \delta_y B^n\,\Delta Q^n = 0(\Delta t^3)$$

which on division by Δt [by which Eq. (112) was previously multiplied] is a second-order-accurate term. For a steady state, this error is zero as $\Delta Q^n = Q^{n+1} - Q^n = 0$. Similar approximate factorization schemes are discussed in Sec. 5, and here, as there, the delta form method is used to automatically subtract the approximate factorization error terms.

A solution algorithm is developed for Eq. (115) with the sequence

$$\left[I + \frac{\Delta t}{2}\,\delta_x A^n - \varepsilon_i(\nabla\Delta)_x\right]Q^* = RHS(112) \quad (116a)$$

$$\left[I + \frac{\Delta t}{2}\,\delta_y B^n - \varepsilon_i(\nabla\Delta)_y\right]\Delta Q^n = Q^* \quad (116b)$$

$$Q^{n+1} = Q^n + \Delta Q^n \quad (116c)$$

The solution of the first sequence [Eq. (116a)] is equivalent to solving the block tridiagonals

$$\left[\left(-\frac{\Delta t}{4\Delta x}A_{j-1,k} - \varepsilon_i I\right), \quad (1 + 2\varepsilon_i)I, \quad \left(\frac{\Delta t}{4\Delta x}A_{j+1,k} - \varepsilon_i I\right)\right]\begin{pmatrix} Q^*_{j-1,k} \\ Q^*_{j,k} \\ Q^*_{j+1,k}\end{pmatrix} = R_{jk} \quad (117)$$

for all j to a given k and for all k. Likewise, the block tridiagonals of Eq. (116b) are of the form

$$\left[\left(-\frac{\Delta t}{4\Delta y}B_{j,k-1}-\varepsilon_i I\right),\quad (1+2\varepsilon_i)I,\quad \left(\frac{\Delta t}{4\Delta y}B_{j,k+1}-\varepsilon_i I\right)\right]\begin{pmatrix}\Delta Q^n_{j,k-1}\\ \Delta Q^n_{j,k}\\ \Delta Q_{jk+1}\end{pmatrix}=Q^*_{jk} \tag{118}$$

In either case the block tridiagonal matrices have 4×4 block elements.

In solving the first sequence [Eq. (116a)], note that Q^* is needed on x boundaries, not ΔQ^n. From Eq. (116b), Q^* is obtained by differencing ΔQ^n on the boundary in y. Using ΔQ^n rather than Q^* results in only a second-order error at the boundary, however, and indeed setting $\Delta Q^n = 0$ at the boundary results in only a first-order error.

The arithmetic operations in solving a block tridiagonal system of equations using L-U decomposition for the block elements is given in Ref. 13 as $N(28M^3+27M^2-M)/6$ where N is the number of rows each with block elements of size M. The inversion work for Eqs. (116) for a grid of $J\times K$ interior points would thus require $2\times J\times K*370$ operations. Here there are K block tridiagonals with J rows with 4×4 block elements produced by Eq. (116a) and J block tridiagonals with K rows produced by Eq. (116b).

3. Navier-Stokes Equations

The unsteady compressible Navier-Stokes equations in two-dimensional flow can be written in conservative nondimensional form as

$$\partial_t Q + \partial_x F + \partial_y G = Re^{-1}(\partial_x F_v + \partial_y G_v)$$

where

$$Q=\begin{pmatrix}\rho\\ \rho u\\ \rho v\\ e\end{pmatrix},\qquad F=\begin{pmatrix}\rho u\\ \rho u^2+p\\ \rho uv\\ u(e+p)\end{pmatrix},\qquad G=\begin{pmatrix}\rho v\\ \rho uv\\ \rho v^2+p\\ v(e+p)\end{pmatrix},$$

$$F_v=\begin{pmatrix}0\\ \tau_{xx}\\ \tau_{xy}\\ f_4\end{pmatrix},\qquad G_v=\begin{pmatrix}0\\ \tau_{xy}\\ \tau_{yy}\\ g_4\end{pmatrix}$$

$$\tau_{xx}=\mu(4u_x-2v_y)/3$$

$$\tau_{xy}=\mu(u_y+v_x)$$

$$\tau_{yy} = \mu(-2u_x + 4v_y)/3$$

$$f_4 = u\tau_{xx} + v\tau_{xy} + \mu Pr^{-1}(\gamma - 1)^{-1}\,\partial_x c^2$$

$$g_4 = u\tau_{xy} + v\tau_{yy} + \mu Pr^{-1}(\gamma - 1)^{-1}\,\partial_y c^2$$

with μ the viscosity coefficient, *Re* and *Pr* are reference Reynolds and Prandtl number, respectively, due to use of nondimensional variables, and the Stokes hypothesis is used.

Compared to solving the inviscid Euler equations, the Navier-Stokes equations offer the following additional difficulties in high-Reynolds-number flow: 1) to accurately resolve viscous terms one generally needs to introduce very fine grid spacings; and 2) viscous shear layers are often turbulent. The viscous terms offer no particular difficulty to numerical solution, although treatment of the cross-derivative terms can be cumbersome and the viscous terms are not readily split for backward and forward space differencing. Viscous surface boundary conditions are typically easier to apply than inviscid surface boundary conditions.

Model Scalar Convection Diffusion Equation

To define a scalar model problem for the one-dimensional inviscid Euler equations, a frozen coefficient form of the equations was diagonalized. One would expect that following a similar approach for the Navier Stokes equations would lead to a simple convection-diffusion model equation. Because the Jacobian matrices of the convection terms do not commute with the Jacobian matrices of the viscous terms, however, such a formal approach is not possible. Nevertheless, experience with the Navier-Stokes and subset equations such as the boundary-layer equations shows that a simple scalar convection-diffusion equation serves as a good model problem. Consequently, in this section differencing schemes for the equation

$$\partial_t u + \lambda\,\partial_x u = \sigma\,\partial_{xx} u \tag{119}$$

are surveyed subject to initial and boundary-value conditions. (The even simpler diffusion alone schemes are discussed under boundary-layer methods.) As before, differencing schemes that are stable for both positive and negative values of λ are needed. The coefficient σ will always be taken as positive. Although flux-splitting like concepts can be used to one-side difference the convection terms, simple one-sided differencing of the viscous term is unstable.

The Lax-Wendroff class of differencing has been used previously for convection alone. To incorporate diffusion into this type of predictor-corrector sequence, an explicit Euler predictor followed by a modified Euler corrector is used. For MacCormack differencing the predictor-corrector sequence for Eq. (119) is given by

$$\bar{u} = u^n - \lambda\,\Delta t\,\nabla_x u^n + \sigma\,\Delta t\,\nabla_x\Delta_x u^n \tag{120a}$$

$$u^{n+1} = \tfrac{1}{2}(u^n + \bar{u} - \lambda\,\Delta t\,\Delta_x \bar{u} + \sigma\,\Delta t\,\nabla_x\Delta_x \bar{u}) \tag{120b}$$

where ∇_x and Δ_x are two-point backward and forward difference operators as used throughout. The convection part of this sequence is recognized as the usual MacCormack scheme. The diffusion part is being updated with a Euler and modified Euler differencing that has been somewhat rearranged. The usual sequence is

Euler:

$$\bar{u} = u^n + h\dot{u}^n \tag{121a}$$

Modified Euler:

$$u^{n+1} = u^n + h\left(\frac{\dot{u}^n + \dot{\bar{u}}}{2}\right) \tag{121b}$$

where $du/dt \equiv \dot{u}$ is differenced in time.

In Eqs. (120) this sequence has been arranged as

$$\bar{u} = u^n + h\dot{u}^n \tag{122a}$$

$$u^{n+1} = \tfrac{1}{2}(u^n + \bar{u} + h\dot{\bar{u}}) \tag{122b}$$

and these are seen to be identical by using the predictor to eliminate $\bar{u}$ of the corrector. These schemes are equivalent to a second-order Runge-Kutta technique.

For real coefficients, the stability bound for the second-order Runge-Kutta scheme is the same as the explicit Euler scheme. For diffusion alone the stability bound is (see Sec. 4)

$$\frac{\sigma \Delta t}{(\Delta x)^2} \leqslant 1/2$$

The stability bound for Eqs. (120) is approximately

$$\Delta t \leqslant \left(\frac{1}{\dfrac{|\lambda|}{\Delta x} + \dfrac{2\sigma}{(\Delta x)^2}}\right) \tag{123}$$

In general the explicit stability bound for diffusion alone is considered to be stringent because $\sigma\, \Delta t \sim (\Delta x)^2$. However, σ corresponds to a viscosity coefficient in the Navier-Stokes equations, and for air, this number can be quite small. As a result the diffusion term alone does not necessarily impose a small time step. But for viscous flow past an obstacle, high gradients must be resolved just near the body surface (caused by the no-slip boundary condition), and there the grid spacing must be fine to resolve the high gradients. Because the grid spacing is small, the stability limit imposed by the convection term becomes stringent. Consequently, the explicit stability limit for convection-diffusion is generally found to be restrictive. Although splitting methods can somewhat alleviate this problem as alluded to later, the use of implicit methods is generally preferred for convection-diffusion problems.

As a representative implicit scheme consider the use of trapezoidal-rule time differencing and three-point second-order-accurate central space differencing

$$\nabla_t u^{n+1} + (\lambda\delta_x - \sigma\delta_{xx})(u^{n+1} + u^n)/2 = 0 \tag{124}$$

with $\delta_x = (\Delta_x + \nabla_x)/2$ and $\delta_{xx} = \nabla_x \Delta_x$. Multiply through by Δt and arrange as

$$(I + \lambda h\delta_x - \sigma h\delta_{xx})u^{n+1} = (I - \lambda h\delta_x + \sigma h\delta_{xx})u^n \tag{125}$$

where $h = \Delta t/2$. Alternately this can be arranged in the delta form:

$$(I + \lambda h\delta_x - \sigma h\delta_{xx})(u^{n+1} - u^n) = -\Delta t(\lambda\delta_x - \sigma\delta_{xx})u^n \tag{126}$$

The differencing is unconditionally stable for convection alone (see Sec. 2) and for diffusion alone (see the discussion for the model problem in Sec. 4). Given periodic or boundary-value conditions in x, the combined equation has an amplification factor of

$$E = \frac{[1 - 2\beta\sin^2(\theta_j/2)] - (\nu/2)i\sin\theta_j}{[1 + 2\beta\sin^2(\theta_j/2)] + (\nu/2)i\sin\theta_j} \tag{127}$$

where $\beta = \sigma\Delta t/(\Delta x)^2$ and $\nu = \lambda\,\Delta t/(\Delta x)$.

Then, as $1 - 2\beta\sin^2(\theta_j/2) \leqslant 1 + 2\beta\sin^2(\theta_j/2)$, the modulus of E is always less than or equal to one and is given by

$$|E| = \left[\frac{[1 - 2\beta\sin^2(\theta_j/2)]^2 + (\nu^2/4)\sin^2\theta_j}{(1 + 2\beta\sin^2(\theta_j/2)) + (\nu^2/4)\sin^2\theta_j}\right]^{\frac{1}{2}} \tag{128}$$

Consequently, the implicit scheme given by Eq. (126) is unconditionally stable for the convection-diffusion problem. Use of three-point backward time differencing leads to the same result. It can be commented that the *inversion* work part of the implicit algorithm is not increased for convection-diffusion over what it is for just convection alone.

Numerical Solution of the Navier-Stokes Equations

In differencing the two-dimensional compressible Navier-Stokes equations, it is convenient to split the viscous terms that form cross derivatives from those that do not. The equations are therefore written as

$$\partial_t Q + \partial_x F + \partial_y G - Re^{-1}(\partial_x F_{vx} + \partial_x F_{vy} + \partial_y G_{vx} + \partial_y G_{vy}) \tag{129}$$

where Q, F, and G were defined previously:

$$F_{vx} = \begin{bmatrix} 0 \\ (4/3)\mu u_x \\ \mu v_x \\ (\mu/2)\,\partial_x[(4/3)u^2 + v^2] + \mu\beta\,\partial_x c^2 \end{bmatrix}, \quad F_{vy} = \begin{bmatrix} 0 \\ -(2/3)\mu v_y \\ \mu u_y \\ -(2/3)\mu u v_y + \mu v u_y \end{bmatrix}$$

$$G_{vx} = \begin{bmatrix} 0 \\ \mu v_x \\ -(2/3)\mu u_x \\ \mu u v_x - (2/3)\mu v u_x \end{bmatrix}, \quad G_{vy} = \begin{bmatrix} 0 \\ \mu u_y \\ (4/3)\mu v_y \\ (\mu/2)\,\partial_y[u^2 + (4/3)v^2] + \mu\beta\,\partial_y c^2 \end{bmatrix} \tag{130}$$

with $\beta \equiv Pr^{-1}(\gamma - 1)^{-1}$ and c is the sound speed.

In this section the fractional step MacCormack scheme and the approximate factorization implicit scheme of Beam and Warming will be used to illustrate differencing of Eq. (129).

Explicit Time Differencing

Although in the preceding section it was noted that implicit time differencing is generally preferred for convection-diffusion equations, explicit schemes have been extensively utilized for the Navier-Stokes equations. Explicit schemes are advocated because they can be more easily programmed on vector computers, and computational efficiency can be somewhat improved by using the fractional step method. A MacCormack fractional step procedure that was previously sketched for multidimensional inviscid flow is widely used.

Following the differencing scheme for the model convection-diffusion equation, central space differencing is used for the viscous terms. A typical term of $\partial_x F_{vx}$ is $(4/3)\,\partial_x \mu\,\partial_x u$, which can be second-order-accurate central differenced as

$$\begin{aligned} \partial_x \mu\,\partial_x u &\simeq \bar{\delta}_x \bar{\mu} \bar{\delta}_x u \\ &= \frac{\mu_{j+1/2}(u_{j+1} - u_j) - \mu_{j-1/2}(u_j - u_{j-1})}{(\Delta x)^2} \\ &\simeq \frac{(\mu_{j+1} + \mu_j)(u_{j+1} - u_j) - (\mu_j + \mu_{j-1})(u_j - u_{j-1})}{2(\Delta x)^2} \end{aligned} \tag{131}$$

Here midpoint operators are used to obtain the most compact second-order-accurate central space differencing. A typical cross-derivative term is $\partial_x \mu\,\partial_y u$ of $\partial_x F_{vy}$, and this can be differenced as

$$\begin{aligned} \partial_x \mu\,\partial_y u &= \delta_x \mu \delta_y u \\ &= \frac{\mu_{j+1}(u_{j+1,k+1} - u_{j+1,k-1}) - \mu_{j-1}(u_{j-1,k+1} - u_{j-1,k-1})}{4\Delta x\,\Delta y} \end{aligned} \tag{132}$$

Throughout this section the notation $\bar{\delta}_x F_{vx}$ implies that each term of $\partial_x F_{vx}$ is differenced according to the format of Eq. (131), whereas $\delta_x F_{vy}$ implies differencing according to the format Eq. (132). That is,

$$\bar{\delta}_x F_{vx} \rightarrow \text{difference like Eq. (131)}$$

$$\delta_x F_{vy} \rightarrow \text{difference like Eq. (132)}$$

Similar operators are defined for $\partial_x G_v$.

With these difference operators defined, the fractional step method of Sec. 2 can be combined with the explicit differencing for the convection-diffusion equation. For the two-dimensional unsteady Navier-Stokes equations, the explicit fractional step algorithm is obtained:

$$\bar{Q}^{(1)} = Q^n - h\,\nabla_x F^n + h\,Re^{-1}(\bar{\delta}_x F^n_{vx} + \delta_x F^n_{vy}) \tag{133a}$$

$$\bar{Q}^{(2)} = \tfrac{1}{2}[Q^n + \bar{Q}^{(1)} - h\,\Delta_x \bar{F}^{(1)} + h\,Re^{-1}(\bar{\delta}_x \bar{F}^{(1)}_{vx} + \delta_x \bar{F}^{(1)}_{vy})] \tag{133b}$$

$$\bar{Q}^{(3)} = \bar{Q}^{(2)} - h\,\nabla_y \bar{G}^{(2)} + h\,Re^{-1}(\delta_y \bar{G}^{(2)}_{vx} + \bar{\delta}_y \bar{G}^{(2)}_{vy}) \tag{133c}$$

$$Q^{n+1} = \tfrac{1}{2}[\bar{Q}^{(2)} + \bar{Q}^{(3)} - h\,\Delta_y \bar{G}^{(3)} + h\,Re^{-1}(\delta_y \bar{G}^{(3)}_{vx} + \bar{\delta}_y \bar{G}^{(3)}_{vy})] \tag{133d}$$

where $h = \Delta t$.

As with the inviscid equations, the ∇ and Δ operators can be permuted with alternate time steps, and the y operators can alternately be carried out before the x operators—all of which can improve accuracy as indicated previously.

Imposing correct boundary conditions on the fractional step sequence given by Eqs. (133) is not trivial. In the x-corrector step of Eqs. (133), for example, $\bar{Q}^{(1)}$ is needed on the upper boundary surface. But $\bar{Q}^{(1)}$ on the upper surface depends on $\nabla_x F^n$, which can readily be evaluated, and $\bar{\delta}_x F^n_{vx}$ and $\delta_x F^n_{vy}$, which are not readily evaluated because they use central difference operators. If the upper surface boundary is a freestream condition, then the viscous derivatives can be safely set to zero. On the lower surface, however, the viscous terms still require $\bar{Q}^{(1)}$. The $\bar{Q}^{(1)}$ needed on the lower surface for the viscous terms may be approximated with Q^n or $(Q^n + Q^{n+1})/2$, but this approximation will degrade the local accuracy to first order.

In viscous flow the grids in one direction are frequently much refined compared to the other direction. It is this fine grid direction that will determine this overall stepsize. If the grid is much refined in y, then it is desirable to advance the y-directional fractional step, at a small Δt, several times for each advance of the x-directional fractional step, at a larger Δt. Adopting this idea to the algorithm [Eqs. (133)] gives

$$\bar{Q}^{(1)} = Q^n - \Delta t\,\nabla_x F^n + \Delta t\,Re^{-1}(\bar{\delta}_x F^n_{vx} + \delta_x F^n_{vy} \tag{134a}$$

$$\bar{Q}^{(2)} = \tfrac{1}{2}[Q^n + \bar{Q}^{(1)} - \Delta t\,\Delta_x \bar{F}^{(1)} + \Delta t Re^{-1}(\bar{\delta}_x \bar{F}^{(1)}_{vx} + \delta_x \bar{F}^{(1)}_{vy})] \tag{134b}$$

and for $m = 1,M$,

$$\bar{Q}^{(2m+1)} = \bar{Q}^{(2m)} - h\,\nabla_y \bar{G}^{(2m)} + h\,Re^{-1}(\delta_y \bar{G}_{vx}^{(2m)} + \bar{\delta}_y \bar{G}_{vy}^{(2m)}) \tag{134c}$$

$$\bar{Q}^{(2m+2)} = \tfrac{1}{2}[\bar{Q}^{(2m)} + \bar{Q}^{(2m+1)} - h\,\Delta_y \bar{G}^{(2m+1)} + h\,Re^{-1}(\delta_y \bar{G}_{vx}^{(2m+1)} + \delta_y G_{vy}^{(2m+1)})] \tag{134d}$$

where $h = \Delta t/M$.

Implicit Time Differencing

In applying implicit differencing to the inviscid compressible flow equations in multidimensions, tricks of local linearization and approximate factorization were employed to maintain computational efficiency. These ideas carry over to differencing the compressible Navier-Stokes equations.

For this illustration, three-point central second-order space differencing and trapezoidal time differencing is used. Extension to higher-order space differencing or alternate time differencing follows in a relatively straightforward way from the previous developments. With $h \equiv \Delta t/2$, Eq. (129) is differenced as

$$\begin{aligned} Q^{n+1} - Q^n + h\delta_x(F^{n+1} + F^n) + h\delta_y(G^{n+1} + G^n) \\ = h\,Re^{-1}[\bar{\delta}_x(F_{vx}^{n+1} + F_{vx}^n) + \delta_x(F_{vy}^{n+1} + F_{vy}^n) \\ + \delta_y(G_{vx}^{n+1} + G_{vx}^n) + \bar{\delta}_y(G_{vy}^{n+1} + G_{vy}^n)] \end{aligned} \tag{135}$$

where δ_x, δ_y are three-point central difference operators and $\bar{\delta}_x$ and $\bar{\delta}_y$ are midpoint operators.

To improve computational efficiency in inviscid flow, the basic nonlinear implicit difference equations were locally linearized and approximately factored. The same steps are taken here. The viscous cross-derivative terms, however, present a difficulty. If the viscous cross-derivative terms are treated implicitly, they will prevent approximate factorization into simple x and y operators. Implicit approximate factorizations that include cross derivatives and are easily invertible are available, but they are not unconditionally stable. However, Beam and Warming[14] have verified that, for the scalar model convection-diffusion equation, the cross-derivative terms can be lagged in time without affecting unconditional stability. Thus, the simplest approach is also the most stable! Consequently, the cross terms will be lagged as

$$(F_{vy}^{n+1} + F_{vy}^n)/2 \doteq (3F_{vy}^n - F_{vy}^{n-1})/2 + 0(\Delta t^2) \tag{136a}$$

$$(G_{vx}^{n+1} + G_{vx}^n)/2 \doteq (3G_{vx}^n - G_{vx}^{n-1})/2 + 0(\Delta t^2) \tag{136b}$$

Although it will not be obvious until one attempts otherwise, lagging the cross derivatives will also lead to a less awkward local linearization expression resulting from energy equation terms such as $\partial_x \mu v\, \partial_y u$.

To avoid an iterative solution process, the inviscid flux terms F and G are simply linearized as before:

$$F^{n+1} = F^n + A^n(Q^{n+1} - Q^n) \tag{137a}$$

$$G^{n+1} = G^n + B^n(Q^{n+1} - Q^n) \tag{137b}$$

The viscous terms, however, contain derivatives and coefficients of viscosity and thermal conductivity that can be complex functions of Q. In particular, the functional relation of μ with respect to Q can be quite involved, especially for turbulent flow. Indeed, for turbulent flow the linearization of μ with Q would have to be performed numerically, and this is costly. As a result, this functional variation is generally ignored and the coefficients are taken at the old time level. If necessary, a single iterative correction as used later with the boundary-layer equations [or similar to that of Eqs. 65a–c] could be employed to improve matters.

Consider a typical viscous term, for example, $\mu\,\partial_y u$ of G_{vy}. Freezing μ for the reasons just discussed, this term is expanded as

$$\mu_0\,\partial_y\left(\frac{\rho u}{\rho}\right) = \mu_0\,\partial_y\left[\frac{\rho_0 u_0}{\rho_0} - \frac{\rho_0 u_0}{\rho_0^2}(\rho - \rho_0) + \frac{1}{\rho_0}(\rho u - \rho_0 u_0)\right] \tag{138}$$

All of the F_{vx} and G_{vy} elements can be expanded term by term in a similar way and written as

$$\partial_x F_{vx} = \partial_x\{F_{vx}|_0 + \mu_0\,\partial_x[A_{vx}|_0(Q - Q_0)]\} \tag{139a}$$

$$\partial_y G_{vy} = \partial_y\{G_{vy}|_0 + \mu_0\,\partial_y[B_{vy}|_0(Q - Q_0)]\} \tag{139b}$$

where the viscous Jacobian matrices A_{vx} and B_{vy} are given by

$$A_{vx} = \begin{bmatrix} 0 & 0 & 0 & 0 \\ -(4/3)(u/\rho) & (4/3)\rho^{-1} & 0 & 0 \\ -(v/\rho) & 0 & \rho^{-1} & 0 \\ a_{x1} & a_{x2} & a_{x3} & (\beta/\rho) \end{bmatrix} \tag{140a}$$

$$B_{vy} = \begin{bmatrix} 0 & 0 & 0 & 0 \\ -(u/\rho) & \rho^{-1} & 0 & 0 \\ -(4/3)(v/\rho) & 0 & (4/3)\rho^{-1} & 0 \\ b_{y1} & b_{y2} & b_{y3} & (\beta/\rho) \end{bmatrix} \tag{140b}$$

and

$$a_{x1} = -(4/3)\left(\frac{u^2}{\rho}\right) - \left(\frac{v^2}{\rho}\right) + \beta\left[-\left(\frac{e}{\rho}\right) + \frac{u^2+v^2}{\rho}\right]$$

$$a_{x2} = (4/3 - \beta)\left(\frac{u}{\rho}\right)$$

$$a_{x3} = (1 - \beta)\left(\frac{v}{\rho}\right)$$

$$b_{y1} = -\left(\frac{u^2}{\rho}\right) - (4/3)\left(\frac{v^2}{\rho}\right) + \beta\left[-\left(\frac{e}{\rho}\right) + \frac{u^2+v^2}{\rho}\right]$$

$$b_{y2} = (1 - \beta)\left(\frac{u}{\rho}\right)$$

$$b_{y3} = (4/3 - \beta)\left(\frac{v}{\rho}\right)$$

Applying the expansions given by Eqs. (136), (137), and (139) to the implicit differencing Eq. (135) gives

$$\begin{aligned}[][I + h\delta_x A^n + h\delta_y B^n - h\,Re^{-1}(\bar{\delta}_x\mu^n\bar{\delta}_x A^n_{vx} + \bar{\delta}_y\mu^n\bar{\delta}_y B^n_{vy})](Q^{n+1} - Q^n) \\ = -\Delta t[\delta_x F^n + \delta_y G^n - Re^{-1}(\bar{\delta}_x F^n_{vx} + \bar{\delta}_y G^n_{vy})] \\ + h\,Re^{-1}[\delta_x(3F^n_{vy} - F^{n-1}_{vy}) + \delta_y(3G^n_{vx} - G^{n-1}_{vx})] \qquad (141)\end{aligned}$$

In terms of the unknown $Q^{n+1}_{j,k}$, this system of difference equations is linear; the nonlinearity is lagged to the previous time levels.

To achieve the next economy the difference equations are approximately factored into easily invertible operators. The cross derivatives have already been lagged; hence, the left-hand side of Eq. (141) is simply factored into one-dimensional like operators:

$$\begin{aligned}(I + h\delta_x A^n - h\,Re^{-1}\bar{\delta}_x\mu^n\bar{\delta}_x A^n_{vx}) \\ \times (I + h\delta_y B^n - h\,Re^{-1}\bar{\delta}_y\mu^n\bar{\delta}_y B^n_{vy})(Q^{n+1} - Q^n) = RHS(141) \qquad (142)\end{aligned}$$

This form requires only the solution of block tridiagonal matrices with 4×4 blocks.

4. Boundary-Layer Methods

In high-Reynolds-number flow, the effect of viscosity is usually confined to a thin shear layer near and trailing the vehicle. The flow outside of this region can be regarded as inviscid. Splitting the flowfield between thin shear layers and an outer inviscid region can result in two sets of governing equations, each of which is simpler than the full Navier-Stokes equations. However, this same boundary-layer approximation creates matching or coupling problems between the inviscid and viscous flows. Furthermore, near separation, the viscous boundary-layer equations can have undersirable critical point behavior. For these reasons, the Navier-Stokes equations are sometimes solved even though the more approximate equation sets are physically valid. The boundary-layer approximation is too useful to ignore, however, and in this section some of the difficulties of solving the boundary-layer equations and coupling them to the outer inviscid flow equations are briefly described.

Unsteady Boundary-Layer Equations

The Navier-Stokes equations in nonconservative form can be written as x-, y-, and z-Momentum equations:

$$\rho(u_t + uu_x + vu_y + wu_z) = -p_x + (2\mu u_x + \beta)_x + (\tau_{xy})_y + (\tau_{xz})_z \tag{143a}$$

$$\rho(v_t + uv_x + vv_y + wv_z) = -p_y + (\tau_{xy})_x + (2\mu v_y + \beta)_y + (\tau_{yz})_z \tag{143b}$$

$$\rho(w_t + uw_x + vw_y + ww_z) = -p_z + (\tau_{xz})_x + (\tau_{yz})_y + (2\mu w_z + \beta)_z \tag{143c}$$

where $\beta = \lambda(u_x + v_y + w_z)$ and $\tau_{xy} = \mu(u_y + v_x)$, $\tau_{xz} = \mu(u_z + w_x)$, and $\tau_{yz} = \mu(v_z + w_y)$.

Energy equation:

$$\rho c_p(T_t + uT_x + vT_y + wT_z)$$
$$= p_t + up_x + vp_y + wp_z + (\kappa T_x)_x + (\kappa T_y)_y + (\kappa T_z)_z + \mu\phi \tag{143d}$$

where the dissipation function ϕ is

$$\phi = 2(u_x^2 + v_y^2 + w_y^2) + [(u_y + v_x)^2 + (v_z + w_y)^2 + (w_x + u_z)^2] - \tfrac{2}{3}(u_x + v_y + w_z)^2$$

Continuity equation:

$$\rho_t + (\rho u)_x + (\rho v)_y + (\rho w)_z = 0 \tag{143e}$$

For a thin laminar shear layer along a flat plate, $y = 0$, analysis (e.g., Ref. 15) indicates that

$$u,w,x,z = 0(1)$$
$$v,y = 0(Re^{-1/2})$$

With this information, order-of-magnitude analysis of Eqs. (143) yields the boundary-layer equations

$$\rho(u_t + uu_x + vu_y + wu_z) = -p_x + (\mu u_y)_y \tag{144a}$$

$$0 = -p_y \tag{144b}$$

$$\rho(w_t + uw_x + vw_y + ww_z) = -p_z + (\mu w_y)_y \tag{144c}$$

$$\rho c_p(T_t + uT_x + vT_y + wT_z) = p_t + up_x + wp_z + (\kappa T_y)_y + \mu(u_y^2 + w_y^2) \tag{144d}$$

$$\rho_t + (\rho u)_x + (\rho v)_y + (\rho w)_z = 0 \tag{144e}$$

with

$$\rho = p/(RT)$$

For a body surface with mild curvature the same equations can be used with y taken as the direction normal to the body surface and x and z being distances along the surface.

The differential boundary-layer equations may be solved as written, or often they are integrated in the y direction to yield what are referred to as the "integral" boundary-layer equations. Except in the case of two-dimensional steady flow, the "integral" boundary-layer equations are not ordinary differential equations, but partial-differential equations with derivatives in one less space direction. The exact form of the integral boundary-layer equations is strongly dependent on the assumptions made about the boundary-layer velocity profiles. (See, for example, the work done at the Office National d'Etudes et de Recherches Aéronautiques (ONERA)[16–18] for a thorough description of an integral representation of the boundary-layer equations.)

There are at least two advantages to the integral formulation. First, for steady two-dimensional flows, the integral boundary-layer equations are ordinary differential equations, and relatively little computer time is required to solve them. For three-dimensional flow integral analysis yields partial differential equations with one less space dimension. Second, the integral formulation is often expressed in terms of quantities (such as displacement thickness) that can be measured experimentally. Therefore, it is possible to directly use experimentally observed trends in an integral formulation. In this way, an integral formulation can be "tailored" to yield excellent agreement with experimental data for a specific problem.

Although the integral methods can prosper from a detailed experimental data base, if this data base does not exist (as is the case for most unsteady and three-dimensional flows), it may be difficult to obtain rational and accurate closure models for the integral methods. This has proven to be a particularly difficult problem for three-dimensional flows with strong crossflow components. In contrast, the differential boundary-layer methods require only as much closure information as the Navier-Stokes methods. Here only differential field boundary-layer methods will be discussed.

Differencing Schemes for the Model Equation

The appropriate one-dimensional model problem for the unsteady boundary-layer equations is the convection-diffusion equation discussed previously. However, the heat equation

$$\frac{\partial u}{\partial t} = \sigma \frac{\partial^2 u}{\partial x^2}, \qquad \sigma > 0 \tag{145}$$

is the model problem usually associated with boundary-layer equations. The heat equation serves well as a model problem for steady boundary-layer flow (with t,x, replacing x,y) and less well for unsteady boundary-layer flow. Having already discussed differencing the convection-diffusion equation, several commonly used schemes for the heat equation are briefly described.

The simplest explicit and implicit schemes for approximating Eq. (145) are

Explicit:

$$\Delta_t u_j^n = \sigma \, \nabla_x \Delta_x u_j^n \tag{146a}$$

or

$$u_j^{n+1} - u_j^n = \beta(u_{j+1}^n - 2u_j^n + u_{j-1}^n)$$

Implicit:

$$\nabla_t u_j^{n+1} = \sigma \, \nabla_x \Delta_x u_j^{n+1} \tag{146b}$$

or

$$u_j^{n+1} - u_j^n = \beta(u_{j+1}^{n+1} - 2u_j^{n+1} + u_{j-1}^{n+1})$$

where $\beta = \sigma \, \Delta t/(\Delta x)^2$.

Using similarity transforms to diagonalize the $\nabla_x \Delta_x$ operator, one finds that

Explicit:

$$\tilde{u}_j^{n+1} = [1 - 4\beta \sin^2(\theta_j/2)]\tilde{u}_j^n \tag{147a}$$

Implicit:

$$[1 + 4\beta \sin^2(\theta_j/2)]\tilde{u}_j^{n+1} = \tilde{u}_j^n \tag{147b}$$

or

$$\tilde{u}_j^{n+1} = [1 + 4\beta \sin^2(\theta_j/2)]^{-1}\tilde{u}_j^n$$

For the amplification factor of the explicit scheme to be less than or equal to 1 requires that

$$|1 - 4\beta| < 1$$

or

$$4\beta < 2$$

The explicit scheme can only be stable if

$$\frac{\sigma\,\Delta t}{(\Delta x)^2} \leqslant \frac{1}{2} \tag{148}$$

For the implicit case the amplification factor is always less or equal to 1. The implicit differencing scheme is unconditionally stable and is generally used with boundary-layer methods.

Higher-order-accurate implicit time-marching schemes can be obtained by approximating ∂_t with either trapezoidal rule (second-order Pade) or three-point backward differencing:

Trapezoidal (Crank-Nicolson scheme):

$$\left(\frac{2}{1+E_t^{-1}}\right)\nabla_t u_j^{n+1} = \sigma\,\nabla_x \Delta_x u_j^{n+1} \tag{149}$$

or

$$\nabla_t u_j^{n+1} = \sigma\left(\frac{\nabla_x \Delta_x u_j^{n+1} + \nabla_x \Delta_x u_j^n}{2}\right)$$

Three-point backward:

$$\nabla_t\left(\frac{3-E_t^{-1}}{2}\right)u_j^{n+1} = \sigma\,\nabla_x \Delta_x u_j^{n+1} \tag{150}$$

Both schemes are unconditionally stable for $\sigma > 0$. The Crank-Nicolson scheme has a lower truncation error than the three-point backward scheme, but it is not as robust. High-frequency noise can be propagated along in the time-march direction with Crank-Nicolson that is filtered out with the three-point backward scheme. The three-point backward scheme, however, requires either two levels of initial data, or it must be started with a two-level scheme such as Crank-Nicolson. Both schemes can be programmed using two levels of storage.

The heat equation can also be written as a first-order system. Let $u_x = w$ so that

$$u_t - \sigma w_x = 0 \tag{151a}$$

$$w - u_x = 0 \tag{151b}$$

A variety of centered numerical schemes have been devised for the equations in this form. One of these, the box scheme (its stencil uses points of a box) is a popular boundary-layer algorithm. The box scheme is usually derived with midpoint differencing and averaging, but it can be readily derived and represented using Pade differencing as

$$\left(\frac{1+E_t^{-1}}{2}\right)^{-1}\nabla_t u_j^n - \sigma\left(\frac{1+E_x^{-1}}{2}\right)^{-1}\nabla_x w_j^n = 0 \tag{152a}$$

$$w_j^n - \left(\frac{E_x + 1}{2}\right)^{-1} \Delta_x u_j^n = 0 \tag{152b}$$

Multiplying Eq. (152a) by $(1 + E_t^{-1})(1 + E_x^{-1})$ and Eq. (152b) by $(E_x + 1)$ and expanding the operators gives

$$\frac{\left(\frac{u_j^u + u_{j-1}^n}{2}\right) - \left(\frac{u_j^{n-1} + u_{j-1}^{n-1}}{2}\right)}{\Delta t} - \sigma \frac{\left(\frac{w_j^n + w_j^{n-1}}{2}\right) - \left(\frac{w_{j-1}^n + w_{j-1}^{n-1}}{2}\right)}{\Delta x} = 0 \tag{153a}$$

$$\left(\frac{w_{j+1}^n + w_j^n}{2}\right) - \frac{u_{j+1}^n - u_j^n}{\Delta x} = 0 \tag{153b}$$

The easy way to analyze the stability of Eqs. (152) is by recombining the difference equations by eliminating w_j^n as

$$\left(\frac{1 + E_t^{-1}}{2}\right)^{-1} \nabla_t u_j - \sigma\left(\frac{E_x + 2 + E_x^{-1}}{4}\right)^{-1} \nabla_x \Delta_x u_j^n = 0 \tag{154}$$

This scheme is unconditionally stable as it has an amplification factor

$$E = \left(\frac{1 + \cos\theta_j - 4\beta \sin^2(\theta_j/2)}{1 + \cos\theta_j + 4\beta \sin^2(\theta_j/2)}\right) \tag{155}$$

with modulus always less than or equal to one.

Equation (154) mimics the trapezoidal rule, Eq. (149), except for the space averaging operator $(E_x + 2 + E_x^{-1})/4$. Although the box scheme is relatively popular, it does not appear to offer any advantage over conventional differencing, and solving a first-order system of equations implicitly requires more work than solving the simple second-order equation implicitly.

Numerical Methods for the Boundary-Layer Equations

Because of the appearance of the reduced vertical momentum equation, $p_y = 0$, the *steady-state* boundary-layer equations with a specified edge pressure can be solved using efficient space-marching procedures for attached flow. The computational efficiency of space marching is lost in unsteady flow, however. Nevertheless, because the pressure gradient forcing terms in the momentum equations are treated as given functions, the boundary-layer equations are weakly coupled and can be solved sequentially at each time step. For example, if p is specified, then clearly Eq. (144a) contains only unknown derivatives in u and Eq. (144c) only those in w. Assuming nonlinear coefficients are available from previous time levels, u and w can each be updated at a time step independently of the other variables. For an implicit solution procedure this means the matrix inversion work is much reduced. Moreover, once u and w are available, the energy equation contains only unknown derivatives of the temperature T; therefore, T can be updated from just the scalar energy equation.

Drawbacks to using the boundary-layer equations are also related to use of the simplified vertical momentum equation. First, there is no longer an equation containing v_t for natural time integration of v. Also, in an interactive mode, pressure must be prescribed to the boundary layer from an edge close enough to the viscous layer that $p_y = 0$ remains valid. Generally, the outer layer will be solved "distinctly" from the viscous layer and inconsistencies can result. Another difficulty arises due to the saddle point behavior of the boundary-layer equations when the forcing function p_x is specified through a separation point. Analysis and numerical experience indicate that critical point behavior will be encountered at a separation point (at least at a steady state) if the pressure is specified. However, this behavior can be suppressed if one of the quantities (such as v, τ_w, and δ^*) which displays singular behavior when the pressure is given is instead specified. Because these quantities are the result of a typical pressure specified (direct) boundary-layer computation, computations with variables such as τ_w or δ^* specified are referred to as inverse computations.

A simple algorithm for the unsteady boundary-layer equations[18] which uses equation uncoupling is illustrated below. Spatial derivatives in the momentum (and energy) equations are approximated with implicit second-order-accurate central difference operators in the direction normal to the body and upwind second-order-accurate operators in the other two directions. That is, central differencing is used in y, and upwind differencing is used in x and z. For example, when differencing a term such as u_x, a backward difference is used if the coefficient ρu is positive and a forward difference is used if the coefficient ρu is negative. In the continuity equation, the x and z derivatives are approximated with second-order-accurate central difference operators, and trapezoidal-rule integration is used in the y direction to yield v.

The time-accurate algorithm is presented below using the conventional operators and the second-order-accurate upwind operator

$$\hat{\alpha}\hat{\delta}_x = \left(\frac{\alpha + |\alpha|}{2}\right)\nabla_x\left(\frac{3 - E_x^{-1}}{2}\right) + \left(\frac{\alpha - |\alpha|}{2}\right)\Delta_x\left(\frac{3 - E_x}{2}\right)$$

where α is the convective coefficient (e.g., ρu). Using a notation that space-time indices are not written unless changed (e.g., $u = u^n_{j,k,l}$, $u^{n+1} = u^{n+1}_{j,k,l}$, etc.), a semi-implicit scheme that takes advantage of the weak coupling between the equations is given as:

Update u and w at the new time step from the x- and z-momentum equations:

$$\rho(\nabla_t u^{n+1} + \hat{u}\hat{\delta}_x u^{n+1} + v\delta_y u^{n+1} + \hat{w}\hat{\delta}_z u) = -p_x + \bar{\delta}_y(\mu\bar{\delta}_y u^{n+1}) \quad (156a)$$

$$\rho(\nabla_t w^{n+1} + \hat{u}\hat{\delta}_x w^{n+1} + v\delta_y w^{n+1} + \hat{w}\hat{\delta}_z w) = -p_z + \bar{\delta}_y(\mu\bar{\delta}_y w^{n+1}) \quad (156b)$$

Each equation is easily updated to time level $n+1$ by inverting scalar-tridiagonal matrices in y and sweeping in x. With u^{n+1} and w^{n+1} available, T is updated from the energy equation using exactly the same approach:

$$\rho c_p(\nabla_t T^{n+1} + \hat{u}\hat{\delta}_x T^{n+1} + v\delta_y T^{n+1} + \hat{w}\hat{\delta}_z T)$$
$$= \nabla_t p + u^{n+1}\delta_x p + w^{n+1}\delta_z p + \bar{\delta}_y(\kappa\bar{\delta}_y T^{n+1}) + \mu(\delta_y u^{n+1})^2 + \mu(\delta_y w^{n+1})^2 \tag{156c}$$

Finally, $\rho^{n+1} = p/T^{n+1}$ and v^{n+1} is obtained by integrating the continuity equation up from the wall using u^{n+1}, w^{n+1} and ρ^{n+1}

$$\nabla_y(\rho v)^{n+1} = -\frac{1+E_y^{-1}}{2}[\delta_x(\rho u) + \delta_z(\rho w) + \nabla_t \rho]^{n+1} \tag{156d}$$

The algorithm presented above is second-order accurate in space and first-order accurate in time. A simple corrector sequence can be used to achieve second-order time accuracy and provide additional stability of the explicitly differenced terms. Letting superscript * denote the values predicted from Eqs. (156) and, for example, using three-point backward time differencing δ_t^b, a corrector step is given by

$$\rho^*(\delta_t^b u^{n+1} + \widehat{u^*}\hat{\delta}_x u^{n+1} + v^*\delta_y u^{n+1} + \widehat{w^*}\hat{\delta}_z u^*) = -p_x + \bar{\delta}_y(\mu^*\bar{\delta}_y u^{n+1}) \tag{157a}$$

$$\rho^*(\delta_t^b w^{n+1} + \widehat{u^*}\delta_x w^{n+1} + v^*\delta_y w^{n+1} + \widehat{w^*}\hat{\delta}_z w^*) = -p_z + \bar{\delta}_y(\mu^*\bar{\delta}_y w^{n+1}) \tag{157b}$$

$$\rho^* c_p(\delta_t^b T^{n+1} + \widehat{u^*}\hat{\delta}_x T^{n+1} + v^*\delta_y T^{n+1} + \widehat{w^*}\hat{\delta}_z T$$
$$= \delta_t^b p + u^{n+1}\delta_x p + w^{n+1}\delta_z p + \bar{\delta}_y(\kappa^*\bar{\delta}_y T^{n+1})$$
$$+ \mu^*(\delta_y u^{n+1})^2 + \mu^*(\delta_y w^{n+1})^2 \tag{157c}$$

$$\nabla_y(\rho v)^{n+1} = -\frac{1+E_y^{-1}}{2}[\delta_x(\rho u) + \delta_z(\rho w) + \delta_t^b \rho]^{n+1} \tag{157d}$$

To solve the boundary-layer equations in the inverse mode, the algorithm must be modified so that the inverse forcing functions can be specified. When solved in primitive variables, it is convenient to specify, for example, the wall shear stress τ_w as the inverse forcing function. To achieve this the momentum equation at the wall is used to replace the pressure term with an expression that contains the wall shear stress. For example, the x-momentum equation evaluated at a wall yields

$$p_x = (\mu u_y)_y|_w = 2\,\frac{\dfrac{\mu_2+\mu_1}{2}\dfrac{u_2-u_1}{y_2-y_1} - \tau_x|_w}{y_2-y_1} \tag{158}$$

This expression allows the eliminations of the pressure term from the momentum equations. However, if the entire inverse forcing term is treated implicitly, a tridiagonal-like matrix with an additional column of nonzero coefficients (due to the u_2 term) must be inverted.

Interaction Schemes

A great deal of progress has been achieved in applying the viscous-inviscid interaction approach to steady transonic problems (e.g., Refs. 16–28). However, relatively few have attempted to apply the interaction approach to unsteady transonic aerodynamics (e.g., Refs. 16–18, 29). In general the interaction methods used for unsteady computations have been those developed for steady flow.

In "direct" interaction the inviscid flow is advanced a time step and then the inviscid pressure distribution is fed to a boundary-layer algorithm. A displacement thickness effect (e.g., a transpiration velocity) is then computed and fed back into the inviscid algorithm. For weak interactions this viscous-inviscid iteration is stable and will converge in steady-state cases. The advantage of the direct interaction scheme is that it is very simple; the disadvantage is that it does not work for steady-state separated flow.

If the boundary-layer equations are solved in the inverse mode, separated flow can be treated. The challenge is to devise an interaction scheme that can supply an inverse forcing function to the boundary-layer algorithm. One such approach is the conceptually simple "inverse" scheme in which the inviscid and viscous flows are computed in the inverse mode. This implies that the surface pressure is specified to the inviscid solver and the effective body thickness (hence the displacement thickness) is a result (which can be fed to the inverse boundary algorithm). However, this approach has seen only limited use for the practical reasons that inverse inviscid computations are not straightforward and, in general, inverse inviscid codes are not as well-developed as their direct counterparts.

To avoid the problems of solving the inverse inviscid problem, the "semi-inverse" interaction procedure was developed. In this scheme, an approximation of the inviscid equations (in the direct, geometry specified mode) is obtained and the resulting pressure distribution is used along with other information to specify an inverse forcing function distribution to the boundary-layer equations, which in turn supplies a displacement thickness distribution that is fed back into the inviscid algorithm. The key to the semi-inverse interaction schemes is the algorithm for obtaining the inverse forcing function distribution fed to the boundary-layer algorithm. A popular example of this type of algorithm is one developed by Carter[30]:

$$\delta^{*n+1} = \delta^{*n} \frac{u_{\text{edge}}\big|_{\text{viscous}}}{u_{\text{edge}}\big|_{\text{inviscid}}} \tag{159}$$

Semi-inverse schemes are fairly simple and have been used extensively in steady flow calculations[23,26–30] to compute flows with mild separation.

Finally, a number of researchers have explored the concept of solving the boundary-layer equations and some subset of the inviscid equations simultaneously. It has been shown numerically that this "semi implicit" procedure suppresses the boundary-layer saddle point behavior.[31-34] This approach has been extended to a variety of flows[17,20,24] and appears to be very promising for computing flows with very strong viscous-inviscid interactions.

5. Unsteady Potential Flow

Governing Equations

If flowfield shock waves are weak, uniform-incoming inviscid compressible flow can be taken as irrotational/isentropic throughout. The Euler equations can then be simplified. For isentropic flow the conservation equations of mass, momentum, and energy can be replaced by equations of continuity, irrotationality, and a Bernoulli equation. In addition, the three velocity variables, u, v, and w, can be replaced with a scalar velocity potential if a proper circulation model is included in the formulation. With introduction of a velocity potential

$$\nabla \Phi = \boldsymbol{q}$$

the irrotationality condition $\nabla \times \boldsymbol{q} = 0$ is identically satisfied, and the governing equations become

Continuity:

$$\partial_t \rho + \partial_x(\rho \Phi_x) + \partial_y(\rho \Phi_y) + \partial_z(\rho \Phi_z) = 0 \tag{160}$$

Bernoulli:

$$\rho = \left[1 + \frac{\gamma - 1}{2}(M_\infty^2 - 2\Phi_t - \Phi_x^2 - \Phi_y^2 - \Phi_z^2)\right]^{1/(\gamma - 1)} \tag{161}$$

where the derivation of the Bernoulli equation assumes a steady far-field flow. The equations have been nondimensionalized throughout with respect to ρ_∞, c_∞, and an arbitrary reference length.

From the Bernoulli equation $\rho = \rho(\Phi(x,y,z,t))$, the continuity equation can be written in nonconservative form as

$$\rho_\Phi \Phi_t + \rho(\Phi_{xx} + \Phi_{yy} + \Phi_{zz}) + \Phi_x \rho_\Phi \Phi_x + \Phi_y \rho_\Phi \Phi_y + \Phi_z \rho_\Phi \Phi_z = 0 \tag{162}$$

where $\rho_\Phi = \partial\rho/\partial\Phi$ is a differential operator:

$$\rho_\Phi = -\rho^{2-\gamma}(\partial_t + \Phi_x\,\partial_x + \Phi_y\,\partial_y + \Phi_z\,\partial_z) \tag{163}$$

Performing the differential operations indicated by ρ_Φ and using the nondimensional isentropic relation $\rho^{\gamma-1} = c^2$ gives the conventional nonlinear potential equation

$$\begin{aligned}\Phi_{tt} + 2(\Phi_x\Phi_{xt} + \Phi_y\Phi_{yt} + \Phi_z\Phi_{zt}) &= (c^2 - \Phi_x^2)\Phi_{xx} + (c^2 - \Phi_y^2)\Phi_{yy} \\ &+ (c^2 - \Phi_z^2)\Phi_{zz} - 2(\Phi_x\Phi_y\Phi_{xy} + \Phi_x\Phi_z\Phi_{xz} + \Phi_y\Phi_z\Phi_{yz})\end{aligned} \tag{164}$$

From this equation transonic small-disturbance equations can be extracted that are useful for approximate solution procedures and that, when linearized further, provide model equations that serve to evaluate numerical schemes.

The small-disturbance transonic low-frequency equation[35] in terms of the disturbance potential can be written as

$$2\kappa M_\infty^2 \phi_{xt} = [(1 - M_\infty^2) - (\gamma + 1)M_\infty^2 \phi_x]\phi_{xx} + \phi_{yy} \tag{165a}$$

or in conservative form

$$2\kappa M_\infty^2 \phi_{xt} = (1 - M_\infty^2)\phi_{xx} - [(\gamma + 1)/2]M_\infty^2[(\phi_x)^2]_x + \phi_{yy} \tag{165b}$$

where κ is the reduced frequency, $\kappa = \bar{\omega}\bar{l}/\bar{U}_\infty \ll 1$. The variables are nondimensionalized as $t = \bar{t}\bar{\omega}$, $\phi = \bar{\phi}/(\bar{U}_\infty \bar{l})$, $x = \bar{x}/\bar{l}$ with l the reference length, and the overbar denotes a dimensional quantity. Equations (165) use the disturbance potential ϕ defined from

$$\Phi_x = U_\infty + \phi_x$$
$$\Phi_y = \phi_y$$
$$\Phi_z = \phi_z$$

Numerical Solution of the Small-Disturbance Equation

In this section implicit numerical algorithms developed for the transonic small-disturbance low-frequency equation are explored. Following the usual pattern, a model linear problem is first studied. The model problem provides a good opportunity to reexamine approximate factorization implicit methods in more detail than previously. Such methods are suitable for unsteady as well as steady applications. From the model problem, algorithms[36] for two-dimensional transonic flow are developed.

Model Problem Differencing

A model linear equation for Eq. (165) is given by

$$\phi_{xt} = \beta\phi_{xx} + \phi_{yy} \tag{166}$$

where $\beta = 1 - M_\infty^2$ and $2M_\infty\kappa$ has been incorporated into the time variable $\tilde{t} = t/(2M_\infty\kappa)$ and the tilde is deleted. An implicit differencing of Eq. (166) is sought to remove any time step restrictions other than those imposed by considerations of accuracy.

A simple first-order accurate in time implicit differencing of Eq. (166) is given by

$$\nabla_t\nabla_x\phi^{n+1} = \beta\,\nabla_x\Delta_x\phi^{n+1} + \nabla_y\Delta_y\phi^{n+1} \tag{167}$$

where $\beta > 0$. Here central difference operators are used to approximate ∂_{xx} and ∂_{yy} because, if a steady state is reached, this differencing is satisfactory for the resulting elliptic equation. The unsteady cross-derivative operator has been backward differencing in space as well as in the time-marching direction.

A simple stability analysis shows that the differencing scheme given by Eq. (167) is unconditionally stable. The difference equations can first be arranged as

$$(\nabla_x - h\beta\ \nabla_x\Delta_x - h\ \nabla_y\Delta_y)\phi^{n+1} = \nabla_x\phi^n \tag{168}$$

where $h = \Delta t$. Assuming spacial periodicity in x and y the orthonormal eigenvectors of a circulant matrix can be used to diagonalize Eq. (168) to its eigenvalues (see the Appendix) as

$$\left[\left(\frac{1-\cos\theta_j + i\sin\theta_j}{\Delta x}\right) + \frac{h\beta 4\sin^2(\theta_j/2)}{(\Delta x)^2} + \frac{4h\sin^2(\theta_j/2)}{(\Delta y)^2}\right]\tilde{\phi}_1^{n+1}$$
$$= \left(\frac{1-\cos\theta_j + i\sin\theta_j}{\Delta x}\right)\tilde{\phi}^n \tag{169}$$

This has the form

$$\tilde{\phi}^{n+1} = \left(\frac{a+ib}{c+ib}\right)\tilde{\phi}^n \equiv E\tilde{\phi}^n$$

where $c \geqslant a \geqslant 0$. Thus, the amplification factor E always has modulus less than or equal to 1, ensuring unconditional stability.

Although the differencing given by Eq. (167) is unconditionally stable, reflection upon Eq. (168) shows that at each time step one must essentially solve a difference equation as complicated as a Laplacian. Because ϕ^{n+1} should not differ much from ϕ^n, an iterative solution process between each time step will likely converge rather quickly, and even one multigrid relaxation sequence may give a close enough result between time steps. However, an efficient solution approach is provided by the method of approximate factorization (AF). In this approach one attempts to approximately factor the left-hand-side operator so that it is easily inverted while accuracy and stability are maintained.

An approximately factored form of Eq. (168) is given by

$$(I - \beta h\ \Delta_x)(\nabla_x - h\ \nabla_y\Delta_y)\phi^{n+1} = \nabla_x\phi^n + \beta h^2\ \Delta_x\nabla_y\Delta_y\phi^n \tag{170}$$

Here the factorization error term

$$\beta h^2\ \Delta_x\nabla_y\Delta_y\phi^{n+1} \simeq \beta(\Delta t)^2\phi_{xyy}$$

was reduced by subtracting the same error at the previous time level. The overall error contribution due to approximate factorization is thus of second-order accuracy and is formally more accurate than the time differencing. Note that the factorization error of Eq. (170) drops out in a steady state when $\phi^{n+1} \to \phi^n$, even for large Δt.

The differenced equations given by Eq. (170) are very easily solved. This is most readily evident by placing Eq. (170) in a two-step algorithm form:

$$(I - \beta h\,\Delta_x)\phi^* = \nabla_x \phi^n + \beta h^2\,\Delta_x \nabla_y \Delta_y \phi^n \tag{171a}$$

$$(\nabla_x - h\,\nabla_y \Delta_y)\phi^{n+1} = \phi^* \tag{171b}$$

The right-hand side of Eq. (171a) is readily formed from known data at level n. The left-hand-side inversion of Eq. (171a) simply requires the solution of upper bidiagonal matrices—one such inversion is required for each point in y. Note that this is a strictly diagonally dominant matrix; hence, the inversion process is exceptionally well behaved. After the upper bidiagonal equations are solved along each $y = \text{const}$ grid line, Eq. (171b) is solved along each $x = \text{const}$ grid line. The left-hand side of Eq. (171b) is also easily solved as it forms a block lower triangular matrix. (There is here an implied reordering of $\phi_{j,k}$ corresponding to solving along $x = \text{const}$ grid lines rather than $y = \text{const}$ grid lines.) Each diagonal block in this case is easily inverted by solving a tridiagonal matrix. This solution process is also well behaved because the tridiagonal matrix is strictly diagonally dominant.

In solving Eq. (170), one can experience problems with boundary conditions which will be somewhat alleviated by using the delta form of the equation. The boundary condition problem consists of the following: In inverting the first factor of Eq. (170) [i.e., Eq. (171a)], it needs as a right-hand-side boundary condition at $j = j_{\max}$, not ϕ^{n+1}, but $(\nabla_x - h\,\nabla_y \Delta_y)\phi^{n+1}$. That is, Δ_x operates on is the ϕ^* of Eq. (171a) and this is the boundary quantity it seeks. If only ϕ^{n+1} is given as a boundary condition, the correct difference equation will not be produced near the boundary and accuracy will be lost. However, supplying ϕ^* on the right-hand-side boundary is not trivial. It is a simple matter to form the $h\,\nabla_y \Delta_y \phi^{n+1}$ part, but the $\nabla_x \phi^{n+1}$ is not easily supplied because its formation requires the unknown $\phi^{n+1}_{j\max - 1}$. What is normally done is to supply $\phi^*_{j\max}$ approximately using some level n data as

$$\phi^*_{j\max} = \nabla_x \phi^n_{j\max} - h\,\nabla_y \Delta_y \phi^{n+1}_{j\max}$$

This often works sufficiently well, but the overall scheme is no longer unconditionally stable.

The *AF* scheme given by Eq. (170) is more readily developed by using the delta or consistency form, although it does not highlight the ideas as well. To maintain the delta term $(\phi^{n+1} - \phi^n)$, Eq. (168) is arranged as

$$[\nabla_x - h(\beta\,\nabla_x \Delta_x + \nabla_y \Delta_y)](\phi^{n+1} - \phi^n) = h(\beta\,\nabla_x \Delta_x + \nabla_y \Delta_y)\phi^n \tag{172}$$

and approximately factored as

$$(I - h\beta\,\Delta_x)(\nabla_x - h\,\nabla_y \Delta_y)(\phi^{n+1} - \phi^n) = RHS(172)$$

Because the factorization is on the delta ϕ quantity, $\phi^{n+1} - \phi^n$, the approximate factorization error term is automatically subtracted out. Thus, the scheme is automatically consistent. The delta form also simplifies the boundary condition problem somewhat. In inverting the first factor one can often assume $\phi^{n+1} \to \phi^n$ at and near the $j = J_{\max}$ boundary. If we do so, only a first-order error in time is created and only at the back boundary.

A second-order time differencing is obtained from Eq. (167) by replacing $\nabla_t \nabla_x \phi^{n+1}$ with

$$\phi_{xt} = \nabla_t \nabla_x \left[\frac{3 - E_t^{-1}(2 - E_x)}{2} \right]$$

This substitution more or less corresponds to using three-point backward implicit time differencing. Alternately, trapezoidal implicit time differencing can be used and is given by

$$\nabla_t \delta_x^b \phi^{n+1} = (\beta\, \nabla_x \Delta_x + \nabla_y \Delta_y) \frac{(\phi^n + \phi^{n+1})}{2} \tag{173}$$

where δ_x^b can be either the two-point first-order-accurate operator in x or the three-point second-order-accurate operator. This scheme can be put into delta form (with $\bar{h} = \Delta t/2$)

$$[\delta_x^b - \bar{h}(\beta\, \nabla_x \Delta_x + \nabla_y \Delta_y)](\phi^{n+1} - \phi^n) = \Delta t(\beta\, \nabla_x \Delta_x + \nabla_y \Delta_y)\phi^n$$

and is approximately factored as

$$[\delta_x^b - \bar{h}\beta\, \nabla_x \Delta_x](\delta_x^b)^{-1}[\delta_x^b - \bar{h}\nabla_y \Delta_y](\phi^{n+1} - \phi^n) = \Delta t(\beta\, \nabla_x \Delta_x + \nabla_y \Delta_y)\phi^n \tag{174}$$

The *AF* scheme, like the unfactored form, is unconditionally stable. The solution algorithm is

$$[\delta_x^b - \bar{h}\beta\, \nabla_x \Delta_x]\, \Delta\phi^* = \Delta t(\beta\, \nabla_x \Delta_x + \nabla_y \Delta_y)\phi^n \tag{175a}$$

$$[\delta_x^b - \bar{h}\, \nabla_y \Delta_y]\, \Delta\phi^n = \delta_x^b\, \Delta\phi^* \tag{175b}$$

$$\phi^{n+1} = \Delta\phi^n + \phi^n \tag{175c}$$

The same sequence can be represented in the equivalent Douglas-Gunn[37] form

$$[\delta_x^b - \bar{h}\beta\, \nabla_x \Delta_x]\tilde{\phi}^{n+1} = (\delta_x^b + \bar{h}\beta\, \nabla_x \Delta_x + \Delta t\, \nabla_y \Delta_y)\phi^n$$

$$[\delta_x^b - \bar{h}\, \nabla_y \Delta_y]\phi^{n+1} = \delta_x^b \tilde{\phi}^{n+1} - \bar{h}\, \nabla_y \Delta_y \phi^n$$

Transonic Flow Algorithms

The preceding algorithms are readily applied to the transonic small-disturbance equation by lagging the nonlinear coefficients at the previous time level and using conventional transonic upwind differencing[38] when the coefficient to ϕ_{xx} becomes negative. Simply lagging nonlinear coefficients at the previous time level will result in first-order time accuracy, but this can be readily overcome in a number of ways, including iterative correction or various forms of extrapolation from previous time levels such as time averaging or Taylor expansion. In steady transonic potential flow, various switches have been developed to transition the x difference operators from central to upwind and back, and these can apply to the unsteady flow equations as well.

Consider, for example, changes to the trapezoidal rule time differencing to account for nonlinearity and upwinding of the term

$$\partial_x[(1 - M_\infty^2)\phi_x - \frac{\gamma + 1}{2} M_\infty^2(\phi_x)^2] \equiv \partial_x f$$

which is to be averaged between time levels n and $n + 1$. Let δ_x be the midpoint operator and difference as

$$\bar{\delta}_x f = \bar{\delta}_x[(1 - M_\infty^2)\bar{\delta}_x\phi - \frac{\gamma + 1}{2} M_\infty^2(\bar{\delta}_x\phi)^2] \tag{176}$$

Averaging in time between n and $n + 1$,

$$\bar{\delta}_x \frac{(f^{n+1} + f^n)}{2}$$
$$= \bar{\delta}_x\left[(1 - M_\infty^2)\bar{\delta}_x\left(\frac{\phi^{n+1} + \phi^n}{2}\right) - \frac{\gamma + 1}{2} M_\infty^2 \frac{(\bar{\delta}_x\phi^{n+1})^2 + (\bar{\delta}_x\phi^n)^2}{2}\right]$$

or

$$\bar{\delta}_x \frac{(f^{n+1} + f^n)}{2} = \bar{\delta}_x[(1 - M_\infty^2)\bar{\delta}_x - (\gamma + 1)M_\infty^2(\bar{\delta}_x\phi^n)\bar{\delta}_x] \frac{\Delta\phi^n}{2}$$
$$+ \bar{\delta}_x[(1 - M_\infty^2)\bar{\delta}_x\phi^n - \frac{\gamma + 1}{2} M_\infty^2(\bar{\delta}_x\phi^n)^2]$$

where the Taylor series is used to expand

$$(\bar{\delta}_x\phi^{n+1})^2 = (\bar{\delta}_x\phi^n)^2 + 2(\bar{\delta}_x\phi^n)\bar{\delta}_x \,\Delta\phi^n$$

Because $\bar{\delta}_x\bar{\delta}_x = \nabla_x\Delta_x$ is the second-order-accurate central three-point second differencing operator, this is a proper differencing for subsonic flow, that is, when the coefficient $[(1 - M_\infty^2) - (\gamma + 1)M_\infty^2(\phi_x)] > 0$. When this coefficient is < 0, backward or upwind space differencing should be used. The simplest such upwind operator is given by $\bar{\delta}_x E^- \bar{\delta}_x = \Delta_x\Delta_x$, which is the first-order-accurate three-point backward difference operator approximation to a second derivative. In switching from one operator to the other, it is important that conservation of the difference operators be maintained, and this can be achieved with the formula

$$\bar{\delta}_x \frac{(f^{n+1}+f^n)}{2}$$
$$= \hat{\delta}_x\left[(1-M_\infty^2)\bar{\delta}_x \frac{\phi^{n+1}+\phi^n}{2} - \frac{\gamma+1}{2} M_\infty^2 \frac{(\bar{\delta}_x\phi^{n+1})^2+(\bar{\delta}_x\phi^n)^2}{2}\right]$$

where

$$\hat{\delta}_x = \bar{\delta}_x(1-\epsilon+\epsilon E_x^-)$$

and the switch $\epsilon = 0$ or 1 as $[(1-M_\infty^2)-(\gamma+1)M_\infty^2(\phi_x)] > 0$ or < 0.

Implementing all of this for the transonic low-frequency small-disturbance equation gives the algorithm

$$[2\kappa M_\infty^2 \delta_x^b - \bar{h}\hat{\delta}_x(1-M_\infty^2-(\gamma+1)\bar{\delta}_x\phi^n)]\,\Delta\phi^*$$
$$= \Delta t\left\{\hat{\delta}_x\left[\left(1-M_\infty^2-\frac{\gamma+1}{2}\bar{\delta}_x\phi^n\right)+\Delta_y\nabla_y\right]\right\}\phi^n \tag{177a}$$

$$[2\kappa M_\infty^2\delta_x^b - \bar{h}\,\nabla_y\Delta_y]\,\Delta\phi^n = 2\kappa M_\infty^2\delta_x^b\,\Delta\phi^* \tag{177b}$$

$$\phi^{n+1} = \Delta\phi^n + \phi^n \tag{177c}$$

Numerical Solution of the Transonic Potential Equation

The nonconservative transonic potential equation as given by Eq. (164) is in a relatively familiar form, and it can be numerically solved using schemes developed for the second-order wave equation. However, for transonic flow the conservative form of the equation, Eq. (160), is needed to give the correct shock jump relation of the equation.

To know how to difference Eq. (160) as an unsteady potential equation like Eq. (164), ρ, $\rho\Phi_x$, etc., are expanded about a nearby state within the operators. Let

$$\rho = \bar{\rho} + \bar{\rho}_\Phi(\Phi-\bar{\Phi})$$
$$\rho\Phi_x = \bar{\rho}\bar{\Phi}_x + \bar{\Phi}_x\bar{\rho}_\Phi(\Phi-\bar{\Phi}) + \bar{\rho}\,\partial_x(\Phi-\bar{\Phi})$$

and so on where the overbar indicates a nearby state. Thus, Eq. (1) is approximated as

$$[\partial_t\bar{\rho}_\Phi + \partial_x(\bar{\Phi}_x\bar{\rho}_\Phi+\bar{\rho}\,\partial_x) + \partial_y(\bar{\Phi}_y\bar{\rho}_\Phi+\bar{\rho}\partial_y) + \partial_x(\bar{\Phi}_z\bar{\rho}_\Phi+\bar{\rho}\,\partial_z)](\Phi-\bar{\Phi})$$
$$= -[\partial_t\bar{\rho} + \partial_x\bar{\rho}\bar{\Phi}_x + \partial_y\bar{\rho}\bar{\Phi}_y + \partial_z\bar{\rho}\bar{\Phi}_z] \tag{178}$$

with conservation form maintained because the expansions are kept within the operators. Equation (178) can be rearranged as

$$\begin{aligned}
&\left[\partial_t \frac{\bar{\rho}}{\bar{c}^2}\partial_t + \partial_t \frac{\bar{\rho}}{\bar{c}^2}\bar{\Phi}_x \partial_x + \partial_t \frac{\bar{\rho}}{\bar{c}^2}\bar{\Phi}_y \partial_y + \partial_t \frac{\bar{\rho}}{\bar{c}^2}\bar{\Phi}_z \partial_z\right. \\
&\quad + \partial_x \frac{\bar{\rho}}{\bar{c}^2}\bar{\Phi}_x \partial_t + \partial_y \frac{\bar{\rho}}{\bar{c}^2}\bar{\Phi}_y \partial_t + \partial_z \frac{\bar{\rho}}{\bar{c}^2}\bar{\Phi}_z \partial_t \\
&\quad + \partial_x\bar{\rho}\left(\frac{\bar{\Phi}_x^2}{\bar{c}^2} - 1\right)\partial_x + \partial_y\bar{\rho}\left(\frac{\bar{\Phi}_y^2}{\bar{c}^2} - 1\right)\partial_y + \partial_z\bar{\rho}\left(\frac{\bar{\Phi}_z^2}{\bar{c}^2} - 1\right)\partial_z \\
&\quad + \partial_x\bar{\rho}\frac{\bar{\Phi}_x\bar{\Phi}_y}{\bar{c}^2}\partial_y + \partial_x\bar{\rho}\frac{\bar{\Phi}_x\bar{\Phi}_z}{\bar{c}^2}\partial_z + \partial_y\bar{\rho}\frac{\bar{\Phi}_x\bar{\Phi}_y}{\bar{c}^2}\partial_x + \partial_y\bar{\rho}\frac{\bar{\Phi}_y\bar{\Phi}_z}{\bar{c}^2}\partial_z \\
&\quad \left. + \partial_z\bar{\rho}\frac{\bar{\Phi}_x\bar{\Phi}_z}{\bar{c}^2}\partial_x + \partial_z\bar{\rho}\frac{\bar{\Phi}_y\bar{\Phi}_z}{\bar{c}^2}\partial_y\right](\Phi - \bar{\Phi}) \\
&\quad = \partial_t\bar{\rho} + \partial_x\bar{\rho}\bar{\Phi}_x + \partial_y\bar{\rho}\bar{\Phi}_y + \partial_z\bar{\rho}\bar{\Phi}_z
\end{aligned} \tag{179}$$

In principle this scheme should be solved for $\Delta\Phi$, but this is not commonly done because the differencing of the left-hand-side terms will not fully mimic the differencing of the right-hand-side terms. Consequently, current algorithms seem to back away from solving Eq. (178) and solve less accurate forms, but forms that are more familiar because of steady flow developments. For example, if only first-order time accuracy is needed, then the terms such as $\partial_x(\bar{\Phi}_x\bar{\rho}_\Phi)$, etc, can be ignored, so that Eq. (178) is replaced by

$$\begin{aligned}
&\left[\partial_t \frac{\bar{\rho}}{\bar{c}^2}\partial_t + \partial_t \frac{\bar{\rho}}{\bar{c}^2}\bar{\Phi}_x \partial_x + \partial_t \frac{\bar{\rho}}{\bar{c}^2}\bar{\Phi}_y \partial_y + \partial_t \frac{\bar{\rho}}{\bar{c}^2}\bar{\Phi}_z \partial_z - \partial_x\bar{\rho}\,\partial_x + \partial_y\bar{\rho}\,\partial_y + \partial_z\bar{\rho}\,\partial_z\right] \\
&\qquad \times (\Phi - \bar{\Phi}) = [\partial_t\bar{\rho} + \partial_x\bar{\rho}\bar{\Phi}_x + \partial_y\bar{\rho}\,\bar{\Phi}_y + \partial_z\bar{\rho}\bar{\Phi}_z]
\end{aligned} \tag{180}$$

In this case upwinding of the spacial derivative terms can be accomplished using density biasing,[39,40] which was developed for steady-state problems.

Letting $\bar{\Phi}$ be Φ of the previous time level, Eq. (180) can be implicitly differenced in delta form to first-order in time as

$$\begin{aligned}
&\left[\left(\frac{\rho}{c^2}\right)^n(\delta_t + \Phi_x^n\delta_x + \Phi_y^n\delta_y + \Phi_z^n\delta_z) - \Delta t(\bar{\delta}_x\rho^n\bar{\delta}_x + \delta_y\rho^n\bar{\delta}_y + \bar{\delta}_z\rho^n\bar{\delta}_z)\right] \\
&\times (\Phi^{n+1} - \Phi^n) = \left(\frac{\rho}{c^2}\right)^{n-1}(\delta_t + \Phi_x^{n-1}\delta_x + \Phi_y^{n-1}\delta_y + \Phi_z^{n-1}\delta_z)(\Phi^n - \Phi^{n-1}) \\
&\qquad + (\rho^n - \rho^{n-1}) + \Delta t(\bar{\delta}_x\rho^n\bar{\delta}_x + \bar{\delta}_y\rho^n\bar{\delta}_y + \bar{\delta}_z\rho^n\bar{\delta}_z)\Phi^n
\end{aligned}$$

Differencing the remaining δt operator and clearing $(\rho/c^2)^n$ gives

$$\begin{aligned}
&\left[I + \Delta t(\Phi_x^n\delta_x + \Phi_y^n\delta_y + \Phi_z^n\delta_z) - \Delta t^2\left(\frac{c^2}{\rho}\right)^n(\bar{\delta}_x\rho^n\bar{\delta}_x + \bar{\delta}_y\rho^n\bar{\delta}_y + \bar{\delta}_z\rho^n\bar{\delta}_z)\right] \\
&\qquad\qquad \times (\Phi^{n+1} - \Phi^n) \\
&= \Delta t\left(\frac{c^2}{\rho}\right)^n\left[\left(\frac{\rho}{c^2}\right)^{n-1}(\delta_t + \Phi_x^{n-1}\delta_x + \Phi_y^{n-1}\delta_y + \Phi_z^{n-1}\delta_z)(\Phi^n - \Phi^{n-1})\right. \\
&\quad \left. + (\rho^n - \rho^{n-1}) + \Delta t(\bar{\delta}_x\rho^n\bar{\delta}_x + \bar{\delta}_y\rho^n\bar{\delta}_y + \bar{\delta}_z\rho^n\bar{\delta}_z)\Phi^n\right] + (\Phi^n - \Phi^{n-1}) \qquad (181)
\end{aligned}$$

Finally, the left-hand side can be approximately factored to give

$$\begin{aligned}
&\left[I + \Delta t\Phi_x^n\delta_x - \Delta t^2\left(\frac{c^2}{\rho}\right)^n\bar{\delta}_x\rho^n\bar{\delta}_x\right]\left[I + \Delta t\Phi_y^n\delta_y - \Delta t^2\left(\frac{c^2}{\rho}\right)^n\bar{\delta}_y\rho^n\bar{\delta}_y\right] \\
&\quad \times\left[I + \Delta t\Phi_z^n\delta_z - \Delta t^2\left(\frac{c^2}{\rho}\right)^n\bar{\delta}_z\rho^n\bar{\delta}_z\right](\Phi^{n+1} - \Phi^n) = RHS(181) \qquad (182)
\end{aligned}$$

Iteration can be used to bring this algorithm to higher-order accuracy, and various strategies are possible.

Circulation Treatment

For lifting cases allowance must be made for a jump of potential across a wakelike cut. In unsteady flow, vorticity is continuously shed from the wing. In a potential formulation, this phenomenon can be approximately modeled by convecting a jump in potential measured at the wing trailing edge along a cut aligned with the shear layer. A further idealization is generally employed by keeping the cut in the wing mean chord plane.

The jump in potential along the cut, $\Phi_u - \Phi_l = \Gamma$, will vary with time. By imposing the usual shear layer assumptions, an equation for $\Gamma(x,y,z,t)$ along the cut can be derived. Across an inviscid shear layer, normal velocity and pressure are continuous, but tangential velocity and density can be discontinuous. However, in flow taken as isentropic throughout, density, like pressure, is continuous across the cut.

The condition that the velocity normal to the shear layer is continuous gives

$$(\Phi_z)_u = (\Phi_z)_l$$

where the cut is taken in the $z = \text{const}$ plane (y is the span direction). The Bernoulli relation together with the continuity of density further requires

$$(2\Phi_t + \Phi_x^2 + \Phi_y^2)_u = (2\Phi_t + \Phi_x^2 + \Phi_y^2)_l$$

Using $\Phi_u = \Phi_l + \Gamma$ gives a relation for convection of Γ along the cut[41]:

$$2\Gamma_t + 2(\Phi_x)\Gamma_x + 2(\Phi_y)\Gamma_y + \Gamma_x^2 + \Gamma_y^2 = 0 \tag{183}$$

Consistent with the thin airfoil relation, this reduces to

$$\Gamma_t + M_\infty \Gamma_x = 0 \tag{184}$$

and for low-frequency flow

$$\Gamma_x = 0 \tag{185}$$

In this case the jump in potential measured at each span station at the wing trailing edge should immediately be convected downstream.

Because the potential can jump across the cut, the differencing expressions must be altered in the vicinity of the cut. This is not particularly difficult, but it requires bookkeeping and special programming which does become tiresome. For example, in differencing Φ_z at point l just above the cut (so the point $l-1$ is just below the cut), the central difference approximation is altered as

$$\delta_z \Phi_l = \frac{(\Phi_{l+1}) - (\Phi_{l-1} + \Gamma)}{2\Delta z}$$

Thus, $-(\Gamma/2\Delta z)$ is simply added to the first difference approximation for points differenced just above the cut, and in fact, the same correction is needed for points just below the cut. Similar corrections are needed for second difference expressions, although the correction adds or subtracts, depending on which side of the cut the difference is taken. Additional corrections should also be made for second derivatives, because, although Φ_z is continuous across the cut, in unsteady or three-dimensional flow, Φ_{zz} is not continuous. From continuity of density and the conservation of mass equation applied across the cut, one can find that

$$\rho[[\Phi_{zz}]] = -\partial(\rho\Gamma_x) - \partial(\rho\Gamma_y)$$

6. Equations in Transformed Coordinates

The previous algorithms were all developed in Cartesian variables, and it was assumed that the grid increments Δx, Δy, and Δz were uniform. The easiest way to extend the previous finite-difference procedures to treat complex boundary surfaces and nonuniform grids is to use curvilinear coordinates and to transform and solve the equations in uniform computational space (Fig. 5). In this way, virtually all of the previously described numerical procedures can then be effectively applied to the governing equations.

Coordinate transformations are used to simplify the application of boundary conditions, maintain well orderness, and to adapt to flowfield

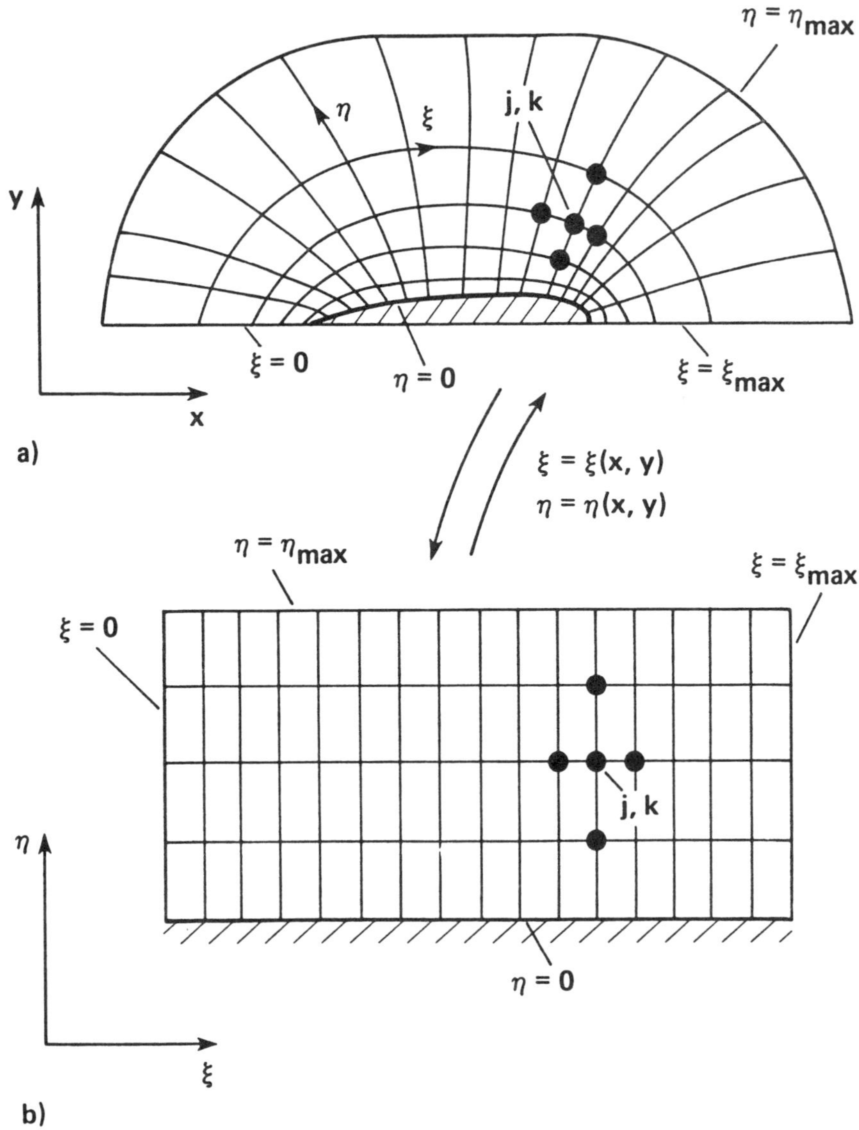

Fig. 5 Mapping from physical space to computational space.

action regions. The use of body-fitted coordinates can also allow use of simplified governing equations, the boundary-layer equations, and the thin-layer Navier-Stokes equations being examples.

This section indicates how the governing equations are altered by coordinate transformations in order to apply the numerical algorithms on curvilinear coordinates. Although not developed here, it is noted that the finite-volume method is closely related to the finite-difference method if the transformed governing equations are in divergence or conservative form. Indeed, the schemes can be made identical.

Coordinate Transformations

Consider the introduction of new independent variables, ξ, η, and τ:

$$\xi = \xi(x,y,t) \tag{186a}$$

$$\eta = \eta(x,y,t) \tag{186b}$$

$$\tau = t \tag{186c}$$

where to simplify the discussion the ideas are developed in only two dimensions. Given a curvilinear grid of the type shown in Fig. 5, relations between derivatives with respect to x,y,t and derivatives with respect to ξ, η, and τ are readily found using the chain rule. Find from the chain rule expansion

$$\partial_\tau = \partial_t + x_\tau\,\partial_x + y_\tau\,\partial_y \tag{187a}$$

$$\partial_\xi = x_\xi\,\partial_x + y_\xi\,\partial_y \tag{187b}$$

$$\partial_\eta = x_\eta\,\partial_x + y_n\,\partial_y \tag{187c}$$

which can be written in matrix form:

$$\begin{bmatrix} \partial_\tau \\ \partial_\xi \\ \partial_\eta \end{bmatrix} = \begin{bmatrix} 1 & x_\tau & y_\tau \\ 0 & x_\xi & y_\xi \\ 0 & x_\eta & y_\eta \end{bmatrix} \begin{bmatrix} \partial_t \\ \partial_x \\ \partial_y \end{bmatrix} \tag{188}$$

Likewise derivatives of t, x, and y can be expanded in terms of τ, ξ, and η to obtain

$$\begin{bmatrix} \partial_t \\ \partial_x \\ \partial_y \end{bmatrix} = \begin{bmatrix} 1 & \xi_t & \eta_t \\ 0 & \xi_x & \eta_x \\ 0 & \xi_y & \eta_y \end{bmatrix} \begin{bmatrix} \partial_\tau \\ \partial_\xi \\ \partial_\eta \end{bmatrix} \tag{189}$$

Solving Eq. (188) for the Cartesian derivatives in terms of the curvilinear derivatives by inverting the coefficient matrix yields

$$\begin{bmatrix} \partial_t \\ \partial_x \\ \partial_y \end{bmatrix} = J \begin{bmatrix} (x_\xi y_\eta - y_\xi x_\eta) & (-x_\tau y_\eta + y_\tau x_\eta) & (x_\tau y_\xi - y_\tau x_\xi) \\ 0 & y_\eta & -y_\xi \\ 0 & -x_\eta & x_\xi \end{bmatrix} \begin{bmatrix} \partial_\tau \\ \partial_\xi \\ \partial_\eta \end{bmatrix} \tag{190}$$

where $J^{-1} = (x_\xi y_\eta - x_\eta y_\xi)$.

Comparison of the elements of Eq. (189) to those of Eq. (190) gives the metric relations

$$\begin{aligned} &\xi_x = Jy_\eta, \qquad &&\xi_y = -Jx_\eta, \qquad &&\xi_t = -x_\tau \xi_x - y_\tau \xi_y \\ &\eta_x = -Jy_\xi, \qquad &&\eta_y = Jx_\xi, \qquad &&\eta_t = -x_\tau \eta_x - y_\tau \eta_y \end{aligned} \tag{191a}$$

where J is defined as the metric Jacobian:

$$J = (\xi_x \eta_y - \eta_x \xi_y) = (x_\xi y_\eta - y_\xi x_\eta)^{-1} \tag{191b}$$

Given a curvilinear grid with $\xi = \text{const}$ and $\eta = \text{const}$ mesh lines, relations (191) show that all of the metric derivatives $\xi_x, \eta_x, \ldots$ needed for the transformed equations can be easily evaluated numerically by differencing values of x and y on the uniform computational space. That is, no functional mapping is needed, only a curvilinear grid from which all needed derivative relations can be obtained numerically.

The conservative form of the equations

$$\partial_t Q + \partial_x F + \partial_y G = 0 \tag{192}$$

can be put into the chain rule form

$$\partial_\tau Q + \xi_t \, \partial_\xi Q + \eta_t \, \partial_n Q + \xi_x \, \partial_\xi F + \eta_x \, \partial_\eta F + \xi_y \, \partial_\xi G + \eta_y \, \partial_\eta G = 0$$

where Q, F, and G can represent Euler flux terms, inviscid and viscous flux terms, or nonlinear potential flux terms.

By scaling the equations by $J^{-1} = \xi_x \eta_y - \xi_y \eta_x$ and using the identities [verified using Eqs. (191)],

$$\partial_\tau (J^{-1}) + \partial_\xi (\xi_t / J) + \partial_\eta (\eta_t / J) = 0 \tag{193a}$$

$$\partial_\xi (\xi_x / J) + \partial_\eta (\eta_x / J) = 0 \tag{193b}$$

$$\partial_\xi (\xi_y / J) + \partial_\eta (\eta_y / J) = 0 \tag{193c}$$

the chain rule form of Eq. (192) is brought into strong conservation law form:

$$\partial_\tau \hat{Q} + \partial_\xi \hat{F} + \partial_\eta \hat{G} = 0 \tag{194}$$

where

$$\begin{aligned} \hat{Q} &= Q/J \\ \hat{F} &= (\xi_t Q + \xi_x F + \xi_y G)/J \\ \hat{G} &= (\eta_t Q + \eta_x F + \eta_y G)/J \end{aligned}$$

Although the transformed equations appear to be considerably more complex than their untransformed Cartesian counterparts, they are not. In particular, if unscaled contravariant velocity components are used as defined by

$$U = \xi_t + \xi_x u + \xi_y v, \qquad V = \eta_t + \eta_x u + \eta_y v \tag{195a}$$

or (with $\dot{x} = x_\tau$ and $\dot{y} = y_\tau$)

$$U = \xi_x(u - \dot{x}) + \xi_y(v - \dot{y}), \qquad V = \eta_x(u - \dot{x}) + \eta_y(v - \dot{y}) \tag{195b}$$

then the inviscid Euler flux terms can be written to resemble their Cartesian counterparts with

$$\hat{Q} = J^{-1}\begin{bmatrix} \rho \\ \rho u \\ \rho v \\ e \end{bmatrix}, \qquad \hat{F} = J^{-1}\begin{bmatrix} \rho U \\ \rho u U + \xi_x p \\ \rho v U + \xi_y p \\ U(e + p) - \xi_t p \end{bmatrix},$$

$$\hat{G} = J^{-1}\begin{bmatrix} \rho V \\ \rho u V + \eta_x p \\ \rho u V + \eta_y p \\ V(e + p) - \eta_t p \end{bmatrix}$$

Note that here only the independent variables are transformed; the momentum equations thus remain Cartesian momentum equations.

The viscous flux terms $\hat{E}_v = J^{-1}(\xi_x E_v + \xi_y F_v)$ and $\hat{F}_v = J^{-1}(\eta_x E_v + \eta_y F_v)$ are more complex in curvilinear coordinates, but they can often be simplified for high-Reynolds-number flow. The stress terms are also transformed in terms of the ξ and η derivatives as

$$\tau_{xx} = \mu[4(\xi_x u_\xi + \eta_x u_\eta) - 2(\xi_y v_\xi + \eta_y v_\eta)]/3$$
$$\tau_{xy} = \mu(\xi_y u_\xi + \eta_y u_\eta + \xi_x v_\xi + \eta_x v_\eta)$$
$$\tau_{yy} = \mu[-2(\xi_x u_\xi + \eta_x u_\eta) + 4(\xi_y v_\xi + \eta_y v_\eta)]/3$$

Boundary conditions are readily imposed in body-conforming curvilinear coordinates. Along a solid boundary surface that conforms with $\eta = \text{const}$, the condition of no flow through (or tangency) for a fixed boundary is

$$\boldsymbol{q} \cdot \nabla\eta = 0 \tag{196}$$

where $\boldsymbol{q}$ is the velocity vector. If the boundary is moving with velocity $\boldsymbol{q}_s$,

$$(\boldsymbol{q} - \boldsymbol{q}_s) \cdot \nabla\eta = 0 \tag{197}$$

In either case, if $\eta = \text{const}$ always conforms to the boundary, the tangency condition is expressed as

$$V = 0 \tag{198}$$

For viscous flow the no-slip condition is expressed as

$$V = 0 \tag{199a}$$

$$U = 0 \tag{199b}$$

and this again applies even if the body surface is moving provided that this surface corresponds to an $\eta = \text{const}$.

Differencing Transformed Equations

The transformed Eq. (194) can be differenced directly on a computational plane which for convenience is chosen to have uniform spacing ($\Delta\xi$ and $\Delta\eta$ are generally set to 1 and ξ and η are thus the grid point index numbers j and k). Because the computational space is uniform, simple difference formula can be used. However, in deriving Eq. (194), use was made of the metric identities (193) to bring the metrics inside the operators; thus, difference approximations to Eq. (194) can accumulate truncation error due to differencing the flow variables ρ, ρu, ρv, and e as well as the metrics $\xi_x, \eta_x, \ldots$ Especially in the far field of external aerodynamic problems, this metric differencing error can be appreciable. In these cases the outer boundary is often taken far away to where the flow variables approach constant freestream values. For constant-flow variables, Eq. (194) degenerates to the relations (193), and with the grid spacing usually stretched out to reach the far-field boundary, the metric truncation error can be very large. This problem, however, can be avoided if the metrics are differenced consistently with the difference operators of Eq. (194) such that the relations (193) are numerically satisfied. For example, if ∂_ξ and ∂_η are central differenced, and if the metrics are evaluated with central differencing, then Eqs. (193) are numerically satisfied. For example, Eq. (193b) using identities (191a) is rewritten as

$$\partial_\xi y_\eta - \partial_\eta y_\xi = 0$$

and is numerically satisfied as

$$\delta_\xi \delta_\eta y - \delta_\eta \delta_\xi y = 0$$

But if ∂_ξ in Eq. (194) is differenced using three-point backward differencing, δ_ξ^b, then Eq. (193b) is satisfied to second-order accuracy:

$$\delta_\xi^b \delta_\eta y - \delta_\eta \delta_\xi y = O(\Delta^2)$$

In the far field with large grid spacing this error may be appreciable.

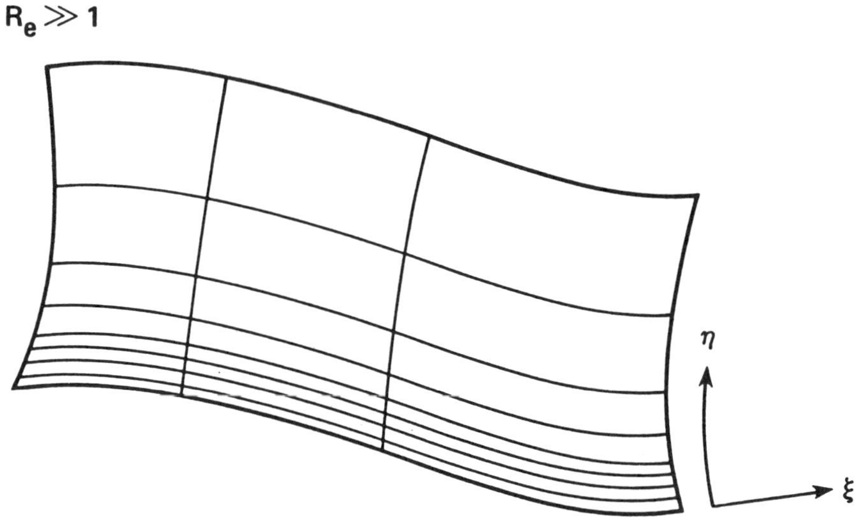

Fig. 6 Sketch suggestive of a grid clustering used for resolving a thin viscous layer near a body surface.

In three dimensions even the use of all central differencing to evaluate metrics and derivatives of Eq. (194) is not sufficient to numerically maintain the metric identities that are now more complex than Eqs. (193), and a simple averaging formula must be used in differencing the metrics.[42] The finite-volume method for equations in divergence form tends to avoid the metric truncation error problem and is akin to a difference method using midpoint operators and averaging.

Thin-Layer Navier-Stokes Approximation[43,44]

In high-Reynolds-number viscous flows the effects of viscosity are concentrated near rigid boundaries and in wake regions. Typically in computations we only have enough grid points available (due to computer limits) to concentrate grid lines near the rigid surfaces. The resulting grid systems usually have fine grid spacing in directions nearly normal to the surfaces and coarse grid spacing along the surface (see Fig. 6).

Even though the full Navier-Stokes equations may be programmed, the viscous terms associated with derivatives along the body will not be resolved in high-Reynolds-number flow. In boundary-layer theory, appropriate scaling arguments show that streamwise components of the viscous terms can be neglected relative to the normal terms. In the thin-layer approximation, we rely on similar arguments and drop these same terms. Therefore, like boundary-layer theory, the thin-layer approximation requires that the Reynolds number be high and that the coordinate surfaces be body conforming. Unlike boundary-layer theory, the normal momentum equation is solved; hence, pressure can vary through the boundary layer, and of course, there are no matching conditions to be satisfied.

Applying the thin-layer approximation to Eqs. (194), where all the viscous terms associated with ξ derivatives are neglected, we obtain

$$\partial_\tau \hat{Q} + \partial_\xi \hat{E} + \partial_\eta \hat{F} = Re^{-1}\, \partial_\eta \hat{S} \tag{200}$$

where

$$\hat{S} = J^{-1} \begin{bmatrix} 0 \\ \eta_x m_1 + \eta_y m_2 \\ \eta_x m_2 + \eta_y m_3 \\ \eta_x(um_1 + vm_2 + m_4) + \eta_y(um_2 + vm_3 + m_5) \end{bmatrix}$$

with

$$\begin{aligned} m_1 &= \mu(4\eta_x u_\eta - 2\eta_y v_\eta)/3 \\ m_2 &= \mu(\eta_y u_\eta + \eta_x v_\eta) \\ m_3 &= \mu(-2\eta_x u_\eta + 4\eta_y v_\eta)/3 \\ m_4 &= \mu Pr^{-1}(\gamma - 1)^{-1}\eta_x\, \partial_\eta(c^2) \\ m_5 &= \mu Pr^{-1}(\gamma - 1)^{-1}\eta_y\, \partial_\eta(c^2) \end{aligned}$$

It must be remarked that the thin-layer equations just given are not rigorously derived. A correct high-Reynolds-number perturbation analysis would give a somewhat different viscous term whose differences would be measurable only for surfaces of small radius of curvature.[45,46]

Three-Dimensional Thin Layer Equations

The full three-dimensional Navier-Stokes equations in strong conservation law form are reduced to the thin-layer form under the same restrictions and assumptions as in two dimensions. The equations in generalized curvilinear coordinates are (with the body conforming with $\zeta = \text{const}$)

$$\partial_\tau \hat{Q} + \partial_\xi \hat{E} + \partial_\eta \hat{F} + \partial_\zeta \hat{G} = Re^{-1}\, \partial_\zeta \hat{S}$$

where now

$$\hat{Q} = J^{-1} \begin{bmatrix} \rho \\ \rho u \\ \rho v \\ \rho w \\ e \end{bmatrix}, \qquad \hat{E} = J^{-1} \begin{bmatrix} \rho U \\ \rho u U + \xi_x p \\ \rho v U + \xi_y p \\ \rho w U + \xi_z p \\ U(e + p) - \xi_t p \end{bmatrix}$$

$$\hat{F} = J^{-1} \begin{bmatrix} \rho V \\ \rho u V + \eta_x p \\ \rho v V + \eta_y p \\ \rho w V + \eta_z p \\ V(e + p) - \eta_t p \end{bmatrix}, \qquad \hat{G} = J^{-1} \begin{bmatrix} \rho W \\ \rho u W + \zeta_x p \\ \rho v W + \zeta_y p \\ \rho w W + \zeta_z p \\ W(e + p) - \zeta_t p \end{bmatrix}$$

with

$$U = \xi_t + \xi_x u + \xi_y v + \xi_z w$$
$$V = \eta_t + \eta_x u + \eta_y v + \eta_z w$$
$$W = \zeta_t + \zeta_x u + \zeta_y v + \zeta_z w$$

with

$$\hat{S} = J^{-1} \begin{bmatrix} 0 \\ \mu m_1 u_\zeta + (\mu/3) m_2 \zeta_x \\ \mu m_1 v_\zeta + (\mu/3) m_2 \zeta_y \\ \mu m_1 w_\zeta + (\mu/3) m_2 \zeta_z \\ \mu m_1 m_3 + (\mu/3) m_2 (\zeta_x u + \zeta_y v + \zeta_z w) \end{bmatrix}$$

where

$$m_1 = \zeta_x^2 + \zeta_y^2 + \zeta_z^2, \qquad m_2 = \zeta_x u_\zeta + \zeta_y v_\zeta + \zeta_z w_\zeta$$

and

$$m_3 = (u^2 + v^2 + w^2)/2 + Pr^{-1}(\gamma - 1)^{-1}(c^2)_\zeta$$

Pressure is related to the conservative flow variables Q by the equation of state:

$$p = (\gamma - 1)[e - \tfrac{1}{2}\rho(u^2 + v^2 + w^2)]$$

The metric terms are defined as

$$\begin{aligned}
\xi_x &= J(y_\eta z_\zeta - y_\zeta z_\eta), & \eta_x &= J(z_\xi y_\zeta - y_\xi z_\zeta) \\
\xi_y &= J(z_\eta x_\zeta - z_\zeta x_\eta), & \eta_y &= J(x_\xi z_\zeta - z_\xi x_\zeta) \\
\xi_z &= J(x_\eta y_\zeta - y_\eta x_\zeta), & \eta_z &= J(y_\xi x_\zeta - x_\xi y_\zeta) \\
\zeta_x &= J(y_\xi z_\eta - z_\xi y_\eta), & \xi_t &= -x_\tau \xi_x - y_\tau \xi_y - z_\tau \xi_z \\
\zeta_y &= J(z_\xi x_\eta - x_\xi z_\eta), & \eta_t &= -x_\tau \eta_x - y_\tau \eta_y - z_\tau \eta_z \\
\zeta_z &= J(x_\xi y_\eta - y_\xi x_\eta), & \zeta_t &= -x_\tau \zeta_x - y_\tau \zeta_y - z_\tau \zeta_z
\end{aligned}$$

with

$$J^{-1} = x_\xi y_\eta z_\zeta + x_\zeta y_\xi z_\eta + x_\eta y_\zeta z_\xi - x_\xi y_\zeta z_\eta - x_\eta y_\xi z_\zeta - x_\zeta y_\eta z_\xi$$

The thin-layer equations have been nondimensionalized with respect to ρ_∞, c_∞, and μ_∞.

Transonic Full Potential Equation

The transformed continuity equation in terms of the velocity potential is given by

$$\left(\frac{\rho}{J}\right)_\tau + \left[\frac{\rho(\alpha_{\xi\xi}\Phi_\xi + \alpha_{\xi\eta}\Phi_\eta + \alpha_{\xi\zeta}\Phi_\zeta)}{J}\right]_\xi + \left[\frac{\rho(\alpha_{\xi\eta}\Phi_\xi + \alpha_{\eta\eta}\Phi_\eta + \alpha_{\eta\zeta}\Phi_\zeta)}{J}\right]_\eta$$
$$+ \left[\frac{\rho(\alpha_{\xi\zeta}\Phi_\xi + \alpha_{\eta\zeta}\Phi_\eta + \alpha_{\zeta\zeta}\Phi_\zeta)}{J}\right]_\zeta = 0$$

where

$$\alpha_{\xi\xi} = \xi_x\xi_x + \xi_y\xi_y + \xi_z\xi_z$$
$$\alpha_{\eta\eta} = \eta_x\eta_x + \eta_y\eta_y + \eta_z\eta_z$$
$$\alpha_{\zeta\zeta} = \zeta_x\zeta_x + \zeta_y\zeta_y + \zeta_z\zeta_z$$
$$\alpha_{\xi\eta} = \xi_x\eta_x + \xi_y\eta_y + \xi_z\eta_z$$
$$\alpha_{\xi\zeta} = \xi_x\zeta_x + \xi_y\zeta_y + \xi_z\zeta_z$$
$$\alpha_{\eta\zeta} = \eta_x\zeta_x + \eta_y\zeta_y + \eta_z\zeta_z$$

Appendix: Circulants and Stability

If periodic boundary conditions are imposed on a spacial difference operator defined on a uniform grid, the matrix generated by the difference operator will be circulant. Circulant matrices are quite special. All circulant matrices commute, are unitarily similar to a diagonal matrix, and have l_2 norm equal to their spectral radius. The sum or product of a circulant matrix is circulant, the inverse of a circulant matrix is circulant, and circulant matrices have known eigenvalues and eigenvectors.

The special properties of circulant matrices follow because all circulant matrices are made up as powers of the elementary periodic normal matrix

$$E = \begin{bmatrix} 0 & 1 & & & & & \\ & 0 & 1 & & & & \\ & & 0 & 1 & & & \\ & & & \cdot & \cdot & & \\ & & & & \cdot & \cdot & \\ & & & & & \cdot & \cdot \\ 1 & & & & & & 0 \end{bmatrix} \tag{A.1}$$

An $N \times N$ circulant matrix has the form (with I the identity matrix)

$$C = c_0 I + c_1 E + c_2 E^2 + c_3 E^3 + \cdots + c_{N-1} E^{N-1} \tag{A.2}$$

or

$$C = \begin{bmatrix} c_0 & c_1 & c_2 & \cdot & \cdot & \cdot & c_{N-2} & c_{N-1} \\ c_{N-1} & c_0 & c_1 & c_2 & \cdot & \cdot & \cdot & c_{N-2} \\ c_{N-2} & c_{N-1} & c_0 & c_1 & \cdot & \cdot & \cdot & \cdot \\ \cdot & \cdot & \cdot & \ddots & & & & \\ \cdot & \cdot & \cdot & & \ddots & & & \\ \cdot & \cdot & \cdot & & & \ddots & & \\ c_2 & \cdot & \cdot & & & & c_0 & c_1 \\ c_1 & c_2 & \cdot & & & & c_{N-1} & c_0 \end{bmatrix} \tag{A.3}$$

The elementary circulant $N \times N$ matrix E has the properties

$$E^N = I = E^0$$

$$E^{N-1} = E^N E^{-1} = E^{-1}$$

$$E^m E^{-m} = I$$

where superscript m or N implies a power, not an index, e.g., $E^2 = E \cdot E$. For example, with the 4×4 matrix E,

$$E = \begin{bmatrix} 0 & 1 & 0 & 0 \\ 0 & 0 & 1 & 0 \\ 0 & 0 & 0 & 1 \\ 1 & 0 & 0 & 0 \end{bmatrix}, \qquad E^2 = \begin{bmatrix} 0 & 0 & 1 & 0 \\ 0 & 0 & 0 & 1 \\ 1 & 0 & 0 & 0 \\ 0 & 1 & 0 & 0 \end{bmatrix}$$

$$E^3 = \begin{bmatrix} 0 & 0 & 0 & 1 \\ 1 & 0 & 0 & 0 \\ 0 & 1 & 0 & 0 \\ 0 & 0 & 1 & 0 \end{bmatrix}, \qquad E^4 = \begin{bmatrix} 1 & 0 & 0 & 0 \\ 0 & 1 & 0 & 0 \\ 0 & 0 & 1 & 0 \\ 0 & 0 & 0 & 1 \end{bmatrix}$$

Thus, for $N = 4$, $E^{-1} = E^3$, $E^3 E = E^4 = I$, and $E^0 = I$.

The eigenvalues of E are given by the roots of unity, i.e.,

$$\lambda_n(E) = 1^{1/N} = e^{2\pi i n/N}, \qquad i = \sqrt{-1} \tag{A.4}$$

as

$$Det(E - \lambda I) = (-\lambda)^N - (-1)^N = 0 \tag{A.5}$$

where expansion of the 4×4 determinant about elements (1, 1) and (4, 1) readily illustrates Eq. (A.5):

$$\begin{vmatrix} -\lambda & 1 & 0 & 0 \\ 0 & -\lambda & 1 & 0 \\ 0 & 0 & -\lambda & 1 \\ 1 & 0 & 0 & -\lambda \end{vmatrix} = -\lambda \begin{vmatrix} -\lambda & 1 & 0 \\ 0 & -\lambda & 1 \\ 0 & 0 & -\lambda \end{vmatrix} - (-1)^4 \begin{vmatrix} 1 & 0 & 0 \\ -\lambda & 1 & 0 \\ 0 & -\lambda & 1 \end{vmatrix}$$

$$= (-\lambda)^4 - (-1)^4$$

The eigenvalues of E^m are given by

$$\lambda_n(E^m) = e^{2\pi i m n/N} \tag{A.6}$$

Thus, the eigenvalues of a circulant matrix are given as

$$\lambda_n(C) = c_0 + c_1 e^{2\pi in/N} + c_2 e^{2(2\pi in/N)} + c_3 e^{3(2\pi in/N)} + \cdots + c_{N-1} e^{(N-1)(2\pi in/N)} \tag{A.7}$$

A complete set of orthonormal eigenvectors corresponding to E is given by

$$\xi_n(E) = \sqrt{\frac{1}{N}} \begin{bmatrix} e^{2\pi in/N} \\ e^{2(2\pi in/N)} \\ e^{3(2\pi in/N)} \\ \vdots \\ e^{(N-1)(2\pi in/N)} \end{bmatrix}, \qquad n = 1,2,3,\ldots,N \tag{A.8}$$

Any circulant matrix C has the same eigenvectors as E because C is made up as a power series of E, and the eigenvectors of E^m are equal to those of E.

The symbol E was chosen for the elementary circulant because it behaves exactly like the shifting operator applied on an N-point uniform grid with periodicity imposed. Moreover, with $c_{-m} \equiv c_{N-m}$ the shifting operator notation holds identically for the eigenvalues as

$$c_{N-1} e^{(N-1)(2\pi in/N)} = c_{-1} e^{-(2\pi in/N)}, \qquad c_{-1} \equiv c_{N-1}$$

$$c_{N-2} e^{(N-2)(2\pi in/N)} = c_{-2} e^{-2(2\pi in/N)}, \qquad c_{-2} \equiv c_{N-2}$$

For grids with periodic data, the known properties of a circulant matrix can be used to analyze difference schemes for stability by transforming the system of difference equations into uncoupled scalar equations using the known eigenvalues and eigenvectors. Consider, for example, the scalar wave equation $u_t - u_x = 0$. The Euler explicit scheme with central spacial differencing and scaled by Δt can be written as

$$u^{n+1} = u^n + \Delta t \delta_x u^n$$

or

$$\boldsymbol{u}^{n+1} = \boldsymbol{u}^n + \Delta t \left(\frac{-E^{-1} + E}{2\Delta x} \right) \boldsymbol{u}^n \tag{A.9}$$

or (for J points)

$$\begin{pmatrix} u_1 \\ u_2 \\ u_3 \\ \vdots \\ \vdots \\ u_J \end{pmatrix}^{n+1} = \begin{pmatrix} u_1 \\ u_2 \\ u_3 \\ \vdots \\ \vdots \\ u_J \end{pmatrix}^{n} + \frac{\Delta t}{2\Delta x} \begin{bmatrix} 0 & 1 & & & \ldots & -1 \\ -1 & 0 & 1 & & & \\ & -1 & 0 & 1 & & \\ & & \ddots & \ddots & \ddots & \\ & & & \ddots & \ddots & \ddots \\ 1 & & & & -1 & 0 \end{bmatrix} \begin{pmatrix} u_1 \\ u_2 \\ u_3 \\ \vdots \\ \vdots \\ u_J \end{pmatrix}^{n}$$

The eigenvalues of the circulant matrix corresponding to δ_x are

$$\lambda_j = \frac{(-e^{-2\pi ij/J} + e^{2\pi ij/J})}{2\Delta x} = \frac{i\,\sin(2\pi j/J)}{\Delta x}$$

as $e^{\pm i\theta} = \cos\theta \pm i\,\sin\theta$. The matrix corresponding to δ_x is diagonalized to its eigenvalues by the (inverse) matrix made up of the circulant eigenvectors

$$\Xi = (\xi_1, \xi_2, \xi_3, \ldots, \xi_J) = \Xi^{-1}$$

Hence, Eq. (A.9) can be transformed into

$$\Xi^{-1}\boldsymbol{u}^{n+1} = \Xi^{-1}\boldsymbol{u}^n + \Delta t\,\Xi^{-1}\left(\frac{-E^{-1} + E}{2\Delta x}\right)\Xi\Xi^{-1}\boldsymbol{u}^n$$

or

$$\Xi^{-1}\boldsymbol{u}^{n+1} = \Xi^{-1}\boldsymbol{u}^n + \Delta t\,\Lambda\Xi^{-1}\boldsymbol{u}^n \tag{A.10}$$

with

$$\Lambda = \begin{bmatrix} \lambda_1 & & & & \\ & \lambda_2 & & & \\ & & \lambda_3 & & \\ & & & \ddots & \\ & & & & \lambda_J \end{bmatrix}$$

Equation (A.10) is an uncoupled set of equations

$$\tilde{u}_j^{n+1} = \tilde{u}_j^n + \Delta t\lambda_j \tilde{u}_j^n \tag{A.11}$$

where $\tilde{u}_j$ is the jth component of $\Xi^{-1}\boldsymbol{u}$. Solving recursively,

$$\tilde{u}_j^{n+1} = (1 + \Delta t\lambda_j)^n \tilde{u}_j^0$$

where u^0 is the initial state. Here the amplification factor

$$(1 + \Delta t\lambda_j)$$

is greater than one in absolute value because λ is imaginary so this scheme is always unstable. Because Ξ is orthonormal, the largest eigenvalue of Λ in absolute value, the spectral radius, is also the l_2 norm of the matrix corresponding to δ_x.

A multidimensional wave equation with constant coefficients, for example $u_t = u_x + u_y$, can be analyzed in the same way for periodic data in x and y. In multidimensions the eigenvectors must be applied in block diagonal form and permutation matrices are needed to diagonalize to the eigenvalues, but the end result is simply the eigenvalues of each operator. This approach applies to any partial differential equation with constant coefficients differenced on a uniform grid and subject to periodicity.

References

[1]Garabedian, P. *Partial Differential Equations*, Wiley, New York, 1964.

[2]Gear, G. W., *Numerical Initial Value Problems in Ordinary Differential Equations*, Prentice-Hall, Englewood Cliffs, NJ, 1971.

[3]Dahlquist, G. and Bjorck, A., *Numerical Methods*, Prentice-Hall, Englewood Cliffs, NJ, 1974.

[4]Richtmyer, R. D. and Morton, K. W., *Difference Methods for Initial-Value Problems* (2nd ed.), Interscience, New York, 1967.

[5]Smith, G. D., *Numerical Solution of Partial Differential Equations: Finite Difference Methods* (2nd ed.), Clarendon, Oxford, UK, 1978.

[6]MacCormack, R. W., "The Effect of Viscosity in Hypervelocity Impact Cratering," AIAA Paper 69-354, 1969.

[7]Steger, J. L. and Warming, R. F., "Flux Vector Splitting of the Inviscid Gasdynamic Equations with Application to Finite Difference Methods," NASA TM-78605, July 1979.

[8]Chakravarthy, S. R., Anderson, D. A., and Salas, M. D., "The Split Coefficient Matrix Method for Hyperbolic Systems of Gas-Dynamic Equations," AIAA Paper 80-0268, 1980.

[9]Moretti, G., "An Old Integration Scheme for Compressible Flows Revisited, Refurbished and Put to Work," *Computational Fluids*, Vol. 7, 1979, p. 191.

[10]Kovenya, V. M and Yanenko, N. N., "The Splitting Method for the Numerical Solution of Problems in Gas Dynamics," *Numerical Methods in Fluid Dynamics*, edited by N. N. Yanenko and Y. I. Shokin, Mir, Moscow, USSR, 1985.

[11]MacCormack, R. W. and Warming, R. F., "Survey of Computational Methods for Three-Dimensional Supersonic Inviscid Flows with Shocks," AGARD Lecture Series 64, Advances in Numerical Fluid Dynamics, 1974.

[12]Beam, R. M. and Warming, R. F., "An Implicit Finite-Difference Algorithm for Hyperbolic Systems in Conservation-Law-Form," *Journal of Computational Physics*, Vol. 22, Sept. 1976, pp. 87–110.

[13]Merriam, M. L., "On the Inversion of Block—Tridiagonals without Storage Constraints," NASA TM-84228, March 1982.

[14]Beam, R. M. and Warming, R. F., "Alternating Direction Implicit Method for Parabolic Equations with a Mixed Derivative," *Society for Industrial and Applied Mathematics. J. Sci. Stat. Comput.* Vol. 1, No. 1, March 1980.

[15]White, F. M., *Viscous Fluid Flow*, McGraw-Hill, New York, 1974.

[16]Le Balleur, J. C., "Numerical Viscid-Inviscid Interaction in Steady and Unsteady Flows," *Proceedings of the 2nd Symposium on Numerical and Physical Aspects of Aerodynamic Flows*, 1983.

[17]Le Balleur, J. C. and Girodroux-Lavigne, P., "A Semi-Implicit and Unsteady Numerical Method of Viscous-Inviscid Interaction for Transonic Separated Flows," *La Recherche Aérospatiale*, 1984, pp. 15–37.

[18]Le Balleur, J. C., "Viscous-Inviscid Interaction Solvers and Computation of Highly Separated Flows," *ICASE Conference on Vortex Dominated Flows*, 1985.

[19]Van Dalsem, W. R. and Steger, J. L., "Efficient Simulation of Separated Three-Dimensional Viscous Flows Using the Boundary-Layer Equations," *AIAA Journal*, Vol. 25, March 1987, pp. 395–400.

[20]Klineberg, J. M. and Steger, J. M., "Calculation of Separated Flows at Subsonic and Transonic Speeds," *Lecture Notes in Physics 19, Proceedings of the Third International Conference on Numerical Methods in Fluid Mechanics*, Vol. 2, Springler-Verlag, New York, 1972, pp. 161–168.

[21]Melnik, R. E., Chow, R., and Mead, H. R., "Theory of Viscous Transonic Flow Over Airfoils at High Reynolds Number," AIAA Paper 77-680, June 1977.

[22]Collyer, M. R., "An Extension to the Method of Garabedian and Korn for the Calculation of Transonic Flow Past an Aerofoil to Include the Effects of a Boundary Layer and Wake," Royal Aircraft Establishment, TR 77104, July 1977.

[23]Whitfield, D. L., Swafford, T. W., and Jacocks, J. L., "Calculation of Turbulent Boundary Layers with Separation, Reattachment and Viscous-Inviscid Interaction," AIAA Paper 80-1439, July 1980.

[24]Fujii, K., "An Approximate Numerical Approach to Separated Flows at Transonic Speeds," Institute of Space and Aeronautical Science, Univ. of Tokyo, Rept. 595, March 1981.

[25]Lock, R. C. and Firmin, M. C. P., "Survey of Techniques for Estimating Viscous Effects in External Aerodynamics," Royal Aircraft Establishment, TR-1900, April 1981.

[26]Melnik, R. E. and Mead, H. R., "A Multi-Grid Method for the Computation of Viscid/Inviscid Interaction on Airfoils," AIAA Paper 83-0234, Jan. 1983.

[27]Wigton, L. and Yoshihara, H., "Viscous-Inviscid Interactions with a Three-Dimensional Inverse Boundary-Layer Code," *Proceedings of the 2nd Symposium on Numerical and Physical Aspects of Aerodynamic Flows*, Jan. 17–20, 1983.

[28]Van Dalsem, W. R. and Steger, J. L., "Finite-Difference Simulation of Transonic Separated Flow Using a Full-Potential Boundary-Layer Interaction Approach," AIAA Paper 83-1689, July 1983.

[29]Rizzetta, D. P., "Procedures for the Computation of Unsteady Transonic Flows Including Viscous Effects," NASA CR-166249, Jan. 1982.

[30]Carter, J. E., "A New Boundary-Layer Interaction Technique for Separated Flows," NASA TM-78690, June 1978.

[31]Abbott, D. E., Holt, M., and Nielsen, J. N., "Investigation of Hypersonic Flow Separation and Its Effects on Aerodynamic Control Characteristics," Vidya Rept. 81, 1962; also "Studies of Separated Laminar Boundary Layers at Hypersonic Speed With Some Low Reynolds Number Data," AIAA Paper 63-172, 1963.

[32]Nielsen, J. N., Lynes, L. L., and Goodwin, F. K., "Calculation of Laminar Separation With Free Interaction by the Method of Integral Relations: Part I—Two-Dimensional Supersonic Adiabatic Flows," AFFDL TR-65-107, Air Force Flight Dynamics Lab., Wright-Patterson AFB, OH, Oct. 1965.

[33]Nielsen, J. N., Lynes, L. L., and Goodwin, F. K., "Calculation of Laminar Separation With Free Interaction by the Method of Integral Relations: Part II—Two-Dimensional Supersonic Nonadiabatic Flow and Axisymmetric Supersonic Adiabatic and Nonadiabatic Flows," AFFDL TR-65-107, Air Force Flight Dynamics Lab., Wright-Patterson AFB, OH, Jan. 1966.

[34]Lees, L. and Reeves, B. L., "Supersonic Separated and Reattaching Laminar Flows: I. General Theory and Application to Adiabatic Boundary-Layer/Shock-Wave Interactions," *AIAA Journal*, Vol. 2, Nov. 1964, pp. 1907–1920.

[35]Landahl, M., *Unsteady Transonic Flow*, Pergamon, New York, 1961.

[36]Ballhaus, W. F. and Steger, J. L., "Implicit Approximate-Factorization Schemes for the Low-Frequency Transonic Equation," NASA TM X-73082, Nov. 1975.

[37]Douglas, J. and Gunn, J. E., "A General Formulation of Alternating Direction Methods," *Numerical Mathematics*, Vol. 6, 1964, pp. 428–453.

[38]Murman, E. M., "Analysis of Embedded Shock Waves Calculated by Relaxation Methods," AIAA CFD Conf. Proceeding, 1973, pp. 27–40.

[39]Holst, T. L. and Ballhaus, W. F., "Fast Conservative Schemes for the Full Potential Equation Applied to Transonic Flows," *AIAA Journal*, Vol. 17, Feb. 1979, pp. 145–152.

[40]Hafez, M., South, J., and Murman, E. "Artificial Compressibility Methods for Numerical Solutions of Transonic Full Potential Equation," *AIAA Journal*, Vol. 17, Aug. 1979, pp. 838–844.

[41]Steger, J. L. and Caradonna, F. X., "A Conservative Implicit Finite Difference Algorithm for the Unsteady Transonic Full Potential Equation," AIAA Paper 80–1368, July 1980.

[42]Pulliam, T. H. and Steger, J. L., "Implicit Finite-Difference Simulations of Three-Dimensional Compressible Flow," *AIAA Journal*, Vol. 18, Feb. 1980 pp. 159–167.

[43]Baldwin, B. S. and Lomax, H., "Thin Layer Approximation and Algebraic Model for Separated Turbulent Flows," AIAA Paper 78-257, 1978.

[44]Steger, J. L., "Implicit Finite Difference Simulation of Flow About Arbitrary Two Dimensional Geometries," *AIAA Journal*, Vol. 16, July 1978.

[45]Mehta, U. and Lomax, H., "Reynolds Averaged Navier-Stokes Computations of Transonic Flows—the State-of-the-Art," *Transonic Aerodynamics*, Vol. 81, edited by D. Nixon, Progress in Astronautics and Aeronautics Series, AIAA, New York, 1982, pp. 297–375.

[46]Steger, J. L., Pulliam, T. H., and Chima, R. V., "An Implicit Finite Difference Code for Inviscid and Viscous Cascade Flow," AIAA Paper 80-1427, 1980.

Chapter 5

Computational Methods for Unsteady Transonic Flows

John W. Edwards* and James L. Thomas†
NASA Langley Research Center, Hampton, Virginia

Abstract

Computational methods for unsteady transonic flows are surveyed with emphasis on applications to aeroelastic analysis and flutter prediction. Computational difficulty is discussed with respect to type of unsteady flow: attached, mixed (attached/separated), or separated. Significant early computations of shock motions, aileron buzz, and periodic oscillations are discussed. The maturation of computational methods toward the capability of treating complete vehicles with reasonable computational resources is noted, and a survey of recent comparisons with experimental results is compiled. The importance of mixed attached and separated flow modeling for aeroelastic analysis is discussed, and recent calculations of periodic aerodynamic oscillations for an 18% thick circular arc airfoil are given.

Introduction

In the past decade there has been much activity in the development of computational methods for the analysis of unsteady transonic aerodynamics about airfoils and wings. Advances have paralleled developments in steady computational fluid dynamics (CFD) with a lag of approximately five years[1] due to the additional requirement of time accuracy. Figure 1, taken from the specification document for U.S. military aircraft,[2] illustrates significant features that must be addressed in the treatment of computational transonic unsteady aerodynamics. On the plot of equivalent airspeed vs Mach number, lines of constant altitude are straight lines through the origin with decreasing altitudes represented by steeper slopes. The flight envelope is typically set by the maximum limit speed, and a typical flutter boundary curve, characterized by the flutter speed gradually dropping to a minimum in the transonic speed range followed by a rapid upward rise, is

*Head, Unsteady Aerodynamics Branch, Loads and Aeroelasticity Division.

†Senior Research Scientist, Analytical Methods Branch, Transonic Aerodynamics Division.

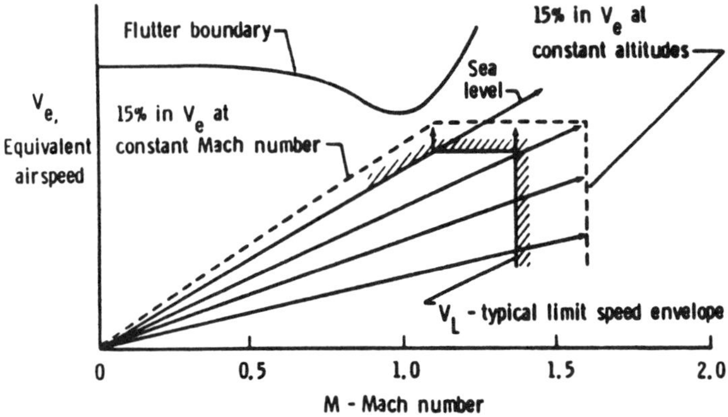

Fig. 1 Graphic representation of minimum required flutter margin.[2]

shown. The ability to predict this minimum, termed the transonic flutter dip, is of great importance in design, since the flutter boundary must be shown by a combination of analysis and flight test to be outside the flight envelope by a margin of at least 15% in equivalent airspeed; i.e., the flutter boundary must be outside the dashed line boundary in Fig. 1. Subsonic linear unsteady aerodynamic theories have been reasonably successful in predicting this flutter boundary for Mach numbers up to 0.6–0.7, but linear theory is unable to account for the effect of aerodynamic shape and maneuvering condition on unsteady airloads. At higher Mach numbers linear analysis has been used with more or less success depending on the severity of local transonic effects. The occurrence of flutter within the flight envelope of an aircraft usually leads to structural failure and loss of the vehicle, highlighting the necessity for careful validation of computational methods intended for use in this area. This is a key difference in the utilization of steady and unsteady computational methods that must be clearly understood.

This paper attempts to provide an assessment of the current situation regarding computational methods for transonic unsteady aerodynamics. It is written from the perspective of aeroelastic applications and the prediction of aircraft flutter. Thus, attention is devoted to assessing the state of the art for predicting unsteady airloads on oscillating lifting surfaces. Also, there is much current interest in airflows over rigid surfaces that exhibit periodic unsteadiness over narrow Mach number regions. The interaction of these two types of flows is likely to be of great interest in the near future as computational aeroelasticity matures. We have restricted this survey to computational methods that solve time-accurate transonic flow equations. Harmonic perturbation and other approximate solution methods have been an active research area, and an assessment of their relative merits vis-à-vis time-accurate solutions would be valuable. Also not covered are unsteady three-dimensional vortex flows, since this field is currently undergoing a rapid growth.[3]

This field received an initial impetus in the mid-1970's from three sources: Tijdeman's[4] pioneering experimental work on transonic unsteady pressure measurements, Magnus and Yoshihara's demonstration of key transonic flow features for an airfoil with an oscillating flap,[5] and the introduction of an economical transonic finite-difference solution algorithm by Ballhaus and Goorjian.[6] Ballhaus[7] gives a survey of the field from this period. The AGARD Structures and Materials Panel Subcommittee on Aeroelasticity has selected experimental unsteady pressure data sets and defined two- and three-dimensional Standard Aeroelastic Configurations[8,9] to provide reference computational test cases for the development and validation of improved computational methods. The data sets were obtained from rigid models undergoing pitch and control surface oscillations and include both conventional and supercritical airfoil geometries.[10–12] In addition to these data sets, Sandford et al.[13] summarize a series of unsteady pressure tests made at NASA Langley, and Tijdeman[14] presents a much-used data set for a fighter wing configuration.

Computational methods have been pursued at a number of differing levels of physical approximation to the flow equations. Magnus and Yoshihara[5,15,16] used an explicit algorithm to solve the Euler (EE) equations. Steger and Bailey[17] reported a significant early application to the problem of aileron buzz using an implicit approximate factorization solution algorithm for the Navier-Stokes (NS) equations. Chyu and his coauthors[18,19] have pursued further applications of derivatives of this code. Most of the nonlinear unsteady computations to date have been made by solving the potential equations, both with and without interacted viscous effects. For example, the alternating-direction implicit (ADI) algorithm embodied in the LTRAN2 code of Ballhaus and Goorjian[6] enabled efficient solutions of the two-dimensional low-frequency transonic small-disturbance (TSD) potential equation through the use of large time steps. Extensions of this ADI algorithm have been widely used by many researchers. A semi-implicit form of the ADI algorithm is used in the three-dimensional (3D) XTRAN3S code[20,21] developed for the aeroelastic analysis of wings. Other TSD and full potential (FP) equation codes are described in Refs. 22–27. There is a growing trend, especially for steady flows, toward use of the Euler equations rather than the potential equations. Euler equation codes treating two-dimensional (2D) oscillating airfoils are reported in Refs. 28–32, whereas Salmond[33] and Belk[34] show results from 3D Euler codes.

Over this same time period, several experimental investigations of periodic aerodynamic flows about rigid airfoils have been reported. McDevitt et al.[35,36] documented such conditions for a very narrow range of Mach number for an 18% thick circular arc airfoil, and Levy[37] reproduced the effect with calculations from an NS code. Subsequently, Mabey[38] studied these oscillations for circular arc airfoils with thicknesses of 10–20%. References 39 and 40 give details for a 14% circular arc airfoil. Related information regarding the interaction of unsteady airloads caused by transitional boundary layers with structural oscillations is given by Mabey et al.[41] Another class of separation-induced periodic flow problems, vortex shedding about rigid cylinders and airfoils at high angle of attack, has been studied using NS codes for a variety of Reynolds numbers in Refs. 42–44.

Unsteady aerodynamics has been the theme of four recent AGARD conferences,[45–48] whose proceedings contain a wealth of information. Survey papers focusing on computational requirements and resources are given by Peterson[49] and McCroskey et al.[1] Summary papers of the 1984 and 1985 AGARD conferences are given by Mykytow[50] and by Mabey and Chambers.[51] The latter reference makes recommendations regarding computational and experimental methods for unsteady flow phenomena and draws particular attention to the need to pay careful attention to the nature of shock motions. The periodic oscillations about circular arc airfoils are recommended as benchmark computational cases for all time-dependent transonic viscous flow theories. Zwaan[52] surveys aeroelastic problems in transonic flow, and Deiwert[53] reviews the numerical simulation of unsteady interactive flows. Finally, Mabey[54] gives a review of pertinent experimental research on time-dependent aerodynamics.

In the following sections, the distinction between attached and separated flow conditions and the available experimental data sets are discussed. A brief discussion of the hierarchy of available flow solvers is given with attention to details specific to unsteady flows. Then, computational capability for two- and three-dimensional unsteady aerodynamics is illustrated, followed by a brief review of aeroelastic applications. Finally, computations of unsteady periodic flow about circular arc airfoils are discussed and the origin of the oscillations is investigated.

Transonic Flow Phenomena

It will be helpful to distinguish the main features of steady transonic flow in order to organize the discussion of unsteady aerodynamics. Figure 2, from Ref. 4, indicates various regions of transonic flow development for the NLR 7301 airfoil, a 16% cambered supercritical-type section. With increasing Mach number and moderate angle of attack, the upper surface becomes critical between $M = 0.4$ and 0.7, with the first shock forming at an increase in Mach number of approximately 0.1. Pearcy et al.[55] have classified several types of flow separation that may occur. For conventional airfoils the typical pattern, termed type A, involves the growth of a local separation bubble induced by boundary-layer separation at the shock foot, spreading rapidly to the trailing edge as Mach number increases. This condition is often accompanied by unsteady phenomena such as buffet and aileron buzz.[4] The steep aft pressure gradients of modern airfoils, such as the NLR 7301, can lead to an alternate pattern, termed type B, in which separation progresses from the trailing edge toward the shock. Figure 2 illustrates this type B separation, with fully separated flow aft of the shock occurring along the line of maximum lift. Note the small "shock-free" design condition occurring over a small isolated range of lift coefficient and Mach number just before the onset of trailing-edge separation. Tijdeman[4] noted that flow conditions in the region between the onset of trailing-edge separation and fully separated flow are very sensitive to Reynolds number and the location of transition from laminar to turbulent flow.

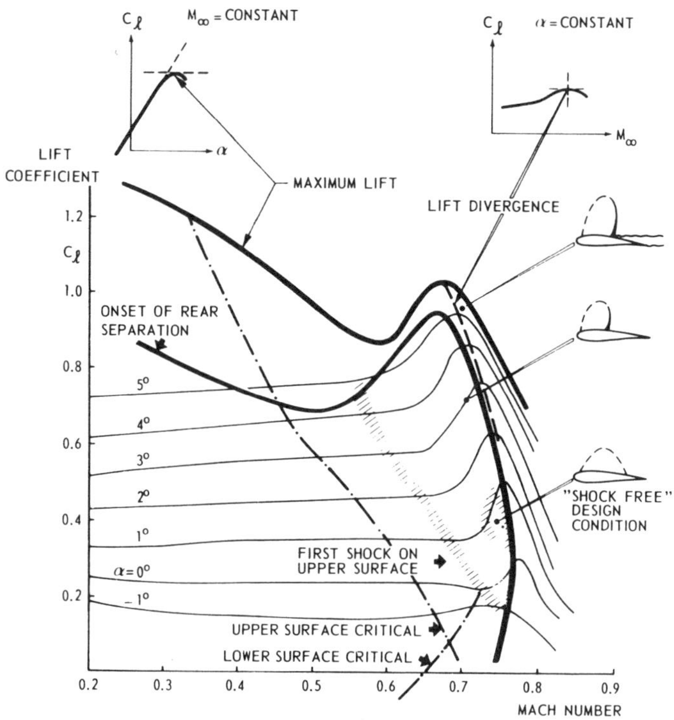

Fig. 2 Main characteristics and boundaries for the onset of separation for the NLR 7301 supercritical airfoil.[4]

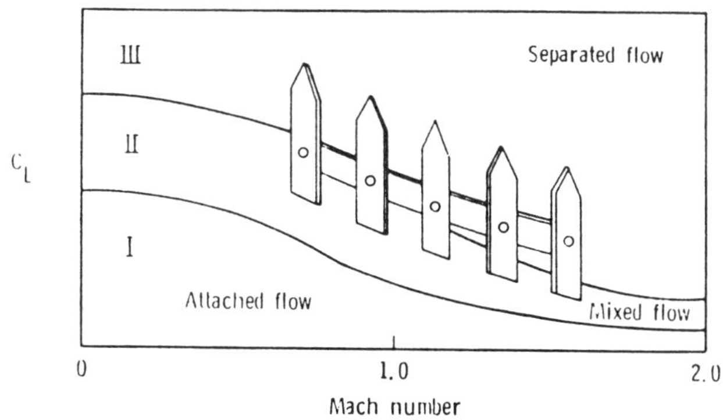

Fig. 3 Characteristics of attached and separated flow for complete aircraft.[56]

Figure 3 shows a similar diagram, derived from Ref. 56, of flow regions for a complete vehicle. Flows that are predominantly attached or separated are designated as type I and III flows, respectively, and mixed attached and separated flows are designated type II. These flow regions will be used to distinguish the relative difficulty of computational problems. For unsteady flows the boundaries between these flow types are not distinct, since an airfoil or wing may exhibit more than one type of flow during the unsteady motion. The "picket fence" in the mixed flow region has been added to emphasize the possibility of "nonclassical" aeroelastic effects in this region. Classical linear aeroelastic analysis methods cannot reliably model results in this region, although they have been surprisingly durable.

Farmer et al.[57] provided early test results documenting the effect of airfoil shape on flutter boundaries. Figure 4 shows their comparison of flutter boundaries for two structurally and geometrically similar wings of the same planform. The supercritical wing was a reduced stiffness model of the modified TF-8A wing while the conventional wing had a symmetrical section. The two wings had leading-edge sweep angles of 44.5 deg. Design cruise Mach number was 0.90 for the conventional wing and 0.99 for the supercritical wing. The supercritical wing was shown to have a 25% lower minimum flutter dynamic pressure near Mach 1.0, where type II mixed flow would be expected. Current transonic computational methods are beginning to address this important area, which will be a key topic for computational aeroelasticity in the future. Other reports of aeroelastic model tests relevant to this area are single mode flutter of a low-aspect-ratio wing studied by Erickson,[58] supercritical wing flutter tests performed at the National Aerospace Laboratory (NLR),[59,60] and torsional buzz of aeroelastic wings tested at the Royal Aircraft Establishment.[61] It should be noted that the present paper deals only with the computational methods for predicting types A and B flow separation. Separated flows involving 3D unsteady vortex flow structures are also of great importance in this field but are outside the scope of this paper.

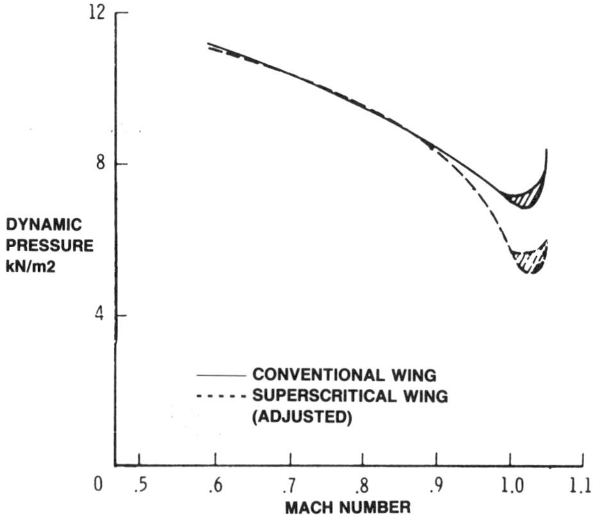

Fig. 4 Comparison of conventional wing flutter boundary with adjusted supercritical wing boundary.[57]

Experimental Data Sets

In this section, the airfoil geometries and wing planforms that have been most frequently studied are summarized. In addition to the AGARD standard configurations, several other model tests have been popular for comparison with computational results. Figures 5 and 6 show the profiles and planforms of the 2D[8] and 3D[9] AGARD configurations, respectively. Data sets for all of these configurations except the 6% parabolic arc, DO Al, and MBB-A3 airfoils are given in Refs. 10, 11, and 70. Tables 1 and 2 tabulate selected references for these and other configurations in which comparisons of experimental and calculated unsteady pressures are given. The entries are grouped by the equation level of the physical modeling used for the calculations. The references are not exhaustive but are an attempt to indicate publication of significant experimental/computational comparisons or new capability.

The first three airfoils in Table 1 are conventional airfoils with 6, 10, and 12% thickness ratios. Tijdeman[4] tested the NACA 64A006 airfoil with an oscillating quarterchord trailing-edge control surface. Interpretations of these tests[4] have provided insights into the underlying mechanisms of unsteady transonic flows. Tijdeman identified three types of shock motion, denoted types A, B, and C. In type A shock motion, the shock wave remains distinct during the oscillation cycle, with a periodic variation of shock location and shock strength. In type B shock motion, the shock wave weakens and disappears during a portion of the cycle, generally during the forward propagation of the shock along the surface. For type C motion, the shock wave on the airfoil remains distinct and propagates forward along the airfoil chord and off the airfoil leading edge.

Davis and Malcolm[62] tested the NACA 64A010A airfoil for pitching oscillations. Two cases from this test have been widely studied: a case with a moderate shock wave at $M = 0.8$ and $\alpha = 0$ deg and a case with steady

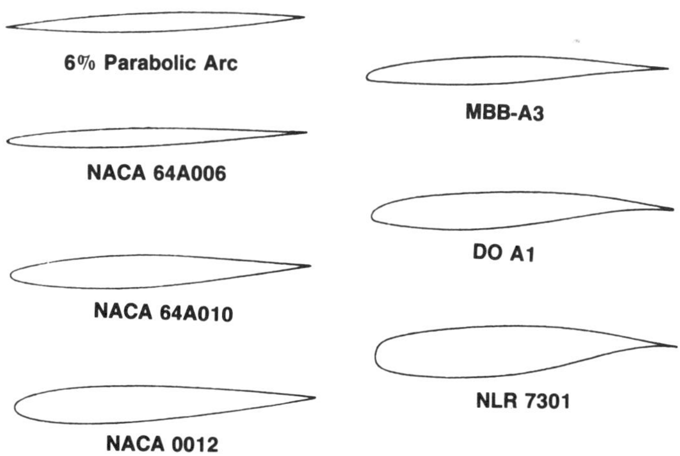

Fig. 5 AGARD Structures and Materials Panel two-dimensional standard aeroelastic configurations.[8]

Table 1 References giving comparisons of experimental and calculated two-dimensional unsteady aerodynamics

	TSD	FP	EE	NS
NACA 64A006	6, 103,[a] 88, 89, 87[a]	102, 116	5, 15	
NACA 64A010A	80, 82, 138, 88, 89, 112,[a] 87,[a] 93, 97,[b] 114,[a] 105[a]–108[a]	117, 116, 79, 119	15, 30, 32	19, 18
NACA 0012	88, 90,[a] 109,[a] 97,[b] 96,[b] 93	119, 100[b]	29, 28, 31	152, 44, 148, 43, 154
NLR 7301	89, 88, 107,[a] 114,[a] 97[b]	117	16, 30, 32	
MBB-A3	112,[a] 89, 82, 138, 114,[a] 63			
Supercritical airfoils	154[a], 109,[a] 65,[a] 66			
Circular arc airfoils	106,[a] 107,[a] 108[a]			35, 37, 36, 40
Other			151	17, 42, 43, 148

[a]Interacted boundary-layer model.
[b]Nonisentropic corrections.

shock-induced separation at $M = 0.8$ and $\alpha = 4$ deg. The NACA 0012 airfoil, tested by Landon,[10] differs from the other entries in Table 1 in that it was tested for larger dynamic pitching amplitudes and for transient ramping motions, making it suitable for dynamic stall computational studies. McDevitt and Okuno[63] have reported measurements of periodic shock-induced oscillations for this airfoil.

Data sets for the 16% thick supercritical NLR 7301 airfoil are given by both Tijdeman and Davis,[10] and the shock-free condition for this supercritical airfoil has been a challenging computational case. The 8.9% thick MBB-A3 airfoil has been tested by Zimmerman[64] and represents a less severe supercritical airfoil computational case. Other supercritical airfoils tested for oscillatory motions or exhibiting unsteady behavior are a 12% thick airfoil tested for pitching, heaving, and flap rotation by den Boer and Houwink,[65] the RA16SC1 airfoil tested by ONERA,[66] and the cryogenic test of a supercritical SC(2)-0714 airfoil by Hess et al.[67] Reference 65 reported large dynamic responses of airloads on the supercritical airfoil for both oscillating and static motions at type II flow conditions and introduced the concept of "aerodynamic resonance." Similar periodic shock-induced oscillations are reported for the RA16SC1 airfoil.[66]

Tests of rigid circular arc airfoils have been reported by McDevitt et al.,[35] McDevitt,[36] Mabey,[38] and Mabey et al.[39] References 35 and 36 give details of tests of an 18% airfoil for Reynolds numbers of 1–17 million, covering laminar to fully developed turbulent flows. The wind-tunnel walls were contoured to approximate the inviscid streamlines over an airfoil at $M = 0.775$. Periodic unsteady airflows were observed over a narrow Mach range whose extent depended on whether Mach number was increasing or decreasing. For increasing Mach numbers, oscillations oc-

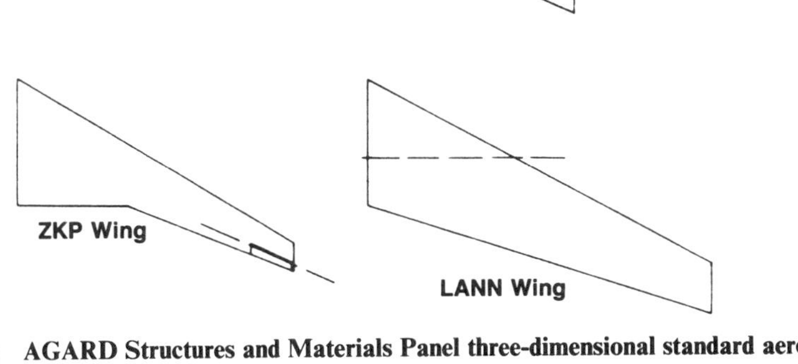

Fig. 6 AGARD Structures and Materials Panel three-dimensional standard aeroelastic configurations.[9]

Table 2 References giving comparisons of experimental and calculated three-dimensional unsteady aerodynamics and aeroelasticity

	TSD	FP	EE
F-5 model	115, 157, 137, 23	24, 136, 153, 102	
NORA	76, 22	64, 26	
LANN	100, 156	45, 40, 24, 76	
RAE wing A		64, 43, 26, 136	
RSW	116, 75	40, 66, 76, 25	96, 34
RAE tail plane	114, 85,[a] 155, 98[a]		75, 33
Other	118,[b] 113,[b] 71,[b] 73[b]	64,[b] 65,[b] 26,[b] 70[b]	

[a]Nonisentropic corrections.
[b]Aeroelastic and flutter comparisons.

curred for $0.76 < M < 0.78$, whereas for decreasing Mach number the range was wider ($0.73 < M < 0.78$). The frequency of the oscillations was 188 ± 3 Hz (reduced frequency $k = 0.48$ based on semichord). Mabey[38] studied similar periodic flows for a series of circular arc airfoils ranging in thickness from 10 to 20% at Reynolds numbers of 0.4–0.6 million. In Ref. 39, further investigations on a larger 14% thick biconvex wing at Reynolds numbers of 1–7 million are reported. Two necessary criteria evident from the experimental results for the existence of the periodic unsteady flow are given: thickness-to-chord ratio greater than 12% and local Mach number upstream of the terminal shock wave in the range of

$$1.24 < M < 1.40$$

McDevitt[36] identifies the predominant shock motion for the 18% thick airfoil as type C, whereas Mabey et al.[39] argue that it is type B motion.

The smaller number of entries in Table 2 reflects the situation regarding 3D testing in that there are fewer experimental data sets widely available and fewer comparisons of experimental and calculated results have been published. Tijdeman et al.[14] tested a model of the F-5 fighter wing including external tanks and stores. This wing has an aspect ratio of 2.98, a taper ratio of 0.31, and a leading-edge sweep of 32 deg. The relatively thin wing section, a modified NACA 64A004.8, has made this a popular computational case, since it is well within the capability of TSD codes. Transonic and low supersonic test conditions are available. Of the AGARD standard configuration models shown in Fig. 6, the NORA model is the most extensively tested. It is a model of the Mirage F-1 horizontal tail, which has been tested in four European wind tunnels.[9,10]

The AGARD rectangular wing and the RAE wing A model have symmetric airfoil sections,[9,10,12] whereas the ZKP wing and LANN wing have supercritical airfoil sections.[9,11] Additional models tested for oscillatory pitching are the NASA Rectangular Supercritical Wing (RSW) model[68,69] and the RAE AGARD tailplane model.[70] The former had a 12%

supercritical airfoil section, whereas the latter had an NACA 64A010A section, the same as one of the AGARD 2D configurations.

Also included in Table 2 are references to several other published comparisons with experimental data. These cases are of interest since the models were aeroelastic and some comparisons of experimental and computed transonic flutter boundaries (or aeroelastic response) are given. Isogai gives comparisons for a high-aspect-ratio supercritical transport wing in Ref. 26 and for the supercritical wing flutter model of Farmer et al.[57] in Ref. 71. Bennett et al.[72] give static aeroelastic comparisons for an aspect ratio 10.3 supercritical wing that was extensively instrumented for unsteady pressure measurements.[73] Finally, Guruswamy and Goorjian[74] present calculations for a rectangular parabolic arc flutter model.

Computational Methods

A variety of methods are available to address transonic unsteady computations. The choice of an appropriate method calls for assessment of the difficulty of the aerodynamic problem being addressed. One possible classification of level of difficulty is indicated in Fig. 3 in which type I problems involve attached flows, type II problems involve mixed flows, and type III problems involve fully separated flows. Type I flows include one of the most important aeroelastic analysis conditions: cruise at high dynamic pressure. Classical linear aeroelastic analysis has primarily focused on this condition. The transition from type I to type II conditions can be induced by maneuvering flight and can occur with little decrease in dynamic pressure. Thus, aeroelastic response and stability of type II flows can be quite important, although they have only recently been brought within the range of transonic computational methods. Type III flows occur as larger maneuvers are performed and can involve the onset of vortex flows, buffet, and dynamic stall. It must also be recognized that the nature of the airflow may change significantly in going from steady to unsteady conditions. For instance, airflows may exhibit intermittently type I and type II conditions while undergoing structural vibrations induced by atmospheric turbulence or by flutter.

Computational methods available for transonic unsteady aerodynamics include potential equation (both TSD and FP), EE, and NS equation solvers. The thin-layer Navier-Stokes (TL-NS) approximation is obtained by neglecting components of the viscous terms in the coordinate direction normal to the body surface. Also, viscous flow capability may be added to the potential and Euler equations by means of interacted boundary-layer models.

Issues that have been central to unsteady CFD have been the choice of implicit vs explicit algorithms, the stability of alternative solution algorithms, and the treatment of computational grids. Explicit schemes are simple to code and easily vectorizable but are limited in allowable time step by the stability limit imposed by the signal propagation time over the smallest grid cell. Faced with the requirement of maintaining time accuracy throughout the entire field for aeroelastic computations, this easily leads to excessive computation times, especially for viscous flow calculations, where a very fine mesh near the surface is required to resolve the boundary layer. The alternative implicit solution algorithms thus are favored, and attention must be given to their relative stability and accuracy characteristics. Grid generation for unsteady problems in which the body boundary moves, such as for an oscillating control surface or an aeroelastic deformation, raises new issues over those involved in steady flows. To maintain accuracy, the body-conforming grid must be realigned with the body at each time step. Schemes for accomplishing this have been studied as well as the necessity of moving the grid at all. When body motions are small with perturbations mainly normal to the surface, imposing boundary conditions on the mean surface location may be an acceptable approximation. Finally, the nature of unsteady calculations means that the solution is not allowed to achieve a steady state; thus, the dynamic response of numerical calculations on the computational grid is more important. For example, grid cell stretching in the near and far field will affect the computational impedance of the grid for unsteady calculations.

In the sections that follow, the status of transonic unsteady aerodynamics is reviewed. It has been helpful to regard the evolution of this computational capability as following four broad stages: 1) early computational demonstrations, 2) maturation of computational methods, 3) application to realistic configurations, and 4) type II mixed flow computation.

A thread that may be discerned in reviewing this field is the continued evolution of computational methods, with applications and evaluations by comparison with experiment, to successively more difficult cases in order to define the boundaries of validity of the computational method. Thus, for instance, an algorithm that was introduced to treat simple type I attached flow cases is upgraded in capability to enable treatment of more difficult type I cases (closer to the type II boundary in Fig. 3) and possibly even type III cases.

Early Computational Demonstration

Three early computations gave impetus to transonic unsteady CFD: demonstration of type A, B, and C shock motions, calculation of periodic aerodynamic oscillations about a rigid airfoil, and calculation of the aileron buzz boundary for the P-80 aircraft. All three of these cases involve 2D computations. A fourth significant early development was the 3D TSD code described by Borland et al.,[20] which was the first time-accurate code designed for the aeroelastic analysis of swept wings.

Computation of Shock Motions

Tijdeman's[4] type A, B, and C shock motions observed for the NACA 64A006 airfoil with oscillating flap were demonstrated computationally by Magnus and Yoshihara,[5] Ballhaus and Goorjian,[6] and Magnus.[15] Figure 7, from Ref. 6, gives these three calculations from the LTRAN2 TSD code with comparisons of Euler code[5,15] results for types A and B. Magnus[15] gives the corresponding Euler code results for type C shock motion. The

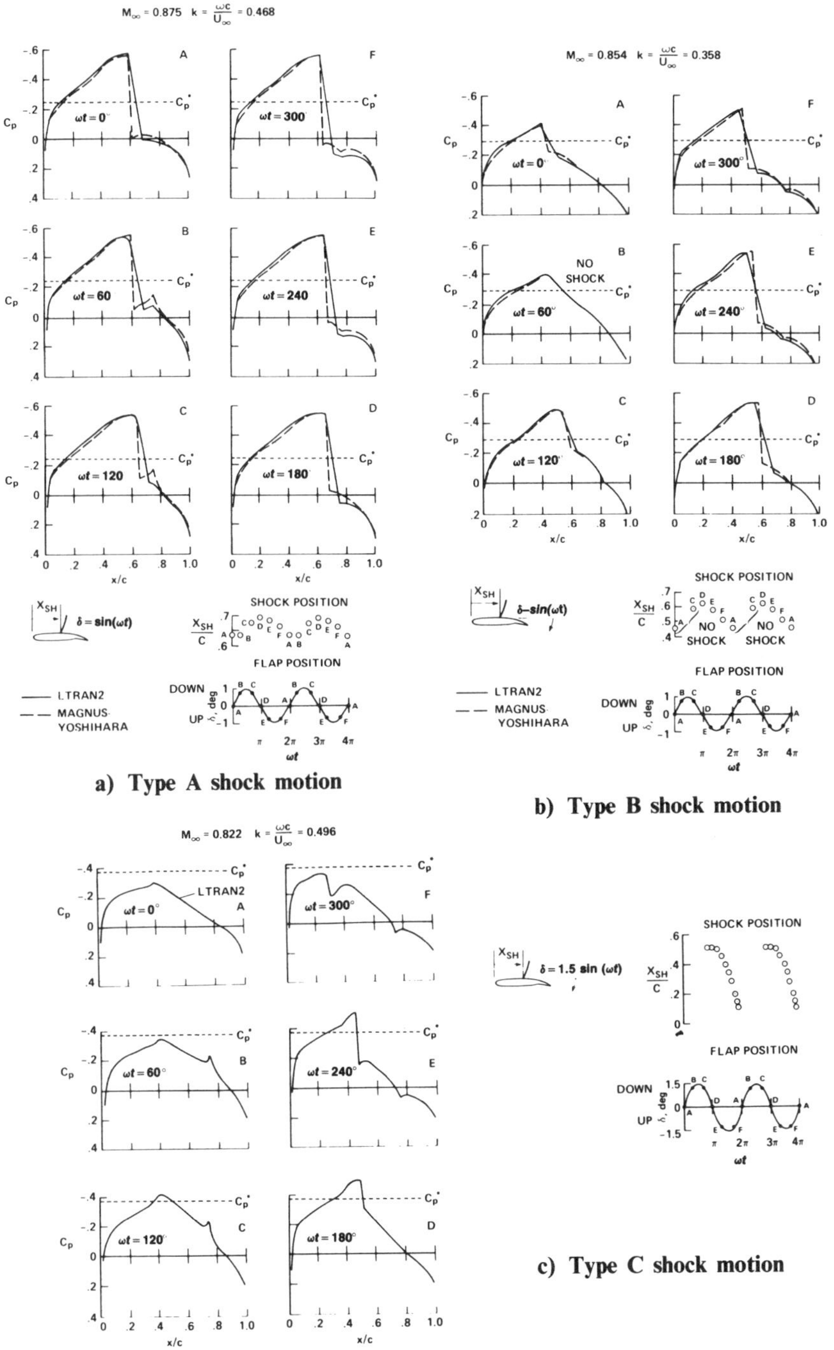

Fig. 7 Upper surface pressure coefficients of an NACA 64A006 airfoil with oscillating trailing-edge flap showing types A, B, and C shock motions.[6]

computational conditions for these three cases are given in Table 3. These conditions are 0.021–0.028 Mach lower than Tijdeman's test conditions, very likely due to wind-tunnel wall interference. The Euler code used by Magnus and Yoshihara was a Lax-Wendroff explicit differencing solution algorithm, and a Cartesian grid with an embedded fine mesh around the moving shocks was used. The boundary conditions were applied at the mean airfoil position. The LTRAN2 code incorporated the ADI algorithm for the solution of the low-frequency TSD potential equation. The Euler code used 5484 grid points and required 1500 s/cycle, whereas the LTRAN2 code required 8 s/cycle (CDC 7600 computer).[6] This significant reduction brought the expense of 2D unsteady transonic CFD calculations within the reach of many researchers.

Periodic Aerodynamic Oscillations

To provide experimental data for validation of viscous flow CFD computer codes, McDevitt et al.[35] conducted tests on a rigid 18% thick circular arc airfoil. Figure 8 illustrates the parameters of the experiment, which was designed to encounter both trailing-edge and shock-induced separations at high Reynolds numbers within the wind-tunnel operating limits. Over a narrow range of Mach numbers, $0.73 < M < 0.78$, oscillatory flow separation was observed (Fig. 9).[37] McDevitt[36] states that the oscilla-

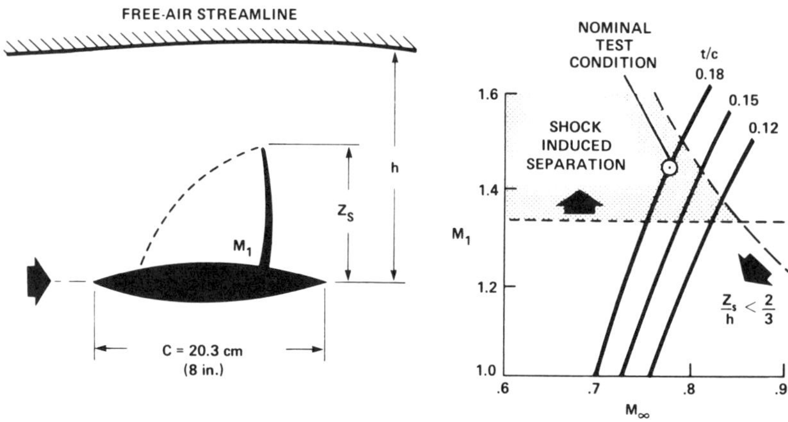

Fig. 8 The 18% thick circular arc airfoil experimental conditions.[36]

Table 3 Shock oscillation conditions for NACA 64A006 airfoil

Type	Mach	Reduced frequency	Flap amplitude
A	0.875	0.234	1.0 deg
B	0.854	0.179	1.0 deg
C	0.822	0.248	1.5 deg

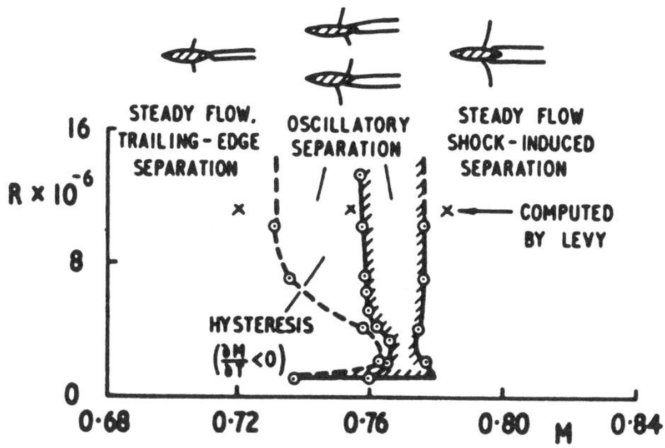

Fig. 9 Steady and oscillatory flow domains for an 18% thick circular arc airfoil.[37]

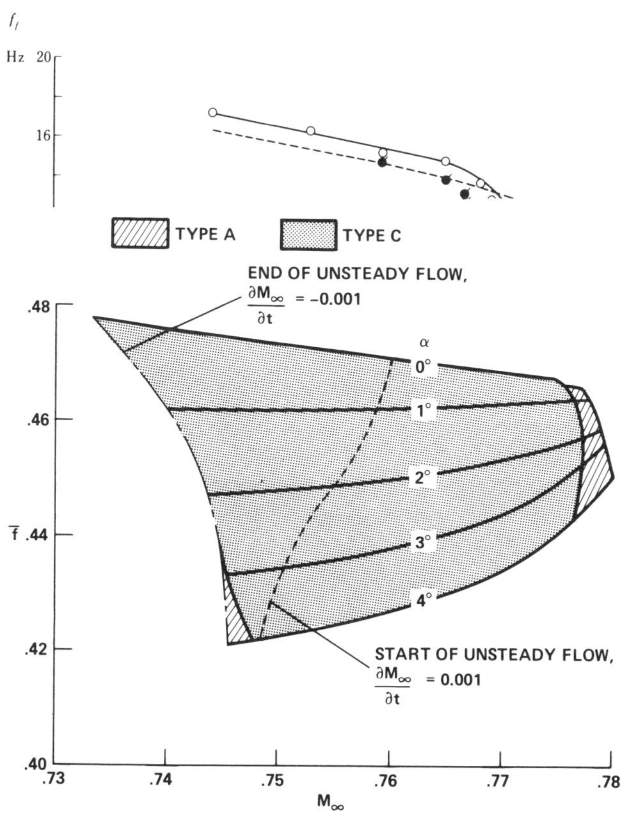

Fig. 10 Reduced frequency of periodic oscillations on an 18% thick circular arc airfoil for varying angles of attack.[36]

tions involve predominantly type C shock motion with small regions of type A motion near the onset of the periodic oscillations (Fig. 10). The reduced frequency of the oscillations is $k \approx 0.48$ for $\alpha = 0$ deg and varies little with angle of attack.

Levy[37] successfully computed such oscillations for this airfoil using an NS flow solver. Levy's code was a modification of the code of Ref. 75 and used MacCormack's explicit solution scheme with an algebraic eddy-viscosity model. Levy modified the code to simulate the contoured wind-tunnel walls. Figure 11 shows steady computed Mach contours for Mach numbers of 0.72 and 0.78, corresponding to trailing-edge and shock-induced separations, respectively, and unsteady flow with oscillatory trailing-edge/shock-induced separation for $M = 0.754$. The reduced frequency of the computed oscillations is 0.40, about 20% lower than the measured frequency. Note particularly the lower surface Mach contours of the third frame for $M = 0.754$. The few lines indicate the collapse of the supersonic region for this portion of the cycle. Also note the dimpled nature of these Mach lines under the airfoil surface. These features will be discussed in more detail later.

Subsequent tests on circular arc airfoils of thicknesses from 10 to 20% were performed by Mabey,[38] who obtained similar periodic oscillations.

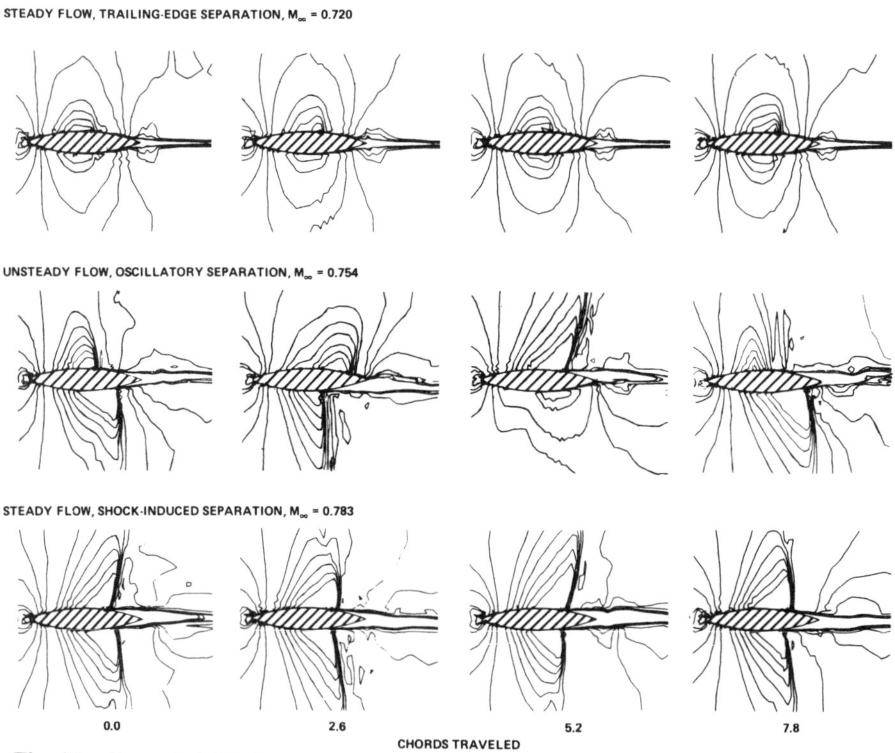

Fig. 11 Computed Mach contours for 18% thick circular arc airfoil; $M = 0.72, 0.75$, and 0.78; $Re_c = 11 \times 10^6$.[37]

The Mach number range of the oscillations increases with decreasing thickness, as does the oscillation frequency, remaining in the range of $0.4 < k < 0.55$ depending on thickness and wind-tunnel wall condition. Mabey et al.[39] and Levy[40] give detailed comparisons of NS calculations with experiment for a 14% thick airfoil, reproducing qualitatively the details of the oscillatory flow.

Calculation of Aileron Buzz

Aileron buzz is a one-degree-of-freedom aeroelastic instability, usually of limited amplitude, which may be encountered under conditions of transonic flow. Steger and Bailey[17] studied such a case for the P-80 aircraft that had been tested in a wind tunnel. They simulated a rigid 2D wing section strip including a free-floating aileron using a thin-layer NS flow solver. They implemented the Beam-Warming implicit approximate factorization (AF) solution algorithm using an algebraic eddy-viscosity turbulence model. A novel treatment of the computational grid was used to follow the aileron motion with a conforming grid. A simple shearing transformation in the coordinate normal to the aileron was used.

Figure 12 shows the limit amplitude "aileron buzz" oscillation that was calculated for $M = 0.82$, $Re = 20 \times 10^6$, and $\alpha = -1$. The calculation was initiated with a 4-deg aileron offset. This and other calculations successfully reproduced the experimental buzz boundary. The computed reduced frequency was $k = 0.38$. The shock motion observed in the calculations was type B, and type II intermittent flow separation is shown in Ref. 17. The code was capable of being run in an inviscid mode (EE mode), and several such calculations were made. Below $M = 0.84$ the aileron exhibited damped oscillations of about $k = 0.36$, whereas divergent oscillations ($k = 0.39$)

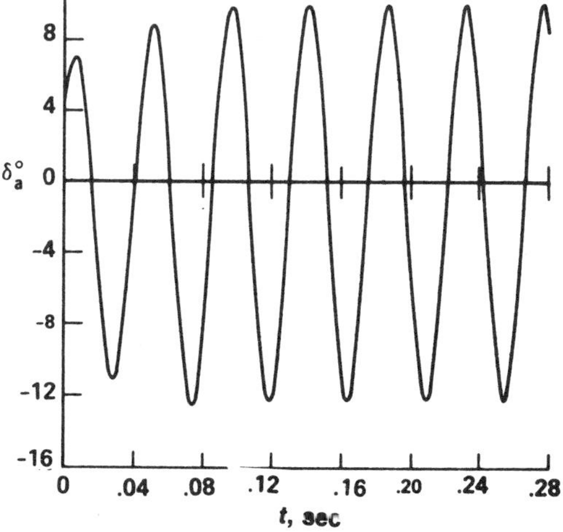

Fig. 12 Computed variation of aileron angle with time showing buzz condition; $Re = 20.7 \times 10^6$.[17]

were calculated at $M = 0.84$. Hence, the tendency to oscillate at a given frequency derives from the inviscid flow equations, and the viscosity apparently plays the key role of limiting the amplitude of oscillation.

These calculations were performed on a 76×42 grid and required approximately 1.5 s of CDC 7600 computer time per time step or 460 μs per grid point per iteration. Nondimensional time steps of 0.005–0.01 were used (based on chord).

Swept-Wing Aeroelastic Analysis

Borland et al.[20] and Borland and Rizzetta[21] describe the development of the XTRAN3S TSD code for aeroelastic analysis of swept wings. Reference 20 describes the extension of the LTRAN2 ADI solution algorithm to three dimensions for the solution of the 3D low-frequency TSD equation. Several cross-derivative terms required explicit treatment yielding a semi-implicit algorithm. Results for an oscillating swept wing are given in Ref. 20. Computations on a $60 \times 20 \times 40$ grid required 3 s of CPU time per time step on the CDC 7600 computer (62 μs per grid point per iteration).

Reference 21 describes the extension of this code to solve the complete TSD equation and its application to coupled structural dynamic/unsteady aerodynamic analysis. Samples of flutter calculations for a rectangular wing are given. Computations on a $60 \times 20 \times 40$ grid required 4 s of CPU time per time step on the CDC 7600 computer. Static aeroelastic solutions were obtained with 300–400 steps and dynamic aeroelastic solutions in about 1000 steps with nondimensional time steps of 0.1. It is noted, however, that this time step is case dependent, depending on planform, Mach number, etc. This capability is noteworthy because it is the first complete wing flutter analysis accomplished using a time-accurate transonic computational method.

A comparison of calculations from the XTRAN3S code with experimental data was given by Seidel et al.,[76] who performed calculations for pitching oscillations of the NASA RSW. Figure 13a shows steady and unsteady pressure distributions for two span stations at $M = 0.7$ and $\alpha = 2$ deg. For this low transonic condition, the agreement with experiment is good except near the leading edge, where grid refinement is probably needed. The unsteady results in Fig. 13b are in good agreement with experiment over most of the chord. Inboard, there is an improvement in the prediction of the leading-edge suction peak over the linear theory RHOIV result. Outboard, there is an underprediction of the leading-edge peak and evidence of viscous effects in the phase angle in the aft cove region. Calculations for this RSW model have also been reported in Refs. 25, 34, and 77.

Maturation of Computational Methods

In this section, the further development and application of the computational methods introduced in the last section are described. These developments have typically been introduced in 2D codes. Hence, the exposition emphasizes 2D methods with parallel 3D results interspersed.

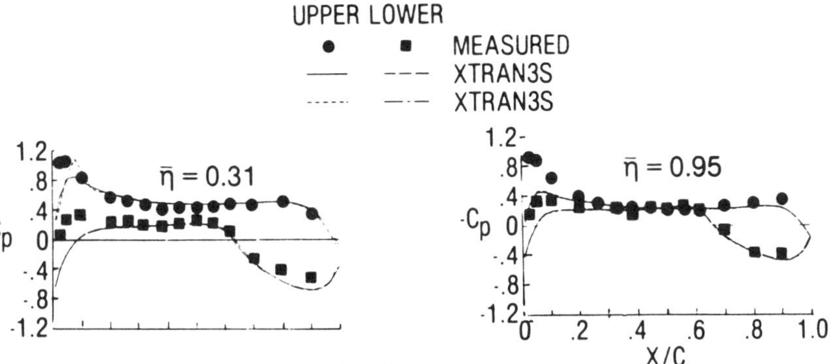

a) Steady pressures

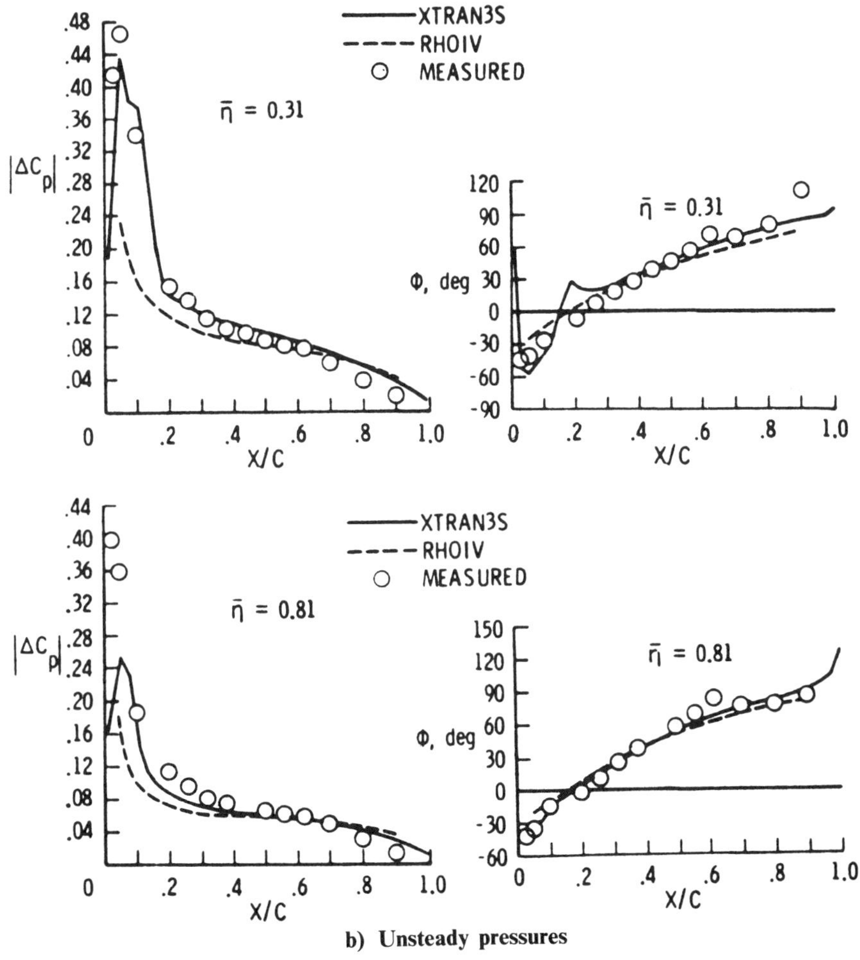

b) Unsteady pressures

Fig. 13 Comparison of measured and calculated steady and unsteady pressures for the NASA rectangular supercritical wing.[76]

Augmentations to Potential Methods

Many researchers have participated in augmenting the capability of unsteady potential codes. These developments have extended the range of applicability of transonic potential codes by improved physical modeling and by improved numerical stability or accuracy.

The low-frequency LTRAN2 ADI algorithm was limited in accuracy above $k \approx 0.1$, and additions of the neglected time derivative terms are described in Refs. 78–83. Chow and Goorjian[82] describe the addition of high-frequency terms and special treatment of the far-field boundary conditions to account for the physical domain of dependence for supersonic freestream conditions. This enabled calculation of 2D supersonic results and enhanced the stability of the code. Increased stability was also added to the LTRAN2 code by Goorjian et al.[84] with the replacement of the Murman-Cole differencing method by the monotone differencing method of Engquist and Osher.[85] This method eliminates the nonphysical formation of expansion shocks in the leading-edge region and allowed a large increase in allowable time-step size.

The far-field boundary conditions imposed can have special significance for unsteady calculations, since disturbances emanating from these boundaries may return to the vicinity of the airfoil and contaminate the solution. Kwak[86] implemented nonreflecting boundary conditions in the LTRAN2 code and showed that the computational grid could be made smaller with reduced computer costs. Whitlow[87] implemented similar conditions for the complete TSD potential equation. Reflections from internal grid points can also occur if the grid stretching is excessive. Seidel et al.[76] show results indicating that grid stretching in the direction normal to the airfoil is most critical.

The XTRAN2L TSD code described by Whitlow[87] incorporates the features mentioned earlier, and typical results from this code are shown in Figs. 14 and 15 for the NACA 64A010A airfoil. Figure 14 shows the steady pressure distribution for the moderate shock case of Davis and Malcolm,[62]

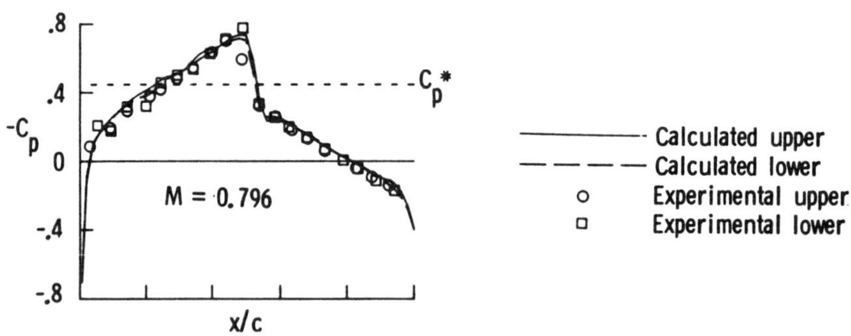

Fig. 14 Steady pressure distribution for the NACA 64A010A airfoil at $M = 0.796$ and $\alpha = 0$ deg.[89]

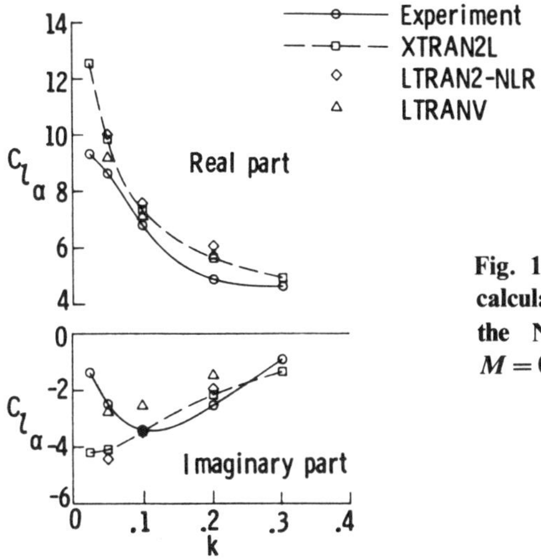

Fig. 15 Comparison of measured and calculated unsteady lift coefficient for the NACA 64A010A airfoil at $M = 0.796$ and $\alpha = 0$ deg.[89]

whereas Fig. 15 compares the unsteady lift coefficient with experiment and the LTRAN2-NLR and LTRANV (with interacted viscous effects) results of Houwink.[88] Agreement is good for moderate frequencies, but there is a characteristic overprediction of the imaginary part for the lowest frequencies. Similar comparisons for a number of the 2D AGARD configurations are given in Refs. 89–93.

Batina[94] studied the effect of airfoil shape, thickness, camber, and angle of attack on unsteady airloads using the XTRAN2L code. An interesting feature shown in Ref. 94 is illustrated in Fig. 16, which shows frequency response functions of pitching moment due to a pulsed airfoil pitching motion for three different airfoils. Fourier transform analysis of such motions and airloads has been shown to be very useful in transonic unsteady aerodynamic calculations,[76] notwithstanding the nonlinear nature of the flow equation. Note the "hump" in the real part and the coincident "wave" of the imaginary part of the pitching moment curves near $k = 0.6$. Such resonance features are familiar to dynamicists as similar to the response of damped mechanical oscillatory systems. Examples from Ref. 94 show that the frequency and strength of this resonance feature are primarily a function of Mach number, with airfoil shape and thickness being secondary effects.

Berry et al.[95] show a similar resonance near $k \approx 0.25$ in the pitching moment due to flap deflection for the NACA 64A006 airfoil at $M = 0.85$. New calculations for this airfoil are shown in Fig. 17 for $M = 0.822$ and 0.854. These are the conditions shown in Fig. 7 for type B and type C shock motion. The symbols in Fig. 17 give the complex valued pitching moment coefficients for these two cases derived from separate harmonic oscillation calculations. Note that the type B shock motion (which is reproduced by the XTRAN2L code) was observed near the "resonance" frequency at $k \approx 0.2$ for $M = 0.854$, whereas the type C shock motion is observed at a frequency approximately one-half the resonance frequency at $M = 0.822$.

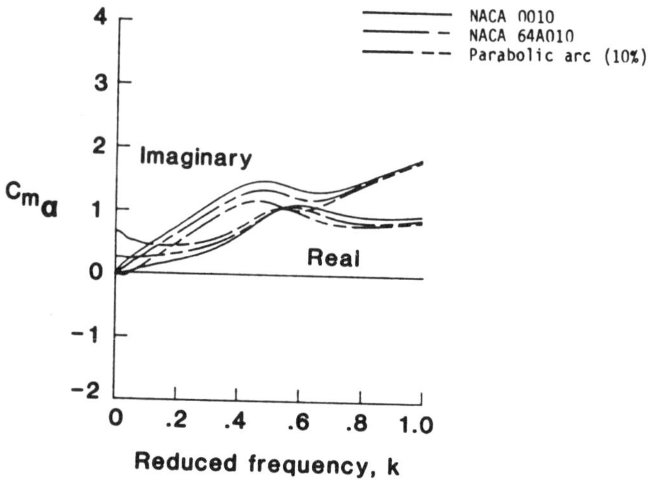

Fig. 16 Comparison of airfoil shape effects on unsteady pitching moment coefficient for pitching oscillations at $M = 0.78$.[94]

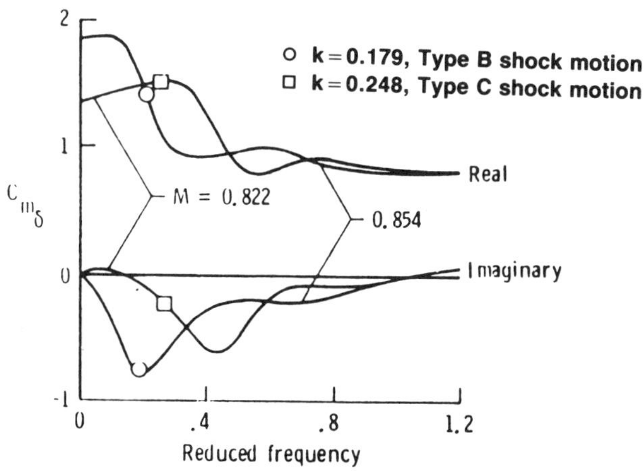

Fig. 17 Pitching moment coefficient due to flap motion for the NACA 64A006 airfoil.

This resonance-type structure is not observed in the pitching moment frequency response function for $M = 0.875$ (type A shock motion), where a strong shock has formed.

Nonunique Solutions of the Potential Equation

Steinhoff and Jameson[96] have demonstrated that the potential equation for transonic flow can have nonunique solutions due to entropy generation within shocks violating the isentropic assumption upon which the theory is based. Williams et al.[97] have studied this effect for the TSD equation.

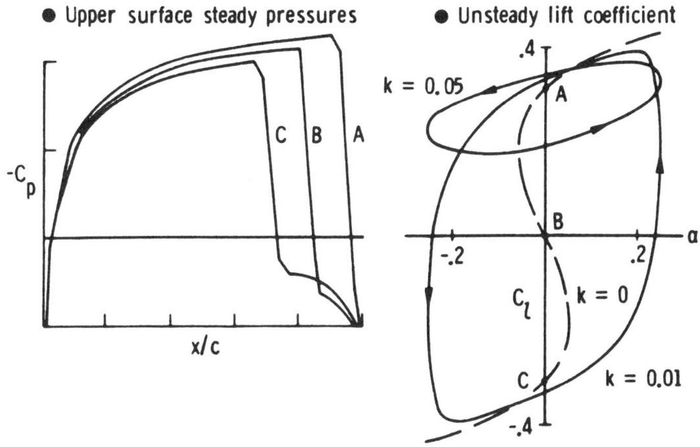

Fig. 18 Anomalous behavior of potential flow solutions for NACA 0012 airfoil at $M = 0.85$ and $\alpha = 0$ deg.[97]

Figure 18 illustrates the multiple solutions obtained for the NACA 0012 airfoil at $M = 0.85$. The unsteady results given on the right of the figure ($k \neq 0$) demonstrate that large phase errors may occur at low reduced frequencies as the solution jumps between the stable equilibrium solutions at A and C. Fuglsang and Williams[98] have developed modifications to the TSD equation to circumvent this problem. These nonisentropic modifications involve a modified flux term, a correction to the pressure coefficient equation, and a correction to the wake jump condition to allow entropy convection. These modifications correct the multiple solutions and enable the use of the TSD equation for a wider range of problems. Nonisentropic cases studied by Fuglsang are noted in Table 1. Gibbons et al.[99] extended this method to 3D using the XTRAN3S TSD code and found evidence of similar multiple solutions. Figure 19 shows time histories of lift coefficient following a pitch pulse excitation. For aspect ratios greater than 24, the lift diverges to a nonzero steady state. Gibbons implemented the 2D modifications discussed earlier stripwise in the XTRAN3S code, obtaining the results in Fig. 20. The unmodified code predicts a lift-curve slope that is twice as large as that obtained with a Euler code, whereas the modified code agrees with the Euler code result. Gibbons also shows results for the RAE AGARD tail plane model, where the nonisentropic modifications resulted in a forward shift of the shock near the wing tip of nearly 10% chord. Further studies of this effect are needed to give a complete assessment of this effect for potential codes. Finally, Osher et al.[100] and Whitlow et al.[101] have also developed entropy condition satisfying approximations for the unsteady FP equation.

Treatment of Moving Grids

For solution of flow equations above the TSD level, body-conforming grids are generally used. For aeroelastic applications in which the structural boundary moves from iteration to iteration, the issue of redefining the grid at each step must be addressed. Generally, the problem of grid definition is of such difficulty that approximate methods have been sought. Chyu and co-workers[18,19] used an interpolation scheme for defining grids at intermediate steps between the extremes of motion for oscillating airfoils. They used Steger's[102] Navier-Stokes code to study the moderate shock case[18] and the shock-induced separation case[19] for the NACA 64A010A airfoil, shown in Figs. 21 and 22. Note that the full and thin-layer Navier-Stokes results in Fig. 21 show no differences except near the shock. This is an important demonstration of capability for types II and III flows with intermittent

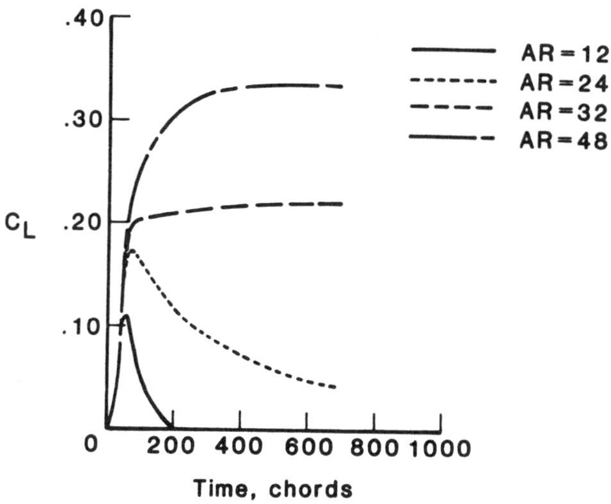

Fig. 19 Lift coefficient time histories due to pulse for NACA 0012 wing at $M = 0.84$ as a function of aspect ratio.[99]

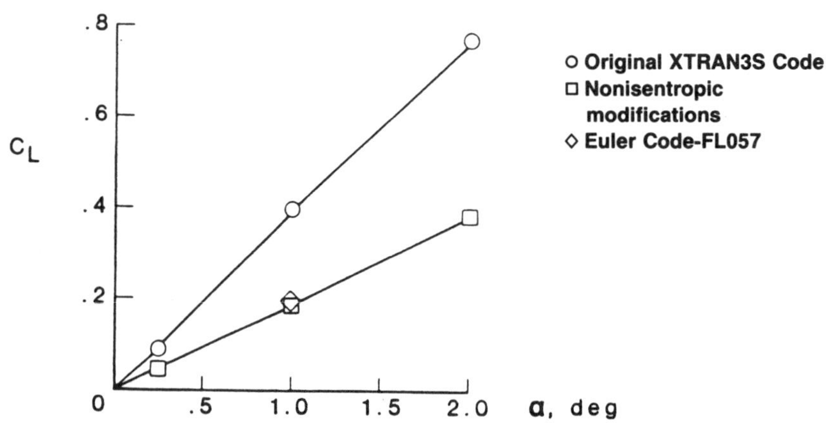

Fig. 20 Comparison of potential and Euler solutions of total lift coefficient for NACA 0012 wing with AR = 12 at $M = 0.82$.[99]

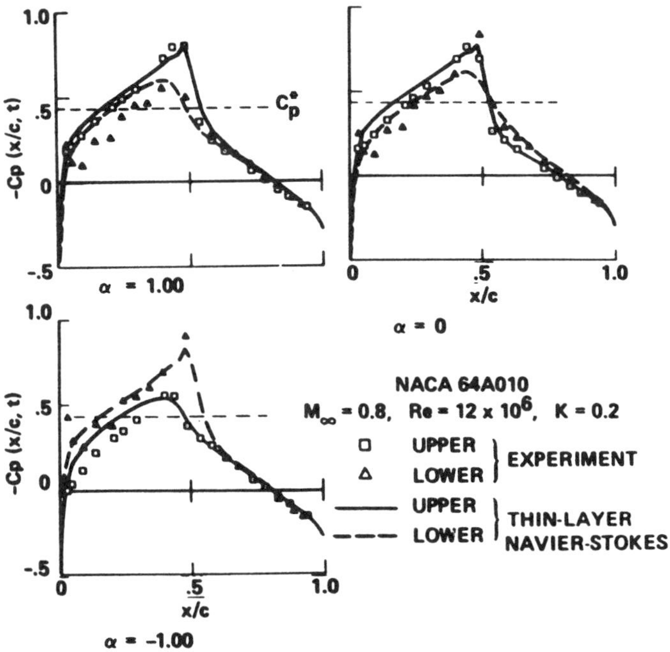

Fig. 21 Comparison of calculated and measured pressure coefficients for NACA 64A010A airfoil for moderate shock condition, $\alpha = 0$ deg + 1 deg cosωt.[18]

separation. These calculations were obtained using 2620 steps per cycle of oscillation, and the times per step on the Cray X-MP computer were 0.33 s, full NS; 0.22 s, TL-NS; and 0.17 s, EE, corresponding to 25–44 μs per grid point per time step.

If the surface motion is small and primarily in the direction normal to the x axis, a suitable engineering approach may be to apply the boundary conditions at the mean surface location with suitable terms added to the boundary condition to simulate surface motion. Sankar et al.[103] have made a preliminary evaluation of this method for unsteady calculations using 2D and 3D FP codes. For the 2D code, a case with an oscillating trailing-edge flap with 1 deg amplitude was studied. They found that results using the exact boundary conditions with a moving grid and the transpiration approach differed by no more than 10% for surface pressures and integrated loads. Further studies are required before the viability of this transpiration boundary condition method can be assessed.

Interactive Viscous Modeling

Within the range of type I unsteady aerodynamic problems (Fig. 3), viscous effects become more important with increasing lift coefficient or Mach number. Further increases lead to the development of type II mixed attached and separated flow where viscous effects must be addressed. This borderline region between type I and type II flows is important since design operating conditions frequently occur here. Thus, simplified methods of accounting for viscous effects without resorting to NS codes have been sought. These interacted boundary-layer methods have most frequently coupled an inviscid potential flow outer region solver with an integral boundary-layer equation inner region viscous flow solver. The two solutions are coupled iteratively through the boundary conditions to ensure consistent coupling. In the "direct" coupling method, the inviscid flow solution provides the pressure coefficient at the edge of the boundary layer, enabling a direct solution for the boundary-layer displacement thickness. This thickness distribution is then added to the airfoil geometry to provide the boundary condition for the next iteration of the inviscid solution. For

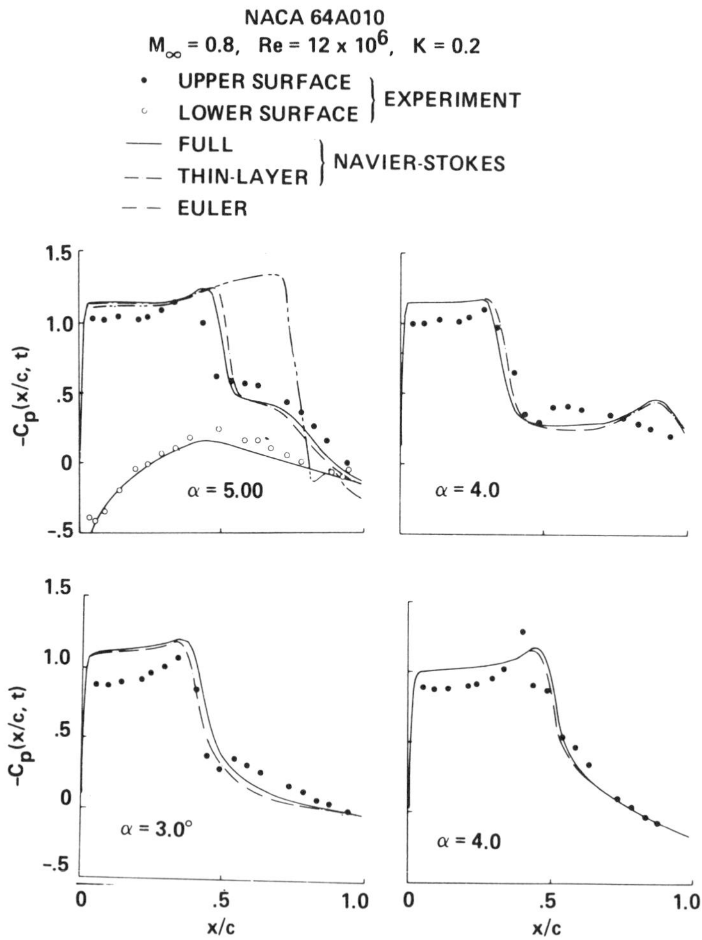

Fig. 22 Comparison of calculated and measured pressure coefficients for NACA 64A010A airfoil at shock-induced separation condition, $\alpha = 4$ deg + 1 deg cosωt.[19]

separated flows, it is necessary to invoke an "inverse" boundary-layer method wherein the displacement thickness is specified and the edge pressure coefficient calculated. Consistency of the inner and outer solutions is maintained by iterating the solution at each step.

Couston et al.[104] published unsteady interacted solutions using a TSD potential code[105] and a viscous defect formulation of the boundary-layer equations. Results for attached flows for the NACA 64A006 airfoil are given. In subsequent calculations, Le Balleur and co-workers, in Refs. 66 and 106–108, have pursued this technique for a wide range of attached and separated flow cases. Cases of airfoils with slats and flaps and airfoils with spoilers are treated. Also, cases of periodic oscillations on the 18% biconvex airfoil[66,109] and the RA16SC1 supercritical airfoil[66] are described.

Similar interacted viscous solutions are given by Houwink[88,110] and by Rizzetta.[111] Applications of Rizzetta's method are given in Refs. 95 and 112–115. Ruo et al.[77] show 3D calculations using 2D strip and 3D integral boundary-layer methods with an FP inviscid code. The viscous defect formulation used by Le Balleur et al. retains time accuracy for the boundary-layer equations, whereas Houwink and Rizzetta's methods use a quasisteady (i.e., time-independent) boundary layer. A comprehensive assessment of flow conditions requiring time-accurate viscous effects would be a valuable addition to the field. Tables 1 and 2 note references in which interacted viscous modeling has been used in comparisons with experimental data.

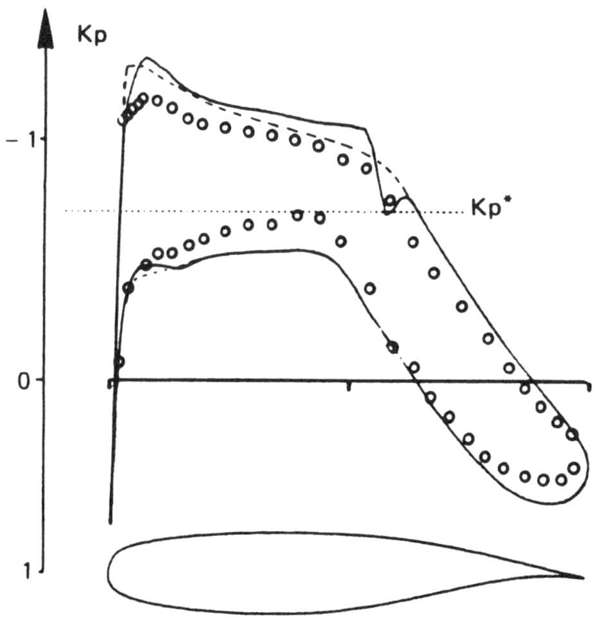

Fig. 23 Pressure coefficient distribution around the NLR 7301 airfoil in steady flow: solid line, Euler calculation; dashed line, hodographic method for $M = 0.721$, $\alpha = -0.19$ deg. Circles, NLR experiment for $M = 0.744$, $\alpha = 0.85$ deg.[30]

Solution Algorithms

Advances in solution algorithms for unsteady CFD have been made for all levels of physical flow modeling. Sidès[30] summarizes work aimed at making time-accurate Euler equation solutions more efficient using an explicit predictor-implicit corrector algorithm. Artificial viscosity is added only in regions of high gradients, and stable solutions for a Courant number of 12 are shown. Results for the NLR 7301 airfoil at its design shock-free condition, $M = 0.721$ and $\alpha = -0.19$ deg, are given in Figs. 23–25. Figure 23 compares calculations from the Euler code and a hodograph method with experimental data. The computational conditions differ from the test conditions due to wind-tunnel corrections and boundary-layer effects.[10] The two calculations agree well except for a weak shock in the Euler solution. It is noted that the Euler solutions were very sensitive to small changes in either Mach number or angle of attack about this condition. Figure 24 shows the steady isobar lines about this airfoil calculated by the Euler code. Note the nearly horizontal isobar pattern shown in the supersonic region over the top of the airfoil. Results are presented for flap oscillations at a reduced frequency of $k = 0.068$, and Fig. 25 shows instantaneous isobar patterns for $\omega t = 0$ and 180 deg. Note the dimpled isobar pattern on the upper surface for $\omega t = 0$ deg. Very good agreement of computed and measured unsteady pressures is shown in Ref. 30 for this case. These calculations were performed on a 188 × 24 grid and required 2180 steps per cycle, using 315 s/cycle on a Cray 1S computer, corresponding to 32 μs per grid point per time step. A factor of 10 increase in speed over the explicit method was achieved with this semi-implicit method.

Another semi-implicit solution algorithm is described by Isogai[26] for the FP equation. The USTF3 code solves the nonconservative 3D FP equation and also implements a quasisteady 2D strip viscous boundary-layer method. Several results for the AGARD standard wings are shown, and significant aeroelastic flutter calculations are compared to experimental flutter results, which will be discussed later.

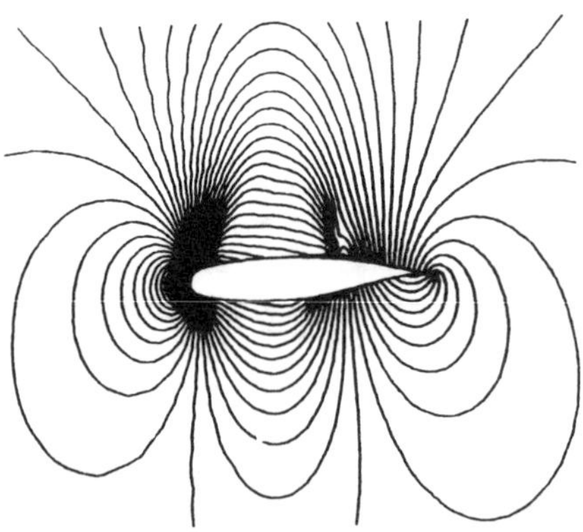

Fig. 24 Isobar lines for NLR 7301 airfoil in steady flow, under design conditions; $M = 0.721$, $\alpha = -0.19$ deg.[30]

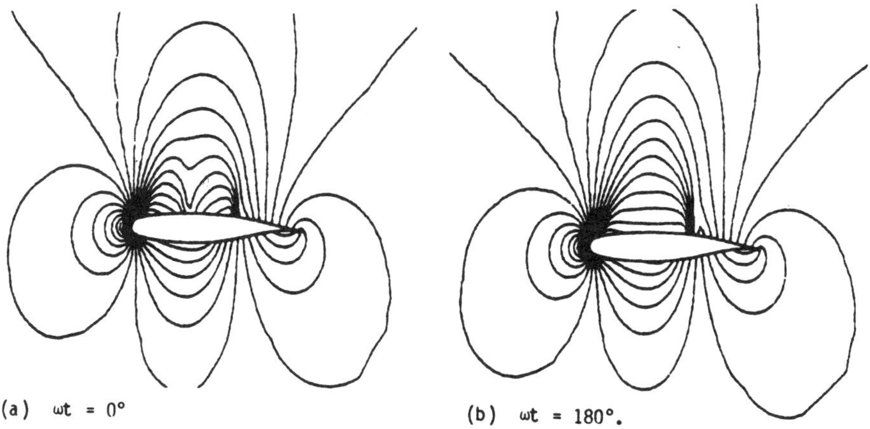

Fig. 25 Pressure field around the NLR 7301 airfoil with oscillating flap; $M = 0.721$, $\alpha = -0.19$ deg, $\delta = 1$ deg sinωt, $k = 0.068$.[30]

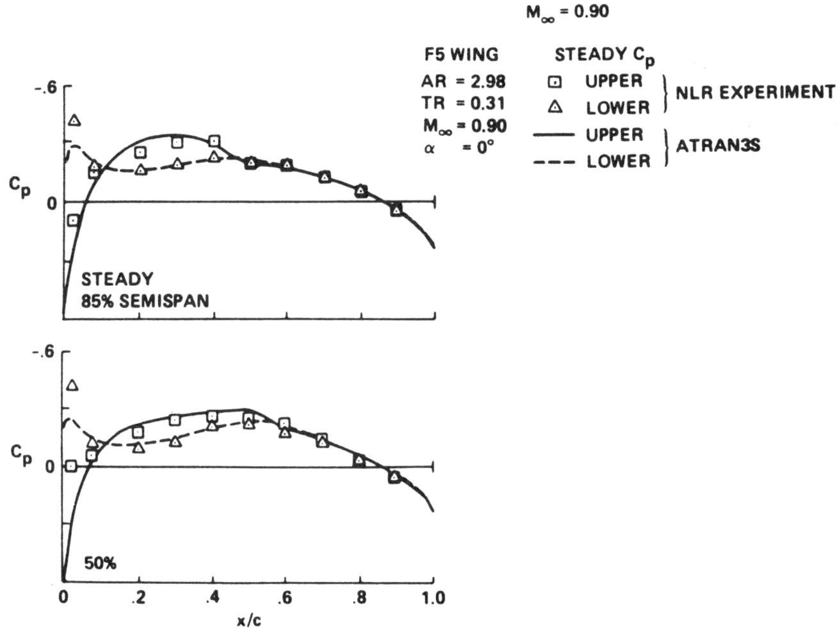

Fig. 26 Comparison of calculated and measured steady pressures for the F-5 wing model; $M = 0.90$.[116]

The semi-implicit ADI solution algorithm used in the XTRAN3S TSD code required increasingly small time steps to maintain stability as wing sweep and/or taper was increased. Guruswamy and Goorjian[116] replaced the grid shearing transformation of the code with a modified shearing transformation that mapped from a rectangular physical domain onto a rectangular computational domain. This alleviated the stability problem, allowing time steps of 0.01 for the F-5 wing model, whereas time steps an order of magnitude smaller had been marginal with the original method. Figures 26 and 27 show the steady and unsteady pressures calculated for this aspect ratio 2.98 wing. The leading-edge sweep angle is 32 deg, and 1200 steps per cycle were used. Both steady and unsteady computations agree closely with experimental results and are representative of other applications of both TSD and FP codes listed in Table 2.

Two advances in solution algorithms for the unsteady FP equation are reviewed next. The first was the development by Malone and Sankar[117] of a strongly implicit AF scheme for the 2D conservative potential equation using body-fitted coordinates. Upwind density biasing is used for stability in supersonic zones. Results from the 2D SUNTANS code are given in Refs. 117 and 118 for several of the AGARD 2D configurations. Malone et al.[24] have also developed a 3D version of the strongly implicit AF algorithm and have published comparisons of unsteady pressures with data for the F-5 wing model,[24] the LANN wing model,[77] and the NASA RSW model.[77] Figures 28 and 29 show steady and unsteady pressures at two span stations for the F-5[24] at $M = 0.95$. The agreement with experiment is quite good. This USIPWING code is implemented on the VAX 11/780 computer. Three-dimensional unsteady results are typically obtained with 240 steps per cycle using nondimensional time steps of 0.03–0.06.

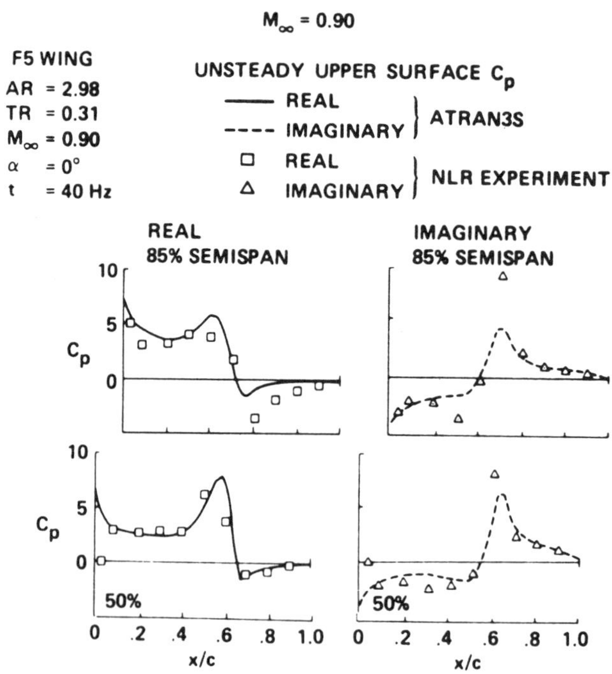

Fig. 27 Comparison of calculated and measured unsteady pressures for the F-5 wing model; $M = 0.90$.[116]

The second FP algorithm development to be noted was the development of the flux-biased AF method by Osher, Hafez, and Whitlow.[100,101,119] The method satisfies an entropy inequality, ruling out expansion shocks and accurately tracks sonic conditions, requiring no empirical constants to specify the amount of artificial viscosity needed for stability. Whitlow et al.[101] give details of the method implemented in a 2D FP code and show examples for the NACA 0012 airfoil. Shankar et al.[120] and Shankar and Ide[25] have used the flux-biased AF technique in a Newtonian linearization method with internal iterations to reduce the linearization and factorization errors and efficiently treat unsteady flows with large time steps. Reference 120 presents results for the NACA 0012 airfoil for nondimensional time steps of 0.1, 0.2, and 0.3 (corresponding to CFL numbers of 35, 70, and 105). The number of internal iterations is increased with the larger time steps to maintain accuracy. For unsteady calculations with two internal iterations, the 2D code required 0.09 s per iteration on a Cray X-MP computer with a 181×30 grid, corresponding to 16 μs per grid point per time step. In Ref. 25 an extension of this flux-biased AF method to 3D is described. Sample calculations for the NASA RSW model are given, and excellent stability characteristics are quoted. On a $100 \times 16 \times 20$ grid, calculations were obtained for 100 time steps per cycle (Courant numbers greater than 100) in about 60 s on a Cray X-MP computer, corresponding to 18 μs per grid point per time step.

Batina[23] has implemented a 3D AF algorithm for the TSD potential equation that is similar to Shankar's[120] method in using a Newtonian linearization in time coupled with internal iterations. Results for the F-5 wing model at $M = 0.9$ were calculated for 100, 200, 300, and 400 steps per cycle. The results for 100 steps per cycle indicated that the time step was too large for engineering accuracy. Comparison calculations with the XTRAN3S ADI algorithm required 2000 steps per cycle. An advantage of such an efficient 3D TSD code is that body-conforming grids are not required and codes may be configured to treat complex geometries. Of course, this advantage must be balanced against the loss of some physical modeling fidelity with the small-disturbance approximation. It seems certain that a variety of unsteady computational methods, with a range of physical modeling approximations, will eventually be used within the design process.

The time-dependent Euler equations form a hyperbolic system of equations, and much of the recent progress in algorithm development[121–126] has hinged on the incorporation of the signal propagation features of the differential equation into the numerical algorithm. There are several methods of incorporating this information into a difference scheme, for example, flux-vector or flux-difference splitting, and an excellent review of the current developments in the field is given by Roe in Ref. 127. The advantages of incorporating an upwind-biased discretization into a numerical algorithm are twofold: 1) the scheme becomes naturally dissipative so that no adjustable constants need to be fine-tuned to a particular application, and 2) improved implicit schemes can be devised for more efficient solutions to both steady and time-dependent problems. Both of these advantages offset the disadvantage that approximately twice as many operations per time

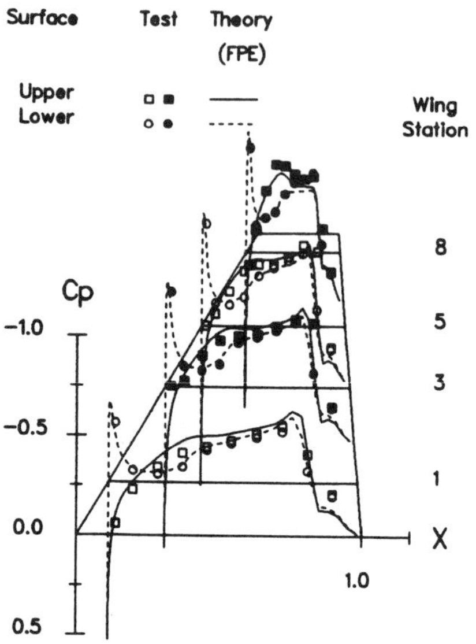

Fig. 28 Comparison of calculated and measured steady pressures for the F-5 wing model; $M = 0.95$.[24]

step are required to implement an upwind scheme as opposed to a central difference scheme.

Most of the calculations made to date with upwind difference schemes, especially for airfoils/wings, have been steady-state applications, for which comparable accuracy can be obtained by central difference methods with added artificial viscosity. The advantages of upwind differencing should be more significant for time-dependent problems, however, where the ability to treat rapid movement of flows with shocks is required. Roe[127] gives several examples of shock-propagation computations in two dimensions, which demonstrate clearly the advantages of a characteristic-based scheme. Viscous effects can also be readily introduced into upwind difference schemes developed for the Euler equations by central differencing the shear-stress/heat-transfer terms.[128,129]

The time-accurate computations made by Steger and Bailey[17] and Chyu et al.[18,19] used a spatially split AF (ADI) scheme, which is unconditionally stable in two dimensions but is at most conditionally stable in three dimensions. Alternate factorizations are possible with the incorporation of an upwind difference discretization in one or more coordinate directions, which can lead to unconditionally stable 3D algorithms.[121] A two-factor eigenvalue-split scheme for the Euler equations has an increased stability limit and fewer operations than the spatially split scheme, although the operations are not completely vectorizable. Belk[34] computed steady and time-dependent inviscid flows for the NASA RSW model with such an algorithm in combination with a blocked-grid strategy. Ying et al.[130] used upwind differences in a single coordinate direction and constructed a two-factor unconditionally stable algorithm for which thin-layer viscous effects are readily incorporated. Applications of the thin-layer NS equations to the high-angle-of-attack unsteady flow over a hemisphere cylinder are

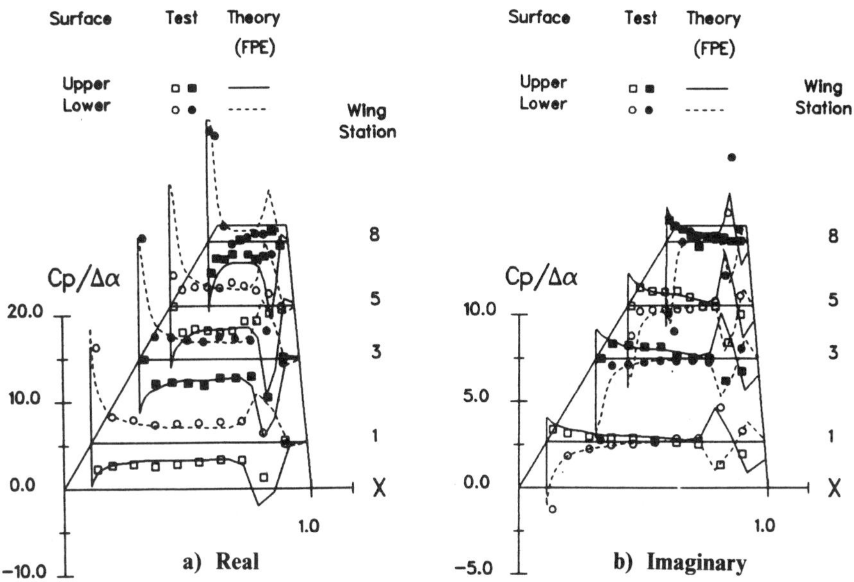

Fig. 29 Comparison of calculated and measured unsteady pressures for the F-5 wing model; $M = 0.95$.[24]

made.[130] Several of these alternate factorizations are investigated in the context of efficient algorithms for three-dimensional steady-state problems by Anderson et al.[131,132]

The use of multigrid techniques to accelerate convergence to the steady state is becoming widespread in the aerodynamic community. These techniques can also be used for time-dependent flows. For instance, multigrid techniques could be used to efficiently solve the large banded matrix equations arising from implicit time discretizations, the solution of which is generally approximated through relaxation and/or factorization methods. Jesperson[133] has demonstrated a time-accurate multiple-grid procedure that was used to overcome the small time-step limitation of an explicit scheme. With the growing memory of today's computers (the Numerical Aerodynamic Simulator has 256 million words of memory), it becomes feasible to solve the banded matrices by direct Gaussian elimination, rather than by approximate techniques. The structure of future implicit algorithms for both steady and time-dependent problems will likely involve a multiple-grid algorithm with direct elimination techniques used on the coarser grid levels.

Treatment of Realistic Configurations

All results presented thus far have dealt with unsteady transonic aerodynamics for isolated lifting surfaces. To realize necessary improvements over existing aeroelastic analysis methods, computational aeroelastic analysis will be required to provide reliable predictions for complex configura-

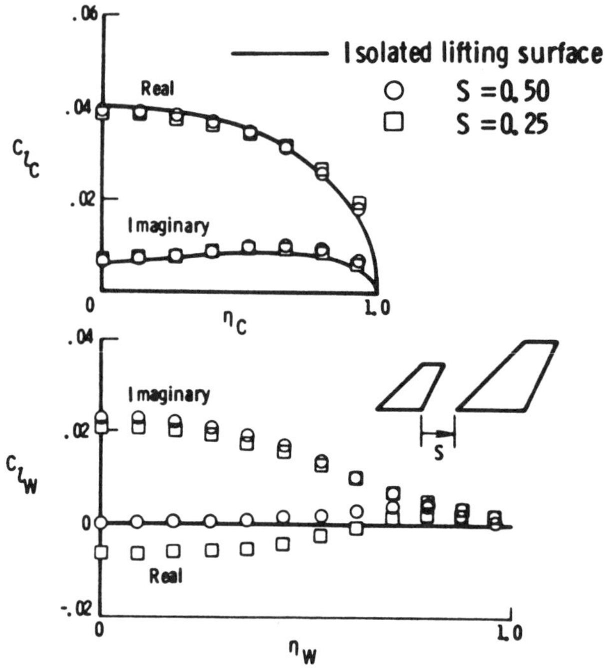

Fig. 30 Unsteady sectional lift coefficients on canard-wing configuration due to canard pitching at $M = 0.90$, $\alpha = 2$ deg, and $k = 0.3$.[134]

tions. The capability to treat such details as wing-fuselage interference and wing-pylon-store effects on flutter boundaries is needed. It is encouraging that studies directed at building the capability of treating these more realistic configuration details are progressing.

Batina has added to the XTRAN3S code separate capabilities for treating interfering lifting surfaces[134] and fuselage-wing combinations.[135] The first study is a direct extension of the ADI algorithm to treat two lifting surfaces. Figure 30[134] illustrates a case of canard-wing interference in which unsteady loads are induced on the wing by the oscillating canard. This effect is obviously a function of the separation distance between the surfaces and the Mach number and frequency. Shankar and Goebel[136] have also published a treatment of canard/wing modeling for the unsteady FP equation. Steady results are shown for a canard/wing research model and the HIMAT vehicle, but no unsteady results are given. Batina[135] describes the treatment of the fuselage that was implemented in the XTRAN3S code. Boundary conditions are applied on a computational fuselage surface aligned with coordinate lines and are formulated in terms of projected areas and arc lengths of the true fuselage. Comparisons with existing steady data are favorable. Figure 31 shows the calculated interference effect of the RAE wing/fuselage model for an assumed wing bending mode. The interference effect on the integrated generalized force, important for aeroelastic analysis, is approximately 5% of the total. Malone and Sankar[137] have also ad-

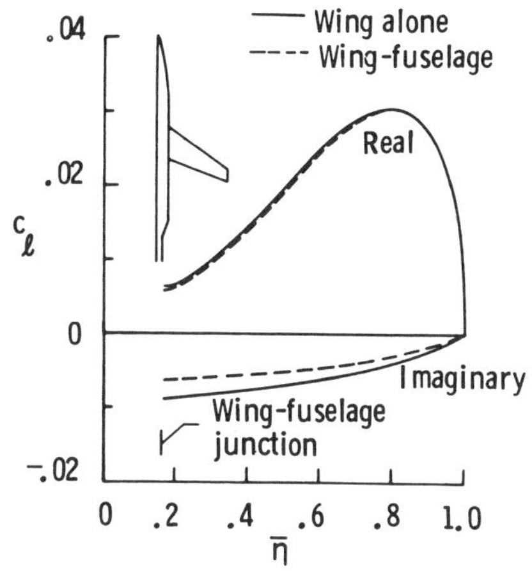

Fig. 31 Unsteady sectional lift coefficients due to assumed wing first bending mode for the RAE wing-fuselage at $M = 0.91$, $\alpha = 1$ deg, and $k = 0.25$.[135]

dressed the treatment of wing-fuselage modeling. They used a body-fitted coordinate system with the USIPWING code and present comparisons with steady data from a research transport wing, the F-14 airplane, and an RAE wing/fuselage model. While unsteady results are shown, there are no experimental data for comparison.

Heavy, concentrated masses such as engines, stores, and tanks can have a large effect on flutter boundaries, and sometimes placement of these components is critical to meeting design objectives. Guruswamy et al.[138] have augmented the ATRAN3S code to treat a wing with a tip store. The store is modeled within the TSD approximation by treating the tip store as an extension of the wing planform. Comparisons with data from the F-5 wing model with a tip store[14] are presented.

Thus, a promising start has been made in the treatment of realistic configurations. The trend to be observed in these examples is that the capability for an additional component has been added to existing codes that treat isolated wings. Based on the recent progress in devising steady flow solution methodology for complex configurations, one might anticipate the development of codes aimed ab initio at more complete configurations in the near future. Another evident requirement is for the necessary experimental data to validate codes for such configuration details. Such tests are quite expensive and require much time for planning and execution. In the meantime, careful attention to comparisons with available steady data is a necessary starting point in any case.

Aeroelastic Applications

The major intended use of unsteady aerodynamic calculations is for prediction of aeroelastic response of aircraft and, more specifically, flutter. This section will review, very briefly, the available publications. There have been numerous published calculations of two- and three-degree-of-freedom 2D flutter studies that are devoid of experimental data comparisons since realistic 2D flutter models are very difficult to fabricate. On the other hand, there are only a small number of published 3D flutter calculations that are compared with experimental results. An important reason for this is the detail and effort required to perform a valid flutter analysis of a flutter model. Vibration mode shapes and masses must be accurately calculated or measured and surface coordinates measured. Another major factor has been the difficulty of the unsteady aerodynamic computation for configurations of current interest. For instance, much interest tends to be focused on cases whose transonic flutter boundaries occur coincidentally with the onset of mixed flow. Hence, aeroelastic applications for such cases have had to await the maturation of computational methods discussed earlier.

A restricted selection of the many publications of 2D flutter calculations includes the works of Isogai,[80] Edwards et al.,[139] Bland and Edwards,[83] and Berry et al.[95] References 80, 83, and 139 document the somewhat surprising locally linear nature of transonic potential flows. That is, about the steady mean flow condition, which is a nonlinear function of Mach number and angle of attack, unsteady perturbation airloads behave very linearly for reasonable airfoil motions. Figure 32[83] illustrates this by showing the normalized unsteady lifting pressure on the NACA 64A010A airfoil for oscillatory pitch amplitudes of 0.25–2.0 deg. Away from the shock pulse little effect of amplitude is seen, and the integrated effect of the shock pulse also varies only slightly. Reference 139 introduced the static twisting of the airfoil due to the steady pitching moment into the flutter problem and demonstrated a marked effect on flutter boundaries, particularly for the supercritical MBB A-3 airfoil. Reference 95 documented the utility of s-plane Padé curve fits of transonic airloads (which rely on the concepts of linearity and superposition) for aeroelastic analysis. Viscous effects are

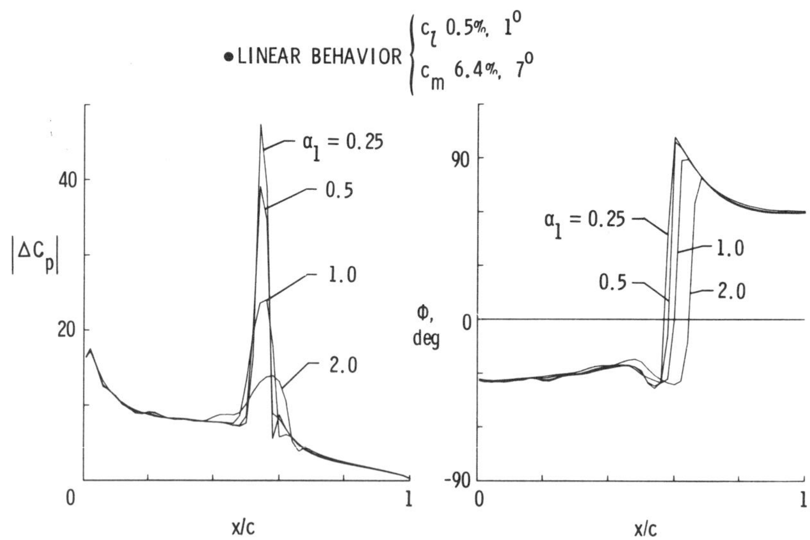

Fig. 32 Pitch amplitude effect on normalized pressure distribution for NACA 64A010A airfoil at $M = 0.78$, $\alpha = 0$ deg, and $k = 0.15$.[83]

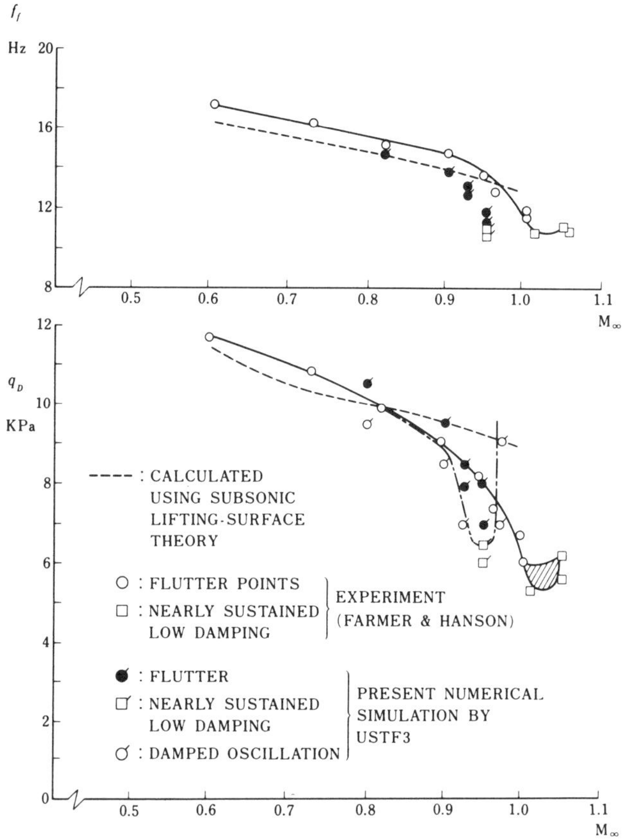

Fig. 33 Comparison of calculated and experimental flutter boundaries for a TF-8A supercritical wing flutter model.[71]

shown to generally result in larger values of flutter speed since transonic effects are alleviated by the boundary layer.

Comparisons of calculated and experimental flutter boundaries for wings have been given by Guruswamy and Goorjian,[140] Isogai,[26,71] and Myers et al.[141] Reference 140 gives comparisons for a rectangular wing flutter model using an early low-frequency version of the XTRAN3S code. Reasonable agreement of flutter speeds is shown at a subcritical speed, $M = 0.715$, but the flutter speed is overpredicted by 75% at $M = 0.9$. Isogai[71] studied the supercritical wing of Farmer et al.[57] using the nonconservative FP USTF3 code and an interacted boundary-layer model. His comparison of flutter boundaries is shown in Fig. 33. The trend of the transonic flutter dip is very nicely predicted, although the dip occurs about 0.08 low in Mach number. The subcritical flutter boundary is very well matched by the calculated flutter point at $M = 0.8$. The premature flutter dip and the subsequent premature rise of the calculated boundary are of concern, since Myers,[141] presenting XTRAN3S calculations for a transport wing, also shows such a premature rise in the boundary. Finally, Isogai[26] shows flutter comparisons for a different supercritical transport wing that agree nicely with the experimental flutter dip.

Since the definition of these minimum flutter speeds very likely involves type II mixed flow, it is not surprising that few calculations are available. It can be anticipated that, with the recent advances in computational algorithms and interactive viscous modeling, such cases will be treated in the near future.

Type II Mixed Flow Computation

As indicated earlier, the unsteady periodic flows encountered over a limited range of Mach numbers and triggered by oscillating trailing-edge/shock-induced boundary-layer separation are just recently coming into the range of computational methods. The weakest link for this capability is the uncertainty in the turbulence modeling of complex separating flows, since rapid progress continues to be made in the development of improved algorithms and faster computers.

It is well known that separated flows depart strongly from equilibrium-type behavior, so that at a minimum some account of the nonequilibrium "upstream history" effects should be included in the computations. Some encouraging results along this line have been obtained by Le Balleur and Girodroux-Lavigne[109] with an integral boundary-layer model and Johnson[142] with an eddy-viscosity Reynolds shear-stress closure model. Simpson[143] recently reviewed calculation methods for turbulent separated flows, and Coakley[144] compared several methods for airfoil applications.

On the other hand, Levy[37] was able to reproduce the unsteady periodic flow behavior of the 18% circular arc airfoil using an equilibrium two-layer algebraic model. The steady flow at Mach numbers below the range of periodic flow, characterized by trailing-edge separation, was predicted accurately. Levy demonstrated that the influence of the channel walls had a substantial impact on the comparisons with experiment, especially at Mach numbers away from the design point. This effect was not considered in the earlier comparisons of Deiwert[75] with the experimental results. The steady flow at a Mach number above the range of periodic flow, characterized by shock-induced separation, was not accurately predicted, since the calculation demonstrated a normal shock pattern (Fig. 11) with trailing-edge pressure recovery, whereas the experiment indicated an oblique shock pattern and a constant pressure region downstream of the shock. Coakley[145] demonstrated that the discrepancy was partly attributable to the turbulence model but more so to the blockage introduced downstream by the separated wake flow, which forces the downstream static pressure to be lower than the upstream static pressure. By modeling the tunnel walls with boundary conditions more appropriate for an internal flow and using a modified two-equation turbulence model, Coakley[145,147] obtained good agreement with the steady experimental results, as shown in Fig. 34. The preceding discussion illustrates some of the difficulties that can be encountered in applying and validating computational methods for turbulent separated flows.

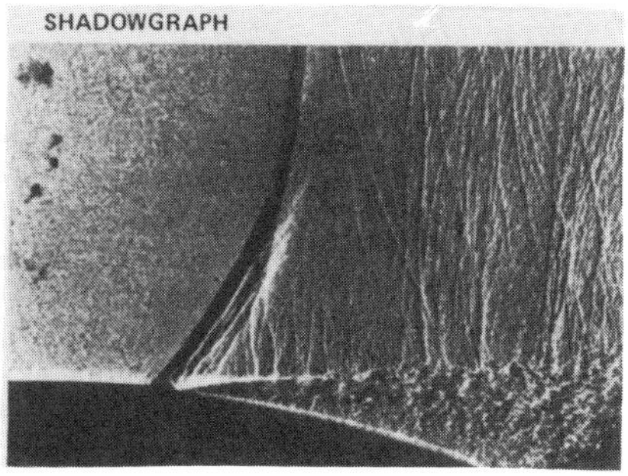

a) Shadowgraph

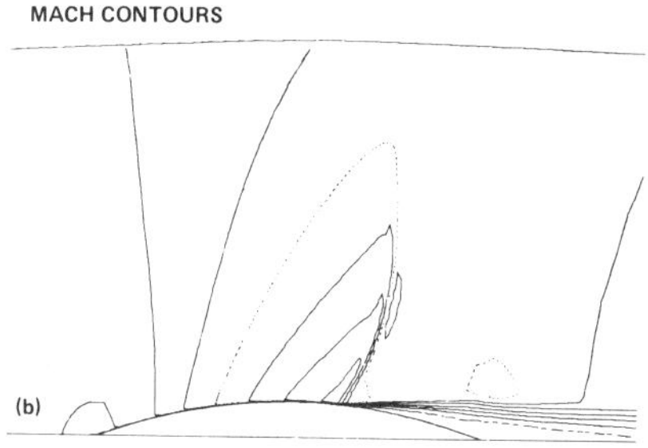

b) Mach contours

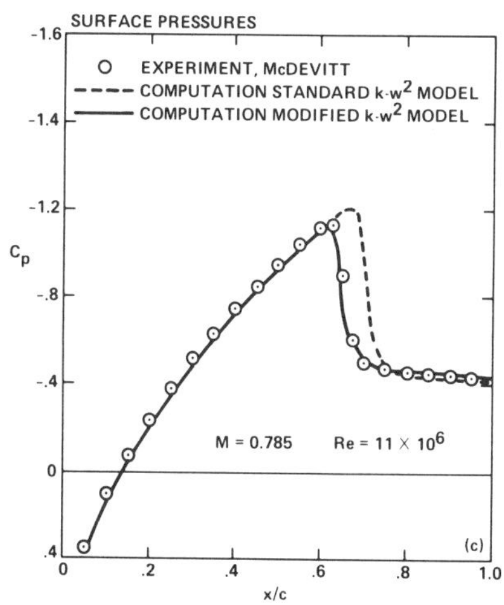

c) Surface pressures

Fig. 34 Comparison of Navier-Stokes computations of Coakley with experiments of McDevitt for steady shock-induced boundary separation; $M = 0.785$, $Re = 11 \times 10^6$.[145]

In addition to Levy's calculations, the unsteady periodic behavior for the 18% biconvex airfoil has also been computed by Steger[102] and by Le Balleur and Girodroux.[109] Steger's calculation was for an airfoil in free air with an implicit NS code using the Baldwin-Lomax[146] algebraic model. The unsteady flow occurred at a higher Mach number ($M = 0.783$) than that of Levy ($M = 0.754$), which can partly be attributed to the free air boundary conditions. The computed reduced frequency (0.41) was remarkably close to that of Levy (0.40), although both are low in comparison to experiment

(0.48). Le Balleur and Girodroux's recent calculations were also made in free air with a small-disturbance potential method including an interacted two-equation integral viscous model. Steady shock-induced separation was computed at $M = 0.788$ and unsteady periodic flow at $M = 0.76$. The reduced frequency (0.34) was lower than either of the two NS solutions.

Some calculations have been made during the preparation of this paper using the implicit upwind-biased NS algorithm described in Ref. 149 using an algebraic turbulence model.[146] The tunnel walls were modeled, and boundary conditions appropriate for internal flow were used; i.e., the downstream pressure and upstream enthalpy, entropy, and flow direction were specified. The results indicated unsteady flow at a higher Mach number than Levy; steady trailing-edge separation occurred at $M = 0.754$ and unsteady periodic flow at $M = 0.78$, although the Mach number for onset of the unsteadiness was sensitive to whether or not the divergence of the tunnel boundary to account for boundary-layer growth was included. Figure 35 shows Mach contours through one half-cycle of oscillation (near-maximum lift to minimum lift) indicating the forward movement, disappearance, and subsequent formation near the trailing edge of the lower surface shock. The reduced frequency of the type B unsteady motion was 0.406, in close agreement with the calculations of both Levy (compare Fig. 11) and Steger. The implicit calculations were made with a time step of 0.01 and a computational time of 18 μs per grid point per time step on the Cyber 205 computer.

The unsteady periodic flow for the 18% circular arc is triggered by the oscillating trailing-edge/shock-induced boundary-layer separation on the airfoil. The unsteady motion can also occur in purely inviscid flows because

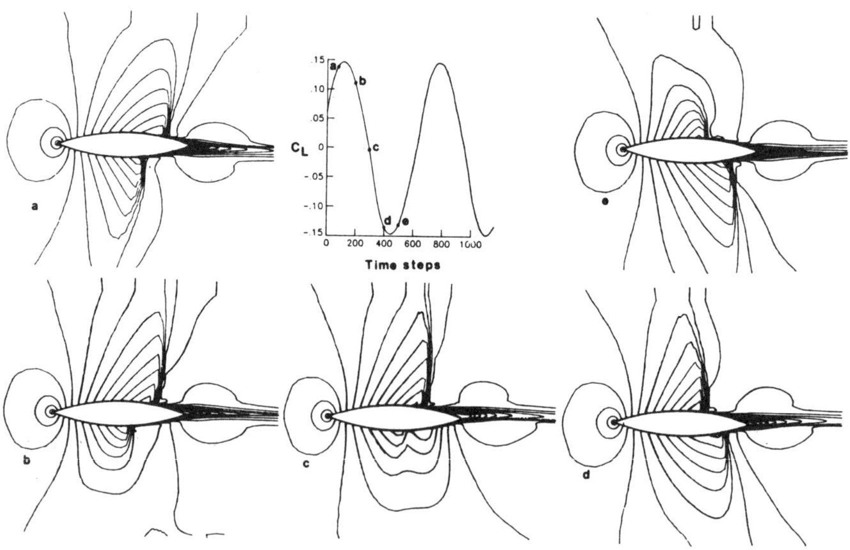

Fig. 35 Calculated periodic oscillation of 18% biconvex airfoil with implicit thin-layer Navier-Stokes code; $M = 0.78$, $Re = 11 \times 10^6$, $k = 0.406$.

of the vorticity introduced at a curved shock. The flow past a circular cylinder at $M = 0.5$ is well known, for example, to be unsteady,[148,150] exhibiting a type B oscillatory shock motion at a reduced frequency of 0.50. Barton and Pulliam[151] have also shown for airfoils at high angle of attack that the curved shocks near the leading edge can lead to shock-induced vorticity and periodic self-induced oscillatory flow. To what extent the inviscid mechanism is playing a role in the unsteady circular arc cases is not known, and a parametric study in terms of thickness and Mach number to establish the range of the inviscid mechanism would be very informative.

The calculation of the unsteady periodic flow boundaries for airfoils is a fruitful area for the development and validation of computational methods. Experimental pressure data[35,36,38,39] over a wide range of Reynolds numbers are available, although detailed boundary-layer measurements are not. For the 18% biconvex experiments of McDevitt, a substantial hysteresis effect in the unsteady flow boundary was found. This aspect has not been demonstrated with computational methods as yet, but it would be expected, based on the preceding discussion, that computational modeling as close as possible to that of the experimental conditions will be a critical consideration. The most interesting behavior, and the most challenging from the computational viewpoint, occurs in the transitional region from laminar to turbulent flow. In the experiment of McDevitt et al.,[35] the Mach number range for the observed unsteady flow diminished near a Reynolds number of 3×10^6 (Fig. 9), and in the experiments of Mabey et al.,[39] it disappeared completely in the range of Reynolds numbers from about 3×10^6 to 5×10^6. Scale effects in either experiment were not significant once turbulent flow was fully established ahead of the shock.

The frequency of these oscillations is of interest in that the flow mechanism causing the unsteadiness might be identified via the characteristic time constants of signal propagation within the various flow regions. Tijdeman[4] noted an almost linear relation between the phase lag of the shock motion and the airfoil motion for type A shock motion with a well-developed shock (for pitch oscillations of the NLR 7301 airfoil). He related this to the signal propagation time from the trailing edge to the shock. Mabey et al.,[39] commenting on characteristic time constants for the 14% circular arc airfoil periodic oscillations, note that this reasoning leads to reduced frequency parameters of 1.15–1.8, much higher than the observed frequencies.

Three items mentioned earlier, which are germane to this discussion, are Steger and Bailey's[17] inviscid EE calculations for the aileron buzz case, Le Balleur and Girodroux-Lavigne's[109] interacted viscous transonic small disturbance (TSD) code result for periodic oscillations, and Batina's[94] demonstration of the possibility of aerodynamic resonance with an inviscid TSD code. The occurrence of damped and undamped oscillations observed for inviscid flows at nearly the same frequency as the oscillations in viscous flow[17] ($k \approx 0.36$–0.39) implies that the flow mechanism determining the oscillation frequency derives from the dynamics of the inviscid flow region. Furthermore, the results of Refs. 94 and 109 give impetus to studying this effect with a TSD code. Accordingly, calculations were made with the XTRAN2L code of the aerodynamic response for the 18% thick circular arc airfoil due to trailing-edge 25% chord flap motions. The nonisentropic

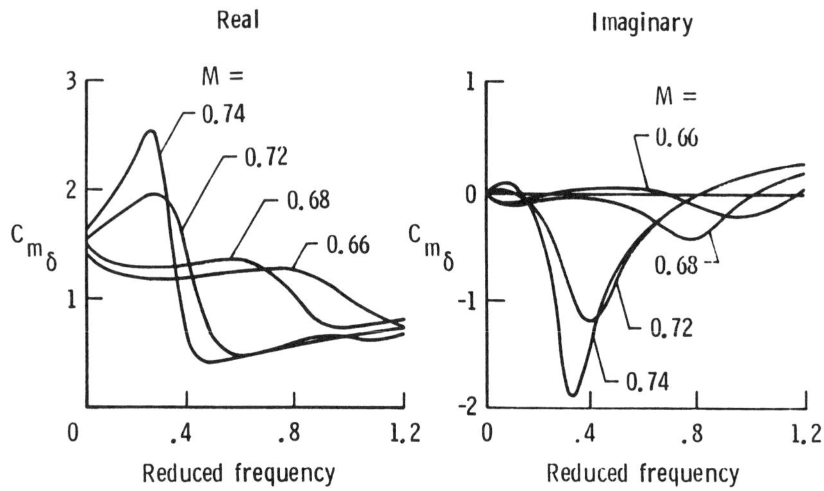

Fig. 36 Calculated pitching moment coefficient for 18% biconvex airfoil with nonisentropic TSD code.

modifications described in Ref. 98 were used to obtain solutions for this strong shock case. Figure 36 presents the resulting $c_{m\delta}$ frequency response function for Mach numbers of 0.66–0.74. There is a very marked development of an aerodynamic resonance effect as Mach number increases. For $M = 0.74$ the airfoil resonance frequency is $k = 0.32$, very close to the computational conditions of Ref. 109 for periodic oscillations ($M = 0.76$ and $k = 0.36$). Steady calculated flow conditions for this case gave a shock location at $x/c \approx 0.73$ with a preshock minimum pressure coefficient of $C_p = -1.27$ (the experimental values were $x/c \approx 0.65$ and $C_p = -1.1$).[36] Although this case is certainly outside the bounds of a TSD code, the occurrence of such a strong resonance feature as shown in Fig. 36 is of interest. Experimental evidence of such aerodynamic resonances is given by den Boer and Houwink,[65] who tested a 12% thick supercritical airfoil. For pitching oscillations, a large resonance effect was observed near $k \approx 0.25$ at conditions corresponding to the onset of type II mixed flow.

The calculations of Fig. 36 were obtained using a 1-deg flap pulse. For the tests of Ref. 36, the inclination angle of the shock-separated region to the airfoil surface ranged up to 7 deg. Thus, harmonic oscillation calculations for this condition with a flap oscillation amplitude of 7 deg (not shown) were made, which show the development of the characteristic dimpled isobar pattern seen in the NS code results of Fig. 35c. These dimpled isobar patterns have been noted in Levy's[37] NS calculations (Fig. 11), Sidès' EE calculations[30] (Fig. 25), and the present NS calculations (Fig. 35) and appear to be a variant of type B and type C shock motions. That is, during a portion of the oscillation two regions of high suction pressure occur simultaneously. The dynamic behavior of the boundaries of these regions (i.e., moving shocks and pressure waves) is an interesting topic for further study.

Concluding Remarks

We have attempted to summarize the status of computational methods for transonic unsteady aerodynamics with a focus on methods for aeroelastic calculations and the prediction of flutter. Three-dimensional vortex flows have not been included in this survey because this is a fast-evolving topic by itself. The subject has been differentiated with respect to the difficulty of flow modeling and computational requirements by distinguishing the status of the unsteady flows as type I, attached flow; type II, mixed attached and separated flows; and type III, separated flow.

1) For type I flows, computational methods have matured with a steady progression of improved techniques for flow simulation. Significant efforts have been devoted to understanding the effects of equation level, computational grid, boundary conditions, and viscous boundary-layer modeling. Extensive comparisons with experimental data sets have been made with potential, EE, and NS equation solvers for two-dimensional airfoil cases, and an understanding of the range of validity of the various methods can be made. Less extensive comparisons are available for three-dimensional wings, and more evaluations are required before definite conclusions can be made.

2) Progress in solution algorithms for unsteady aerodynamics has been significant with large decreases in required computer resources due to the larger time steps allowed by more stable solution algorithms. Development of programs treating more realistic configurations, such as multiple lifting surfaces and wing appendages, is anticipated in the near future. The treatment of body-conforming grids for deforming elastic vehicles must be addressed in order to fully utilize these emerging methods.

3) There is a need for experimental data sets with which to validate unsteady aerodynamic codes simulating these more complete configurations.

4) Many more transonic flutter calculations are needed before an assessment of the range of validity of the computational methods can be made. Of the few reported flutter calculations, a tendency is noted of predicting a premature upturn in the flutter boundary, very possibly due to violation of flow modeling assumptions (e.g., onset of type II unsteady flow).

5) Type II periodic aerodynamic oscillations about an 18% thick circular arc airfoil have been calculated with an upwind differenced NS code using flux splitting, duplicating earlier calculations by Levy. More insight into this case is provided by a TSD potential code using nonisentropic modifications. Calculations of flap oscillations for this case illustrate the frequency dependence of the oscillations and give further evidence of an aerodynamic resonance phenomenon in the inviscid flow about airfoils at transonic speeds. Interaction of such type II flows with structural vibration characteristics may be expected to lead to nonclassical aeroelastic responses.

References

[1]McCroskey, W. J., Kutler, P., and Bridgeman, J. O., "Status and Prospects of Computational Fluid Dynamics for Unsteady Transonic Viscous Flows," "Transonic Unsteady Aerodynamics and its Aeroelastic Application," AGARD CP-374, Jan. 1985.

[2]Anonymous, "Military Specification; Airplane Strength and Rigidity; Flutter, Divergence, and Other Aeroelastic Instabilities," MIL-A-008870A (USAF), March 1971.

[3]Newsome, R. W. and Kandil, O. A., "Vortical Flow Aerodynamics—Physical Aspects and Numerical Simulation," AIAA Paper 87-0205, Jan. 1987.

[4]Tijdeman, H., "Investigations of the Transonic Flow Around Oscillating Airfoils," National Aerospace Lab., The Netherlands, NLR TR-77090U, 1978.

[5]Magnus, R. and Yoshihara, H., "The Transonic Oscillating Flap," AIAA Paper 76-327, July 1976.

[6]Ballhaus, W. F. and Goorjian, P. M., "Implicit Finite-Difference Computations of Unsteady Transonic Flows about Airfoils," *AIAA Journal*, Vol. 15, Dec. 1977, pp. 1728–1735.

[7]Ballhaus, W. F., "Some Recent Progress in Transonic Flow Computations," *VKI Lecture Series: Computational Fluid Dynamics*, von Kármán Institute for Fluid Dynamics, Rhode-St-Genese, Belgium, March 1976.

[8]Bland, S. R., "AGARD Two-Dimensional Aeroelastic Configurations," AGARD-AR-156, 1979.

[9]Bland, S. R., "AGARD Three-Dimensional Aeroelastic Configurations," AGARD-AR-167, 1982.

[10]Lambourne, N. C., "Compendium of Unsteady Aerodynamic Measurements," AGARD Rept. 702, 1982.

[11]Dau, K., Vogel, S., and Zimmerman, H., "Compendium of Unsteady Aerodynamic Measurements Addendum No. 1," AGARD Rept. 702, 1985.

[12]Mabey, D. G., Welsh, B. L., and Pyne, C. R., "Measurements of Steady and Oscillatory Pressure on a Rectangular Wing," Royal Aircraft Establishment TR-86040, July 1986.

[13]Sandford, M. C., Ricketts, R. H., and Hess, R. W., "Recent Transonic Unsteady Pressure Measurements at the NASA Langley Research Center," Paper 85-23, April 1985.

[14]Tijdeman, H., van Nunen, J. W. G., Kraan, A. N., Persoon, A. J., Poestkoke, R., Schippers, P., and Siebert, C. M., "Transonic Wind Tunnel Tests on an Oscillating Wing with External Stores," AFFDL TR-78-194, Pts. I–IV, Dec. 1978.

[15]Magnus, R. J., "Calculations of Some Unsteady Transonic Flows about the NACA 64A006 and 64A010 Airfoils," AFFDL TR-77-46, July 1977.

[16]Magnus, R. J., "Some Numerical Solutions of Inviscid, Unsteady, Transonic Flows Over the NLR 7301 Airfoil," Convair Division of General Dynamics, San Diego, CA, CASD/LVP 78-013, Jan. 1978.

[17]Steger, J. L. and Bailey, H. E., "Calculation of Transonic Aileron Buzz," *AIAA Journal*, Vol. 18, March 1980, pp. 249–255.

[18]Chyu, W. J., Davis, S. S., and Chang, K. S., "Calculation of Unsteady Transonic Flow over an Airfoil," *AIAA Journal*, Vol. 19, June 1981, pp. 684–690.

[19]Chyu, W. J. and Davis, S. S., "Numerical Studies of Unsteady Transonic Flow Over an Oscillating Airfoil," "Transonic Unsteady Aerodynamics and its Aeroelastic Application," AGARD CP-374, Jan. 1985.

[20]Borland, C. J., Rizzetta, D. P., and Yoshihara, H., "Numerical Solution of Three-Dimensional Unsteady Transonic Flow Over Swept Wings," *AIAA Journal*, Vol. 20, March 1982, pp. 340–347.

[21]Borland, C. J. and Rizzetta, D. P., "Nonlinear Transonic Flutter Analysis," *AIAA Journal*, Vol. 20, Nov. 1982, pp. 1606–1615.

[22]Laurent, A., "Calcul D'Ecoulement Transsonique Instationnaire Authour D'Aile A Forte Fleche," "Transonic Unsteady Aerodynamics and its Aeroelastic Application," AGARD CP-374, Jan. 1985.

[23]Batina, J. T., "An Efficient Algorithm for Solution of the Unsteady Transonic Small-Disturbance Equation," AIAA Paper 87-0109, Jan. 1987.
[24]Malone, J. B., Sankar, L. N., and Sotomayer, W. A., "Unsteady Aerodynamic Modeling of a Fighter Wing in Transonic Flow," *Journal of Aircraft*, Vol. 23, Aug. 1986, pp. 611–620.
[25]Shankar, V. and Ide, H., "Treatment of Steady and Unsteady Flows Using a Fast, Time-Accurate Full Potential Scheme," AIAA Paper 85-4060, Oct. 1985.
[26]Isogai, K., "The Development of Unsteady Transonic 3-D Full Potential Code and Its Aeroelastic Applications," "Transonic Unsteady Aerodynamics and its Aeroelastic Application," AGARD CP-374, Jan. 1985.
[27]Mulak, P. and Angelini, J., "Amelioration et Extension D'une Methode de Calcul d'Ecoulements Transsoniques Tridimensionnels," "Transonic Unsteady Aerodynamics and its Aeroelastic Application," AGARD CP-374, Jan. 1985.
[28]Belk, D. M., Janus, J. M., and Whitfield, D. L., "Three-Dimensional Unsteady Euler Equations Solutions on Dynamic Grids," AIAA Paper 85-1704, July 1985.
[29]Janus, J. M., "The Development of a Three-Dimensional Split Flux Vector Euler Solver with Dynamic Grid Applications," M.S. Thesis, Mississippi State Univ., Aug. 1984.
[30]Sidès, J., "Computation of Unsteady Transonic Flows with an Implicit Numerical Method for Solving the Euler Equations," *Recherche Aerospatiale*, Vol. 2, 1985, pp. 17–39.
[31]Jameson, A. and Venkatakrishnan, V., "Transonic Flows About Oscillating Airfoils Using the Euler Equations," AIAA Paper 85-1514, July 1985.
[32]Smith, G. E., Whitlow, W., Jr., and Hassan, H. A., "Unsteady Transonic Flows Past Airfoils Using the Euler Equations," AIAA Paper 86-1764-CP, June 1986.
[33]Salmond, D. J., "Calculation of Harmonic Aerodynamic Forces on Aerofoils and Wings from the Euler Equations," "Transonic Unsteady Aerodynamics and its Aeroelastic Application," AGARD CP-374, Jan. 1985.
[34]Belk, D. M., "Unsteady Three-Dimensional Euler Equations Solutions on Dynamic Blocked Grids," Ph.D. Dissertation, Mississippi State Univ., Aug. 1986.
[35]McDevitt, J. B., Levy, L. L., Jr., and Deiwert, G. S., "Transonic Flows about a Thick Circular-Arc Airfoil," *AIAA Journal*, Vol. 14, May 1976, pp. 606–613.
[36]McDevitt, J. B., "Supercritical Flow About a Thick Circular-Arc Airfoil," NASA TM-78549, Jan. 1979.
[37]Levy, L. L., Jr., "Experimental and Computational Steady and Unsteady Transonic Flows about a Thick Airfoil," *AIAA Journal*, Vol. 16, June 1978, pp. 564–572.
[38]Mabey, D. G., "Oscillatory Flows from Shock-Induced Separations on Biconvex Aerofoils of Varying Thickness in Ventilated Wind Tunnels," AGARD CP-296, 1980.
[39]Mabey, D. G., Welsh, B. L., and Cripps, B. E., "Periodic Flows on a Rigid 14% Thick Biconvex Wing at Transonic Speeds," Royal Aircraft Establishment TR-81059, May 1981.
[40]Levy, L. L., Jr., "Predicted and Experimental Steady and Unsteady Transonic Flows About a Biconvex Airfoil," NASA TM-81262, Feb. 1981.
[41]Mabey, D. G., Ashill, P. R., and Welsh, B. L., "Aeroelastic Oscillations Caused by Transitional Boundary Layers and Their Attenuation," AIAA Paper 86-0736, March 1986.
[42]Osswald, G. A., Ghia, K. N., and Ghia, U., "An Implicit Time-Marching Method for Studying Unsteady Flow with Massive Separation," AIAA Paper 85-1489, July 1985.
[43]Lecointe, Y. and Piquet, J., "Unsteady Viscous Flow Round Moving Circular Cylinders and Airfoils," AIAA Paper 85-1490, July 1985.

[44]Anderson, W. K., Thomas, J. L., and Rumsey, C. L., "Application of Thin Layer Navier-Stokes Equations Near Maximum Lift," AIAA Paper 84-0049, Jan. 1984.
[45]Anonymous, "Unsteady Aerodynamics," AGARD CP-227, 1978.
[46]Anonymous, "Boundary Layer Effects on Unsteady Airloads," AGARD CP-296, 1981.
[47]Anonymous, "Transonic Unsteady Aerodynamics and its Aeroelastic Application," AGARD CP-374, 1985.
[48]Anonymous, "Unsteady Aerodynamics—Fundamentals and Applications to Aircraft Dynamics," AGARD CP-386, 1985.
[49]Peterson, V. L., "Trends in Computational Capabilities for Fluid Dynamics," "Transonic Unsteady Aerodynamics and its Aeroelastic Application," AGARD CP-374, Jan. 1985.
[50]Mykytow, W. J., "Transonic Unsteady Aerodynamics and its Aeroelastic Applications," "Unsteady Aerodynamics—Fundamentals and Applications to Aircraft Dynamics," AGARD CP-386, 1985.
[51]Mabey, D. G. and Chambers, J. R., "Unsteady Aerodynamics—Fundamentals and Applications to Aircraft Dynamics," AGARD-AR-222, May 1985.
[52]Zwaan, R. J., "Aeroelastic Problems of Wings in Transonic Flow," presented at VKI Lecture Series, Unsteady Airloads and Aeroelastic Problems in Separated and Transonic Flow, Rhode-Saint-Genese, Belgium, March 1981; also NLR MP-81005 U.
[53]Deiwert, G. S., "Finite Difference Simulation of Unsteady Interactive Flows," *Computational Methods in Viscous Flow*, edited by W. G. Habashi, Pineridge Press, Swansea, UK, 1984, pp. 281–308.
[54]Mabey, D. G., "A Review of Some Recent Research on Time-Dependent Aerodynamics," *The Aeronautical Journal*, Feb. 1984, pp. 23–37.
[55]Pearcy, H. H., Haines, A. B., and Osborne, J., "The Interaction Between Local Effects at the Shock and Real Separation," "Transonic Aerodynamics," AGARD CP-35, 1968.
[56]Bradley, R. G., "Practical Aerodynamic Problems—Military Aircraft," *Transonic Aerodynamics*, edited by D. Nixon, Vol. 81, Progress in Astronautics and Aeronautics Series, AIAA, New York, 1982.
[57]Farmer, M. G., Hanson, P. W., and Wynne, E. C., "Comparison of Supercritical and Conventional Wing Flutter Characteristics," NASA TM-X-72837, May 1976.
[58]Erickson, L. L., "Transonic Single-Mode Flutter and Buffet of a Low Aspect Ratio Wing Having a Subsonic Airfoil Shape," NASA TN D-7346, Jan. 1974.
[59]Houwink, R., Kraan, A. N., and Zwaan, R. J., "A Wind-Tunnel Study of the Flutter Characteristics of a Supercritical Wing," AIAA Paper 81-0651; also *Journal of Aircraft*, Vol. 19, May 1982, pp. 400–405.
[60]Persoon, A. J., Horsten, J. J., and Meijer, J. J., "On Measuring Transonic Dips in the Flutter Boundaries of a Supercritical Wing in the Wind Tunnel," AIAA Paper 83-1031-CP, May 1983; also National Aerospace Lab. NLR MP-83008 U, Amsterdam, The Netherlands, Jan. 1983.
[61]Moss, G. F. and Pierce, D., "The Dynamic Response of Wings in Torsion at High Subsonic Speeds," AGARD CP-226, 1977.
[62]Davis, S. D. and Malcolm, G. N., "Transonic Shock-Wave/Boundary-Layer Interactions on an Oscillating Airfoil," *AIAA Journal*, Vol. 18, Nov. 1980, pp. 1306–1312.
[63]McDevitt, J. B. and Okuno, A. F., "Static and Dynamic Pressure Measurements on a NACA 0012 Airfoil in the Ames High Reynolds Number Facility," NASA TP-2485, June 1985.

[64]Zimmerman, H., "The Application of Transonic Unsteady Methods for Calculation of Flutter Airloads," "Transonic Unsteady Aerodynamics and its Aeroelastic Application," AGARD CP-374, Jan. 1985.

[65]den Boer, R. G. and Houwink, R., "Analysis of Transonic Aerodynamic Characteristics for a Supercritical Airfoil Oscillating in Heave, Pitch and with Oscillating Flap," "Transonic Unsteady Aerodynamics and its Aeroelastic Application," AGARD CP-374, Jan. 1985.

[66]Le Balleur, J. C. and Girodroux-Lavigne, P., "Prediction of Buffeting and Calculation of Unsteady Boundary Layer Separation Over Airfoils," presented at IUTAM Symposium on Boundary Layer Separation, Univ. College, London, Aug. 1986.

[67]Hess, R. W., Seidel, D. A., Igoe, W. B., and Lawing, P. L., "Highlights of Unsteady Pressure Tests on a 14 Percent Supercritical Airfoil at High Reynolds Number, Transonic Condition," AIAA Paper 87-0035, Jan. 1987.

[68]Sandford, M. C., Ricketts, R. H., and Hess, R. W., "Recent Transonic Unsteady Pressure Measurements at the NASA Langley Research Center," AIAA Paper 85-23, April 1985.

[69]Ricketts, R. H., Sandford, M. C., Watson, J. J., and Seidel, D. A., "Subsonic and Transonic Unsteady- and Steady-Pressure Measurements on a Rectangular Supercritical Wing Oscillated in Pitch," NASA TM-85765, Aug. 1984.

[70]Mabey, D. G., Welsh, B. L., and Cripps, B. E., "Measurements of Steady and Oscillatory Pressures on a Low Aspect Ratio Model at Subsonic and Supersonic Speeds," Royal Aircraft Establishment, TR-84095, Sept. 1984.

[71]Isogai, K. and Suetsugu, K., "Numerical Simulation of Transonic Flutter of a Supercritical Wing," National Aerospace Lab., NAL Rept. TR-726T, Aug. 1982.

[72]Bennett, R. M., Seidel, D. A., and Sandford, M. C., "Transonic Calculations for a Flexible Supercritical Wing and Comparison with Experiment," AIAA Paper 85-0665-CP, April 1985.

[73]Seidel, D. A., Sandford, M. C., and Eckstrom, C. V., "Measured Unsteady Transonic Aerodynamic Characteristics of an Elastic Supercritical Wing with an Oscillating Control Surface," AIAA Paper 85-0598-CP, April 1985.

[74]Guruswamy, P., Goorjian, P. M., and Tu, E. L., "Transonic Aeroelasticity of Wings with Tip Stores," AIAA Paper 86-1007-CP, May 1986.

[75]Deiwert, G. S., "Computation of Separated Transonic Turbulent Flows," AIAA Paper 75-829, June 1975.

[76]Seidel, D. A., Bennett, R. M., and Ricketts, R. H., "Some Recent Applications of XTRAN3S," AIAA Paper 83-1811, July 1983.

[77]Ruo, S. Y., Malone, J. B., and Sankar, L. N., "Steady and Unsteady Full Potential Calculations for High and Low Aspect Ratio Supercritical Wings," AIAA Paper 86-0122, Jan. 1986.

[78]Rizzetta, D. P. and Chin, W. C., "Effect of Frequency in Unsteady Transonic Flows," *AIAA Journal*, Vol. 12, July 1979, pp. 779–781.

[79]Houwink, R. and van der Vooren, J., "Improved Version of LTRAN2 for Unsteady Transonic Flow Computations," *AIAA Journal*, Vol. 18, Aug. 1980, pp. 1008–1010.

[80]Isogai, K., "Numerical Study of Transonic Flutter of a Two-Dimensional Airfoil," NAL TR-716T, July 1980.

[81]Hessenius, K. A. and Goorjian, P. M., "Validation of LTRAN2-HI by Comparison with Unsteady Transonic Experiment," *AIAA Journal*, Vol. 20, May 1982, pp. 731–732.

[82]Chow, L. J. and Goorjian, P. M., "Implicit Unsteady Transonic Airfoil Calculations at Supersonic Freestreams," AIAA Paper 82-0934, June 1982.

[83]Bland, S. R. and Edwards, J. W., "Airfoil Shape and Thickness Effects on Transonic Airloads and Flutter," *Journal of Aircraft*, Vol. 21, March 1984, pp. 209–217.

[84]Goorjian, P. M., Meagher, M. E., and Buskirk, R. V., "Monotone Switches in Implicit Algorithms for Potential Equations Applied to Transonic Flows," *AIAA Journal*, Vol. 23, April 1985, pp. 492–498.

[85]Engquist, B. E. and Osher, S. J., "Stable and Entropy Satisfying Approximations for Transonic Flow Calculations," *Journal of Computation*, Vol. 34, No. 149, Jan. 1980, pp. 45–75.

[86]Kwak, D., "Nonreflecting Far-Field Boundary Conditions for Unsteady Transonic Flow," *AIAA Journal*, Vol. 19, Nov. 1981, pp. 1401–1407.

[87]Whitlow, W., Jr., "XTRAN2L: A Program for Solving the General-Frequency Unsteady Transonic Small Disturbance Equation," NASA TM-85723, Nov. 1983.

[88]Houwink, R., "Unsteady Viscous Transonic Flow Computations Using the LTRAN2-NLR Code Coupled with Green's Lag-Entrainment Method," *Numerical and Physical Aspects of Aerodynamic Flows*, Vol. II, edited by T. Cebeci, Springer-Verlag, New York, 1984, Chap. 15.

[89]Edwards, J. W., Bland, S. R., and Seidel, D. A., "Experience with Transonic Unsteady Aerodynamic Calculations," "Transonic Unsteady Aerodynamics and its Aeroelastic Application," AGARD CP-374, Jan. 1985.

[90]Bland, S. R. and Seidel, D. A., "Calculation of Unsteady Aerodynamics for Four AGARD Standard Aeroelastic Configurations," NASA TM-85817, May 1984.

[91]Goorjian, P. M. and Guruswamy, G. P., "Unsteady Transonic Aerodynamic and Aeroelastic Calculations About Airfoils and Wings," "Transonic Unsteady Aerodynamics and its Aeroelastic Application," AGARD CP-374, Jan. 1985.

[92]Malone, J. B., Ruo, S. Y., and Sankar, N. L., "Computation of Unsteady Transonic Flows About Two-Dimensional and Three-Dimensional AGARD Standard Configurations," "Transonic Unsteady Aerodynamics and its Aeroelastic Application," AGARD CP-374, Jan. 1985.

[93]Edwards, J. W., "Applications of Potential Theory Computations to Transonic Aeroelasticity," *Fifteenth Congress of the International Council of the Aeronautical Sciences*, Paper ICAS-86-2.9.1, Sept. 1986.

[94]Batina, J. T., "Effects of Airfoil Shape, Thickness, Chamber, and Angle of Attack on Calculated Transonic Unsteady Airloads," NASA TM-86320, March 1985.

[95]Berry, H. M., Batina, J. T., and Yang, T. Y., "Viscous Effects on Transonic Airfoil Stability and Response," *Journal of Aircraft*, Vol. 23, May 1986, pp. 361–369.

[96]Steinhoff, J. and Jameson, A., "Multiple Solutions of the Transonic Potential Flow Equation," *AIAA Journal*, Vol. 20, Nov. 1982, pp. 1521–1525.

[97]Williams, M. H., Bland, S. R., and Edwards, J. W., "Flow Instabilities in Transonic Small-Disturbance Theory," *AIAA Journal*, Vol. 23, Oct. 1985, pp. 1491–1496.

[98]Fuglsang, D. F. and Williams, M. H., "Non-Isentropic Unsteady Transonic Small Disturbance Theory," AIAA Paper 85-0600, April 1985.

[99]Gibbons, M. D., Whitlow, W., Jr., and Williams, M. H., "Nonisentropic Unsteady Three Dimensional Small Disturbance Potential Theory," AIAA Paper 86-0863, May 1986.

[100]Osher, S., Hafez, M., and Whitlow, W., Jr., "Entropy Condition Satisfying Approximations for the Full Potential Equation of Transonic Flow," *Mathematics of Computations*, Vol. 44, No. 169, Jan. 1985, pp. 1–29.

[101]Whitlow, W., Jr., Hafez, M. H., and Osher, S. J., "An Entropy Correction Method for Unsteady Full Potential Flows with Strong Shocks," AIAA Paper 86-1768-CP, June 1986.

[102]Steger, J. L., "Implicit Finite-Difference Simulation of Flow about Arbitrary Two-Dimensional Geometries," *AIAA Journal*, Vol. 16, July 1978, pp. 679–686.

[103]Sankar, L. M., Ruo, S. Y., and Malone, J. B., "Application of Surface Transpiration in Computational Aerodynamics," AIAA Paper 86-0511, Jan. 1986.

[104]Couston, M., Angelini, J. J., Le Balleur, J. C., and Girodroux-Lavigne, P., "Prise en Compte Deffets de Couche Limite Instationnaire dans un Calcul Bidimensionnel Transsonique," "Boundary Layer Effects on Unsteady Airloads," AGARD CP-296, 1981.

[105]Couston, M., Angelini, J. J., and Mulak, P., "Application de l'Equation des Petites Perturbations Transonniques aux Calculs d'Ecoulements Bidimensionnels Instationnaires," *Recherche Aerospatiale*, No. 1975-5, 1975, pp. 325–340.

[106]Le Balleur, J. C., "Numerical Viscid-Inviscid Interaction in Steady and Unsteady Flows," *Numerical and Physical Aspects of Aerodynamic Flows II*, edited by T. Cebeci, Springer-Verlag, New York, 1984, Chap. 14.

[107]Le Balleur, J. C. and Girodroux-Lavigne, P., "A Semi-Implicit and Unsteady Numerical Method of Viscous-Inviscid Interaction of Transonic Separated Flows," *Recherche Aerospatiale*, No. 1984-1, 1984, pp. 14–37.

[108]Girodroux-Lavigne, P. and Le Balleur, J. C., "Calcul d'Ecoulements Instationnaires Transsoniques avec Decollements par Interaction Visqueux-Non Visqueux," "Transonic Unsteady Aerodynamics and its Aeroelastic Application," AGARD CP-374, Jan. 1985.

[109]Le Balleur, J. C. and Girodroux-Lavigne, P., "A Viscous-Inviscid Interaction Method for Computing Unsteady Transonic Separation," *Proceedings of the 3rd Symposium on Numerical and Physical Aspects of Aerodynamic Flows*, edited by T. Cebeci, Springer-Verlag, New York, 1986.

[110]Houwink, R., "Computations of Separated Subsonic and Transonic Flow About Airfoils in Unsteady Motion," National Aerospace Lab., NLR MP-84-0944, Amsterdam, The Netherlands, 1984.

[111]Rizzetta, D. P., "Procedures for the Computation of Unsteady Transonic Flows Including Viscous Effects," NASA CR-166249, Jan. 1982.

[112]Rizzetta, D. P. and Borland, C. J., "Unsteady Transonic Flow Over Wings Including Inviscid/Viscous Interaction," *AIAA Journal*, Vol. 21, March 1983, pp. 363–371.

[113]Guruswamy, P. and Goorjian, P. M., "Effects of Viscosity on Transonic-Aerodynamic and Aeroelastic Characteristics of Oscillating Airfoils," *Journal of Aircraft*, Vol. 21, Sept. 1984, pp. 700–707.

[114]Marstiller, J. W., Guruswamy, P., Yang, T. Y., and Goorjian, P. M., "Effects of Viscosity and Modes on Transonic Flutter Boundaries of Wings," AIAA Paper 84-0870, May 1984.

[115]Howlett, J. T., "Efficient Self-Consistent Viscous-Inviscid Solutions for Unsteady Transonic Flow," AIAA Paper 85-0482, Jan. 1985.

[116]Guruswamy, G. P. and Goorjian, P. M., "Efficient Algorithm for Unsteady Transonic Aerodynamics of Low-Aspect-Ratio Wings," *Journal of Aircraft*, Vol. 22, March 1985, pp. 193–199.

[117]Malone, J. B. and Sankar, N. L., "Numerical Simulation of Two-Dimensional Unsteady Transonic Flows Using the Full-Potential Equation," *AIAA Journal*, Vol. 22, Aug. 1984, pp. 1035–1041.

[118]Malone, J. B. and Sankar, N. L., "Numerical Solutions of 2-D Unsteady Transonic Flows Using Coupled Potential-Flow/Boundary-Layer Methods," AIAA Paper 84-0268, Jan. 1984.

[119]Hafez, M. M., Osher, S., and Whitlow, W., Jr., "Improved Finite Difference Schemes for Transonic Potential Calculations," AIAA Paper 84-0092, Jan. 1984.

[120]Shankar, V., Ide, H., Gorski, J., and Osher, S., "A Fast, Time Accurate Unsteady Full Potential Scheme," AIAA Paper 85-1512-CP, July 1985.

[121]Steger, J. L. and Warming, R. F., "Flux Vector Splitting of the Inviscid Gasdynamic Equations with Application to Finite-Difference Methods," *Journal of Computational Physics*, Vol. 40, April 1981, pp. 263–293.

[122]Harten, A., Lax, P. D., and Van Leer, B., "On Upstream Differencing and Godunov-Type Schemes for Hyperbolic Conservation Laws," *SIAM Review*, Vol. 25, 1983, pp. 35–61.

[123]Chakravarthy, S. and Osher, S., "A New Class of High Accuracy TVD Schemes for Hyperbolic Conservation Laws," AIAA Paper 85-0363, Jan. 1985.

[124]Thomas, J. L., Van Leer, B., and Walters, R. W., "Implicit Split-Flux Schemes for the Euler Equations," AIAA Paper 85-1680, July 1985.

[125]Yee, H. C. and Harten, A., "Implicit TVD Schemes for Hyperbolic Conservation Laws in Curvilinear Coordinates," AIAA Paper 85-1513CP, July 1985.

[126]Chakravarthy, S. R., "Relaxation Methods for Unfactored Implicit Schemes," AIAA Paper 84-0165, Jan. 1984.

[127]Roe, P. L., "Characteristics Based Schemes for the Euler Equations," *Annual Review of Fluid Mechanics*, Vol. 18, 1986, pp. 337–365.

[128]Coakley, T. J., "Implicit Upwind Methods for the Compressible Navier-Stokes Equations," AIAA Paper 83-1958, July 1983.

[129]Thomas, J. L. and Walters, R. W., "Upwind Relaxation Algorithms for the Navier-Stokes Equations," AIAA Paper 85-1501, July 1985.

[130]Ying, S. X., Steger, J. L., Schiff, L. B., and Baganoff, D., "Numerical Simulation of Unsteady, Viscous, High-Angle-of-Attack Flows Using a Partially Flux-Split Algorithm," AIAA Paper 86-2179, Aug. 1986.

[131]Anderson, W. K., Thomas, J. L., and Whitfield, D. C., "Multigrid Acceleration of the Flux-Split Euler Equations," AIAA Paper 86-0274, Jan. 1986.

[132]Anderson, W. K., "Implicit Multigrid Algorithms for the Three-Dimensional Flux-Split Euler Equations," Ph.D. Dissertation, Mississippi State Univ., MS, Aug. 1986.

[133]Jesperson, D. C., "A Time-Accurate Multiple-Grid Algorithm," AIAA Paper 85-1493-CP, July 1985.

[134]Batina, J. T., "Unsteady Transonic Flow Calculations for Interfering Lifting Surface Configurations," NASA TM-86432, May 1985.

[135]Batina, J. T., "Unsteady Transonic Flow Calculations for Wing-Fuselage Configurations," AIAA Paper 86-0862, May 1986.

[136]Shankar, V. and Goebel, T., "Computation of Canard-Wing Transonic Flows Using the Full Potential Equation," AIAA Paper 86-1795, June 1986.

[137]Malone, J. B. and Sankar, L. N., "Unsteady Full Potential Calculations for Complex Wing-Body Configurations," AIAA Paper 85-4062, Oct. 1985.

[138]Guruswamy, G. P., Goorjian, P. M., and Tu, E. L., "Unsteady Transonics of a Wing with Tip Store," *Journal of Aircraft*, Vol. 23, Aug. 1986, pp. 662–672.

[139]Edwards, J. W., Bennett, R. M., Whitlow, W., Jr., and Seidel, D. A., "Time-Marching Transonic Flutter Solutions Including Angle-of-Attack Effects," *Journal of Aircraft*, Vol. 20, Nov. 1983, pp. 899–906.

[140]Guruswamy, P. and Goorjian, P. M., "Computations and Aeroelastic Applications of Unsteady Transonic Aerodynamics About Wings," *Journal of Aircraft*, Vol. 21, Jan. 1984, pp. 37–43.

[141]Myers, M. R., Guruswamy, P., and Goorjian, P. M., "Flutter Analysis of a Transport Wing Using XTRAN3S," AIAA Paper 83-0922, May 1983.

[142]Johnson, D. A., "Predictions of Transonic Separated Flow with an Eddy-Viscosity/Reynolds-Shear Stress Closure Model," AIAA Paper 85-1683, July 1985.

[143]Simpson, R., "A Review of Two-Dimensional Turbulent Separated Flow Calculation Methods," IUTAM Symposium on Boundary Layer Separation, University College, London, Aug. 1986.

[144]Coakley, T. J., "Impact of Turbulence Modeling on Numerical Accuracy and Efficiency of Compressible Flow Simulations," *10th International Conference on Numerical Methods in Fluid Mechanics*, Lecture Notes in Physics No. 264, Springer-Verlag, New York, June 1986, pp. 186–191.

[145]Coakley, T. J., "Numerical Method for Gas Dynamics Combining Characteristic and Conservation Concepts," AIAA Paper 81-1257, June 1981.

[146]Baldwin, B. and Lomax, H., "Thin-Layer Approximation and Algebraic Model for Separated Turbulent Flows," AIAA Paper 78-257, Jan. 1978.

[147]Coakley, T. J., "A Compressible Navier-Stokes Code for Turbulent-Flow Modeling," *Workshop on Computational Fluid Dynamics*, Univ. of Tennessee Space Inst. Publ. No. E02-4005-023-84, Tullahoma, TN, March 1984, Chap. 12.

[148]Buning, P. G. and Steger, J. L., "Solution of the Two-Dimensional Euler Equations with Generalized Coordinate Transformation Using Flux Vector Splitting," AIAA Paper 82-0971, June 1982.

[149]Rumsey, C. L., Thomas, J. L., Warren, G. P., and Liu, G. C., "Upwind Navier-Stokes Solutions for Separated Periodic Flows," AIAA Paper 86-0247, Jan. 1986.

[150]Thomas, J. L., Walters, R. W., Van Leer, B., and Anderson, W. K., "Implicit Finite-Volume Algorithms for the Flux-Split Euler Equations," *Proceedings, 6AMM Workshop on the Numerical Simulation of Compressible Euler Flows* (to appear in *Notes on Numerical Fluid Mechanics*, Vieweg, Braunschweig).

[151]Barton, J. T. and Pulliam, T. H., "Airfoil Computation at High Angles of Attack, Inviscid and Viscous Phenomena," *AIAA Journal*, Vol. 24, May 1986, pp. 705–712.

[152]Shamroth, S. J., "Calculation of Steady and Oscillating Airfoil Flow Fields Via the Navier Stokes Equations," AIAA Paper 84-525, Jan. 1984.

[153]Sankar, L. N., Malone, J. B., and Schuster, D., "Full Potential and Euler Solutions for the Unsteady Transonic Flow Past a Fighter Wing," AIAA Paper 85-4061, Oct. 1985.

[154]Houwink, R. and Veldman, A. E. P., "Steady and Unsteady Separated Flow Computations for Transonic Airfoils," AIAA Paper 84-1618, June 1984.

[155]Bennett, R. M., Wynne, E. C., and Mabey, D. G., "Calculation of Transonic Steady and Oscillatory Pressures on a Low Aspect Ratio Model and Comparison with Experiment," NASA TM-86449, June 1985.

[156]Ruo, S. Y., Malone, J. B., Horsten, J. J., and Houwink, R., "Steady and Unsteady Transonic Airloads on a Supercritical Wing," *Journal of Aircraft*, Vol. 22, Jan. 1985, pp. 28–36.

[157]Borland, C. J. and Sotomayer, W. A., "An Algorithm for Unsteady Transonic Flow About Tapered Wings," AIAA Paper 84-1567, June 1984.

Chapter 6

Application of Transonic Flow Analysis to Helicopter Rotor Problems

F. X. Caradonna*
Ames Research Center, Moffett Field, California

Introduction

A rotor blade in edgewise motion experiences a large variation in dynamic pressure due to the varying component of forward flight velocity in the blade chordwise direction. This variation would result in large hub rolling moments if the blade were not hinged to permit free flapping. The resulting flow on a rotor is one in which the advancing blade operates at a high Mach number and low lift coefficient (C_L), whereas the retreating blade operates at a low Mach number and high C_L (Fig. 1). Therefore, the two major constraints on rotor forward flight performance are the initiation of stall on the retreating rotor and the occurrence of excessive compressibility effects on the advancing side. Of these two problems, stall is sometimes considered to be the more tractable because it can be delayed by designing rotors with an increased blade area (at no small cost in weight, however). On the advancing side of the rotor, however, there is no such avoiding of high tip Mach number if the object is to maximize the useful forward flight capability of the helicopter. Therefore, a major goal in the development of advanced helicopters is an ability to design rotors that can operate efficiently (i.e., with minimum drag, pitching moment, and noise) at the highest possible Mach numbers. This goal entails a development of the ability to predict the transonic flow of a rotor operating in a highly unsteady environment.

Unlike fixed wings, rotors are intrinsically unsteady, and this fact is an important determinant of the choice of computational methods. The unsteadiness of the rotor environment comes from a number of sources. The blade chordwise Mach number, flapping, and pitch control have large amplitude variations that occur at the frequency of the blade rotation. These variations are especially significant when the blade is at azimuthal angles between 90 and 180 deg (0 deg is defined as the downwind direction)

---

*Research Scientist, U.S. Army, Aeroflightdynamics Directorate.

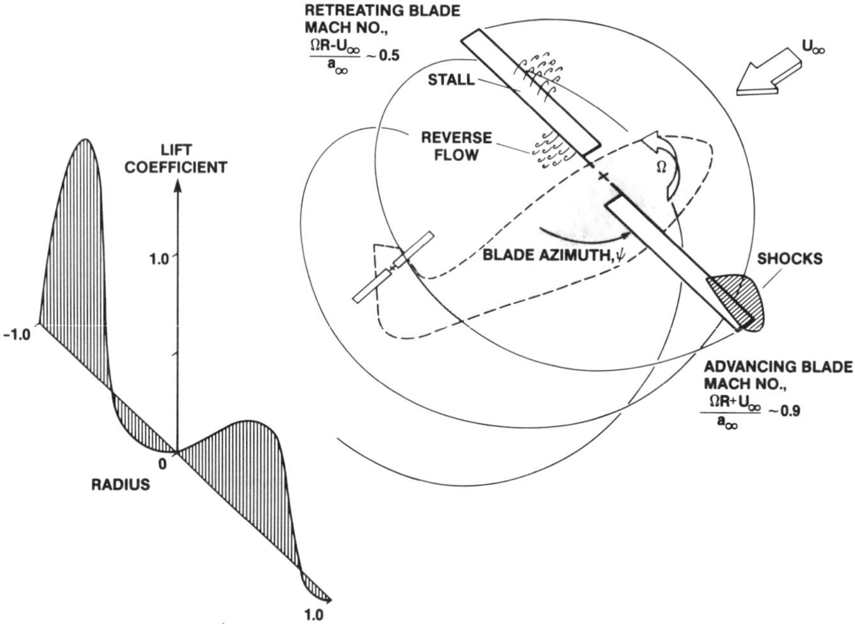

Fig. 1 Aerodynamics of the advancing rotor.

where the flow is transonic and decelerating. Superimposed on these variations are an assortment of high-frequency, high-amplitude unsteady flows that result from the near passage of vortices from previous blades.

Although the main emphasis of this review is the transonic aspect of helicopter flow analysis, it must be recognized that the rotor wake determines much of the basic flow environment, and a truly reliable flow prediction will not be complete without adequate wake models. Therefore, a prime issue—beyond the description of transonic flow models and algorithms—is the incorporation of rotor wakes into these transonic models. The more basic issue of the development and accuracy of wake models is beyond the scope of this review. Furthermore, a validated model of the advancing rotor wake simply does not exist at present.

The development of flow analysis methodologies for rotorcraft lags far behind its fixed-wing counterparts. One reason for this situation is the great complexity of rotor flows. However, another factor is that, in the past, the helicopters greatest problems have been mechanical rather than aerodynamic. At present, many of these problems have been solved due to the development of the small turbine engine, composite structures, bearingless rotor technology, and a general understanding of rotor dynamics and stability. It is clear now that a renewed emphasis must be placed on rotor aerodynamic development if helicopters are to meet increasing productivity demands. A central facet of this work must be the development of new aerodynamic computational tools.

Equations of Motion and Their Implementations

The basic complexity of rotor flows is such that the simplest flow models hold a treasured position in the analysis hierarchy. However, when detailed analyses are necessary, these are nearly always based in an integral solution of the incompressible potential flow equation. Usually this is a lifting line method with the wake being modeled by a vortex lattice that is either free to convect or constrained to an assumed path. Various attempts to extend the applicability of these methods (e.g., by implementing a lifting surface or including linear compressibility) have not yet proven to be worth the extra effort. Therefore, real flow effects are currently treated by using two-dimensional experimental airfoil tables in which the section drag, lift, and moment are given as functions of local Mach number and angle of attack. The result of this approach is that unsteady, three-dimensional transonic effects are not accounted for. Furthermore, this approach limits the analyst to treating only those profiles and configuration types that have been physically tested. These limitations are finally being overcome by the use of differential (i.e., finite-difference/finite-volume) methods.

Potential methods are currently the primary finite-difference approaches to high-speed rotor computations. The starting point for the various potential formulations is the mass conservation and Bernoulli's equation (shown here for an inertial reference frame):

$$\rho_t + \nabla \cdot (\rho \nabla \phi) = 0 \tag{1}$$

$$\rho/\rho_\infty = \left\{1 - \frac{\gamma - 1}{\alpha_\infty^2}\,[\phi_t + \tfrac{1}{2}(\nabla\phi)^2]\right\}^{1/(\gamma - 1)} \tag{2}$$

The conservation form (i.e., the divergence form) of Eq. (1) is essential because it guarantees mass conservation through shocks. However, this model does not conserve momentum through shocks and is isentropic. This results in an approximate shock drag mechanism and an approximate jump condition that are incorrect for strong shocks. It will be shown later that these approximations are correctable. This is fortunate because potential theory provides the simplest and most computationally efficient flow model.

For helicopter computations, it is first necessary to transform Eqs. (1) and (2) to a translating and rotating coordinate system fixed to a blade. That is,

$$\boldsymbol{r}' = \boldsymbol{U}_\infty t + \boldsymbol{\Omega} t \times \boldsymbol{r}$$

$$t' = t$$

where $r' = (x',y',z')$ and $r = (x,y,z)$ are the inertial and body-fixed coordinates (Fig. 2) and $\boldsymbol{U}_\infty$ and $\boldsymbol{\Omega}$ are the translational and angular velocities of the rotor. Under this transformation Eqs. (1) and (2) become

$$\rho_t + \nabla \cdot [\rho \mathbf{V}] = 0 \tag{3}$$

$$\rho/\rho_\infty = \left\{1 - \frac{\gamma - 1}{\alpha_\infty^2}\,[\phi_t + \tfrac{1}{2}(V^2 - V_\infty^2)]\right\}^{1/(\gamma - 1)} \tag{4}$$

where $\boldsymbol{V} = \boldsymbol{V}_\infty + \nabla\phi$ and $\boldsymbol{V}_\infty = \boldsymbol{U}_\infty + \boldsymbol{\Omega} \times \boldsymbol{r}$ is the local freestream velocity seen by an observer in the rotating frame. A common misunderstanding concerns the validity of potential theory where rotary motion is involved. This question arises because the fixed-wing community traditionally includes the freestream velocity in the potential function. In the rotor case, $\boldsymbol{V}_\infty$ is rotational and cannot be described by a potential. However, $\boldsymbol{V}_\infty$ arises from a coordinate transformation rather than the basic flow, which is certainly irrotational in the inertial reference frame. In the blade fixed frame, ϕ should be considered to describe a perturbation field about $\boldsymbol{V}_\infty$.

Equation (3) is commonly written for a generalized, moving coordinate system ξ,η,ζ,t as

$$\partial_t(\rho/J) + \partial_\xi(\rho U/J) + \partial_\eta(\rho V/J) + \partial_\zeta(\rho W/J) = 0 \tag{5}$$

where U, V, and W are contravariant velocity components and J is the Jacobian. Bernoulli's equation is similarly expressed in general coordinates:

$$\rho/\rho_\infty = \left\{1 - \frac{\gamma - 1}{\alpha_\infty^2}[\phi_t + (U + \xi_t)\phi_\xi + (V + \eta_t)\phi_\eta + (W + \zeta_t)\phi_\zeta]\right\}^{1/(\gamma-1)} \tag{6}$$

In this formulation, the blade motion is specified by the coordinate terms ξ_t,η_t,ζ_t. Equations (5) and (6) are probably the ideal formulation of the potential problem because they are conservative and geometrically general.

However, simpler formulations have been and continue to be important to the development of rotorcraft applications. The first of these simpler approaches is the small-disturbance equation. This equation is obtained by substituting Eq. (4) into Eq. (3) to eliminate ρ. A scaling and limiting process is then used to eliminate all but the lowest-order nonlinear term. The resulting equation has the form[1]

$$A\phi_{tt} + B\phi_{xt} = F_x + \phi_{zz} + C\phi_{yy} + D\phi_{xy} \tag{7}$$

where the nonlinearity is contained in the flux F, which has the form $\alpha\phi_x + \beta\phi_x^2$. Equation (7) is not unique, and various modifications and additions have been made to improve its stability and accuracy.[2,3] An interesting feature of Eq. (7) is that it maintains conservation form. The maintenance of conservation is a result of the small-disturbance limiting process. If all terms are retained after Eqs. (3) and (4) are combined, the following equation results (shown here in Cartesian coordinates for a rotor in pure edgewise motion):

$$\begin{aligned}\phi_{tt} + 2V_x\phi_{xt} + 2V_zV_{zt} = (a^2 - V_x^2)\phi_{xx} + (a^2 - V_z^2)\phi_{zz} + (a^2 - V_y^2)\phi_{yy}\\ -2V_xV_y\phi_{xy} - 2V_xV_z\phi_{xz} - 2V_yV_z\phi_{yz} + (\Omega^2 x - 2\Omega U_\infty \cos\psi)\phi_x\\ + (\Omega^2 z + 2\Omega U_\infty \sin\psi)\phi_z\end{aligned} \tag{8}$$

Equation (8) is essentially the familiar textbook form of the potential equation. It cannot be put in conservative form. Since Eqs. (7) and (8) have the same general form, it is to be expected that they share the same general solution methods. Equation (7) is stably discretized using a "Murman mixed difference scheme" (i.e., central/backward differences in the subsonic/supersonic flow regions) in the streamwise direction. The counterpart to the Murman scheme for Eq. (8) is the "Jameson rotated method" in which mixed differences in all directions are used. Equation (7) has been implemented in the finite-difference rotor code, FDR,[4,5] which uses the Douglas-Gunn Alternating Direction Implicit solution scheme. Equation (8) has been implemented in several codes. The first such implementation is the steady code ROT22,[6] a derivative of FLO22, which uses successive line overrelaxation (SLOR). Another steady code for Eq. (8) is TFAR1,[7] which used an approximate factorization solution method (rather than SLOR). The fully unsteady form of Eq. (8) is implemented in the code TFAR2,[8] a derivative of TFAR1.

The previously mentioned codes are limited either by small-disturbance or conservation considerations. The implementation of a conservative full potential method is complicated by the inability to either directly eliminate ρ (which destroys the conservation form) or to stably solve Eqs. (5) and (6) as a two-equation system. The spacial terms in Eq. (5) present no problem, since these are easily treated using central differences (with stability being maintained by upstream biasing of density), where density is evaluated at the previous time step or iteration. The real problem is to express the term ρ_t in a form that does not depend on ρ at the new time step $n + 1$. This can be achieved[9,10] by expanding ρ^{n+1} *as*

$$\rho^{n+1} \cong \rho^n + \partial\rho/\partial\phi|^n(\phi^{n+1} - \phi^n) \tag{9}$$

where $\partial\rho/\partial\phi$ is a differential operator obtained from the Bernoulli equation. When Eq. (9) is substituted into the discretized form of Eq. (5), there results an equation of the form

$$\begin{aligned} &\delta_{tt}\phi^{n+1} + \phi_\xi^n\delta_{\xi t}\phi^{n+1} + \phi_\eta^n\delta_{\eta t}\phi^{n+1} + \phi_\zeta^n\delta_{\zeta t}\phi^{n+1} \\ &+ \delta_\xi(\rho^n\delta_\xi\phi)^{n+1} + \delta_\eta(\rho^n\delta_\eta\phi)^{n+1} + \delta_\zeta(\rho^n\delta_\eta\phi)^{n+1} = C^n \end{aligned} \tag{10}$$

where δ denotes a difference operator. This equation can be considered to be a hybrid between Eqs. (5) and (8). The spacial flux operators correspond to those in Eq. (5). However, the time derivatives have the same form as the nonconservative time terms of Eq. (8). The resulting conservation error is canceled by the term C^n, which is a conservation correction term evaluated at time n. Equation (10) has been implemented for rotors in the code FPR[11,12] (a derivative of TUNA[10]), in which an approximate factorization solution method is employed. In a similar implementation (called RFS2), Eq. (10) has been solved using a strongly implicit procedure.[13]

All of the previously mentioned rotor codes (ROT22, FDR, TFAR1, TFAR2, FPR, RFS2) have seen use in industry. It may be that ROT22 has seen the widest use because its simplicity and robustness do not put great demands on the user. However, its inability to handle unsteady effects and accurate shock solutions (because it is nonconservative) is a considerable handicap. Of all the codes mentioned earlier, FPR probably stands out as being the most available and highly developed (both from a technical and user viewpoint). It also has a sizable user community.

The emphasis of the previous discussion on potential methods is an accurate reflection of current rotorcraft industry activities and needs. It should be noted that, unlike fixed-wing requirements, rotor requirements are intrinsically unsteady, and this imposes a computational burden, which is best met by potential schemes. However, potential methods are approximations that give incorrect solutions for strong shocks. This error also appears to cause nonuniqueness in various two-dimensional solutions (such effects have not been seen in unsteady, three-dimensional rotor computations) and can constitute a major argument for the use of the Euler equations. However, various modifications have been made to correct this problem.[14–16] One of the simplest such modifications has been applied to the FPR code by Bridgeman et al.[17] This consists of applying the non-isentropic density expression

$$\rho = \rho_i e^{-\Delta s/R} \tag{11}$$

where ρ_i is the isentropic density given by Eq. (6), and Δs is the shock generated entropy (computed by the Rankine-Hugoniot relations). It can be shown that it suffices to use this expression only at shock points. Equation (11) is only strictly valid for steady flows. However, it has produced good results for both steady and unsteady solutions.

The alternative to the entropy corrected potential equation is to use the Euler equations, which in Cartesian coordinates are

$$q_t + F_x + G_y + H_z = 0 \tag{12}$$

where

$$\begin{aligned}
q &= \{\rho,\rho u,\rho v,\rho w,e\}^T \\
F &= \{\rho u,\rho u^2 + p,\rho uv,\rho uw,u(e+p)\}^T \\
G &= \{\rho v,\rho uv,\rho v^2 + p,\rho vw,v(e+p)\}^T \\
H &= \{\rho w,\rho uw,\rho vw,\rho w^2 + p,w(e+p)\}^T
\end{aligned}$$

This set of equations is exact for inviscid flows and can be readily extended to treat viscous flows. However, the computational cost is currently far greater than any of the potential methods. The Euler equations were first implemented for rotors by Sankar et al.[18,19] in a semi-implicit scheme that uses the discretization

$$\boldsymbol{\delta}_t q^{n+1} + \delta_\xi F^{n+1} + \delta_\eta G^{n+1} + \delta_\zeta H^n = 0 \tag{13}$$

where $\boldsymbol{\delta}$ is a backward difference and δ is a central difference. This scheme has been implemented using an approximate factorization method. Other implementations include an explicit method by Agarwal and Deese[20] and a

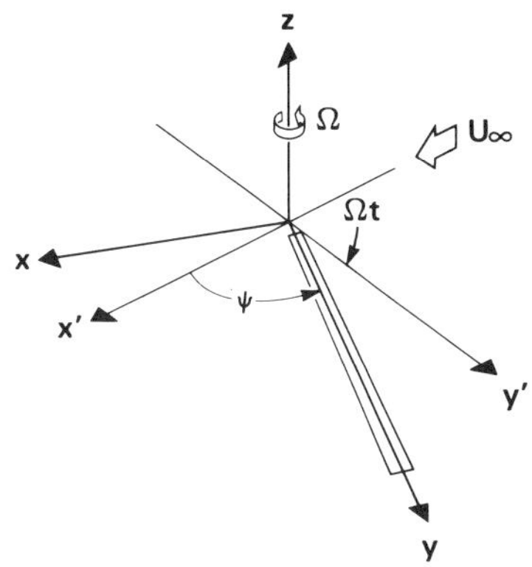

$$\begin{bmatrix} x' \\ y' \\ z' \\ t \end{bmatrix} = \begin{bmatrix} \cos \Omega t & -\sin \Omega t & 0 & 0 \\ \sin \Omega t & \cos \Omega t & 0 & 0 \\ 0 & 0 & 1 & 0 \\ 0 & 0 & 0 & 1 \end{bmatrix} \begin{bmatrix} x \\ y \\ z \\ t \end{bmatrix}$$

Fig. 2 Transformation from inertial to blade-fixed coordinates.

fully implicit method by Chen and McCroskey.[21] However, in view of the recent success of entropy corrected potential codes (and the fact that strong shocks can be avoided by proper design), it may be that the main significance of Euler methods is as a stepping stone to full Navier-Stokes computations.

Computation of Real Rotor Flows

The previous section described a number of very serviceable transonic rotor codes and methods. The next step is to convert these into realistic engineering tools. In fact, all of the previously mentioned codes were only designed to solve for a small portion of the total rotor domain, typically a region extending about 5–10 chords from one blade in all directions. Within this domain there are a number of standard boundary conditions that all finite-difference codes employ. These include surface tangency conditions, vorticity convection from the trailing edge (for potential codes), and radiation or freestream conditions at the outer grid boundary. These grids and boundary conditions are insufficient for helicopters because they do not account for a rotors extensive wake system, only some of which can be contained in a typical grid.

The problem is illustrated in Fig. 3 for a hovering rotor. A rotor sheds a wake that consists of a large system of vorticity sheets and tip vortices. A typical finite-difference grid wrapped around a blade will have several tip vortices passing through. Vortices that pass close to the blade will produce large local lift perturbations. However, these vortices are only a small part of the total wake system, which must be accounted for if lift and power are to be accurately predicted. Therefore, the problem of using local Computational fluid dynamics (CFD) solutions to predict real rotor loading has two parts: 1) how to incorporate nearby vortex elements into the grid region and 2) how to account for the remaining wake vorticity regions that lie outside of or are not included in the grid.

At first glance, the problem of putting vorticity into a solution would appear to be to use Euler or Navier Stokes solutions, which do describe the

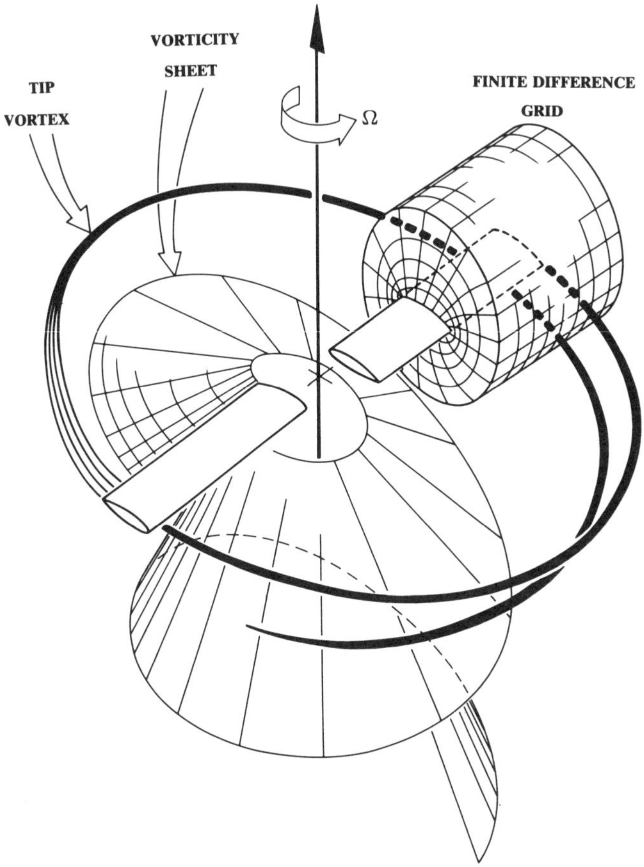

Fig. 3 A blade-fixed finite-difference grid embedded in a global rotor flow.

generation, convection, and dissipation of vorticity. The flaw with this approach is that the numerical (or artificial) dissipation of vorticity is considerable and all but eliminates (presently) the possibility of long preserving a wake structure with a practical grid. The alternative to solving for this wake is to specify it. In fact, the shed circulation sheet in all potential codes is such a specified wake. Accordingly, the first near-wake vortex model entailed a simple modification of the trailing vorticity sheet logic already contained in all potential codes. In this approach an additional potential discontinuity sheet is inserted into the grid. The edge of this sheet constitutes the location of a desired vortex. Although this scheme is simple when a vortex can be aligned with a grid line, it is not easily applied to arbitrary grid/vortex orientations of unsteady flows. An alternate approach is to recast the flow equations to solve for a perturbation about a known vortex solution. For example, the difference equation that results from the conservative full potential equations can be expressed in functional form as

Discretized mass conservation:

$$L(\rho^n,\Phi^{n+1}-\Phi^n)=R(\rho^n,\rho^{n-1},\Phi^n,\Phi^{n-1},\Phi^{n-2}) \tag{14a}$$

Discretized Bernoulli equation:

$$\rho^n=F(\Phi^n) \tag{14b}$$

Now letting $\Phi=G+\phi$, where G is a known potential due to a specified system of vortex elements and ϕ is a perturbation potential, yields

$$L(\rho^n,\phi^{n+1}-\phi^n)=R(\rho^n,\rho^{n-1},\phi^n+G^n,\phi^{n-1}$$
$$+G^{n-1},\phi^{n-2}+G^{n-2})+L(\rho^n,G^{n+1}-G^n) \tag{15a}$$
$$\rho^n=F(G^n+\phi^n) \tag{15b}$$

That is, the left-hand-side operator is unchanged and the right-hand side and Bernoulli equation undergo minor modification. The task of finding G for various time steps for a general vortex geometry can be quite laborious, however. Fortunately, various studies[12,22–24] indicate that in Eqs. (14), the spacial operators acting on G dominate the time operators, which can be neglected. This implies that the effect of the wake on the solution can be imposed as a specified velocity, $V_s=\nabla G$, which is simpler to find than G. It also happens that rotational V_s fields can be used if the vortical region is confined to a small area. Since these approaches express the total velocity as $V=V_s+\nabla\phi$, they are often referred to as "velocity decomposition" methods (the term "split potential method" is also used frequently).

After including the near-vortex segments in the grid by the preceding methods, it is necessary to find the inflow at the rotor which is induced by all of the remaining vorticity elements. One method to find this inflow is to use a Biot-Savart computation. The induced velocities so obtained can be imposed either on the outer grid boundary[25] or the blade surface.[5] The latter approach entails the necessity to excise the grid-contained vorticity from the Biot-Savart computation. However, this approach is convenient in that it amounts to an effective angle-of-attack modification (the chordwise variation of this induced velocity being small). This approach also leaves the outer boundary free for the specification of nonradiative boundary conditions (an important stability concern in unsteady computations).

The preceding discussion is based on the assumption that a rotor loads model is to be built from a CFD code that is purely blade-oriented. However, it is possible now to solve for the entire blade-wake system. Recently, Steinhoff and Ramachandran[26–28] have devised a potential scheme for solving the entire rotor hover problem. Rather than using a grid that wraps around the blade (such as that depicted in Fig. 3), they employ a cylindrical grid (around the axis of rotation) that can resolve both the blade and its wake system. The basis of their approach is the use of velocity decomposition (which, in fact, Steinhoff et al. pioneered in Ref. 29).

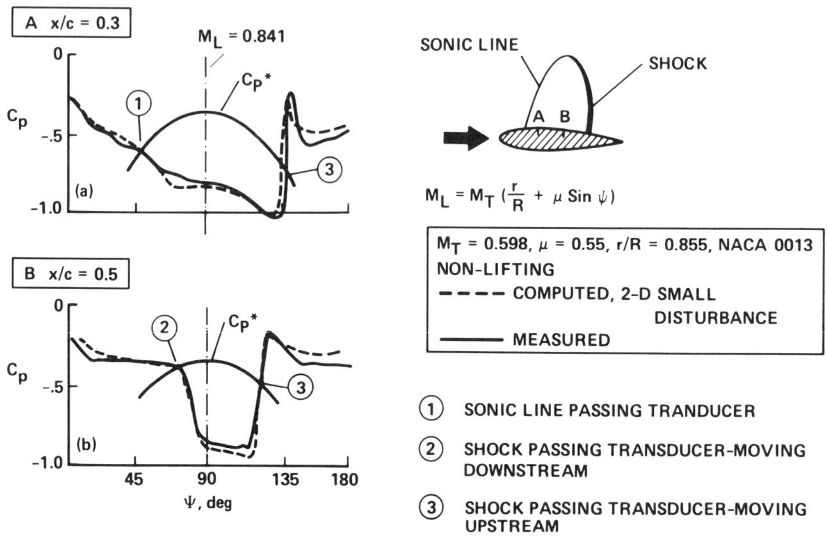

Fig. 4 A comparison of measured and computed surface pressure for a nonlifting rotor.

However, they carry velocity decomposition a step farther by actually solving for V_s in the course of the solution procedure. They accomplish this by replacing the usual shed potential discontinuity sheet (which is not easily convected) by vorticity markers that are free to convect (in the Lagrangian sense). The vorticity that these markers contain is expressed as a normal velocity distribution V_s, whose structure is a fixed function of the local grid and whose magnitude is determined by the circulation that the markers represent. This approach results in a freely convecting wake that compares well with hover wake measurements. This approach is extendable to treating advancing rotors.

An additional problem that arises in forward flight computations concerns the prediction of the blade trim. In a general forward flight computation it is required to find the rotor control inputs that produce a desired thrust and propulsive force. This process involves iterating the control inputs while the blade motion and aerodynamic forces are simultaneously computed. Convergence of this process can require the computation of as much as 10 rotor revolutions—an unacceptable computational burden for any finite-difference scheme. Current trim codes have few computational time problems because they use table lookups for the blade aerodynamics. A scheme[5] has been recently developed that efficiently couples a current trim analysis code with a finite-difference rotor code. The method is to iterate between a finite-difference code and the tabular trim analysis. However, the trim analysis is modified to obtain only a lift correction to the finite-difference solution rather than the entire lift. On convergence, the lift correction (and the effect of the tabular data) disappear. The trim code used is CAMRAD,[30] a comprehensive code that predicts the inflow, rotor dynamics, and trim. This code has currently been coupled with the FDR, FPR, and TFAR2 codes.

Role of Experimental Data

One of the primary reasons for the importance of new computational methods for helicopters is that rotor experiments are far more difficult, expensive, and time consuming than fixed-wing tests. Furthermore, the results of such tests can be much more difficult to interpret. Therefore, a final goal is to be able to design new rotors without an extensive cut-and-try process. This goal entails the need for computational methods whose uncertainties are less than those of typical rotor tests. Paradoxically, the development of new computational tools generates a need for new rotor measurements—for validation rather than design purposes. For rotors, the development of special purpose validation experiments can involve a greater time and expense than the code developments. It is not surprising, therefore, that the sources of suitable validation data are few. The following sections draw from these tests as well as from the previously discussed methods in order to present a picture of current computational capabilities.

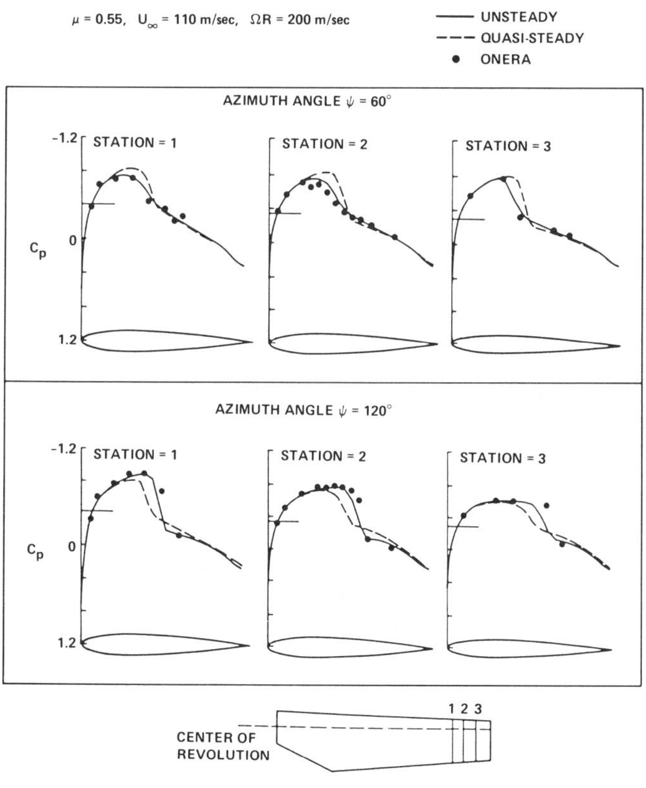

Fig. 5 Effect of unsteadiness on three-dimensional, transonic rotor flows.

These experiment-computation comparisons are arranged in order of increasing complexity and constitute a validation path for any new rotor CFD code.

Nonlifting Rotor Flows

Surface pressure data from nonlifting rotors validate the basic ability of codes to predict transonic flows free from the complications of wake effects. These tests are usually performed on advancing rotors in order to avoid the wake buildup that would occur for a nonlifting hover.

Probably the most extensive surface pressure model rotor data base has been acquired by ONERA.[31–33] These data include nonlifting and lifting data for two- and three-bladed rotors with a variety of blade profiles and planforms. Figure 4 shows an early nonlifting computation of the pressure variation on two surface pressure transducers performed with a two-dimensional small-disturbance code. Subsequent three-dimensional computations compare equally well with the data. However, the point of showing an inboard two-dimensional computation is to show the importance of transonic unsteadiness without any mitigating three-dimensional influences. This unsteadiness is seen in the asymmetry of the pressure about the $\psi = 90$ deg azimuth. Computations of a higher speed case are shown in Fig. 5 using the steady and unsteady full potential codes TFAR1 and TFAR2.[8] The inadequacy of steady computations is clearly shown in this comparison. Also, the slight upstream location of the computed shock with respect to the measured location is typical of nonconservative methods. These conservation discrepancies become more pronounced with increasing shock strength.

Figures 6 and 7 show a nonlifting computation/data comparison,[34] which is unique in that it employs optical methods rather than the usual pressure instrumentation. Figure 6 shows a series of interferograms that are obtained from holograms produced for a range of rotor/laser beam orientations. These interferograms can be used to reconstruct nearly the entire pressure field around the rotor. Figure 7 shows these reconstructed pressure fields at 0.08 chords above the rotor and at several radii. Also included in Fig. 7 are a comparison to small-disturbance computations and to laser velocimetry results. Although several unexplained differences are seen, the overall comparison is quite good. This comparison demonstrates the ability of CFD methods to predict pressure fields away from the surface of the rotor—an essential need for high-speed acoustics.

Nonlifting experimental data have also been used to validate the ability of codes to predict high-speed profile power. Figure 8 shows a comparison of computed and measured torque for a nonlifting two-bladed rotor with a NACA 0012 airfoil. The computation[17] was performed using the FPR code with both the isentropic and nonisentropic density expression [Eq. (11)]. This computation also used a Nash-MacDonald boundary layer. The torque was obtained by integrating the skin-friction and surface pressure. It is seen that the transonic drag divergence is well predicted, whereas the purely isentropic computation overpredicts the torque at higher-tip Mach numbers.

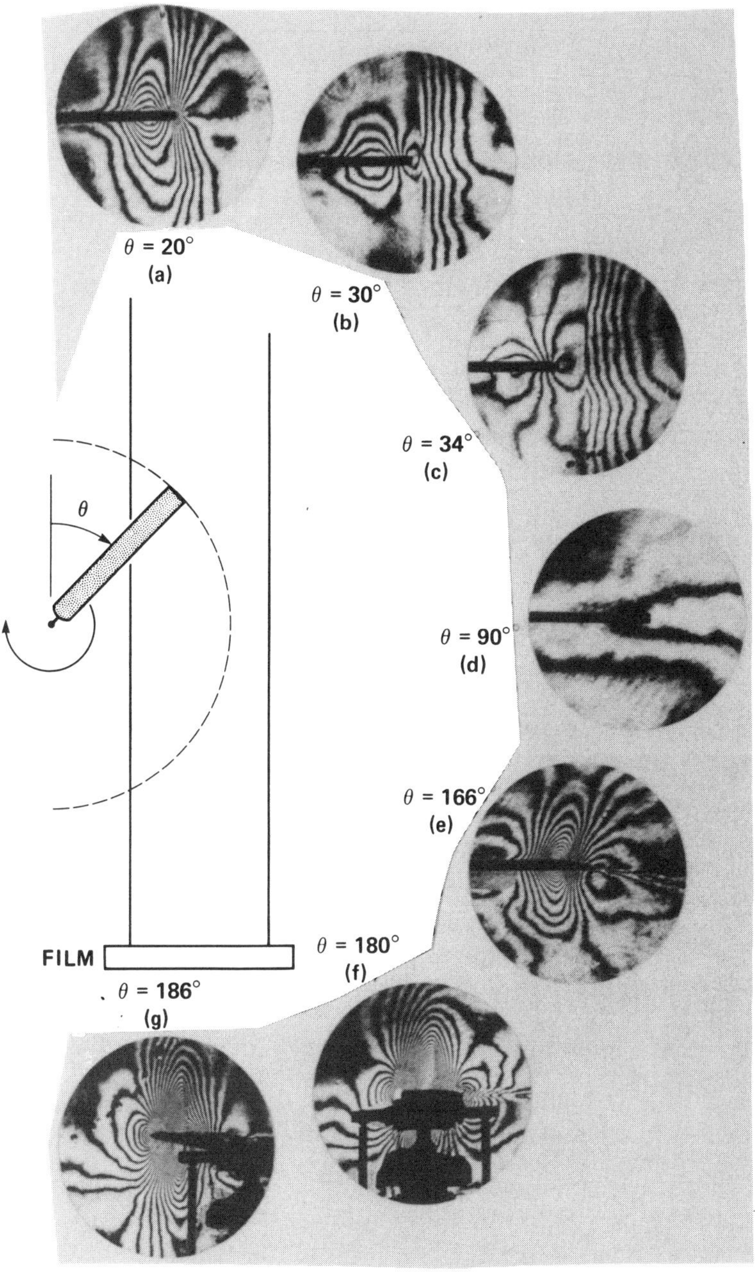

Fig. 6 Holographic interferograms for a rotor at various azimuth angles.

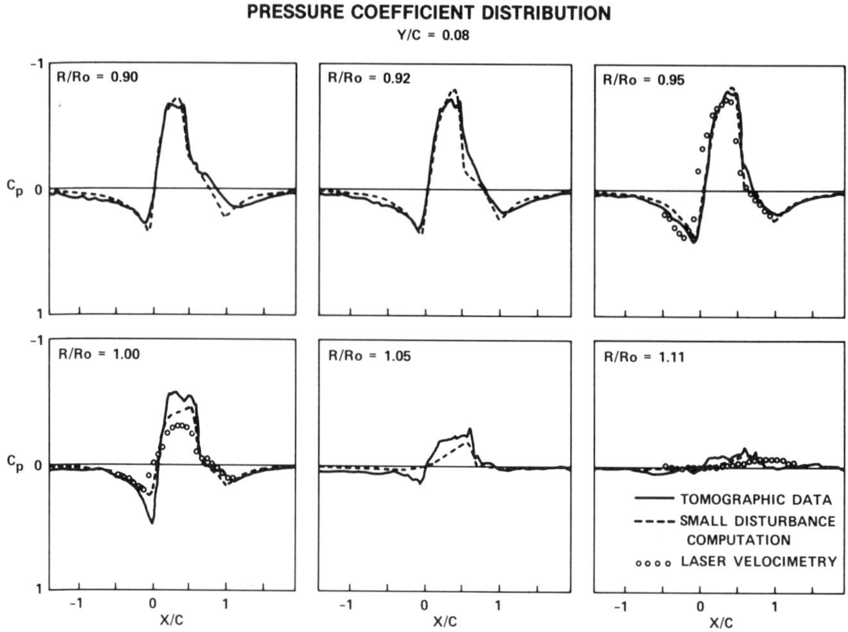

Fig. 7 A comparison of rotor flowfields obtained by computation and optical reconstruction methods.

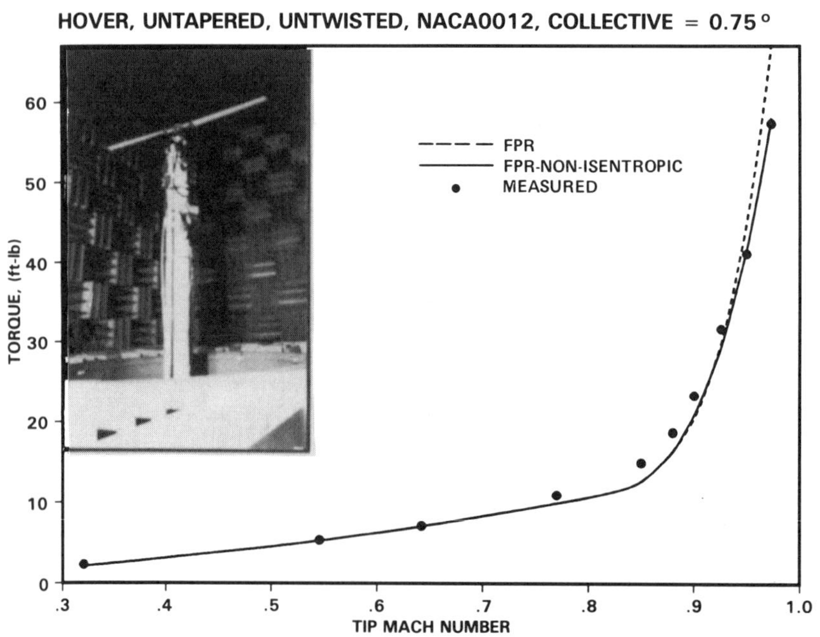

Fig. 8 A comparison of measured and computed torque for a nonlifting, hovering rotor.

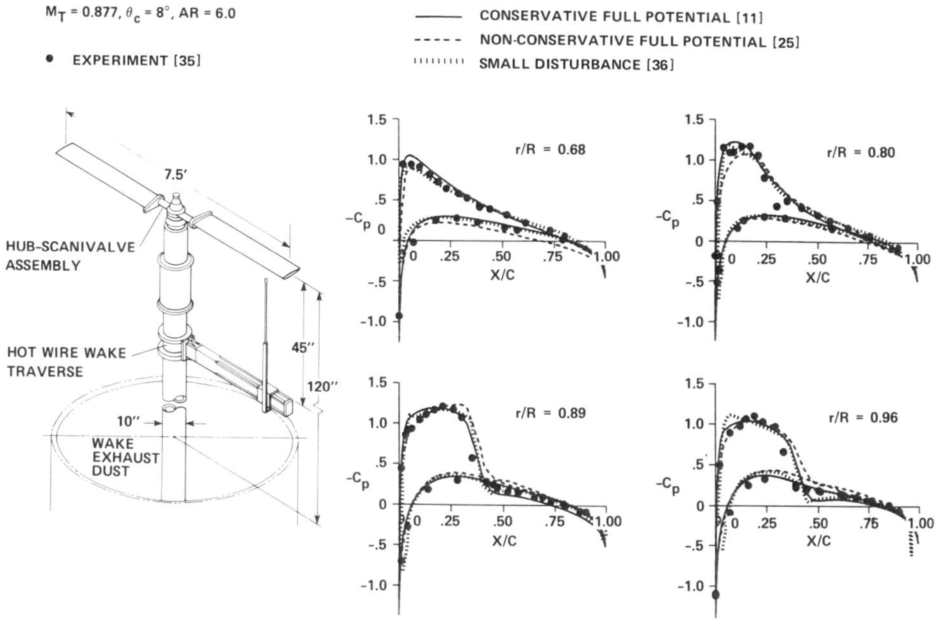

Fig. 9 A comparison of measured and computed flow on a lifting, hovering rotor.

High-Speed Hover

The next step in the validation process is to verify an ability to predict the effect of a known wake on the rotor flow. Initially, such a validation is best performed for a hovering rotor because the hover wake geometry is easily measured.

Toward this end, an extensively pressure instrumented model rotor was hover tested[35] by the (then) U.S. Army Aeromechanics Laboratory. Simultaneous measurements of the wake geometry (see the sketch in Fig. 9) provides the tip vortex location information that is needed for computational vortex modeling studies. To date, these data have been studied using 1) a small perturbation approach with a branch-cut model of the vortex[36] and 2) conservative (FPR[11]) and nonconservative (ROT22[25]) full potential schemes using the perturbation approach to modeling the vortex. These methods assume that the vortex strength equals the maximum blade bound circulation. Figure 9 shows a comparison of all three approaches with the data. Overall, there is a remarkable agreement between all three approaches. In all three methods the predicted shock location is slightly aft of the measured location. This discrepancy probably results from the isentropic flow approximation.[17] For this particular case, the conservative and nonconservative results are in good agreement.

The previously mentioned codes required that the measured tip vortex locations be specified. However, one new code, HELIX-1,[28] employs the previously mentioned vorticity convection scheme, which obviates the need for specified measured vortex locations or any assumptions on the strength of these vortices. This code has produced fair comparisons with the measured pressures and excellent comparisons with the wake measurements, a result of a grid that is presently aimed more at resolving the wake rather than the blade. A further advantage of being able to predict the wake is that it enables the separation of induced power from profile power because it permits the computation of an energy flux integral about the rotor.

It seems clear from this and other similar comparisons that near-wake modeling presents few problems in a steady transonic rotor environment.

High Advance Ratio Lifting Rotor Flows

In forward flight, the wake structure is not always well defined or known. However, for those tests that are run at high forward speeds, the wake-induced inflow is a relatively small percentage of the total inflow, and it becomes possible to use simplified wake models. Data obtained under these

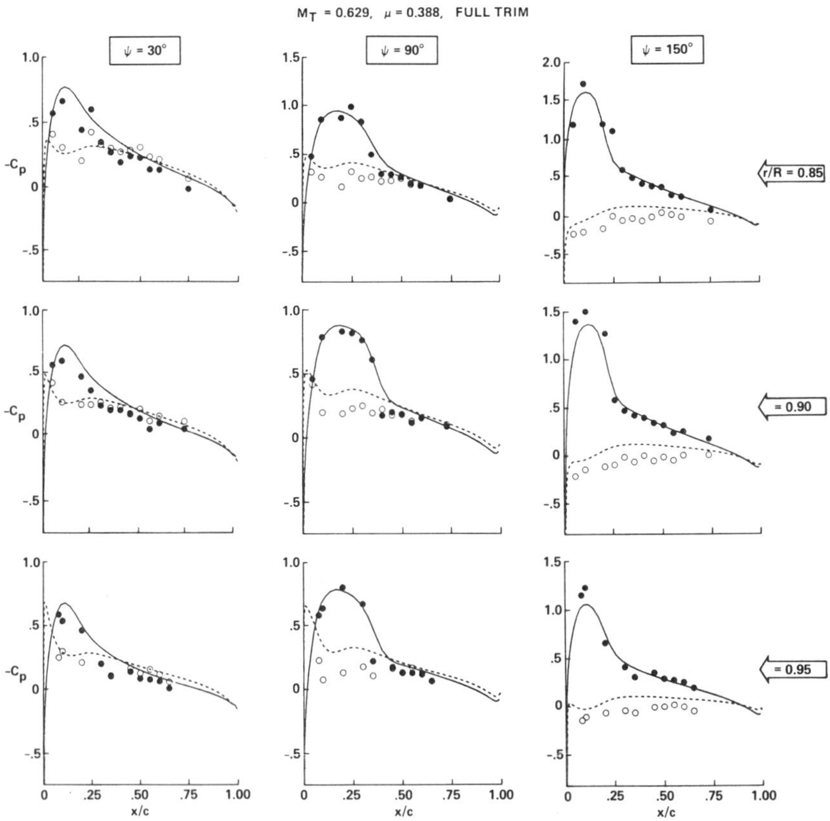

Fig. 10 A comparison of CAMRAD/FPR computations with pressure data from an ONERA advancing, lifting rotor.

conditions provide a first test of our ability to predict realistic operational advancing rotor flows.

The first realistic lifting rotor computations for forward flight were performed using the FDR, small-disturbance code.[36] These computations did not involve any trim determination other than to use measured control inputs and flapping. Furthermore, this early work used an approximate analytic expression for the wake-induced inflow, rather than actually modeling the wake itself. By means of these simplified approaches a number of very good comparisons with the ONERA three-bladed data were obtained. Beginning with Ref. 5, however, forward flight computations are increasingly using dynamic trim computations with full vortex modeling of the wake-induced inflow. However, these computations do not actually insert the vortex into the grid as was done in the hover cases that were previously shown. Rather, a separate vortex element wake model is used to find an induced angle of attack. This is only valid when the blade and shed wake do not approach each other too closely. An example of such a result is shown in Fig. 10, which makes a comparison with one of the ONERA three-bladed test cases. For this case, the tip rotational Mach number is 0.628 and the advance ratio (ratio of forward speed to rotational speed) is 0.387. The airfoils used are S-130XX (a variant of the NACA

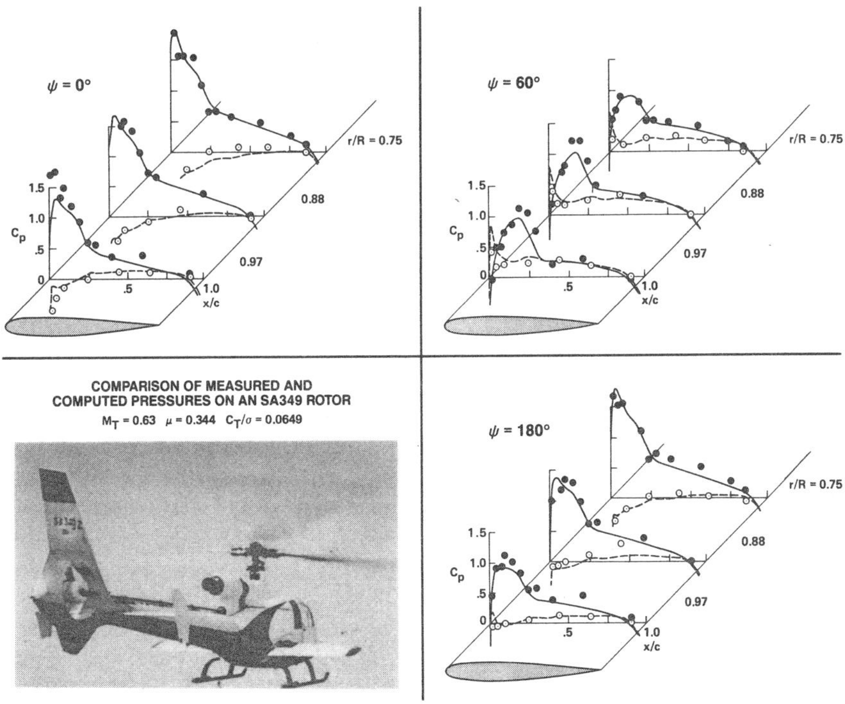

Fig. 11 A comparison of CAMRAD/FPR computations with in-flight pressure data on the SA 349 rotor.

five-digit family). This particular computation is accomplished using the coupled FPR and CAMRAD codes.[12] Overall, the comparison of the data and the computation are quite good. However, the computation does somewhat overpredict the upper pressures early in the first quadrant $|0 < \psi < 90°|$ as well as the lower surfaces in the second quadrant. The shock location and strength seem to be well predicted throughout the computation.

The previous cases involved stiff articulated model rotors; that is, the blade motion was almost pure flapping with very little elastic torsion or bending. Full-scale rotors tend to be much softer than the previously mentioned model blades, and elastic deformations can be important. At present, there has been only one published computational/experimental data comparison for an actual flight vehicle, an Aerospatiale SA 349. A typical data/computation comparison,[37] performed using CAMRAD/FPR, is shown in Fig. 11. Although this comparison of surface pressures is very promising, the differences between the computed and measured lift actually constitute a significant load error. The resolutions of these problems will probably require improvements in our structural and wake models. However, the present coupled codes that are now available do provide the first total rotor analysis to include a complete transonic loads prediction. These codes can constitute a valuable design tool in the hands of a knowledgeable user.

However, several important flow modeling improvements remain. A significant requirement is to improve the modeling of close blade/vortex interactions.

Strong Blade/Vortex Interaction Computations

Recall that, for hover flows (a wake-dominated flow), much of the wake-induced inflow was provided by actually embedding some vortex elements into the finite-difference domain. In contrast, for high advance ratio flows, no vortex elements were inserted in the grid, and the effect of the wake was given solely through its effect on the local angle of attack. However, many forward flight conditions actually have very strong blade-vortex interactions (BVI's). These interactions are much more complex than those in hover cases because they are unsteady and can occur for a wide range of blade-vortex orientations and proximities. Such flows can probably be treated only by split potential (or velocity decomposition) schemes at present.

However, the development of such a scheme for a full rotor computation is presently unrealistic because of the complexity of the wake geometry and the fact that no suitable body of full rotor BVI data exist. Our present requirement is for a body of BVI data in which the vortex has a simple geometry and a known location, strength, and structure. Figure 12 shows a recent experimental setup[38,39] for a scheme to produce a geometrically simple BVI. This test employed an upstream movable wing whose tip vortex interacted with a downstream nonlifting rotor, which was pressure instrumented. This scheme permits the setup of parallel and oblique BVI's of various speeds and blade-vortex proximities. Many of these interactions

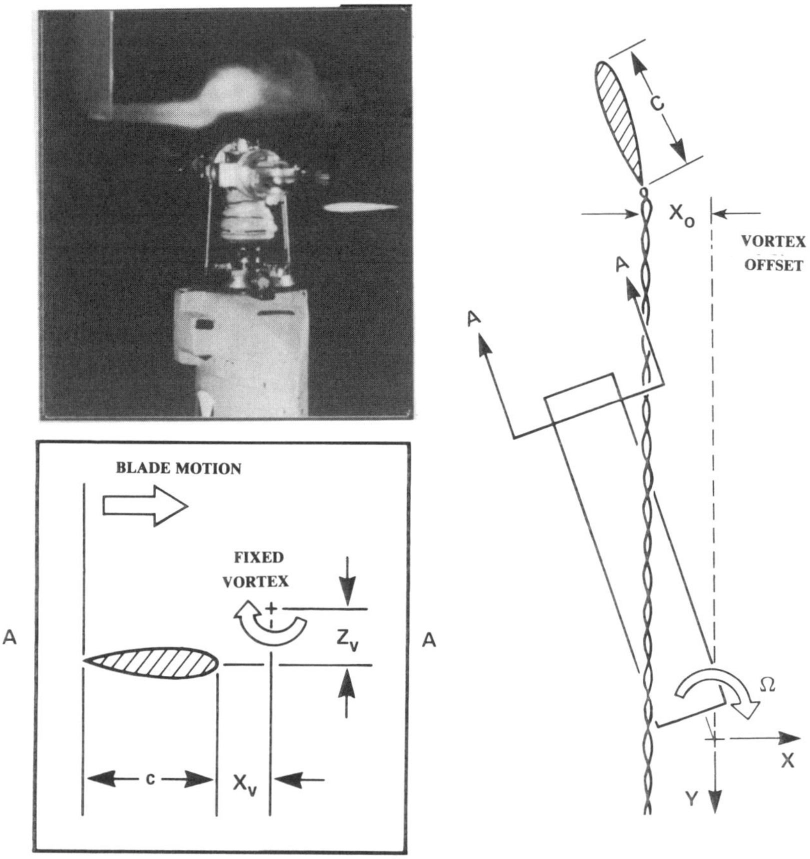

Fig. 12 A method for obtaining blade-vortex interaction data.

have been modeled using a split potential vortex model with the FPR code.[24] Figure 13 shows a comparison of computed and measured surface pressure for a high-speed, oblique (i.e., the vortex generator is offset from the tunnel centerline) interaction in which the blade misses the vortex by 0.25 chords. The timewise and chordwise pressure distributions are very well predicted. Equally good comparisons have been obtained for a variety of conditions, including head-on parallel interactions. This probably constitutes one of the most extreme tests of the potential method because the imposed velocity distribution (i.e., the vortex with its vortical core) violates the irrotational flow assumption. It therefore appears that the velocity decomposition model is suitable for computing most BVI conditions and should permit a comprehensive computation of nearly the entire rotor-wake environment when the wake is specified.

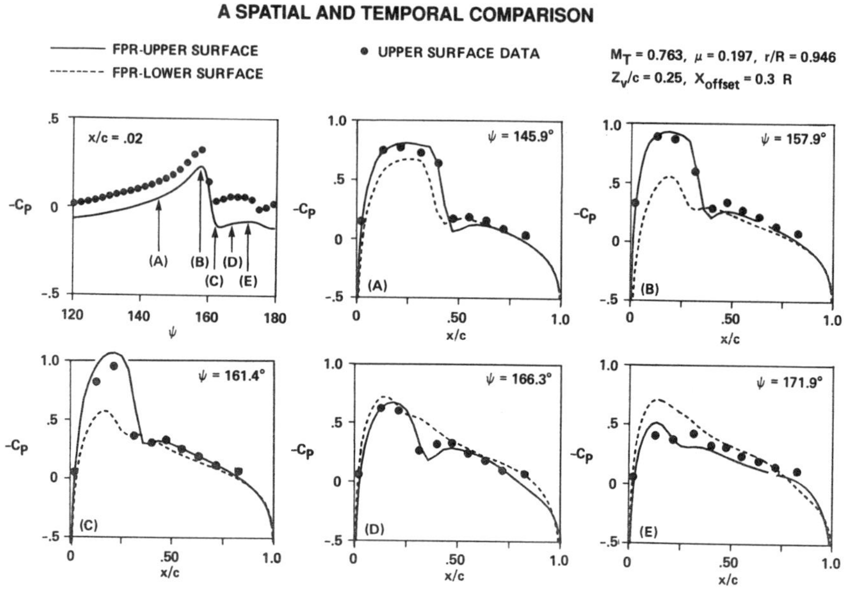

Fig. 13 A comparison of measured and computed surface pressures for an oblique blade-vortex interaction.

Concluding Remarks

At a time when fixed-wing computations are being handled by Euler and Navier-Stokes methods, rotary wing work is mainly represented by potential methods. These methods are aimed at predicting the transonic flows that occur on the advancing side of the rotor disk. These computations must take full account of the blade dynamic motion and various degrees of interactions with shed wakes from previous blades. The latest capabilities include the ability to perform a full-trim solution using full potential methods for flows with weak viscous interactions. This approach is a considerable improvement over present capabilities (which employ lifting line methods with tabular airfoil corrections) and should result in new design activities that aim to capitalize on this capability. Further extensions to present potential methods will enable the development of strong wake interaction capabilities.

Presently, potential methods are considered to be adequate for rotors because the main problem is seen as being the advancing blade transonic flow. Actually, helicopter rotors have several other major flow problems that have yet to be successfully treated. Foremost among these problems is retreating blade stall. An ability to design a higher $C_{L_{\max}}$ into the retreating blade would permit lower rotor tip speeds, permitting an alleviation of the advancing blade shock problems. The key to being able to handle such problems is the development of three-dimensional, unsteady Navier-Stokes methods capable of predicting the onset and evolution of mass separation. Another problem that has not yet been successfully treated is the wake geometry prediction problem. An ability to accurately predict the wake shape would greatly increase the accuracy with which rotor performance could be predicted. Presently, wake geometry can only be determined by experiment. However, the requisite forward flight wake data base is sparse. Although promising hover wake computations have been performed, this work has not yet been extended to forward flight. Finally, this paper has been concerned entirely with rotor flows. Nevertheless, the fuselage drag of helicopters is very high and constitutes the largest high-speed power sink as well as a prime driver of blade stall. Therefore, methods that can accurately predict the separated flow on the fuselage, especially when combined with a main and tail rotor, would promise very large performance gains and greatly diminish the technical risk of a new design.

It is therefore vital and inevitable that the helicopter community will also develop and use various complex geometry Navier-Stokes methods such as is becoming commonplace for fixed-wing work. However, the time when such developments become practical engineering tools is still far off. Until such time much work will be done in the area of learning to apply present CFD tools and incorporating them into our various present and future simpler flow models.

References

[1]Isom, M. P., "Unsteady Subsonic and Transonic Potential Flow Over Helicopter Rotor Blades," NASA CR-2463, Oct. 1974.

[2]vander Vooren, J., Sloof, J. W., Huizing, G. H., and van Essen, A., "Remarks on the Suitability of Various Transonic Small Perturbation Equations to Describe Three-Dimensional Transonic Flow," presented at Proceedings of the Symposium Transonicum II, 1975.

[3]Desopper, A., "Study of the Transonic Flow on a Rotor Blade with Different Tip Shapes," presented at the Tenth European Rotorcraft Forum, Aug. 28–31, 1984, The Hague, The Netherlands.

[4]Chattot, J. J., "Calculation of Three-Dimensional Unsteady Transonic Flows Past Helicopter Blades," NASA TP-1721, Oct. 1980.

[5]Tung, C., Caradonna, F. X., Boxwell, D. A., and Johnson, W. R., "The Prediction of Transonic Flows on Advancing Rotors," American Helicopter Society 40th Annual Forum, May 1984.

[6]Arieli, R. and Tauber, M. E., "Computation of Subsonic and Transonic Flow about Lifting Rotor Blades," AIAA Paper 79-1667, Aug. 1979.

[7]Chang, I. C., "Transonic Flow Analysis for Rotors. Part I: Three Dimensional Quasi-Steady Full Potential Calculation," NASA TP-2375, 1984.

[8]Chang, I. C., "Transonic Flow Analysis for Rotors. Part II: Three Dimensional, Unsteady, Full Potential Calculation," NASA TP-2375, Jan. 1985.

[9]Steger, J. L. and Caradonna, F. X., "A Conservative Implicit Finite Difference Algorithm for the Unsteady Transonic Full Potential Equation," AIAA Paper 80-1368, July 1980.

[10]Bridgeman, J. O., Steger, J. L., Caradonna, F. X., "A Conservative Finite Difference Algorithm for the Unsteady Transonic Potential Equation in Generalized Coordinates," AIAA Paper 82-1388, Aug. 1982.

[11]Strawn, R. C. and Caradonna, F. X., "Numerical Modeling of Rotor Flows with a Conservative Form of the Full Potential Equations," AIAA Paper 86-0079, Jan. 1986.

[12]Strawn, R. C. and Tung, C., "The Prediction of Transonic Loading on Advancing Helicopter Rotors," *AGARD/FDP Symposium on Application of Computational Fluid Dynamics in Aeronautics*, AGARD-CP-412, April 1986.

[13]Sankar, L. N. and Prichard, D. S., "Solution of Transonic Flow Past Rotor Blades Using the Conservative Full Potential Equation," AIAA Paper 85-0120, Jan. 1985.

[14]Siclari, M. J. and Rubel, A., "Entropy Corrections to Supersonic Conical Nonlinear Potential Flows," AIAA Paper 84-1683, June 1984.

[15]Fuglsang, D. F. and Williams, M. H., "Non-Isentropic Unsteady Small Disturbance Theory," AIAA Paper 85-0600, April 1985.

[16]Whitlow, W., "Application of a Nonisentropic Full Potential Method to AGARD Standard Airfoils," AIAA Paper 88-0710, Jan. 1988.

[17]Bridgeman, J. O., Strawn, R. C., and Caradonna, F. X., "An Entropy and Viscosity Corrected Potential Method for Rotor Performance Prediction," *Proceedings of the 42nd Annual Forum of the American Helicopter Society*, 1988.

[18]Sankar, L. N., Wake, B. E., and Lekoudis, S. G., "Solution of the Unsteady Euler Equations for Fixed and Rotary Wing Configurations," AIAA Paper 85-0120, Jan. 1985.

[19]Wake, B. E., Sankar, N. L., and Lekoudis, S. G., "Computation of Rotor Blade Flows Using the Euler Equations," AIAA Paper 85-5010, Oct. 1985.

[20]Agarwal, R. K. and Deese, J. E., "An Euler Solver for Calculating the Flow Field of a Helicopter Rotor in Hover and Forward Flight," AIAA Paper 87-1427, June 1987.

[21]Chen, C. L. and McCroskey, W. J., "Numerical Simulation of Helicopter Multibladed Rotor Flow," AIAA Paper 88-0046, Jan. 1988.

[22]Sankar, L. N. and Malone, J. B., "Unsteady Transonic Full Potential Solutions for Airfoils Encountering Vortices and Gusts," AIAA Paper 85-1710, July 1985.

[23]Jones, H., "Full Potential Modeling of Blade-Vortex Interactions," Ph.D. Thesis, George Washington Univ., Washington, DC, Sept. 1986.

[24]Caradonna, F. X., Strawn R. C., and Bridgeman J. O., "An Experimental and Computational Study of Rotor-Vortex Interactions," *Proceedings of the Fourteenth European Rotorcraft Forum*, Paper 18, 1988.

[25]Egolf, T. A. and Sparks, S. P., "Hovering Rotor Airload Prediction Using a Full Potential Flow Analysis with Realistic Wake Geometry," *41st American Helicopter Society Forum*, 1985.

[26]Steinhoff, J. S. and Ramachandran, K., "A Vortex Embedding Method for Free Wake Analysis of Helicopter Rotor Blades in Hover," *Thirteenth European Rotorcraft Forum*, Paper 2-11, 1987.

[27]Ramachandran, K., "Free Wake Analysis of Helicopter Rotor Blades in Hover Using a Finite Volume Technique," Ph.D. Thesis, Univ. of Tennessee, Knoxville, TN, Dec. 1987.

[28]Ramachandran, K., Tung, C., and Caradonna, F. X., "The Free-Wake Prediction of Rotor Hover Performance Using a Vortex Embedding Method," AIAA Paper 89-0638, Jan. 1989.

[29]Steinhoff, J. S., Ramachandran, K., and Suryanarayana, K., "The Treatment of Convected Vortices in Compressible Potential Flow," *AGARD Symposium on Aerodynamics of Vortical Type Flows in Three Dimensions*, 1983.

[30]Johnson, W., "A Comprehensive Analytical Model of Rotorcraft Aerodynamics and Dynamics. Part 1, Analysis Development," NASA TM-81182, 1980.

[31]Caradonna, F. X. and Philippe, J. J., "The Flow over a Helicopter Blade Tip in the Transonic Regime," *Vertica*, Vol. 2, 1978, pp. 43–60.

[32]Philippe, J. J. and Chattot, J. J., "Experimental and Theoretical Studies on Helicopter Blade Tips at ONERA," *Sixth European Rotorcraft and Powered Lift Aircraft Forum*, Bristol, UK, Paper 46, 1980, pp. 16–19.

[33]Monnerie, B. and Philippe, J. J., "Aerodynamic Problems of Helicopter Blade Tips," *Vertica*, Vol. 2, pp. 217–231.

[34]Yu, Y., Kittleson, J. K., and Becker, F., "Reconstruction of a Three-Dimensional Transonic Rotor Flow Field From Holographic Interferogram Data," *41st American Helicopter Society Forum*, 1985.

[35]Caradonna, F. X. and Tung, C., "Experimental and Analytic Studies of a Model Helicopter Rotor in Hover," NASA TM-81232, Sept. 1981.

[36]Caradonna, F. X., Desopper, A., and Tung, C., "Finite Difference Modeling of Rotor Flows Including Wake Effects," *Journal of the American Helicopter Society*, Vol. 29, No. 2, April 1984.

[37]Strawn, R. C. and Tung, C., "Prediction of Unsteady Transonic Rotor Loads With a Full-Potential Rotor Code," *Proceedings of the 43rd Annual Forum of the American Helicopter Society*, 1987.

[38]Caradonna, F. X., Laub, G. H., and Tung, C., "An Experimental Investigation of the Parallel Blade-Vortex Interaction," *Proceedings of the 10th European Rotorcraft Forum*, 1984; see also NASA TM-86005, Nov. 1984.

[39]Caradonna, F. X., Lautenschlager, J. L., and Silva, M. J., "An Experimental Study of Rotor-Vortex Interactions," AIAA Paper 88-0045, Jan. 1988.

Chapter 7

Unsteady Aerodynamics for Turbomachinery Aeroelastic Applications

Joseph M. Verdon*
United Technologies Research Center, East Hartford, Connecticut

1. Introduction

The development of theoretical methods to predict unsteady flows in turbomachines is a formidable task. The analyst is confronted with determining the time-dependent three-dimensional flow of a viscous compressible fluid through a geometric configuration of enormous complexity. Phenomena of interest include aerodynamic interactions between closely spaced blades and blade rows, intricate shock patterns, and viscous separations. Because of the complexity of the unsteady phenomena occurring within the machine, approaches to design and development have been based generally on steady-state theoretical aerodynamic analyses along with the extensive use of empirical correlations to account for unsteady effects. An important area in which theoretical unsteady aerodynamic analyses have had an impact on design is the prediction of the aeroelastic behavior of the blading. But even in most aeroelastic calculations, a heavy reliance is still placed on empirical correlations. The continuing demands for increased performance are resulting in design trends such as higher rotational speeds, higher pressure ratios per stage, thinner airfoils, and increased operating temperatures, which tend to aggravate dynamic problems in the various components, particularly the blading, and require empirical extrapolations beyond previous experience levels. Thus the development of theoretical aerodynamic analyses to predict the unsteady flow phenomena associated with blade vibrations is becoming more essential to the successful design process.

An aeroelastic prediction system that identifies the conditions under which blades will flutter and determines the vibratory stress amplitudes induced by various external aerodynamic excitations consists of the following elements: a structural analysis to define vibratory mode shapes and natural frequencies; a structural damping model to provide estimates of the material damping due to cyclic straining and the mechanical damping due

---

*Principal Scientist, Theoretical & Computational Fluid Dynamics.

to the energy dissipated by mechanical interactions; methods for estimating the strength, mode, frequency, and attenuation rate of various external aerodynamic excitations; and, finally, an unsteady aerodynamic analysis to predict the unsteady loads acting on the blades due to prescribed structural and aerodynamic excitations. Each of these elements is essential to a successful aeroelastic prediction system; however, because of the complexity of the aerodynamic environment within the machine, the unsteady aerodynamic component is generally regarded as the one requiring the most serious research attention at the present time.

In general, an unsteady aerodynamic analysis intended for turbomachinery aeroelastic applications must be applicable to fan, compressor, and turbine blade rows operating at low subsonic through low supersonic Mach numbers and subjected to moderate through high frequency structural and aerodynamic excitations. In addition, because of the large number of controlling parameters involved, such an analysis must be capable of providing reliable aerodynamic response information efficiently and economically. Unfortunately, these requirements have been met only to a very limited extent. As a result, significant improvements in theoretical unsteady aerodynamic modeling and in the associated analytic and/or numerical solution methods are required to meet the goal of providing engine designers with a comprehensive analysis that they can use in aeroelastic prediction schemes.

In this chapter, a brief description of the vibration problems encountered with axial flow turbomachinery blading will be given. This will be followed by statements of the major assumptions invoked in the development of unsteady aerodynamic analyses intended for turbomachinery aeroelastic design applications and the specific requirements placed on such analyses. Then, theoretical models that have been proposed to address these requirements will be described. The emphasis here will be given to two-dimensional linearized inviscid theories and, in particular, to one that accounts for the effects of blade geometry, mean loading, and transonic phenomena on the unsteady aerodynamic response at a blade surface. Although this theory has been developed primarily for flutter applications, the governing equations presented in Sec. 6 account for the effects of external aerodynamic excitations, i.e., incident entropic, vortical, and acoustic disturbances, as well as those due to self-excited blade vibrations. Example response predictions will be presented to illustrate features of the unsteady aerodynamic response to blade vibrations and to indicate partially the status of solution procedures for linearized unsteady flows. Finally, the future research required to exploit more fully and to extend linearized unsteady aerodynamic formulations will be outlined.

2. Blade Vibrations

Aerodynamically induced vibrations are usually classified into two general categories: flutter and forced vibration. In the former case, the unsteady aerodynamic forces are regarded as being dependent solely on the structural motion, and the airstream supplies the energy to sustain that motion, whereas in the latter, the aerodynamic forces that excite the structural motion are independent of that motion. In both circumstances the vibratory motion of the structure can lead to a fatigue failure. There are fundamental differences between the flutter and forced vibration characteristics of turbomachinery blading and those of aircraft wings. These are associated with the relatively high density of the blade structure and its placement within a cascade. As a result, blade vibration frequencies tend to be higher than those of wings, and the unsteady airloads acting on a given blade depend intimately on the aerodynamic interaction between this blade and its neighbors. Also, because of the large disparity between the density of the blade and the surrounding air, the mode and natural frequency of a blade motion tend to be independent of aerodynamic effects. Important features of the flutter and forced vibration of turbomachinery blading will be described. This information has been summarized from material presented in Refs. 1–8, where the reader can find more complete information.

Flutter

Blades flutter generally at a frequency corresponding to one of the lower blade or coupled blade-disk natural frequencies. This results in high stresses in the blading and leads to a very short-term high-cycle fatigue failure. Compressor blade failure due to flutter has been recognized as a major problem almost from the beginning of the development of the axial flow compressor. In modern engines flutter can be encountered over a wide range of operating conditions and imposes significant constraints on the design of fan and compressor stages. Although to a far lesser extent, turbine blading is also susceptible to flutter, particularly in the high-aspect-ratio aft stages of low-pressure power turbines. In addition, with the performance trend toward higher flow velocities that lead to shocks in turbine blade passages, flutter may also become an increasingly important concern in the turbine stages of jet engines.

Boundaries for the most common types of flutter are shown schematically on a compressor map in Fig. 1. Because of their frequency of occurrence, subsonic/transonic positive incidence flutter and classical supersonic flutter have generally been regarded as the two most important types and have therefore received the most attention from the turbomachinery designers and aerodynamic researchers. However, the distinguishing features and aerodynamic environments associated with each of the various types of flutter indicated in Fig. 1 are summarized below.

1) Subsonic/transonic positive incidence flutter is the oldest, most commonly encountered type of flutter. Traditionally, it has been referred to as subsonic/transonic stalled flutter, but more recent evidence indicates that extensive flow separation is not essential to its occurrence. This type of flutter occurs usually when a compressor is operating near the surge or stall-limit line (Fig. 1) and is characterized by an increase in flutter stress as pressure ratio is increased. It may occur at part speed, i.e., at a fraction of the design rotational speed, in a high-speed fan stage, and at or near design speed in a low- or high-pressure compressor stage. It generally occurs in blading that is highly loaded and operating at high subsonic relative inlet

Mach number. Bending, torsion, and coupled (bending-torsion) blade vibration modes have been observed over a reduced frequency (based on relative inlet flow speed and blade chord) range extending from 0.4 to 1.6. In an unshrouded rotor the blade motions are generally unphased at low amplitudes with the possibility of a constant interblade phase angle at larger amplitudes of vibration. In a shrouded rotor phasing is enforced by part-span mechanical ties.

2) Negative incidence flutter, also referred to as choke flutter, is encountered during part speed operation when the blades are operating either subsonically or transonically and at negative incidence. It occurs primarily in the first bending mode at reduced frequencies extending from 0.3 to 0.5 with vibratory stresses increasing as the pressure ratio is lowered. The causes of choke flutter are not fully understood; both high negative incidence angles and choked flow are important candidates. In the near-choke condition in-passage shocks and associated flow separations are believed to influence blade stability. Analogous to the subsonic positive incidence case, unphased blade motions occur with the possibility that these become phased at higher amplitudes.

3) Classical supersonic flutter generally occurs in thin fan blading when the outer span of each rotor blade is operating at a supersonic relative but subsonic axial inlet Mach number. It occurs at pressure ratios in the vicinity of the compressor operating line, or at lower pressure ratios, and is important because it can impose a limit on the high-speed operation of the machine. The blades vibrate generally in the torsional mode at a common

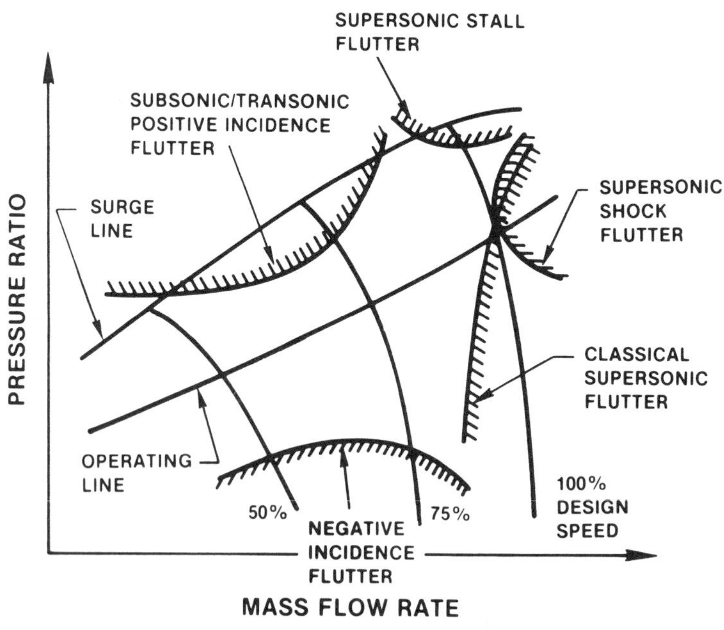

Fig. 1 Schematic of common types of fan/compressor blade flutter.

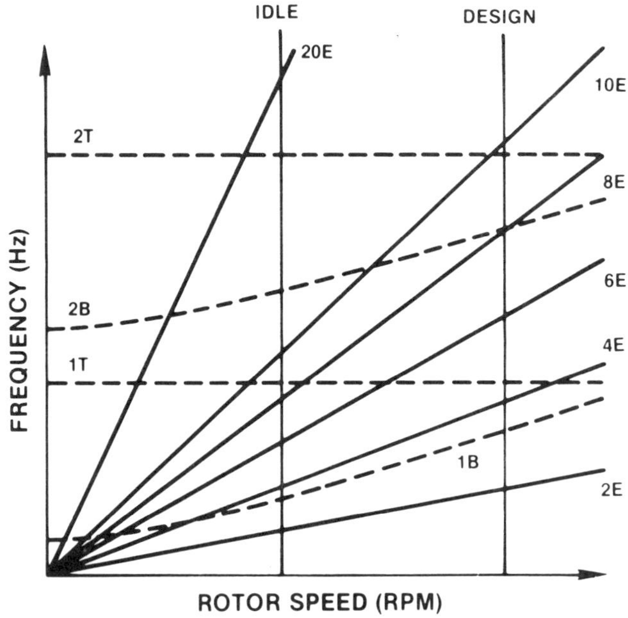

Fig. 2 Typical frequency-speed diagram for a rotating system.

reduced frequency lying approximately between 0.7 and 1.3 and with a constant phase angle between the motions of adjacent blades. The stresses encountered during this type of flutter can be quite large but diminish as one traverses a constant speed line toward higher pressure ratios.

4) Supersonic shock flutter is encountered near the operating line and can also impose a limit on high-speed or over-speed capability. In many ways it is similar to supersonic unstalled flutter. It occurs in fans in which the outer sections operate at supersonic relative Mach number and results in high stresses with all blades vibrating at a common frequency and interblade phase angle. However, supersonic shock flutter is characterized by the existence of a threshold pressure ratio above which flutter is encountered, and an increase in pressure ratio at constant speed can alleviate the flutter. This type of flutter may be caused by strong in-passage shocks that can excite either a bending or torsional blade vibration.

5) Supersonic high-incidence or stalled flutter occurs during high-speed operation when the outer portion of the blading is operating supersonically and the stage is operating near the surge line. Flutter in the first bending mode has generally been associated with these higher pressure ratios. As in the case of supersonic unstalled flutter, all blades vibrate at a common frequency, generally within the range of 0.2–0.5, and at a constant interblade phase angle. Because this type of flutter is associated with high pressure ratios, strong shocks and their motions and flow separations probably contribute to its occurrence.

Forced Vibration

Destructive forced vibrations can occur in fan, compressor, and turbine blading when a periodic aerodynamic forcing function, with frequency close to a system resonant natural frequency, acts on the blades in a given row. Such forcing functions are generated at multiples of the engine rotational frequency and arise from a variety of sources. The latter include aerodynamic disturbances resulting from the presence of upstream and downstream struts, stator vanes, and rotor blades, and disturbances due to inlet flow nonuniformities, rotating stall patterns, and compressor surge. A primary source of resonant forced vibration is the aerodynamic interaction between adjacent blade rows. The two principal types of such interaction are potential flow or static pressure interaction and wake interaction. The former results from the variations in the velocity potential or pressure field associated with the blades of a given row and their effect on the blades of a neighboring row moving at a different rotational speed. This type of interaction is of serious concern when the axial spacings between neighboring blade rows are small or flow Mach numbers are high. Wake interaction is the effect on the flow through downstream blade rows of the wakes shed by one or more upstream rows and persists over considerable distances.

The rotor speeds at which significant forced vibrations may occur are predicted with frequency-speed diagrams (e.g., see Fig. 2) which display the

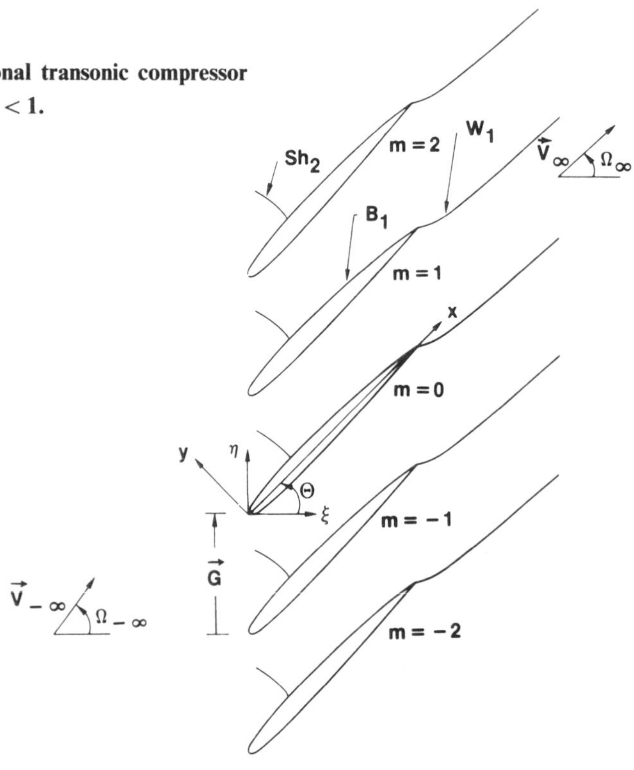

Fig. 3 Two-dimensional transonic compressor cascade; $M_{\infty} < M_{-\infty} < 1$.

natural frequency of each blade vibration mode and the forcing function frequencies as functions of rotor speed. Wherever these curves cross, the possibility of high vibratory stress amplitudes is indicated. It is not possible to eliminate the sources of forced vibration from the operating range of a turbomachine and, as a result, the number of vibratory fatigue failures attributable to forced vibration problems may exceed those due to flutter. Over the lifetime of the engine, the former can often prove to be more significant in terms of cost and repair time than the latter. However, since forced vibration problems are not as crucial to an engine development program, they have not received the same degree of attention from turbomachinery designers as flutter problems.

3. Cascade Unsteady Aerodynamic Analyses

For the most part, the theoretical formulations that have been developed to predict the unsteady aerodynamic phenomena associated with blade flutter or forced vibration consider the blades of an isolated two-dimensional cascade (e.g., see Fig. 3). In such analyses viscous effects are usually neglected at the outset and the unsteady fluctuations are assumed to be of sufficiently small magnitude that a linearized analysis of the inviscid unsteady perturbation is justified. Then, to determine all of the unsteady characteristics of interest for an isolated blade row, such aerodynamic theories must be capable of predicting the unsteady forces acting on the blades, the entropy and vorticity fluctuations convected downstream from the blade row, and the static pressure or acoustic waves radiated away from the blade row for prescribed structural (i.e., blade motions) and externally induced aerodynamic excitations. The latter include variations in total pressure and total temperature ("entropy and vorticity waves") at inlet and variations in static pressure (acoustic waves) entering the blade row from upstream or downstream.[9] In particular, for flutter applications it is only necessary to predict the unsteady loads due to prescribed blade motions, whereas for forced vibration applications predictions of the unsteady loads due to prescribed incident entropic, vortical, and acoustic disturbances are also required.

Until recently, the inviscid unsteady aerodynamic analyses that have been available for turbomachinery aeroelastic applications were based on classical linearized theory.[10,11] Here the departures of the steady and unsteady flow variables from their uniform freestream values are regarded as small and of the same order of magnitude leading to uncoupled linear constant coefficient boundary-value problems for the steady and unsteady disturbance potentials. Thus, unsteady solutions based on the classical linearization are essentially restricted to cascades of unloaded flat-plate blades that operate in an entirely subsonic or an entirely supersonic environment. Very efficient semianalytic solution procedures have been developed for two-dimensional attached subsonic[12–14] and supersonic[15–19] flows and applied with some success, particularly in the supersonic regime, in turbomachinery flutter and forced vibration design calculations. Models based on variations or extensions of the two-dimensional classical linearization have also been developed for treating separated subsonic flows,[20,21] supersonic flows with strong in-passage normal shocks,[22] and high-frequency transonic flows,[23] but these have received only limited application in the design process. We will discuss the classical linearization and the foregoing extensions in more detail in Sec. 8 of this article. Also, although not discussed herein, it should be mentioned that extensive efforts have been made to develop three-dimensional unsteady aerodynamic analyses, based on the classical linearization, for turbomachinery aeroelastic applications (cf. Refs. 24 and 25).

The classical linearization does not account for interactions between steady and unsteady disturbances. Such interactions are crucial to the successful prediction of a wide variety of turbomachinery aeroelastic phenomena, in particular, phenomena associated with transonic flows. The influence of steady disturbances on unsteady response is partially retained in the so-called "time-linearized," transonic small-disturbance approximation.[26] Here, unsteady disturbances are regarded as small relative to steady disturbances caused by airfoil shape and incidence, which are in turn assumed to be small relative to the freestream speed. These assumptions along with appropriate independent variable scalings provide a linearized theory that formally applies at freestream Mach numbers close to 1 but only for low-frequency unsteady motions. Numerical field methods for solving the time-linearized transonic equations have received extensive development since the early 1970's. In particular, time[27] and frequency[28] domain finite-difference methods have been determined for time-linearized transonic flows, with moving shocks, around isolated airfoils. Such developments have been directed primarily toward external aeronautical applications, but solution methods for cascade flows have also appeared.[29–31] It should be noted, however, that the large mean flow gradients and the high-frequency unsteady fluctuations typical of the flows through turbomachines may limit the usefulness of the time-linearized transonic approximation for turbomachinery aeroelastic applications.

The foregoing linearized unsteady aerodynamic theories allow a simple treatment of airfoil, wake, and shock surface conditions. However, the assumptions that permit this simplification place severe restrictions on airfoil geometry and loading and, in the transonic case, on the frequency of the unsteady motion. As such these theories fail to meet the needs of turbomachinery designers over a wide range of practical operating conditions. Therefore, more general linearizations are being developed[32–42] that include the effects of important design features such as real blade geometry, finite mean pressure variation across a blade row, and operation at transonic Mach numbers. Here, unsteady disturbances are regarded as small-amplitude harmonic fluctuations relative to a fully nonuniform steady (in a coordinate frame attached to the blade row) potential flow. The steady flow is determined as a solution of the full potential equation, and the unsteady flow is governed by linear equations with variable coefficients that depend on the underlying steady flow. A number of useful solution algorithms for the nonlinear steady problem are currently available.[43–45] In addition, solution methods (see Ref. 46) for the linearized unsteady problem are reaching the stage where it will be appropriate to consider them for aeroelastic design applications.

The concept of linearizing unsteady fluctuations with respect to a fully nonuniform mean flow has been employed in other applications, e.g., in studies on turbulent flows past obstacles.[47–49] However, with the exception of the unsteady cascade analyses cited earlier, this concept has not been exploited in developing the unsteady aerodynamic analyses needed for aeroelastic predictions. The improvements in physical modeling offered by such a linearization coupled with advances in our ability to numerically resolve the resulting governing equations should make this approach an increasingly attractive one in future considerations. In this chapter we will derive this general linearization for unsteady two-dimensional cascade flows. In addition, in Sec. 9 we will present numerical results to illustrate several effects, associated with the features of nonuniform steady flow, on the unsteady aerodynamic response to self-excited blade vibrations. The reader can find additional useful information on unsteady linearizations relative to nonuniform steady flows in Refs. 10, 46, and 48.

4. Nonlinear Time-Dependent Aerodynamic Formulation

Problem Description

We consider time-dependent two-dimensional adiabatic flow, with negligible body forces, of an inviscid non-heat-conducting perfect gas through an isolated and infinite array of airfoils, such as the one shown in Fig. 3. The time-dependent or unsteady fluctuations in the flow are assumed to be of small amplitude and to arise from one or more of the following sources (cf. Fig. 4): blade motions, upstream and/or downstream acoustic disturbances that carry energy toward the blade row, and upstream entropic and vortical disturbances. It is assumed that, in the absence of these unsteady excitations, the blades are identical in shape, equally spaced, and their chord lines are each inclined at the angle Θ relative to the axial flow (or positive ξ) direction and that beyond some finite distance upstream (say $\xi < \xi_-$) and downstream ($\xi > \xi_+$) from the blade row, the flow is at most a small irrotational steady perturbation (due to the presence of the cascade) from a uniform freestream. Blade shape and orientation relative to the inlet freestream direction, the exit-to-inlet static pressure ratio, and the amplitude frequency and mode of the unsteady excitations are assumed to be such that the flow remains attached to the blade surfaces. Thus, thin vortex sheets (unsteady wakes) emanate from the blade trailing edges and extend infinitely far downstream. Finally, any changes in the entropy or vorticity of a fluid particle as it passes through a shock are regarded as negligible. Thus, we are restricting our consideration to flows containing shocks of only weak to moderate strength.

In the following discussion all variables are dimensionless. Lengths have been scaled with respect to blade chord, time with the ratio of blade chord to upstream freestream speed, and density and velocity with respect to the upstream freestream density and velocity, respectively. The scalings for the remaining flow variables are readily determined from the equations given later. In anticipation of the small unsteady-disturbance approximation, which will be introduced in Sec. 5, the symbol ~ is used below to denote

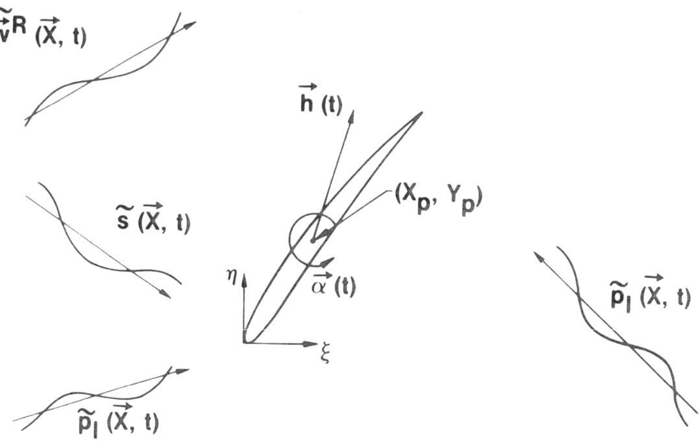

Fig. 4 Unsteady excitations: blade motion (rotation and translation), incident vortical and entropic disturbances from upstream, and incident acoustic disturbances from upstream and downstream.

a time-dependent fluid property, and, with the exception of density, uppercase and lowercase letters will be used to represent the various nonlinear flow variables and their small-disturbance counterparts, respectively. Thus, for example, $\tilde{P}$ and P will refer to the nonlinear time-dependent and steady-state fluid pressure, respectively, and $\tilde{p}$ and p will refer to the time-dependent small-disturbance unsteady pressure and its complex amplitude. The corresponding symbols for the fluid density are $\tilde{\bar{\rho}}$, $\bar{\rho}$, $\tilde{\rho}$, and ρ, respectively. The subscripts $-\infty$ and $+\infty$ will be used to denote the uniform flow conditions that exist far upstream and far downstream, respectively, from the blade row.

In addition to these notations for the fluid dynamic variables, uppercase script letters represent instantaneous surface (blade, wake, and shock) locations, whereas corresponding uppercase block letters refer to steady-state surface positions. The vector $\mathscr{R}$ measures the displacement of a point on a moving surface relative to its mean or steady-state position (see Fig. 5). The unit vectors $\boldsymbol{n}$ and $\boldsymbol{\tau}$ are normal and tangent, respectively, to a surface and directed such that $\boldsymbol{n} \times \boldsymbol{\tau} = \boldsymbol{e}_z$ points out from the page. Finally, as in Fig. 3, the mean or steady-state positions of the blade chord lines are assumed to coincide with the line segments $\eta = \xi \tan\Theta + mG$, $0 < \xi < \cos\Theta$, $m = 0, \pm 1, \pm 2, \ldots$, where ξ and η are the cascade axial and "circumferential" coordinates, m is a blade number index, Θ is the cascade stagger angle, and $\boldsymbol{G}$ is the cascade gap vector that is directed along the η axis with magnitude equal to the blade spacing.

For aeroelastic applications the goal is to predict the unsteady pressures and the global unsteady airloads that act on the blades and result from prescribed periodic (in time and space) unsteady excitations. These excitations are assumed to occur at temporal frequency ω. Thus, we consider blade motions of the form

$$\mathscr{R}(\boldsymbol{X} + m\boldsymbol{G}, t) = \mathrm{Re}\{\boldsymbol{r}(\boldsymbol{X}) \exp[i(\omega t + m\sigma)]\}, \qquad \vec{\boldsymbol{X}} \text{ on } B \tag{1}$$

where $\boldsymbol{X}$ is a position vector, t is time, $\boldsymbol{r}$ is a complex displacement-amplitude vector, σ is the phase angle between the motions of adjacent blades, and Re{ } denotes the real part of { }. Incident disturbances are of the form

$$\tilde{p}_I(\boldsymbol{X}, t) = \mathrm{Re}\{p_{I,\mp\infty} \exp[i(\boldsymbol{k}_{\mp\infty} \cdot \boldsymbol{X} + \omega t)]\}, \qquad \xi \lessgtr \xi_{\mp} \tag{2}$$

where $p_{I,\mp\infty}$ and $\boldsymbol{k}_{\mp\infty}$ are the amplitude and wave number vector, respectively, of the incident pressure fluctuation $\tilde{p}_I(\boldsymbol{X}, t)$ traveling toward the blade row from upstream or downstream.

Governing Equations

The equations governing the fluid motion are obtained from the integral conservation laws for mass, momentum, and energy, and the thermodynamic relations for a perfect gas. The former provide corresponding differential equations, in regions where the flow variables are continuously differentiable, and "jump" conditions at surfaces across which (in the inviscid approximation) the flow variables are discontinuous, i.e., at vortex-sheet wakes and shocks. In continuous regions the energy equation can be replaced by the requirement that the entropy following a fluid particle must remain constant.[50] The differential equations governing the flow in continuous regions then can be written in the form

$$\frac{\partial \tilde{\bar{\rho}}}{\partial t} + \nabla \cdot (\tilde{\bar{\rho}} \tilde{\boldsymbol{V}}) = 0 \tag{3}$$

$$\frac{\partial}{\partial t}(\tilde{\bar{\rho}} \tilde{\boldsymbol{V}}) + \nabla \cdot (\tilde{\bar{\rho}} \tilde{\boldsymbol{V}} \otimes \tilde{\boldsymbol{V}}) + \nabla \tilde{P} = 0 \tag{4}$$

and

$$\frac{\mathrm{D}\tilde{S}}{\mathrm{D}t} = 0 \tag{5}$$

where $\tilde{\boldsymbol{V}}(\boldsymbol{X}, t)$ and $\tilde{S}(\boldsymbol{X}, t)$ are the fluid velocity and entropy, respectively, $\otimes$ denotes the tensor product of two vectors, and $\mathrm{D}/\mathrm{D}t = \partial/\partial t + \tilde{\boldsymbol{V}} \cdot \nabla$ is the material or convective derivative operator. In addition, the entropy is related to the pressure and density by the differential expression

$$\mathrm{d}\tilde{S} = \gamma^{-1}\, \mathrm{d}\tilde{P}/\tilde{P} - \mathrm{d}\tilde{\bar{\rho}}/\tilde{\bar{\rho}} \tag{6}$$

where γ is the specific heat ratio of the fluid (constant pressure to constant volume). Equation (6) can be integrated to relate the change in the entropy between any two equilibrium states to the corresponding changes in pressure and density.

The foregoing equations must be supplemented by boundary conditions at moving blade surfaces $\mathscr{B}_m$ and jump conditions across moving wake, $\mathscr{W}_m$, and shock, $\mathscr{S}\!\mathscr{h}_{m,n}$, surfaces. The subscripts on $\mathscr{S}\!\mathscr{h}$ refer to the nth shock associated with the mth blade. For attached flows, the normal component of the fluid velocity must equal the normal component of the surface velocity at blade surfaces (flow tangency); i.e.,

$$\left(\tilde{\boldsymbol{V}} - \frac{\partial \boldsymbol{\mathscr{R}}}{\partial t}\right) \cdot \boldsymbol{n} = 0 \tag{7}$$

where the $\boldsymbol{\mathscr{R}}$ is prescribed. Since the vortex-sheet unsteady wakes are material surfaces, Eq. (7) also applies at wake surfaces. However, more convenient wake conditions can be specified. These follow from the foregoing kinematic condition and the conservation laws and are given by

$$[\![\tilde{P}]\!] = 0 \tag{8}$$

and

$$[\![\tilde{\boldsymbol{V}}]\!] \cdot \boldsymbol{n} = 0 \tag{9}$$

where $[\![\ ]\!]$ denotes the change in a flow quantity as experienced by an observer in moving across a surface of discontinuity in the $\boldsymbol{n}$ direction. Since we have assumed that entropy and vorticity changes across shocks are negligible, the jump conditions (i.e., the Rankine-Hugoniot conditions) based on the mass, momentum, and energy conservation laws cannot all be enforced at shock surfaces. Therefore, we will follow the usual practice and require only that mass and tangential momentum be conserved at shocks; i.e.,

$$\left[\!\!\left[\tilde{\rho}\left(\tilde{\boldsymbol{V}} - \frac{\partial \boldsymbol{\mathscr{R}}}{\partial t}\right)\right]\!\!\right] \cdot \boldsymbol{n} = 0 \tag{10}$$

and

$$[\![\tilde{\boldsymbol{V}}]\!] \cdot \boldsymbol{\tau} = 0 \tag{11}$$

To complete this formulation we require conditions on the flow far upstream and far downstream from the blade row. Since we are not interested in transient response behavior, a precise knowledge of the initial state of the fluid is not required. Information on the uniform inlet and exit flow conditions, the entropic and vortical disturbances at inlet, and the incident acoustic disturbances at inlet and exit must be specified to satisfy the far-field requirement. The remaining steady and unsteady departures from the uniform inlet and exit conditions (e.g., steady pressure disturbances, entropic and vortical fluctuations at exit, and the outgoing acoustic waves) must be determined as part of the time-dependent nonlinear solution.

The foregoing aerodynamic problem is a very difficult one. It consists of a system of nonlinear time-dependent partial differential equations along with conditions imposed at moving blade, wake, and shock surfaces. Moreover, the instantaneous locations of the wake and shock surfaces must, in principle, be determined as part of the solution. Although numerical solutions to this problem would be of substantial interest, they would be of limited practical value because of the prohibitive expense involved in obtaining the detailed unsteady aerodynamic response information required for turbomachinery aeroelastic applications. Thus, the usual approach is to examine limiting forms of the nonlinear time-dependent equations with the intention of providing efficient methods for predicting the unsteady aerodynamic response to prescribed blade motions and external aerodynamic excitations.

5. Small Unsteady-Disturbance Approximation

We now seek an approximation to the foregoing nonlinear time-dependent equation set that is valid for small-amplitude unsteady excitations; i.e., $|\mathscr{R}(X,t)|$, $|\tilde{p}_I(X,t)|$, etc., $\sim\mathcal{O}(\varepsilon) \ll 1$. For this purpose, the flow variables are each expanded in an asymptotic series in ε; e.g.,

$$\tilde{P}(X,t) = P(X) + \tilde{p}(X,t) + \cdots \tag{12}$$

where $\tilde{P}(X,t)$ and $\tilde{p}(X,t)$ are of order ε^0 and ε^1, respectively, and the ellipses refer to terms that are of higher than first order in ε. In addition, Taylor series expansions, e.g.,

$$\tilde{P}_{\mathscr{S}} = (\tilde{P} + \mathscr{R} \cdot \nabla \tilde{P} + \cdots)|_S \tag{13}$$

are applied to refer information at a moving blade, wake, or shock surface to the mean position of this surface (see Fig. 5). In Eq. (13) the subscripts $\mathscr{S}$ and S refer to the instantaneous and mean surface locations, respectively. The unit tangent and normal vectors at a point on a moving surface are related to the unit tangent and normal vectors at the location of this point on the mean surface as follows:

$$\boldsymbol{\tau}_{\mathscr{S}} = \boldsymbol{\tau}_S + (\boldsymbol{\tau}_S \times \partial\mathscr{R}/\partial\tau_S) \times \boldsymbol{\tau}_S + \cdots \tag{14}$$

and

$$\boldsymbol{n}_{\mathscr{S}} = \boldsymbol{n}_S + (\boldsymbol{\tau}_S \times \partial\mathscr{R}/\partial\tau_S) \times \boldsymbol{n}_S + \cdots \tag{15}$$

After we substitute the foregoing series expansions and surface vector relations into the full governing equations, equate terms of like power in ε, and neglect terms of higher than first order in ε, we obtain nonlinear and linear variable coefficient boundary-value problems, respectively, for the zeroth- and first-order flows.

A significant analytic advantage offered by this linearization is that unsteady effects arising from self-excited blade motions and from the various incident disturbances are not coupled and hence can be determined separately. Indeed, it is sufficient to develop solution procedures only for a

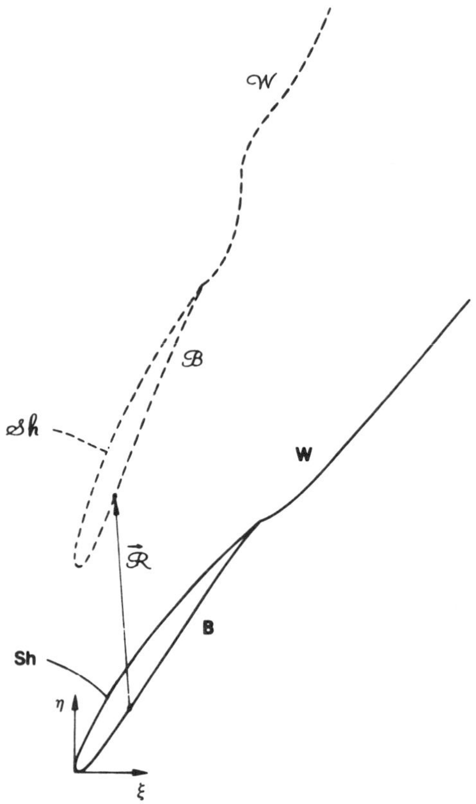

Fig. 5 Steady-state (solid lines) and instantaneous (dashed lines) blade, wake, and shock positions.

single harmonic component of a given disturbance [cf. Eqs. (1) and (2)]. Solutions for arbitrary disturbances and arbitrary combinations of the various disturbances can then be obtained by Fourier superposition. Note, however, that the linearization does limit the unsteady response phenomena that can be analyzed, since nonlinear unsteady effects, such as those resulting from highly distorted inlet flows or large-amplitude blade vibrations, are neglected.

As $\varepsilon \rightarrow 0$ the blade surfaces collapse to their mean positions and the incident entropic vortical and acoustic disturbances vanish. Hence, the zeroth-order terms of the asymptotic expansions (12) are the fluid properties corresponding to the steady flow past a stationary cascade. Since the equations governing the first-order unsteady properties will be linear, the fluctuations in these properties induced by a harmonic (in time) unsteady excitation must also have harmonic time dependence. We take advantage of this feature by introducing a complex representation, e.g., $\tilde{p}(\boldsymbol{X},t) = \mathrm{Re}\{p(\boldsymbol{X})\, e^{i\omega t}\}$, for all first-order flow properties and adopting the convention that the real parts of the various complex parameters represent the actual time-dependent physical quantities. The complex representation serves to remove explicit time dependence from the linearized unsteady problem, thereby facilitating the determination of a solution. In addition, the cascade geometry, the prescribed form of the unsteady excitations [cf. Eqs. (1) and (2)], and the linearity of the first-order equations require that both the steady and unsteady flows exhibit blade-to-blade periodicity. Thus, for example,

$$P(\boldsymbol{X} + m\boldsymbol{G}) = P(\boldsymbol{X}) \tag{16}$$

and

$$p(\boldsymbol{X} + m\boldsymbol{G}) = p(\boldsymbol{X})\, e^{im\sigma} \tag{17}$$

where σ is the phase angle between the motions of adjacent blades or the scalar product of the far-field circumferential wave number and the cascade gap vector $\boldsymbol{G}$ (i.e., $\sigma = \boldsymbol{k}_{-\infty} \cdot \boldsymbol{G}$) for an incident disturbance. Conditions (16) and (17) allow a numerical resolution of the steady and the linearized unsteady flows to be limited to a single extended blade-passage region of the cascade and permit fluid properties at the mth blade or wake surface or the m,nth shock surface to be evaluated in terms of information available at the corresponding reference ($m = 0$) surface. For simplicity we will usually omit the subscript m from the following discussion when referring to a reference surface.

Steady Background Flow

The equations governing the zeroth-order or steady flow follow from Eqs. (3–11) after we replace the time-dependent flow properties $\tilde{\bar{\rho}}$, $\tilde{\boldsymbol{V}}$, $\tilde{P}$, and $\tilde{S}$ by their zeroth-order or steady flow counterparts $\bar{\rho}$, $\boldsymbol{V}$, P, and S and set temporal derivatives and surface displacement vectors equal to zero. As a consequence of the assumptions regarding shocks and the far-upstream or inlet flow conditions, the steady background flow will be isentropic, $S = S_{-\infty}$, and irrotational, $\boldsymbol{V} = \nabla\Phi$. Thus, the mass conservation Eq. (1) can be written as

$$\nabla \cdot (\bar{\rho}\, \nabla\Phi) = 0 \tag{18}$$

Moreover, the momentum equation can be integrated and, after the application of some algebra, it yields the following convenient relations between the mean flow variables:

$$\begin{aligned}(M_{-\infty}V/M)^2 &= (M_{-\infty}A)^2 = \bar{\rho}^{(\gamma-1)} = (\gamma M_{-\infty}^2 P)^{(\gamma-1)/\gamma} \\ &= 1 - \frac{\gamma-1}{2} M_{-\infty}^2[(\nabla\Phi)^2 - 1] = \frac{2 + (\gamma-1)M_{-\infty}^2}{2 + (\gamma-1)M^2}\end{aligned} \tag{19}$$

where $A = (\gamma P/\bar{\rho})^{\frac{1}{2}}$ and $M = V/A$ are the local speed of sound propagation and Mach number, respectively, in the mean or steady flow. Recall that $\bar{\rho}_{-\infty} = V_{-\infty} = 1$, and we can, without loss in generality, set the mean entropy S equal to zero.

Surface conditions for the zeroth-order or steady flow apply at the mean positions of the blade ($\mathcal{B}_m$), wake ($\mathcal{W}_m$), and shock ($\mathrm{Sh}_{m,n}$) surfaces. Blade mean positions are prescribed, but the mean wake, i.e., the stagnation streamlines downstream of the blade row, and shock positions must be determined as part of the steady solution. Since, by assumption, the flow remains attached to the blade surfaces, it follows that

$$\boldsymbol{V} \cdot \boldsymbol{n} = \nabla\Phi \cdot \boldsymbol{n} = 0 \tag{20}$$

for $\boldsymbol{X}$ on B_m. In addition, the steady pressure and normal velocity component must be continuous across blade wakes. Hence, the conditions

$$[\![P]\!] = 0 \qquad \text{or} \qquad [\![\Phi]\!] = [\![\Phi]\!]_{\text{TE}} \tag{21}$$

and

$$[\![\nabla\Phi]\!] \cdot \boldsymbol{n} = 0 \tag{22}$$

where the subscript TE refers to the trailing-edge point of a blade, apply at the mean wake surfaces, W_m. Finally, the conservation laws for mass and tangential momentum at shocks, $\text{Sh}_{m,n}$, require that

$$[\![\bar{\rho}\ \nabla\Phi]\!] \cdot \boldsymbol{n} = 0 \tag{23}$$

and

$$[\![\nabla\Phi]\!] \cdot \boldsymbol{\tau} = 0 \qquad \text{or} \qquad [\![\Phi]\!] = [\![\Phi]\!]_{\text{Ref}} \tag{24}$$

where $[\![\Phi]\!]_{\text{Ref}}$ can be set to zero if the shock terminates at a sonic point, i.e., within the flowfield.

For the cascade flows of practical interest, the axial components of the inlet and exit freestream velocities will be subsonic; i.e., $V_{\xi,\mp\infty} < A_{\mp\infty}$. In this case, three of the far-field uniform velocity components or their equivalents (e.g., Mach number $M_{\mp\infty}$, flow angle $\Omega_{\mp\infty}$, etc.) must be prescribed to completely specify the steady boundary-value problem. The fourth or remaining component can be determined in terms of the three prescribed using the integral or global form of the mass conservation law, which requires that

$$\bar{\rho}_{\infty} V_{\infty} \cos\Omega_{\infty} = \cos\Omega_{-\infty} \tag{25}$$

Numerical solution procedures for determining two-dimensional steady potential flows through cascades have been developed extensively (e.g., see Refs. 7 and 43–45), particularly for flows with subsonic relative inlet and exit Mach numbers (i.e., $M_{\mp\infty} < 1$). In such calculations far-field boundary conditions are imposed at axial stations placed at finite distances upstream and downstream (i.e., at $\xi = \xi_{\mp}$) from the blade row where linearized solutions describing the behavior of the steady potential for $\xi \lessgtr \xi_{\mp}$ can be matched to a nonlinear near-field solution. In addition, conditions are often imposed at the blade edges [e.g., a zero load condition at a sharp leading and/or trailing (Kutta condition) edge] in lieu of prescribing an inlet and/or exit freestream flow property. Finally, the usual practice is to solve the conservative form of the mass balance equation [i.e., Eq. (18)] throughout the entire fluid domain while allowing for a discontinuity in the velocity potential at the blade trailing-edge points and along the arbitrary periodic lines [cf. Eq. (16)] that emanate from these points and extend downstream. Thus, shock jump conditions are usually not imposed in steady potential flow calculations; instead, shock phenomena are captured through the use of special differencing techniques. It appears that such procedures lead to reasonably accurate predictions of mean shock locations and the mean flow behavior in the vicinity of a shock. Mean wake locations are determined a posteriori from the resulting steady solution.

6. Linearized Unsteady Flow

Field Equations

The differential equations governing the first-order or linearized unsteady flow in continuous regions are determined by substituting the asymptotic expansions for the flow variables into the full time-dependent mass, momentum, and entropy transport equations; subtracting out the corresponding steady equations; and neglecting terms of higher than first order in ε. This procedure yields the following system of linearized equations when the mean flow is isentropic and irrotational:

$$i\omega\rho + \nabla \cdot (\bar{\rho}\boldsymbol{v} + \rho\, \nabla\Phi) = 0 \tag{26}$$

$$\bar{\rho}\left(\frac{\bar{\mathrm{D}}\boldsymbol{v}}{\mathrm{D}t} + (\boldsymbol{v} \cdot \nabla)\, \nabla\Phi\right) + \rho\, \nabla[(\nabla\Phi)^2/2] + \nabla p = 0 \tag{27}$$

and

$$\frac{\bar{\mathrm{D}}s}{\mathrm{D}t} = 0 \tag{28}$$

where ρ, $\boldsymbol{v}$, p, and s are the complex amplitudes of the first-order density, velocity, pressure, and entropy, respectively, and $\bar{\mathrm{D}}/\mathrm{D}t = i\omega + \nabla\Phi \cdot \nabla$ is a mean flow convective derivative operator. The additional equation required to complete this system is obtained by integrating the thermodynamic relation (6) from the state occupied by the steady background flow at the point $\boldsymbol{X}$ to the actual state of the fluid at $\boldsymbol{X}$. It follows that

$$s = \gamma^{-1} p/P - \rho/\bar{\rho} \tag{29}$$

Equation (29) can be used to eliminate the first-order density from Eqs. (26) and (27). In particular, it follows that, after introducing Eqs. (28) and (29) and the mean flow Bernoulli relations (19) into the linearized momentum Eq. (27) and performing some algebra, we can express the latter equation in the form

$$\frac{\bar{\mathrm{D}}}{\mathrm{D}t}(\boldsymbol{v} - s\, \nabla\Phi/2) + [(\boldsymbol{v} - s\, \nabla\Phi/2) \cdot \nabla]\, \nabla\Phi = -\nabla(p/\bar{\rho}) \tag{30}$$

In addition, after we combine the steady [Eq. (18)] and linearized unsteady [Eq. (26)] mass conservation equations and introduce Eqs. (28) and (29) into the result, the linearized continuity equation can be written as

$$\frac{\bar{D}}{Dt}[p/(\bar{\rho}A^2)] + \bar{\rho}^{-1}\nabla\cdot(\bar{\rho}\boldsymbol{v}) = 0 \tag{31}$$

It is possible to simplify further the first-order field equations by introducing the Goldstein[48,49] velocity decomposition. That is, we set

$$\boldsymbol{v} = \nabla\phi + \boldsymbol{v}^R \tag{32}$$

where

$$\nabla\cdot\boldsymbol{v}^R = 0 \qquad \text{for} \qquad \xi < \xi_- \tag{33}$$

and

$$\frac{\bar{D}\phi}{Dt} = -p/\bar{\rho} \tag{34}$$

After substituting Eqs. (32) and (34) into Eq. (30), we find that

$$\frac{\bar{D}}{Dt}(\boldsymbol{v}^R - s\,\nabla\Phi/2) + [(\boldsymbol{v}^R - s\,\nabla\Phi/2)\cdot\nabla]\,\nabla\Phi = 0 \tag{35}$$

Note that the rotational component of the linearized unsteady velocity $\boldsymbol{v}^R$ is independent of the pressure fluctuation and therefore depends only on the prescribed upstream rotational velocity and entropy fluctuations and the mean velocity field. Finally, after substituting Eqs. (32) and (34) into Eq. (31), we find that

$$\frac{\bar{D}}{Dt}\left(A^{-2}\frac{\bar{D}\phi}{Dt}\right) - \bar{\rho}^{-1}\nabla\cdot(\bar{\rho}\,\nabla\phi) = \bar{\rho}^{-1}\nabla\cdot(\bar{\rho}\boldsymbol{v}^R) \tag{36}$$

Thus, the linearized unsteady flow can be determined by solving the system of Eqs. (28), (35), and (36) subject to appropriate first-order surface and far-field conditions.

Surface Conditions

Conditions on the linearized unsteady perturbation at blade, wake, and shock mean positions are obtained in a similar fashion, i.e., by substituting the expansions (12–15) into the time-dependent surface conditions, subtracting out the corresponding steady conditions, and eliminating higher-order terms. After we perform some algebra, the following conditions on the linearized unsteady flow are determined. The first-order flow tangency condition has the form

$$\nabla\phi\cdot\boldsymbol{n} = [-\boldsymbol{v}^R + i\omega\boldsymbol{r} + (\nabla\Phi\cdot\boldsymbol{\tau})(\boldsymbol{\tau}\cdot\nabla)\boldsymbol{r} - (\boldsymbol{r}\cdot\nabla)\,\nabla\Phi]\cdot\boldsymbol{n}, \qquad X \text{ on } B_m \tag{37}$$

In addition, since the steady (irrotational) velocity and pressure are continuous and have continuous derivatives across the downstream stagnation streamlines, the conditions of continuity of normal velocity and pressure across blade wakes reduce simply to

$$[\![\nabla\phi]\!]\cdot\boldsymbol{n} = -[\![\boldsymbol{v}^R]\!]\cdot\boldsymbol{n}, \qquad X \text{ on } W_m \tag{38}$$

and

$$\left[\!\!\left[\frac{\bar{\mathrm{D}}\phi}{\mathrm{D}t}\right]\!\!\right] = \left[i\omega + (\nabla\Phi\cdot\boldsymbol{\tau})\frac{\partial}{\partial\tau}\right][\![\phi]\!] = 0, \qquad X \text{ on } W_m \tag{39}$$

Equation (39) can be integrated to yield a condition on the jump in the first-order potential across each wake; i.e.,

$$[\![\phi]\!] = [\![\phi]\!]_{\mathrm{TE}} \exp\left[-i\omega\int_{\tau_{\mathrm{TE}}}^{\tau}(\nabla\Phi\cdot\boldsymbol{\tau})^{-1}\,\mathrm{d}\tau\right] \tag{40}$$

Finally, after making use of the steady mass conservation Eq. (18), the unsteady thermodynamic (29) and pressure-potential (34) relations, and neglecting changes in the entropy and rotational velocity across shocks, we find that the mass and tangential-momentum conservation laws yield the following linearized shock jump conditions:

$$\left[\!\!\left[\bar{\rho}\left(\nabla\phi - A^{-2}\frac{\bar{\mathrm{D}}\phi}{\mathrm{D}t}\nabla\Phi\right)\right]\!\!\right]\cdot\boldsymbol{n} = -[\![\bar{\rho}]\!]\boldsymbol{v}^R\cdot\boldsymbol{n} + [\![\bar{\rho}]\!][i\omega + (\nabla\Phi\cdot\boldsymbol{\tau})\boldsymbol{\tau}\cdot\nabla](\boldsymbol{r}\cdot\boldsymbol{n})$$
$$+ (\boldsymbol{r}\cdot\boldsymbol{n})\boldsymbol{\tau}\cdot\nabla([\![\bar{\rho}]\!]\,\nabla\Phi\cdot\boldsymbol{\tau}), \qquad X \text{ on } \mathrm{Sh}_{m,n} \tag{41}$$

and

$$[\![\nabla\phi]\!]\cdot\boldsymbol{\tau} = -\boldsymbol{\tau}\cdot\nabla(\boldsymbol{r}\cdot\boldsymbol{n}[\![\nabla\Phi]\!]\cdot\boldsymbol{n}), \qquad X \text{ on } \mathrm{Sh}_{m,n} \tag{42}$$

Note that, for a shock that terminates at a sonic point, Eq. (42) can be integrated to yield

$$[\![\phi]\!] = -\boldsymbol{r}\cdot\boldsymbol{n}[\![\nabla\Phi]\!]\cdot\boldsymbol{n}, \qquad X \text{ on } \mathrm{Sh}_{m,n} \tag{43}$$

Equations (41) and either (42) or (43) provide two relations for determining the jump in the unsteady potential $[\![\phi]\!]$ at the mean position of a shock and the shock displacement normal to the mean shock locus, $\boldsymbol{r}\cdot\boldsymbol{n}$. Recall that the mean positions and the unsteady displacements of the blades are prescribed, but the mean positions and unsteady displacements of the wakes and shocks must be determined as part of the steady and unsteady solutions, respectively. However, wake displacements have no direct impact on the solution to the linearized unsteady problem nor on the aerodynamic response to the blade motion.

Far-Field Behavior

We have assumed that in the far-upstream ($\xi < \xi_-$) and far-downstream ($\xi > \xi_+$) regions the irrotational steady flow is at most a small perturbation [also of $\mathcal{O}(\varepsilon)$] from a uniform stream. Therefore, far from the blade row and to within the first-order unsteady approximation considered here, the linearized unsteady field equations can be reduced to constant coefficient equations for which analytical solutions can be determined. These solutions can be matched to near-field numerical solutions, thereby serving to complete the specification of the unsteady problem.

Thus, far from the blade row, the solutions for the linearized entropy and rotational velocity fluctuations have the form

$$s(\boldsymbol{X}) = s_{\mp\infty} \exp[i\boldsymbol{k}_{\mp\infty} \cdot (\boldsymbol{X} - \boldsymbol{X}_{\mp})], \qquad \xi \lessgtr \xi_{\mp} \tag{44}$$

and

$$\boldsymbol{v}^R(\boldsymbol{X}) = \boldsymbol{v}^R_{\mp\infty} \exp[i\boldsymbol{k}_{\mp\infty} \cdot (\boldsymbol{X} - \boldsymbol{X}_{\mp})], \qquad \xi \lessgtr \xi_{\mp} \tag{45}$$

where $s_{\mp\infty}$, $\boldsymbol{v}^R_{\mp\infty}$, and $\boldsymbol{k}_{\mp\infty}$ are the complex amplitudes and wave numbers of the entropic and rotational velocity fluctuations, and $\boldsymbol{X}_{\mp} = (\xi_{\mp}, \eta)$. Recall that $\omega = -\boldsymbol{k}_{\mp\infty} \cdot \boldsymbol{V}_{\mp\infty}$ and $\sigma = \boldsymbol{k}_{\mp\infty} \cdot \boldsymbol{G}$. These equations illustrate that the small-amplitude entropic and vortical gusts are convected without distortion by the uniform mean flows in the far-upstream and far-downstream regions. The wave numbers $\boldsymbol{k}_{\mp\infty}$ and the complex amplitudes $s_{-\infty}$ and $\boldsymbol{v}^R_{-\infty}$ are prescribed quantities with $\boldsymbol{k}_{-\infty} \cdot \boldsymbol{v}^R_{-\infty} = 0$ [see Eq. (33)]. The complex amplitudes s_{∞} and $\boldsymbol{v}^R_{-\infty}$ and $\boldsymbol{v}^R_{\infty}$ must be determined as part of the overall unsteady solution.

In the far field the linearized continuity Eq. (36) can be reduced to

$$A^2_{\mp\infty} \frac{\bar{D}^2\phi}{Dt^2} - \nabla^2\phi = \nabla \cdot \boldsymbol{v}^R \qquad \text{for } \xi \lessgtr \xi_{\mp} \tag{46}$$

where

$$\nabla \cdot \boldsymbol{v}^R = \begin{cases} 0 & \text{for } \xi < \xi_- \\ i\boldsymbol{k}_{\infty} \cdot \boldsymbol{v}^R_{\infty} \exp[i\boldsymbol{k}_{\infty} \cdot (\boldsymbol{X} - \boldsymbol{X}_+)] & \text{for } \xi > \xi_+ \end{cases} \tag{47}$$

We require a solution to this equation subject to the blade-to-blade periodicity condition (17), the wake jump conditions (38) and (40), and the condition that acoustic response waves must either attenuate with increasing distance from the blade row or propagate away from or parallel to the blade row. For this purpose it is convenient to set

$$\phi(\boldsymbol{X}) = \begin{cases} \phi^H(\boldsymbol{X}) & \text{for } \xi < \xi_- \\ \phi^H(\boldsymbol{X}) + \phi^P(\boldsymbol{X}) & \text{for } \xi > \xi_+ \end{cases} \tag{48}$$

where $\phi^H(\boldsymbol{X})$ is a continuous function that accounts for the fluctuations in velocity potential due to acoustic or pressure waves and $\phi^P(\boldsymbol{X})$ is discontinuous at blade wakes and accounts for the fluctuations in the potential due to rotational velocity fluctuations and countervorticity shed from the blades and convected along the wakes. The latter disturbances do not give rise to pressure fluctuations; therefore,

$$\frac{\bar{D}\phi^P}{Dt} = 0 \qquad \text{for } \xi > \xi_+ \tag{49}$$

An expression for the continuous component of the potential, i.e., a solution to the homogeneous equation corresponding to Eq. (46), can be determined using Fourier methods and has the form[32,35,37,40]

$$\phi^H(\boldsymbol{X}) = a_{\mp\infty} \exp[i\boldsymbol{k}_{\mp\infty} \cdot (\boldsymbol{X} - \boldsymbol{X}_{\mp})]$$
$$+ \sum_{j=-\infty}^{\infty} b_{j,\mp\infty} \exp[\pm\beta_{j,\mp\infty}(\xi - \xi_{\mp}) - i\boldsymbol{\kappa}_{j,\mp\infty} \cdot (\boldsymbol{X} - \boldsymbol{X}_{\mp})] \quad \text{for } \xi \lessgtr \xi_{\mp} \tag{50}$$

The first term in Eq. (50) accounts for an incident acoustic excitation that comes from far upstream or far downstream; therefore, the complex amplitudes $a_{\mp\infty}$ and wave numbers $\boldsymbol{k}_{\mp\infty}$ are prescribed quantities. The infinite series accounts for the acoustic response waves. The constants $\beta_{j,\mp\infty}$ and wave numbers $\boldsymbol{\kappa}_{j,\mp\infty}$ for these waves are determined in terms of the far-upstream or far-downstream uniform mean flow conditions, the blade spacing, and the frequency and interblade phase angle of the unsteady motion, but the complex amplitudes $b_{j,\mp\infty}$ of these response disturbances must be determined by matching the analytic far-field solution to a numerical near-field solution for $\phi(\boldsymbol{X})$ at the inlet and exit ($\xi = \xi_{\mp}$) axial stations.

A particular solution, $\phi^P(\boldsymbol{X})$, of Eq. (46) can be determined conveniently in terms of Cartesian coordinates τ_m and n_m, where the τ_m axis is parallel to the far-downstream uniform flow direction, and the origin of the τ_m, n_m axes coincides with the intersection of the axial line $\xi = \xi_+$ and the mean mth wake locus. It then follows from Eq. (49) that

$$\phi^P(\boldsymbol{X}) = F(n_m)\, e^{-i\omega\tau_m/V_\infty} \qquad \text{for } \tau_m > 0,\ 0 < n_m < G \cos\Omega_\infty \tag{51}$$

The function $F(n_m)$ is determined to within arbitrary constants by substituting this expression into Eq. (46) and solving the resulting ordinary differential equation subject to the wake conditions (38) and (40). Once $F(n_m)$ is determined in this manner, the arbitrary constants are determined by matching to the near-field solution.

7. Aerodynamic Response at a Blade Surface

The foregoing unsteady boundary-value problem accounts for the effects of blade geometry, mean blade loading, and transonic, including moving shock, phenomena on the unsteady fluctuations arising from small-amplitude harmonic excitations. The unsteady equations are linear time independent and contain variable coefficients that depend on a fully nonlinear

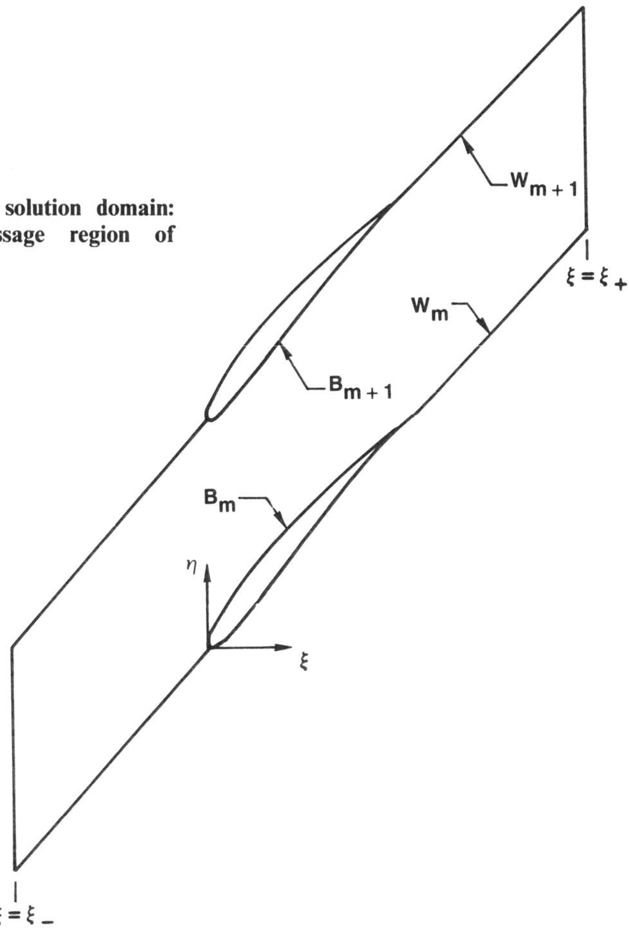

Fig. 6 Numerical solution domain: extended blade-passage region of finite extent.

isentropic and irrotational steady background flow. A numerical resolution of this problem is required to determine the aerodynamic response information needed for aeroelastic predictions, i.e., the unsteady pressures and global unsteady airloads acting on the blades. Because of the assumed form of the unsteady excitations (i.e., periodic in the cascade or η direction), such a resolution is required only over a single extended blade-passage region. In addition, since analytical far-field unsteady solutions have been determined, the numerical solution domain can be restricted further to a single extended blade-passage region of finite extent in the axial flow direction as shown in Fig. 6.

Surface Pressures

Solutions based on the foregoing linearization must be interpreted carefully to provide the correct response information in the vicinity of a moving shock.[51,52] The zeroth- and first-order solutions will be discontinuous at the mean shock location; therefore, they do not account for the fact that an observer situated between the extreme shock positions will experience large-amplitude jumps in the flow variables as the shock passes by. Such local anharmonic effects can be accommodated by analytically continuing the zeroth- and first-order solutions from the upstream and downstream sides of the mean shock locus to the upstream and downstream sides, respectively, of the instantaneous shock locus.[52] This procedure essentially transfers the discontinuities in the flow variables from the mean to the instantaneous shock locations.

The pressure acting at the instantaneous position of the mth blade surface is given by

$$\tilde{P}_{\mathscr{B}_m}(\tau,t) = P_{\mathrm{B}}(\tau) + \mathrm{Re}\{p_{\mathscr{B}}(\tau)\, e^{i(\omega t + m\sigma)}\} + \sum_n P_{\mathscr{S}\!\mathscr{h}_{m,n}}(\tau,t) + \cdots \tag{52}$$

where τ is a coordinate measuring distance in the counterclockwise (or $\boldsymbol{\tau}$) direction along the mean blade surface. The first two terms on the right-hand side of Eq. (52) are the steady and first-harmonic components of the fluid pressure acting at the mth moving blade surface. The third term represents the anharmonic contribution to the unsteady surface pressure caused by the motions of shocks along the surface of the mth blade.

After expanding the pressure P in the manner indicated by Eqs. (12) and (13) and making use of the unsteady pressure-potential relation (34), we find that

$$p_{\mathscr{B}} = \left[-\bar{\rho}\,\frac{\bar{\mathrm{D}}\phi}{\mathrm{D}t} + (\boldsymbol{r}_{\mathrm{B}} \cdot \nabla)P \right]_B \tag{53}$$

where $\bar{\rho}$ and P are determined from the steady solution and $\boldsymbol{r}_{\mathrm{B}}$ is the complex amplitude of the reference blade displacement. The first term on the right-hand side of Eq. (53) represents the harmonic unsteady pressure acting at the mean position of the reference blade surface, and the second term represents the harmonic pressure induced by the motion of the reference blade through a spatially nonuniform steady pressure field.

The local anharmonic effect caused by the motion of a shock is determined by an analytic continuation of the solutions to the steady and the linearized unsteady boundary-value problems.[52] Thus, for example,

$$P_{\mathscr{S}\!\mathscr{h}}(\tau,t) = -\frac{\mathrm{Re}\{r_{\mathrm{Sh,B}}\}}{|\mathrm{Re}\{r_{\mathrm{Sh,B}}\}|}\, H[(\tau - \tau_{\mathrm{Sh}})(\tau_{\mathscr{S}\!\mathscr{h}} - \tau)]Q(\tau,t) \tag{54}$$

where $r_{\mathrm{Sh,B}} = (\boldsymbol{r}_{\mathrm{Sh}} - \boldsymbol{r}_{\mathrm{B}}) \cdot \boldsymbol{\tau}$ is the complex amplitude of the displacement of the shock foot in the counterclockwise or $\boldsymbol{\tau}$ direction along the moving blade surface, the subscripts $\mathscr{S}\!\mathscr{h}$ and Sh refer to the instantaneous and mean shock locations respectively, H is the unit-step function, i.e.,

$$H(\tau) = \begin{cases} 0, \tau < 0 \\ 1, \tau > 0 \end{cases} \tag{55}$$

and

$$Q(\tau,t) = [\![P_B]\!]_{Sh} + (\tau - \tau_{Sh})[\![\partial P_B/\partial\tau]\!]_{Sh} + \mathrm{Re}\{[\![p_{\mathscr{B}}]\!]_{Sh}\, e^{i(\omega t + m\sigma)}\} \tag{56}$$

The discontinuous terms on the right-hand side of Eq. (56) are evaluated at the mean position of the shock foot and by moving across the shock in the τ direction; i.e., $[\![P_B]\!]_{Sh} = P_B(\tau_{Sh}^+) - P_B(\tau_{Sh}^-)$. The first two terms on the right-hand side of Eq. (52) are discontinuous at the undisturbed shock locations. The third term cancels these discontinuities and transfers them to the instantaneous shock locations. This can be seen by setting $\tau = \tau_{\mathscr{S}h} \pm$ in Eqs. (54) and (55).

The relative displacement of a normal shock along a blade surface is readily determined in terms of the prescribed blade motion and the linearized unsteady solution for the shock displacement in the normal direction [cf. Eqs. (41) and (42) or (43)]. It should be noted, however, that the additional steps required to determine $r_{Sh,B}$ for an oblique shock are not apparent at the present time. Finally, although the unsteady pressure disturbance is not everywhere harmonic, its regions of anharmonicity are small. Consequently, the first-order global aerodynamic coefficients are harmonic in time.[28,53]

Unsteady Airloads

Up to this point we have placed no restriction on the mode of the blade motion. But we now limit our consideration to the condition usually treated in turbomachinery aeroelastic calculations wherein each incremental blade section undergoes a rigid-body motion; i.e.,

$$\boldsymbol{r}_B(\boldsymbol{X}) = \boldsymbol{h} + \boldsymbol{\alpha} \times \boldsymbol{R}_P \qquad \text{for } \boldsymbol{X} \text{ on } B \tag{57}$$

For such motions, the unsteady force and moment are the only global response parameters needed to analyze aeroelastic behavior. In Eq. (57) $\boldsymbol{h}$ defines the amplitude and direction of blade translations (positive in the positive coordinate directions), $\boldsymbol{\alpha} = \alpha \boldsymbol{e}_z$ defines the amplitude and direction of blade rotations (positive counterclockwise), and $\boldsymbol{R}_P$ is a position vector extending from the mean position of the reference blade axis of rotation (i.e., from the point X_P, Y_P) to points on the mean position of the reference blade surface. These rigid two-dimensional motions model bending and torsional vibrations of actual rotor blades. The terms h_x, h_y, and α are, in general, complex to permit phase differences between the translational and angular displacements.

The force and moment acting on the mth blade are given by

$$\tilde{\boldsymbol{F}}_m = -\oint_{\mathscr{B}_m} \tilde{P}\boldsymbol{n}\, d\tau = \boldsymbol{F} + \mathrm{Re}\{\boldsymbol{f} e^{i(\omega t + m\sigma)}\} + \cdots \tag{58}$$

and

$$\tilde{M}_m = \oint_{\mathscr{B}_m} \tilde{P}\tilde{\boldsymbol{R}}_P \cdot d\boldsymbol{\tau} = M + \mathrm{Re}\{m\, e^{i(\omega t + m\sigma)}\} + \cdots \tag{59}$$

The components of the force vector $\tilde{\boldsymbol{F}}_m$ are taken as positive in the positive coordinate directions, and the moment $\tilde{M}_m$ is taken about the moving pitching axis of the mth blade and is regarded as positive in the counterclockwise direction. The vector $\tilde{\boldsymbol{R}}_P$ in Eq. (59) extends from this axis to points on the moving mth blade surface. After some algebra, it follows that

$$\boldsymbol{f} = \boldsymbol{\alpha} \times \boldsymbol{F} - \oint_{\mathrm{B}} p_{\mathscr{B}} \boldsymbol{n} \, \mathrm{d}\tau + \sum_n r_{\mathrm{Sh}_n,\mathrm{B}} ([\![P_{\mathrm{B}}]\!] \boldsymbol{n}_{\mathrm{B}})_{\mathrm{Sh}_n} \tag{60}$$

and

$$m = \oint_{\mathrm{B}} p_{\mathscr{B}} \boldsymbol{R}_P \cdot \mathrm{d}\boldsymbol{\tau} - \sum_n r_{\mathrm{Sh}_n,\mathrm{B}} ([\![P_{\mathrm{B}}]\!] \boldsymbol{R}_P \cdot \boldsymbol{\tau}_{\mathrm{B}})_{\mathrm{Sh}_n} \tag{61}$$

where $\boldsymbol{f}$ and m are the complex amplitudes of the linearized unsteady force and moment, respectively, acting on the reference blade, and the summations account for the concentrated loads due to shock motion.

8. Response Predictions: Uniform Mean Flow

In this and the following sections we will present unsteady aerodynamic response predictions for several cascade configurations. These will serve to illustrate partially the relative influence of the various physical parameters that control blade response as well as the status of the procedures used for solving the linearized boundary-value problem derived in Sec. 6. We will restrict our consideration to unsteady flows through two-dimensional fan- or compressor-type cascades for which torsional blade motions are the only source of unsteady excitation. Unless otherwise stated these motions occur at unit amplitude ($|\alpha| = 1$) and relative to axes at the blade midchords ($X_P, Y_P = 0.5, 0$).

Theoretical results including first harmonic unsteady surface pressure $p_{\mathscr{B}}$ and pressure difference [i.e., $\Delta p(x) = p_{\mathscr{B}}(x,y_-) - p_{\mathscr{B}}(x,y_+)$, where the subscripts $-$ and $+$ refer to the lower and upper blade surfaces, respectively] distributions, unsteady aerodynamic moments m, and, where appropriate, unsteady shock displacements $r_{\mathrm{Sh,B}}$ will be presented for blades undergoing pure torsional vibrations with $\mathrm{Im}\{\alpha\} = 0$. For such motions the real and imaginary parts of a complex response parameter represent the response components that are in and out of phase, respectively, with the reference blade displacement. Moreover, the stability of the blade motion depends on the sign of the out-of-phase moment.[54] Thus, if $\mathrm{Im}\{m\} < 0$, the airstream is removing energy from the blade motion; hence, this motion is stable. If $\mathrm{Im}\{m\} = 0$, there is no net transfer of energy from the blade motion to the airstream and the former is neutrally stable. Finally, if $\mathrm{Im}\{m\} > 0$, the airstream is supplying energy to the blade motion, and this motion is unstable according to linearized theory.

First, we will consider cascades of flat-plate airfoils whose mean positions are aligned with the inlet flow direction. In such cases the underlying mean flow is uniform; i.e., $M = M_{-\infty}$, where M is the local steady Mach number, and the unsteady perturbation is governed by the equations of classical linearized subsonic or supersonic flow theory. Our primary purpose here is to familiarize the reader with important aspects of classical subsonic and supersonic unsteady response behavior. However, several extensions of the classical linearization will also be discussed. The subsonic results presented in this section have been determined using the semi-analytic solution procedure developed by Smith[14]; those for supersonic flow, using the procedure developed by Verdon.[15] At supersonic Mach numbers weak oblique shock (or Mach) waves occur. These waves, which originate at the leading and trailing edges of the blade, and their reflections off neighboring blades remain stationary in the classical approximation and lie on the characteristic lines $\mathrm{d}x/\mathrm{d}y = \pm(M^2 - 1)^{\frac{1}{2}} = \pm\mu$. The unsteady potential ϕ is continuous, but, in general, the fluid properties are discontinuous across such weak oblique shocks.

Subsonic Flow

Response predictions for the subsonic flat-plate cascades are shown in Figs. 7–10 for representative values of the cascade parameters. Unsteady pressure difference distributions and aerodynamic moments for $\Theta = 45$ deg, $G = 1.0$, $M = 0.7$, and $\omega = 1.0$ are shown in Fig. 7 for four interblade phase angles: $\sigma = 0$, ± 90, and 180 deg. Recall that Θ and G are the cascade stagger angle and blade spacing, respectively, and ω is the frequency of the unsteady motion. Classical subsonic unsteady pressure difference distributions are singular and behave like $x^{-\frac{1}{2}}$ at blade leading edges and like $x^{\frac{1}{2}}$ at blade trailing edges. The blade motions considered in Fig. 7 are stable (i.e., $\mathrm{Im}\{m\} < 0$), and the response predictions show a strong dependence on the phase angle σ between the motions of adjacent blades.

The effects of variations in Mach number and blade vibration frequency on the unsteady moment response for a cascade with $\Theta = 45$ deg and $G = 1$ are illustrated in Figs. 8 and 9, respectively. The results in Fig. 8 are for unit-frequency ($\omega = 1.0$) blade motions at mean flow Mach numbers of 0.5, 0.7, and 0.8. Those in Fig. 9 are for a steady Mach number of 0.7 and vibration frequencies of 0.5, 1.0, and 1.5. The effects of variations in stagger angle Θ and blade spacing G on the out-of-phase component of the unsteady moment are illustrated in Fig. 10. Here $M = 0.7$ and $\omega = 1.0$. The results in Fig. 10a are for $G = 1$ and $\Theta = 30$, 45, and 60 deg; those in Fig. 10b are for $\Theta = 45$ deg and $G = 0.5$, 1.0, and 1.5. The predictions shown in Figs. 7–10 illustrate the strong dependence of cascade subsonic unsteady response behavior on interblade phase angle, Mach number, vibration frequency, stagger angle, and blade spacing.

The spikes in the foregoing unsteady moment curves are indicative of acoustic resonances. For example, the results in Fig. 8 indicate that such resonances occur at $\sigma = -22.2$ deg and $\sigma = 49.2$ deg for $M = 0.5$, at -29.4 and 107.3 deg for $M = 0.7$, and at -33.0 and 177.0 deg for $M = 0.8$. For an unloaded flat-plate cascade operating at a resonance condition, one acoustic wave persists and carries energy parallel to the blade row in both the far-upstream and far-downstream flowfields. Blade motions at interblade phase angles between the two resonant values are termed superresonant; i.e., they occur at a frequency greater than the resonant frequency

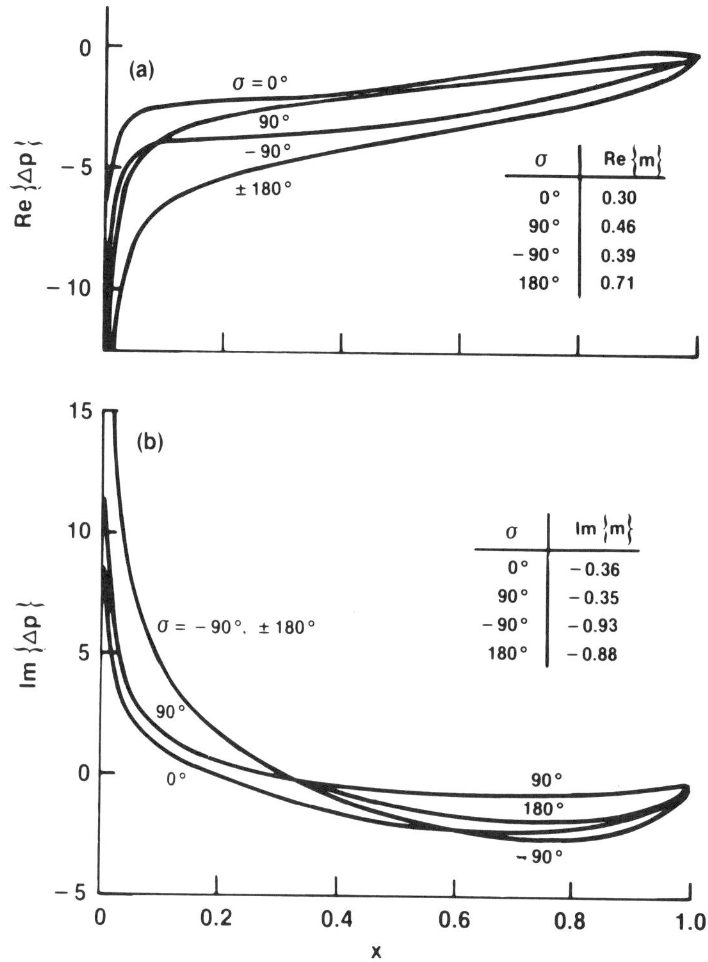

Fig. 7 Unsteady pressure difference distributions and aerodynamic moments due to the torsional blade vibrations of a subsonic flat-plate cascade; $\Theta = 45$ deg, $G = 1.0$, $M = 0.7$, $\omega = 1.0$: a) in-phase component (real part); b) out-of-phase component (imaginary part) of the unsteady response.

corresponding to the prescribed values of Θ, G, M, and σ. Those motions for which σ is less than or greater than the two resonant interblade phase angles are subresonant. For subsonic flat-plate cascades at least one acoustic response wave persists and carries energy away from the blade row (in both the upstream and downstream axial directions) when the blade motion is super-resonant, but all such waves attenuate with distance from the blade row when the motion is subresonant.

Finally, the out-of-phase moment curves in Figs. 8–10 indicate that, except for blade motions occurring at interblade phase angles near resonant values, the torsional blade motions considered are stable. Classical subsonic analyses will predict unstable motions at low vibration frequencies, but those occurring at the frequencies of practical interest are usually found to be stable. As a result, classical unsteady aerodynamic analyses have provided only a limited insight into the causes of flutter in fan or compressor blading operating under subsonic mean inlet and exit conditions.

Supersonic Flow

Modern supersonic fans operate at supersonic relative inlet velocities that have subsonic axial velocity components. Thus, a classical two-dimensional simulation of the relevant flow behavior will contain a system of weak

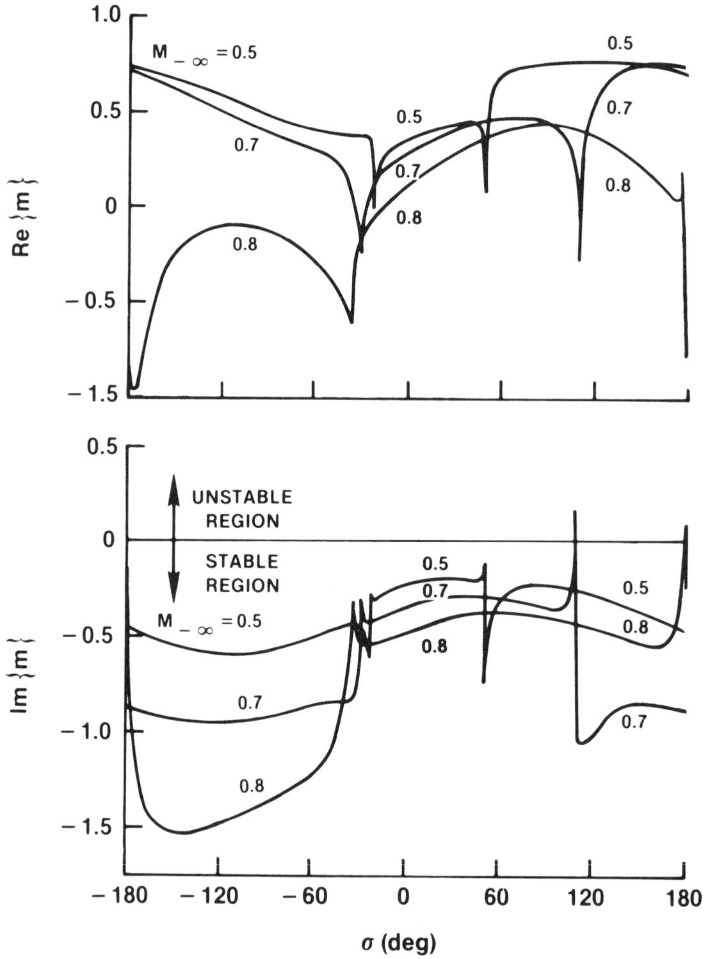

Fig. 8 Effect of Mach number on the unsteady moment response to the torsional blade vibrations of a subsonic flat-plate cascade; $\Theta = 45$ deg, $G = 1.0$, $\omega = 1.0$.

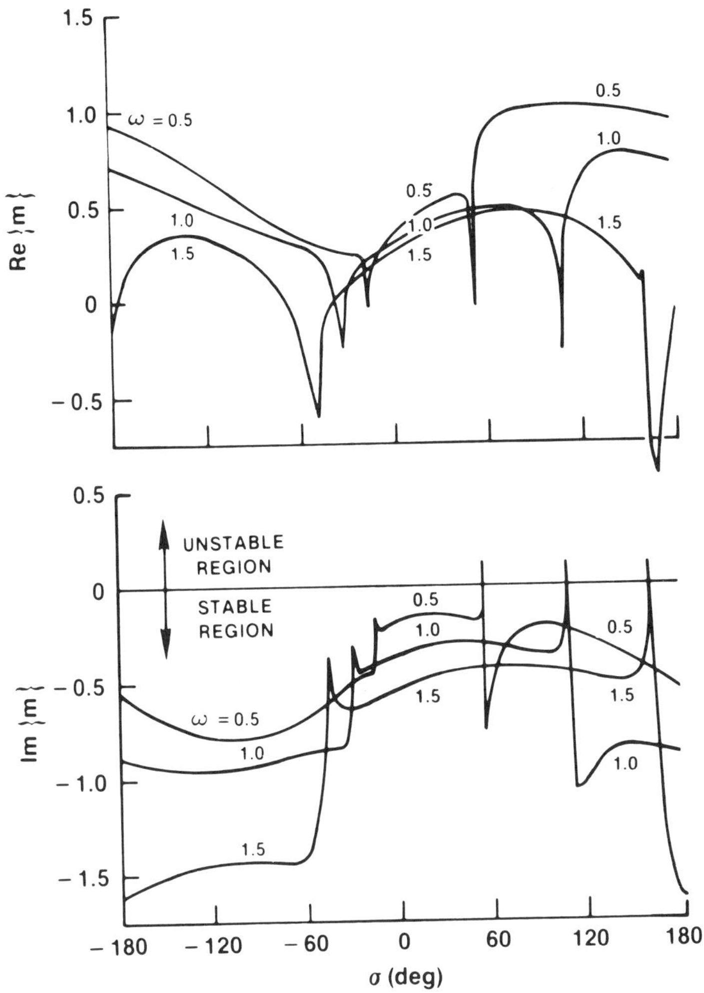

Fig. 9 Effect of vibration frequency on the unsteady moment response to the torsional blade vibrations of a subsonic flat-plate cascade; $\Theta = 45$ deg, $G = 1.0$, $M = 0.7$.

oblique shock (Mach) waves extending infinitely far upstream of the blade row. These waves originate at the leading edges of the blades, on the suction side, and their locations are given by $x - \mu y = m(x_A - \mu y_A)$, $x > mx_A$, $m = 0, \pm 1, \pm 2, \ldots$. Here $x_A = G \sin\Theta$ and $y_A = G \cos\Theta$. There are two such supersonic configurations, as shown in Fig. 11, depending on the freestream Mach number and the blade stagger and spacing, which are representative of the flows at the outer span radial sections of modern fans. In the first configuration (Fig. 11a), $x_A + \mu y_A > 1$, and the leading-edge Mach wave off the lower or pressure surface of each blade passes behind the blades below. In the second configuration (Fig. 11b), $x_A + \mu y_A < 1 <$

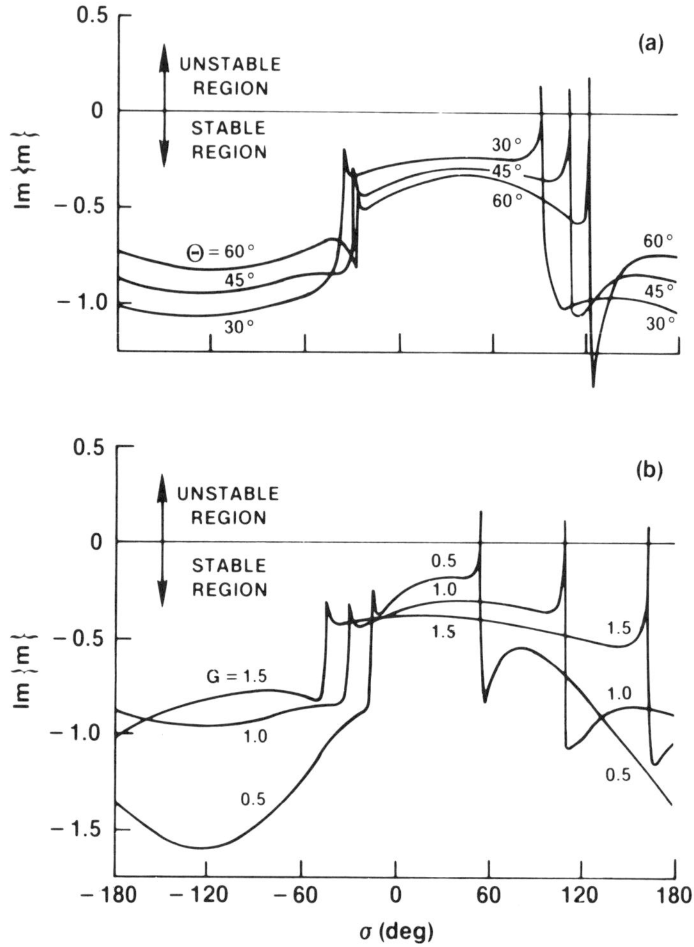

Fig. 10 Effect of cascade stagger angle and blade spacing on the out-of-phase component of the unsteady moment due to the torsional blade vibrations of a subsonic flat-plate cascade; $M = 0.7$, $\omega = 1.0$: a) effect of stagger angle; $G = 1.0$; b) effect of blade spacing; $\Theta = 45$ deg.

$x_A + 3\mu y_A$, and this Mach wave is reflected once by the adjacent blade below. The points labeled 1, 2, and 3 in Fig. 11 denote the locations at which leading- and trailing-edge Mach waves impinge on and are reflected by the reference blade and, as such, the locations along this blade at which the surface pressure is discontinuous.

Unsteady response predictions for specific examples of the foregoing supersonic configurations are given in Figs. 12 and 13. These results have been taken from Ref. 15, where they were reported in terms of a compressible reduced frequency parameter $k = \omega M \mu^{-2}$. In one configuration (cascade A), $x_A = 0.68$, $y_A = 0.4$, $\mu = 0.9$ ($\Theta = 59.5$ deg, $G = 0.789$,

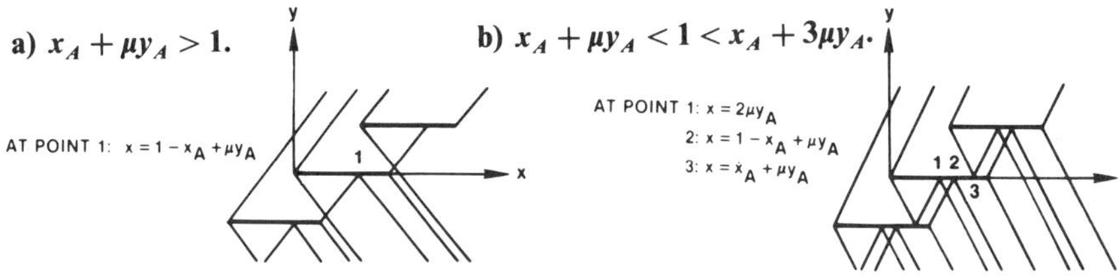

Fig. 11 Two representative classical supersonic cascade configurations.

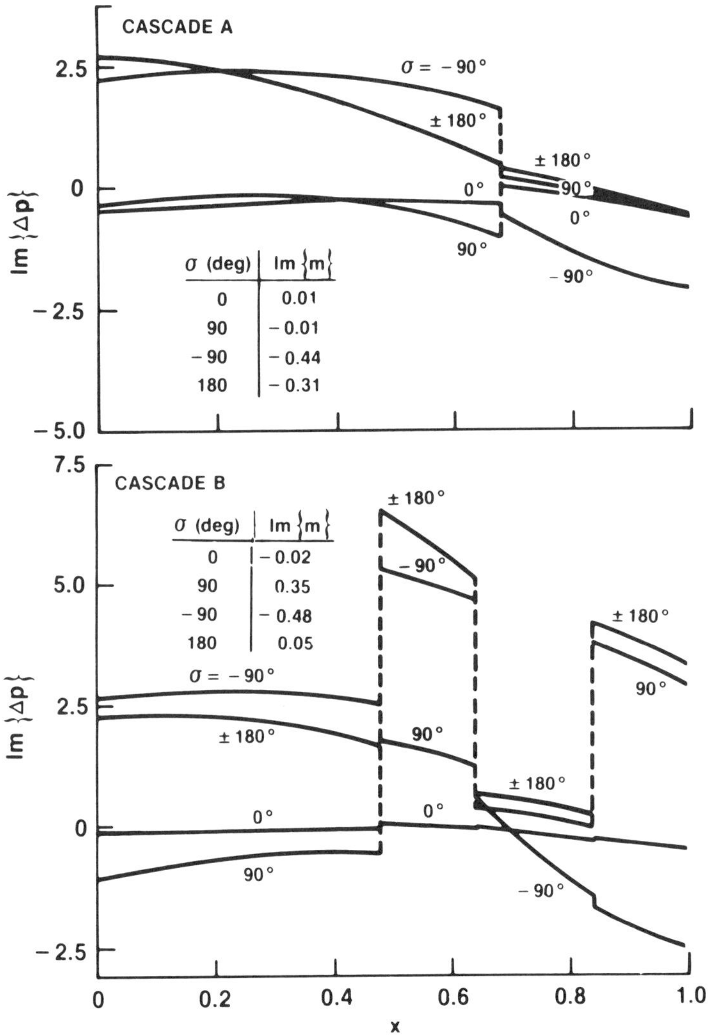

Fig. 12 Out-of-phase unsteady pressure difference distributions and aerodynamic moments due to the torsional blade vibrations at $k = \omega M \mu^{-2} = 1.5$ of supersonic flat-plate cascades A and B.[15]

$M = 1.345$), and the leading-edge Mach waves off the blade pressure surfaces pass behind the lower blades. In the other (cascade B), $x_A = 0.6$, $y_A = 0.3$, $\mu = 0.8$ ($\Theta = 63.4$ deg, $G = 0.671$, $M = 1.281$), and these Mach waves are reflected once by the adjacent blades below. Out-of-phase unsteady pressure difference distributions for $k = 1.5$ and $\sigma = 0$, ± 90, and 180 deg, and out-of-phase moments for $k = 0.5$, 1.0, 1.5, and 2.0 are shown in Figs. 12 and 13, respectively, for pure torsional motions (with $\alpha = 1,0$) about midchord. Note that $\omega = 0.602k$ for cascade A, and $\omega = 0.5k$ for cascade B.

At supersonic freestream Mach numbers, far-field acoustic waves are primarily of propagating type. Blade motions are subresonant when all such waves carry energy away from the blade row, resonant when at least one carries energy along the blade row, and super-resonant when at least one attenuates with increasing axial distance from the blade row. For cascade A, resonances occur at $\sigma = -19.4k$ and at $\sigma = -85.5k$ deg, and for cascade B the resonant interblade phase angles are $\sigma = -12.5k$ and $-75.5k$ deg. As in the subsonic case, blade motions at phase angles between the resonant values are super-resonant, whereas those at higher or lower interblade phase angles are subresonant.

Isolated airfoils are susceptible to supersonic torsional instabilities but only at low vibration frequencies.[55] Thus, the results shown in Figs. 12 and 13 demonstrate that the aerodynamic interactions between airfoils in cascade have a pronounced effect on aeroelastic stability at supersonic Mach numbers. The out-of-phase pressure difference distributions in Fig. 12 indicate that the pressure differences over the forward parts of the blades in cascades A and B tend to support the torsional motions at $\sigma = 0$ and 90 deg. As a result, these motions are unstable or lie close to the stability boundary. The moment curves depicted in Fig. 13 indicate that the torsional blade motions for the two cascades are unstable over a broad range of vibration frequencies and interblade phase angles. In particular, unstable torsional motions of the blades in cascade A occur over a wide band of subresonant interblade phase angles for $k = 0.5$, 1.0, and 1.5 and over a narrow band of super-resonant phase angles for $k = 0.5$ and 1.0. A similar behavior is observed for the blades in cascade B, but here unstable subresonant and super-resonant motions also occur at $k = 2.0$ and 1.5, respectively. Because of this ability to identify possible regions of torsional instability using classical supersonic analyses, such analyses have proved to be very useful to turbomachinery designers in predicting the flutter characteristics of proposed supersonic fan designs.[1,56]

Related Theories

During a period extending from the late 1960's through the mid-1970's, a significant level of research activity was focused on the development of semianalytic solutions of the classical linearized equations for unsteady cascade flows. This work has resulted in the availability of very efficient solutions that are now used extensively in turbomachinery flutter and resonant stress design prediction systems. The development of supersonic methods was motivated by the occurrence of supersonic unstalled torsional

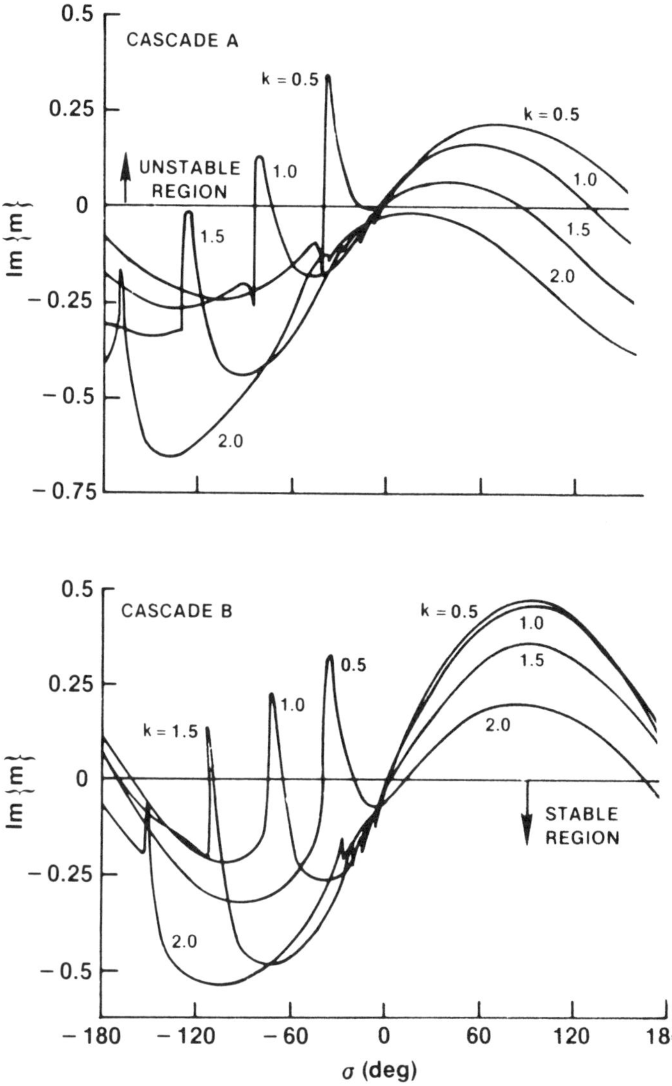

Fig. 13 Effect of frequency on the out-of-phase components of the unsteady moments due to the torsional blade vibrations of supersonic flat-plate cascades A and B.[15]

flutter in fan stages. Since this type of flutter can occur at design operating conditions, it can impose a limit on the high-speed operation of the machine. Classical unsteady supersonic solutions have been found to give conservative estimates for the onset of supersonic unstalled torsional flutter and have therefore provided engine designers with an effective means for predicting and controlling its occurrence.

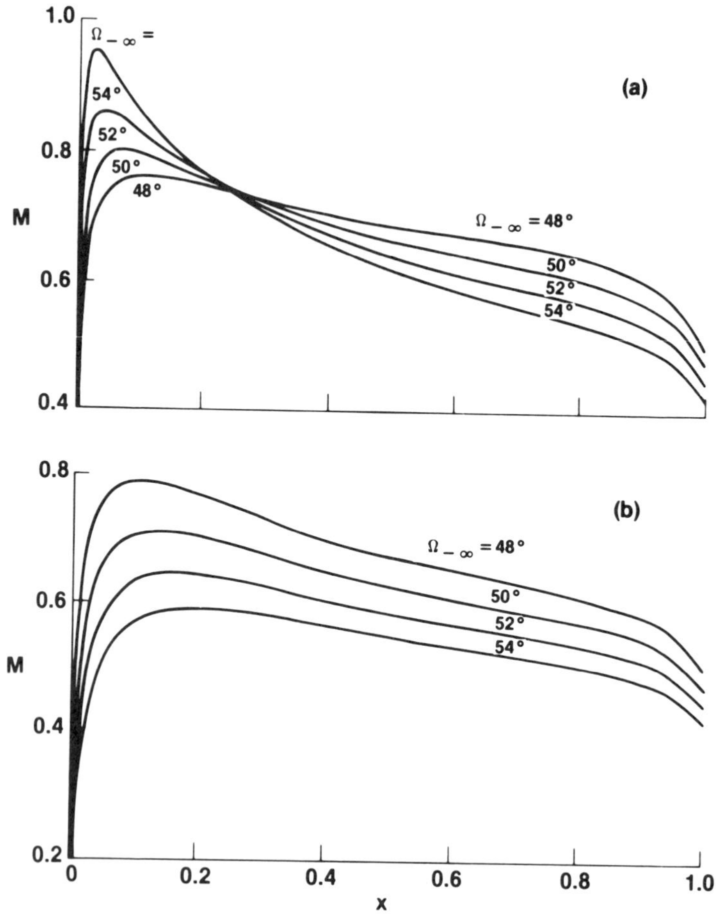

Fig. 14 Steady surface Mach number distributions for the example NACA 0012 cascade; $M_{-\infty} = 0.6$: a) suction surface; b) pressure surface.[10]

Modifications to the classical linearization have also been proposed to treat separated subsonic flows, transonic (i.e., supersonic inlet/subsonic exit) flows containing strong in-passage shocks and high-frequency transonic flows. In the flow-separation model[20,21] the flow is assumed to separate at a prescribed point on each blade suction surface, and a constant surface pressure condition is prescribed, in lieu of a flow tangency condition, along the suction surface of each blade from the point of separation to the trailing edge. Solutions based on this model have been found to be in qualitative agreement with experimental measurements of unsteady lift and moment and tend to predict observed flutter in fans in cases where subsonic attached flow analyses indicate stability.[21] Thus, although this simple model ignores much of the complicated physics of separated flow, it may reveal an essential mechanism for the occurrence of subsonic flutter in cascades of thin, highly loaded blades.

Goldstein et al.[22] have provided an analysis of unsteady flow through a supersonic cascade in which a strong normal shock occurs in each blade passage. These shocks extend from the lower surface near the leading edge of each blade to the upper surface of the adjacent blade below. The resulting flow configuration approximates that observed in the tip region of a loaded supersonic fan rotor. The steady base flow upstream of the normal shocks is uniform and supersonic with subsonic axial velocity component, whereas that downstream is uniform and subsonic. The unsteady perturbation upstream of the strong normal shocks is irrotational, whereas that downstream contains both irrotational and rotational components. The upstream supersonic and the downstream subsonic unsteady solutions are connected through jump conditions applied at the normal shock surfaces. Analytic predictions based on this strong shock model indicate a greater torsional stability margin than those based on the classical supersonic linearization. This is consistent with experimental observations of an increasing supersonic torsional stability margin with increasing backpressure. However, the analytical results also indicate the occurrence of pure bending flutter at low frequencies. This is attributed partially to the destabilizing effect of concentrated shock loads, since a pure bending flutter has not been revealed by classical supersonic analyses.

Finally, Surampudi and Adamczyk[23] have developed a Wiener-Hopf solution technique for unsteady perturbations relative to uniform flows at Mach numbers near 1. Their analysis gives results that are in agreement with classical subsonic predictions at $M = 0.9$ and with classical supersonic predictions at $M = 1.1$. Hence, the unsteady transonic analysis of Ref. 23 appears to bridge the gap between separate subsonic and supersonic unsteady cascade analyses. Moreover, the parametric trends revealed by the transonic solutions are consistent with those revealed by earlier classical supersonic solutions.

9. Response Predictions: Nonuniform Mean Flow

Because of the restrictions on physical modeling that are imposed by classical linearizations, unsteady linearizations relative to nonuniform steady base flows are under active current development for turbomachinery aeroelastic applications. Our purpose in this section is to illustrate, to some extent, the status of numerical procedures for solving the associated nonlinear steady and the linear variable coefficient unsteady boundary-value problems and to demonstrate several effects associated with nonuniform steady flows on the unsteady aerodynamic response to blade vibrations. One such effect is that different acoustic response environments exist in the far-upstream and far-downstream flow regions. These are caused by differences between the uniform inlet and exit flow conditions, and they have an important impact on the unsteady aerodynamic response at a blade surface.

We will restrict our consideration to flows through subsonic fan- or compressor-type ($M_{\infty} < M_{-\infty} < 1$) cascades and to unsteady perturbations caused by pure torsional blade vibrations with $\mathrm{Im}\{\alpha\} = 0$. The steady flows considered here are determined by imposing a Kutta condition, i.e.,

$$\boldsymbol{V} \cdot \mathrm{d}\boldsymbol{\tau}\big|_{\mathrm{B}-} = -\boldsymbol{V} \cdot \mathrm{d}\boldsymbol{\tau}\big|_{\mathrm{B}+} \tag{62}$$

at blade trailing edges; therefore, only the inlet freestream conditions, $M_{-\infty}$ and $\Omega_{-\infty}$, are prescribed. We refer the reader to Refs. 39, 41, 42, and 44 for a description of the numerical methods used to obtain the results presented below. For a description of some alternative methods along with their applications to unsteady flows involving forced aerodynamic excitations, turbine cascades, and supersonic inlet and/or exit conditions, see also Refs. 37, 38, 40, and 46.

We consider first a cascade of NACA 0012 airfoils to illustrate the effects of a relatively thick, blunt-nosed blade geometry and variable mean incidence on unsteady aerodynamic response. We then turn to a cascade of sharp-edged double circular arc (DCA) airfoils to illustrate the effects of mean blade loading and transonic phenomena on the response at high subsonic inlet Mach number. For purposes of comparison, results for flat-plate cascades (also determined by the methods of Refs. 39, 41, and 42) operating in uniform mean flows will be included along with those for the NACA 0012 and the DCA cascades. The example NACA 0012, DCA, and flat-plate cascades all have a stagger angle Θ of 45 deg and a unit gap-to-chord ($G = 1$) ratio, and the torsional blade motions occur at $|\alpha| = 1$ about axes at the blade midchords. Subsequently, we will present response predictions for cascades of NACA 65 series and biconvex airfoils for which experimental measurements are also available.

NACA 0012 Cascade

The mean position of the reference blade for a cascade of (modified) NACA 0012 airfoils is given by

$$y_{\pm}(x) = \pm 5T[0.2969x^{\frac{1}{2}} - 0.1260x - 0.3516x^2 + 0.2843x^3 - 0.1036x^4] \quad \text{for} \quad 0 \leqslant x \leqslant 1 \tag{63}$$

where $T = 0.12$ is the nominal blade thickness and $1.1019T = 0.015867$ is the leading-edge radius. The coefficient of the x^4 term in Eq. (63) differs from the standard one (i.e., -0.1015),[57] so that the example profile will close in a wedge-shaped trailing edge. Steady and unsteady flows through a staggered cascade ($\Theta = 45$ deg, $G = 1$) of such airfoils have been determined for an inlet Mach number ($M_{-\infty}$) of 0.6 and four inlet flow angles ($\Omega_{-\infty}$).

The predicted steady surface Mach number distributions for $\Omega_{-\infty} = 48$, 50, 52, and 54 deg are shown in Fig. 14. The calculated exit Mach numbers are, respectively, 0.595, 0.557, 0.522, and 0.490, and in each case the calculated exit flow angle is approximately 47.7 deg. These steady flows are entirely subsonic with a peak Mach number of 0.789 occurring at $x = 0.113$ on the pressure (lower) surface of the blade for $\Omega_{-\infty} = 48$ deg, and 0.8, 0.86, and 0.96 occurring at $x = 0.07$, 0.05, and 0.03 on the suction (upper)

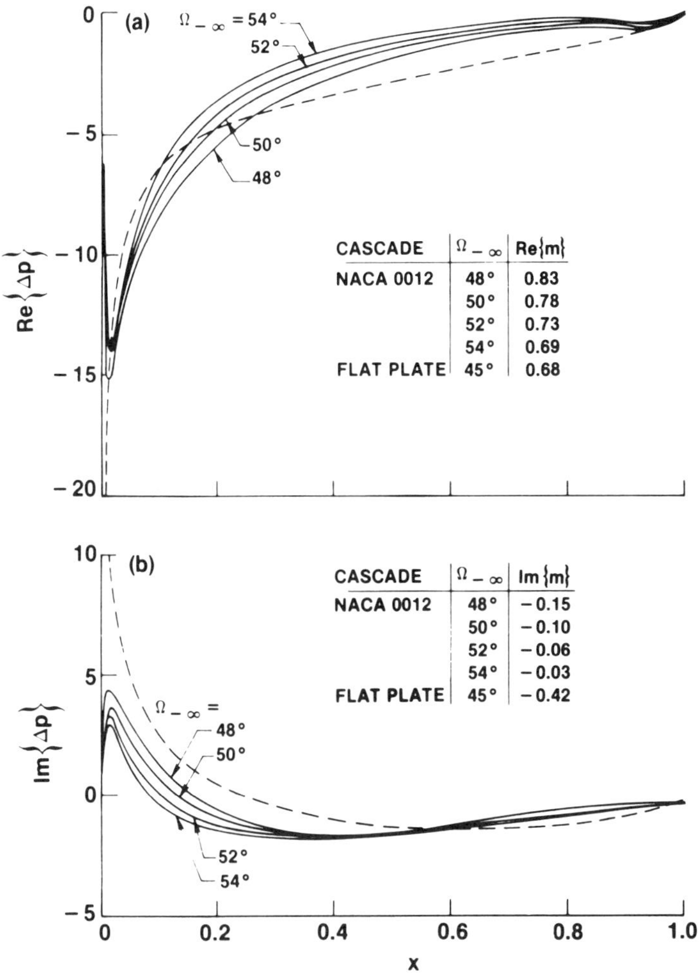

Fig. 15 Unsteady response to torsional blade vibrations; $M_{-\infty} = 0.6$, $\omega = 1.0$, $\sigma = 90$ deg. Solid lines, NACA 0012 cascade; dashed lines, flat-plate cascade[10]: a) in-phase component (real part); b) out-of-phase component (imaginary part) of the unsteady response.

surface for $\Omega_{-\infty} = 50$, 52, and 54 deg, respectively. In each case the mean flow stagnates within 0.2% of blade chord downstream from the leading edge.

Unsteady response predictions for the NACA 0012 cascade are shown in Figs. 15–18. Results are also given in Figs. 15, 17, and 18 for the corresponding flat-plate cascade [$T = 0$ in Eq. (63)] operating at $M_{-\infty} = 0.6$ and $\Omega_{-\infty} = 45$ deg. Unsteady pressure difference distributions and aerodynamic moments for the NACA 0012 and flat-plate blades under-

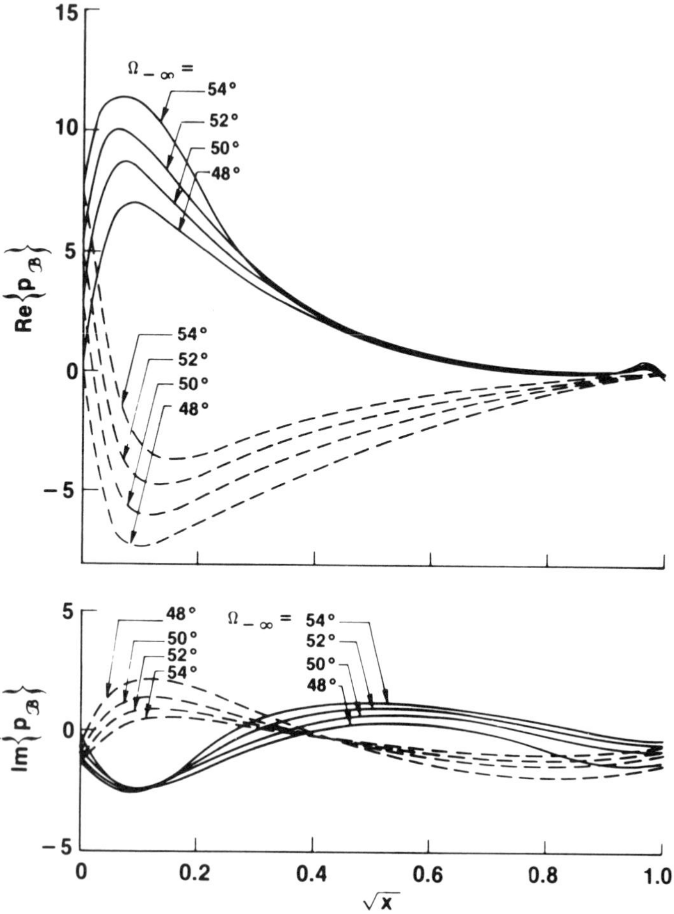

Fig. 16 Unsteady surface pressures due to torsional vibrations of the NACA 0012 blades; $M_{-\infty} = 0.6$, $\omega = 1.0$, $\sigma = 90$ deg. Solid lines, suction surface; dashed lines, pressure surface.[10]

going unit-frequency motions at $\sigma = 90$ deg are shown in Fig. 15. In addition, the unsteady pressures acting on the reference NACA 0012 airfoil are plotted in Fig. 16 vs the square root of the distance along the chord to emphasize the surface pressure behavior near the airfoil leading edge. As mentioned in the previous section, the unsteady pressure difference is singular and behaves like a multiple of $x^{-\frac{1}{2}}$ near the leading edge of the flat-plate airfoil. In contrast, the linearized unsteady pressure is analytic in the vicinity of the rounded NACA 0012 leading edge. In the latter case both the real and imaginary components of the unsteady pressure difference are zero at the leading edge and reach local extrema very close to the leading edge. The results in Figs. 15 and 16 also indicate that the coupling between the steady and unsteady flows, due to blade geometry and increasing mean blade loading, leads to a reduction in the out-of-phase pressure difference, $\mathrm{Im}\{\Delta p(x)\}$, over the forward part of the blade and therefore a reduction in the moment opposing the blade motion.

Unsteady moments acting on the reference blades of the NACA 0012 and the flat-plate cascades undergoing unit-frequency torsional vibrations are shown in Fig. 17 for the entire range of interblade phase angles. Again, the abrupt changes in the moment curves are indicative of an acoustic

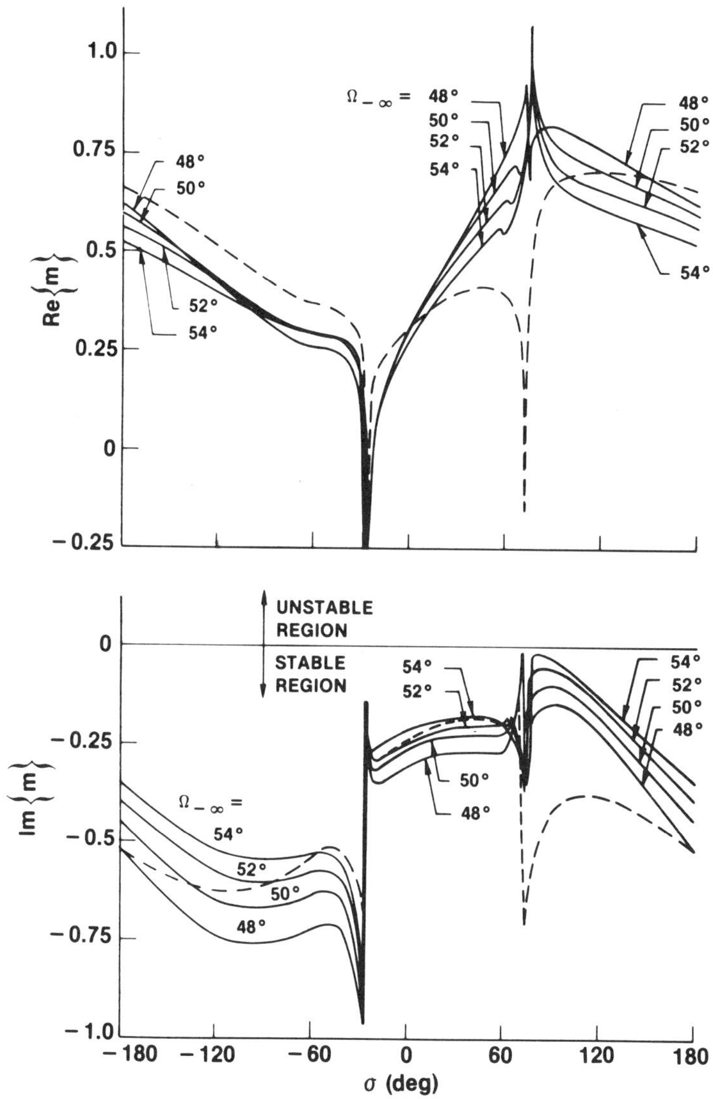

Fig. 17 Unsteady moments due to torsional blade vibrations at $M_{-\infty} = 0.6$ and $\omega = 1.0$. Solid lines, NACA 0012 cascade; dashed lines, flat-plate cascade.[10]

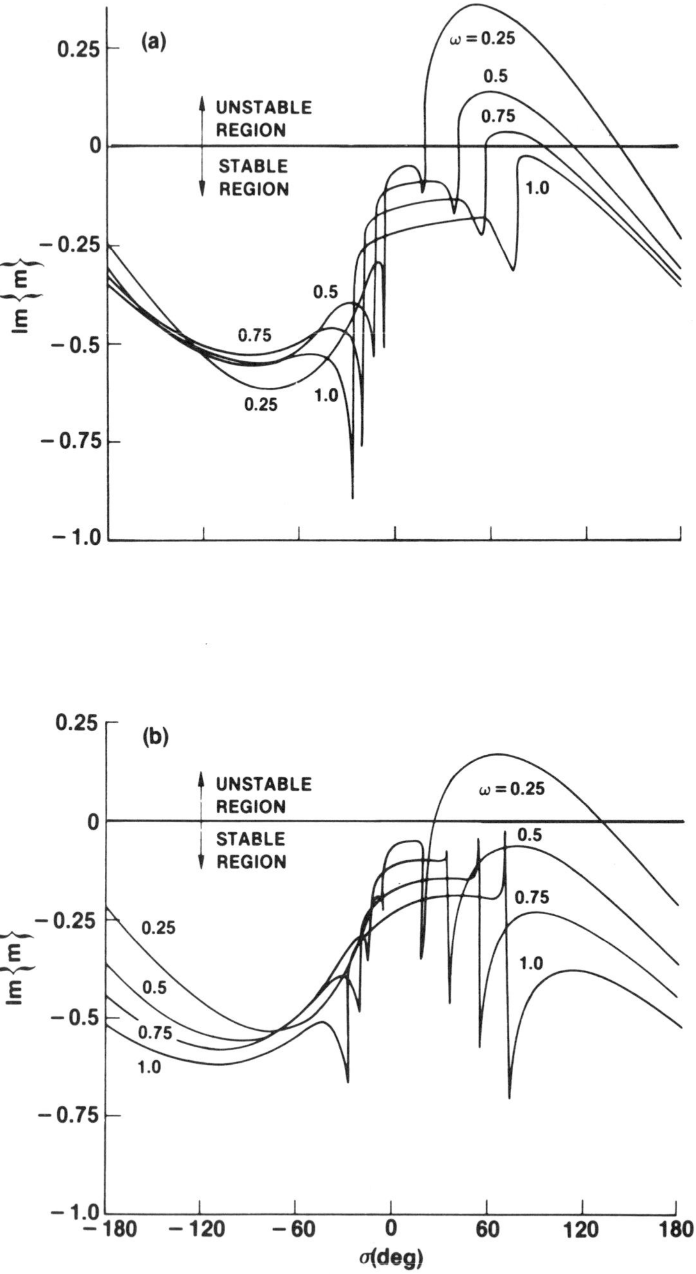

Fig. 18 Effect of vibration frequency on the out-of-phase components of the unsteady moments due to torsional blade vibrations at $M_{-\infty} = 0.6$: a) NACA 0012 cascade, $\Omega_{-\infty} = 54$ deg; b) flat-plate cascade, $\Omega_{-\infty} = 45$ deg.[10]

resonance at inlet and/or exit. For the loaded NACA 0012 blades there are four resonant interblade phase angles: two associated with the uniform inlet flow and two with the uniform exit flow. However, the previous classification for blade motions occurring in subsonic flows can be retained; i.e., such motions are super-resonant (except those occurring at the other resonant phase angles) at phase angles lying between the lowest and the highest resonant phase angles and subresonant at phase angles lying beyond the resonant values. The motions considered in Fig. 17 are stable. But the NACA 0012 results indicate that for a given σ the effect of an increase in mean blade loading (in this case an increase in $\Omega_{-\infty}$) is usually destabilizing since the out-of-phase moment curves move closer to the stability boundary as the inlet flow angle is increased. Also, the overall stability margin for $\omega = 1$ is greater for the unloaded flat-plate cascade than it is for the NACA 0012 cascade operating at the inlet flow angles reported in Fig. 17.

The effect of frequency on the out-of-phase unsteady moment due to torsion about midchord for the NACA 0012 cascade operating at $\Omega_{-\infty} = 54$ deg and for the flat-plate cascade operating at $\Omega_{-\infty} = 45$ deg is illustrated in Fig. 18. The NACA 0012 blades experience a region of subresonant torsional instability for $\omega = 0.25$, 0.5, and 0.75, but the extent of this region decreases with increasing frequency. The subresonant torsional motions of the flat-plate cascade are unstable only at the lowest frequency considered, i.e., $\omega = 0.25$.

DCA Cascade

The mean position of the reference blade of a cascade of sharp-edged double circular arc airfoils is defined by

$$y_{\pm}(x) = \begin{cases} \operatorname{sgn}(H_{\pm})\{|H_{\pm}| - R_{\pm} + [R_{\pm}^2 - (x-0.5)^2]^{\frac{1}{2}}\} & \text{for} \quad H_{\pm} \neq 0 \\ 0 \quad \text{for} \quad H_{\pm} = 0 \end{cases} \tag{64}$$

where $0 \leqslant x \leqslant 1$, H is the y coordinate of a surface at midchord, $R = |H|^{-1}(H^2 + 0.25)/2$ is the radius of curvature of this surface, $\operatorname{sgn}(H) = \pm 1$ for $H \gtrless 0$, and the subscripts + and − refer to the upper and lower surfaces of the blade. Here we consider a cascade (with $\Theta = 45$ deg and $G = 1$) of 5% thick flat-bottomed DCA airfoils (i.e., $H_+ = 0.05$ and $H_- = 0$). Full potential steady and linearized unsteady flows through this example configuration have been determined for prescribed inlet Mach numbers ($M_{-\infty}$) of 0.5, 0.7, 0.8, and 0.9. In each case the requirement (62) has been imposed to eliminate steady loading at both blade leading and trailing edges. As a consequence, only the inlet Mach number is prescribed with the remaining inlet and exit parameters determined as part of the steady solution. Numerical results for this DCA configuration along with those for a corresponding flat-plate ($H_{\pm} = 0$, $\Omega_{-\infty} = 45$ deg) cascade are given in Figs. 19–25.

The predicted surface Mach number distributions for the example DCA cascade are shown in Fig. 19. For the prescribed inlet Mach numbers stated

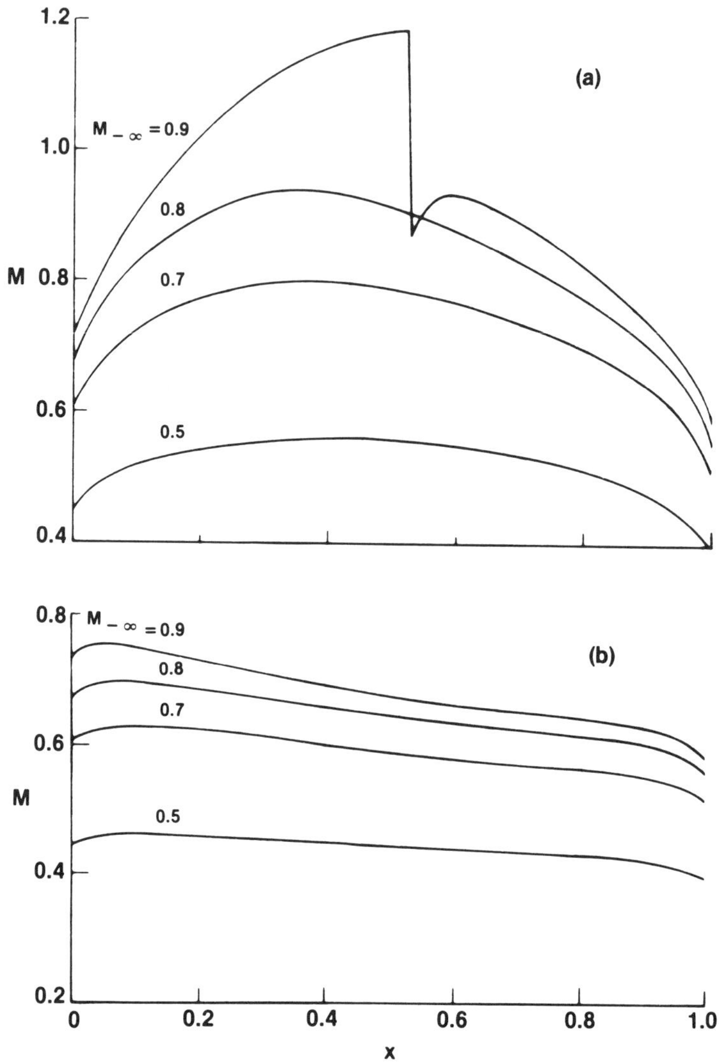

Fig. 19 Steady surface Mach number distributions for the example DCA cascade: a) suction surface; b) pressure surface.[42]

earlier, the calculated exit Mach numbers, M_{∞}, are 0.43, 0.57, 0.62, and 0.64, respectively. In addition, the calculated inlet flow angles are 49.0, 49.2, 49.4, and 49.6 deg, respectively, and in each case the calculated exit flow angle is approximately 43.0 deg. The steady flows at $M_{-\infty} = 0.5$, $M_{-\infty} = 0.7$, and $M_{-\infty} = 0.8$ are entirely subsonic with maximum suction surface Mach numbers of 0.561, 0.804, and 0.941 occurring at, respectively, 40.8, 38.5, and 36.5% of blade chord downstream from the leading edge. The steady flow is transonic for $M_{-\infty} = 0.9$ with a supersonic region extending from 18.5 to 52.5% of blade chord along the suction surface and terminating at a normal shock discontinuity. The Mach numbers at the foot of the shock are 1.193 on the upstream or supersonic side and 0.871 on the downstream or subsonic side.

First-harmonic pressure difference distributions and aerodynamic moments for the example DCA and flat-plate cascades operating at an inlet Mach number of 0.9 are presented in Figs. 20 and 21, respectively. Here the blades are undergoing out-of-phase ($\sigma = 180$ deg) torsional vibrations about their midchords at different prescribed frequencies. Recall that for a discontinuous transonic flow there are two contributions to the unsteady moment: one arising from the harmonic unsteady surface-pressure response and the other from the anharmonic surface pressures associated with shock motion. However, for the DCA cascade at $M_{-\infty} = 0.9$, the mean shock location is only slightly aft of blade midchord; therefore, the anharmonic surface pressures give only a small contribution to the unsteady moment. A comparison of the DCA and flat-plate results depicted in Figs. 20 and 21 indicates the dramatic impact of transonic mean flow phenomena on the unsteady aerodynamic response. A second interesting feature, revealed by these results, is the change in character of the pressure difference curves and in the magnitude of the unsteady moments as the blade vibration frequency is increased from 0.25 to 0.5. This change occurs because the out-of-phase blade motions of the DCA and flat-plate cascades are subresonant for $\omega = 0.1$ and 0.25 and super-resonant for $\omega = 0.5$, 0.75, and 1.0, and the resulting difference in the far-field acoustic wave behavior has an important impact on the aerodynamic response at the blade surface.

The effect of Mach number on the response to in-phase ($\sigma = 0$ deg) unit-frequency torsional vibrations is illustrated in Figs. 22 and 23. These motions are super-resonant; hence, acoustic waves persist in the far field and propagate away from the blade row. For the flows at $M_{-\infty} = 0.5$, 0.7, and 0.8, two such waves persist: one upstream and one downstream. For the DCA cascade operating at $M_{-\infty} = 0.9$, there are three propagating waves: two upstream and one downstream. Finally, for the flat-plate cascade operating at $M_{-\infty} = 0.9$, there are four such waves: two upstream and two downstream. The pressure difference distributions and moments reported for the DCA (Fig. 22) and flat-plate (Fig. 23) blades reflect this change in character of the unsteady flow in the far field, since the trends indicated by the results for $M_{-\infty} = 0.5$, 0.7, and 0.8 are not maintained at $M_{-\infty} = 0.9$. Also, a comparison of the DCA and flat-plate pressure difference curves for in-phase motions suggests that the influence of mean flow gradients on the unsteady response becomes more pronounced with increasing inlet Mach number. The pressure difference distributions for the DCA and flat-plate blades are very similar for the two lower inlet Mach numbers, differ somewhat for $M_{-\infty} = 0.8$, and differ substantially for $M_{-\infty} = 0.9$. The differences at $M_{-\infty} = 0.8$ can be attributed partially to the relatively large gradients in the subsonic mean flow that occur along the suction surface of each DCA blade, whereas the substantial differences at $M_{-\infty} = 0.9$ are caused by the transonic effects associated with the DCA cascade.

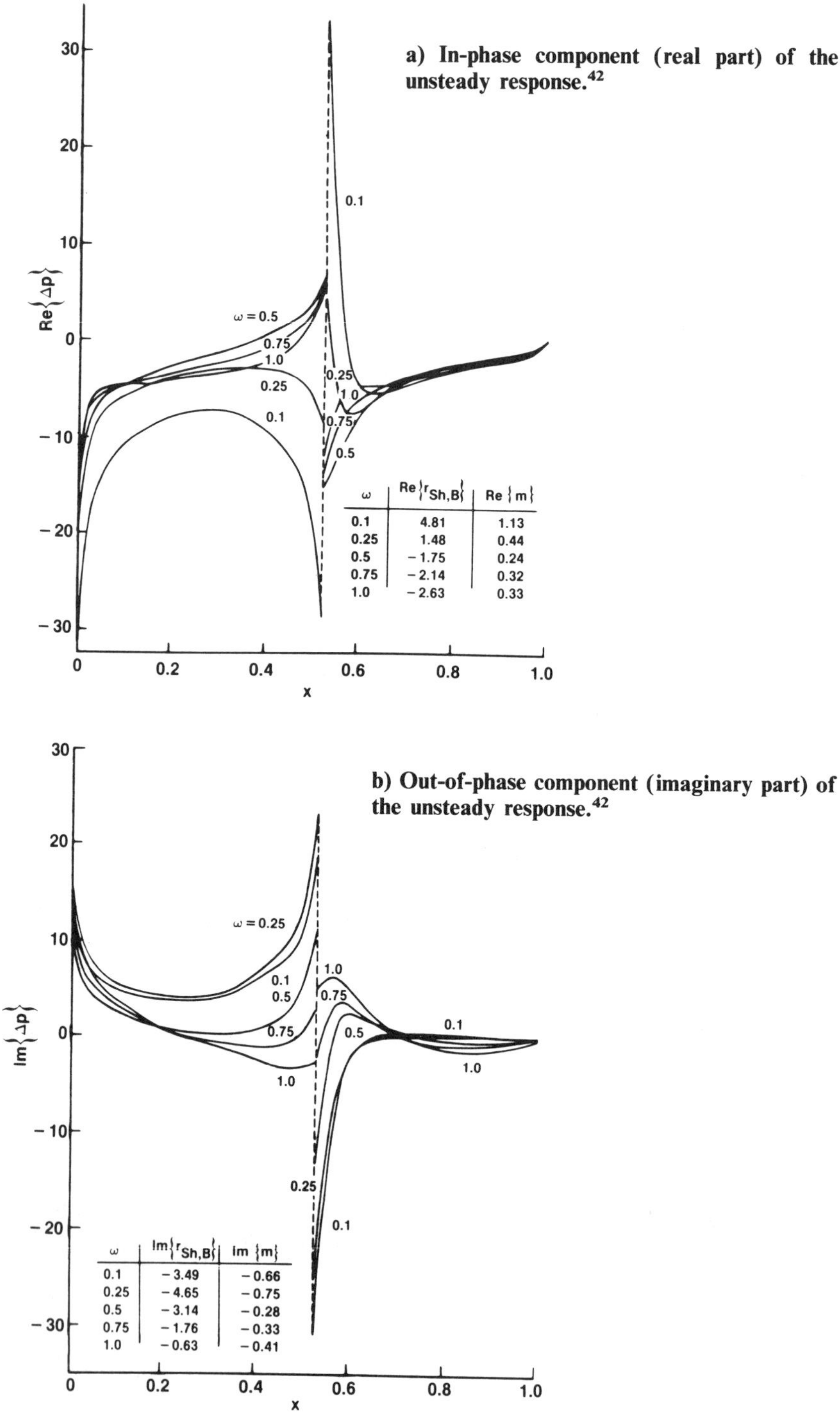

a) In-phase component (real part) of the unsteady response.[42]

ω	$Re\{r_{Sh,B}\}$	$Re\{m\}$
0.1	4.81	1.13
0.25	1.48	0.44
0.5	−1.75	0.24
0.75	−2.14	0.32
1.0	−2.63	0.33

b) Out-of-phase component (imaginary part) of the unsteady response.[42]

ω	$Im\{r_{Sh,B}\}$	$Im\{m\}$
0.1	−3.49	−0.66
0.25	−4.65	−0.75
0.5	−3.14	−0.28
0.75	−1.76	−0.33
1.0	−0.63	−0.41

Fig. 20 Effect of frequency on the unsteady response to the torsional blade vibrations of the DCA cascade; $M_{-\infty} = 0.9$, $\sigma = 180$ deg.

Predictions for the aerodynamic moment vs interblade phase angle for torsional motions of the flat-plate and DCA cascades are shown in Fig. 24 for $M_{-\infty} = 0.7$ and in Fig. 25 for $M_{-\infty} = 0.9$. The unit-frequency torsional motions of the two cascades are stable (i.e., $\text{Im}\{m\} < 0$) at both inlet numbers. Those angles at which acoustic resonances occur are indicated by the arrows at the top of each figure. At $M_{-\infty} = 0.7$ the moment curves for the two cascades are similar except at interblade phase angles lying in the approximate range 74.8 deg $< \sigma <$ 107.3 deg. At such angles the flat-plate ($M_\infty = 0.7$, $\Omega_\infty = 45$ deg) blade motions produce a propagating acoustic wave in the far-downstream region, whereas the far-downstream acoustic response disturbances produced by the DCA ($M_\infty = 0.57$, $\Omega = 43$ deg)

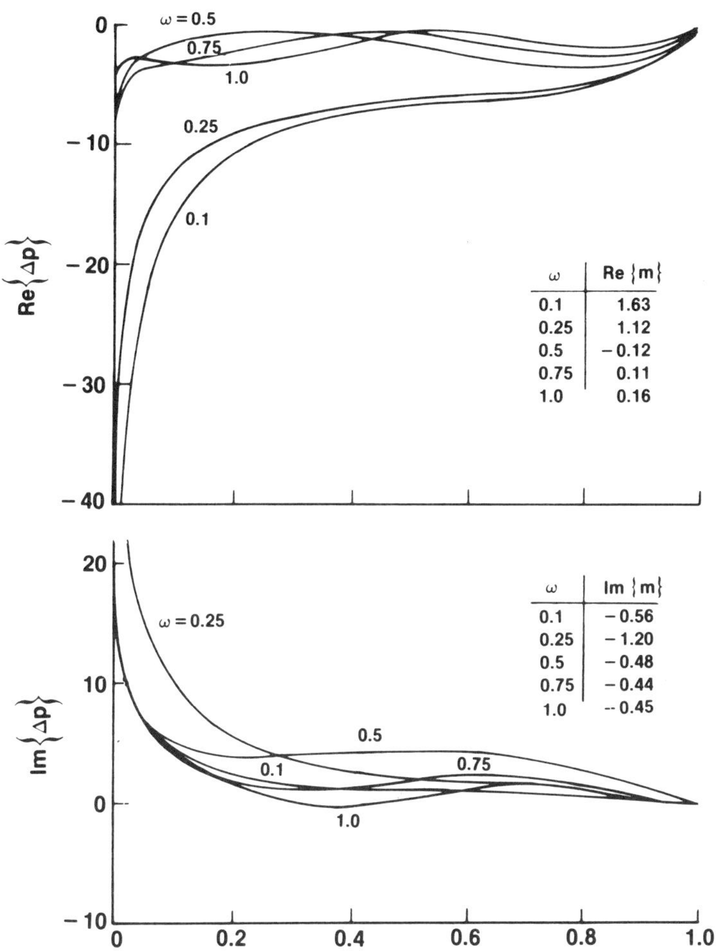

ω	Re{m}
0.1	1.63
0.25	1.12
0.5	−0.12
0.75	0.11
1.0	0.16

ω	Im{m}
0.1	−0.56
0.25	−1.20
0.5	−0.48
0.75	−0.44
1.0	−0.45

Fig. 21 Effect of frequency on the unsteady response to torsional blade vibrations of the example flat-plate cascade; $M_{-\infty} = 0.9$, $\sigma = 180$ deg.[42]

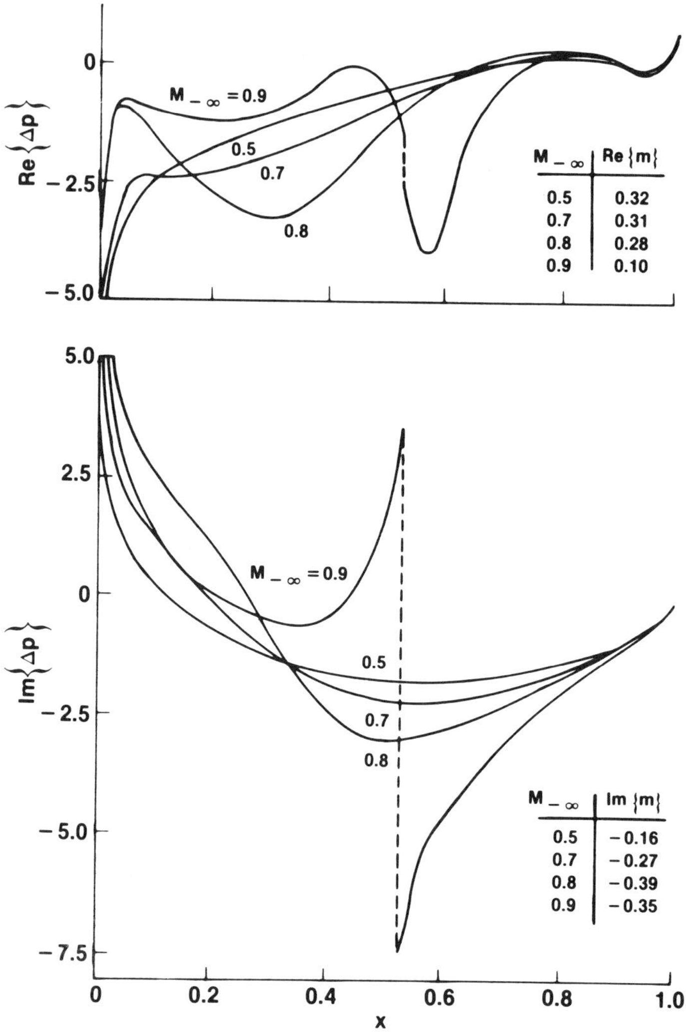

Fig. 22 Effect of Mach number on the unit-frequency response to in-phase ($\sigma = 0$ deg) torsional blade vibrations of the DCA cascade.[42]

blade motions all attenuate. At the higher inlet Mach number the trends exhibited by the flat-plate and DCA moment response curves are quite different over the entire range of interblade phase angles. These differences occur not only because the mean flow through the flat-plate cascade is purely subsonic, whereas that through the DCA cascade is transonic with a shock discontinuity, but also because of the substantial difference between the exit Mach numbers for the flat-plate ($M_\infty = 0.9$) and the DCA ($M_\infty = 0.64$) cascades. This difference implies that the two cascades operate in very different far-downstream acoustic response environments over almost the entire range of interblade phase angles.

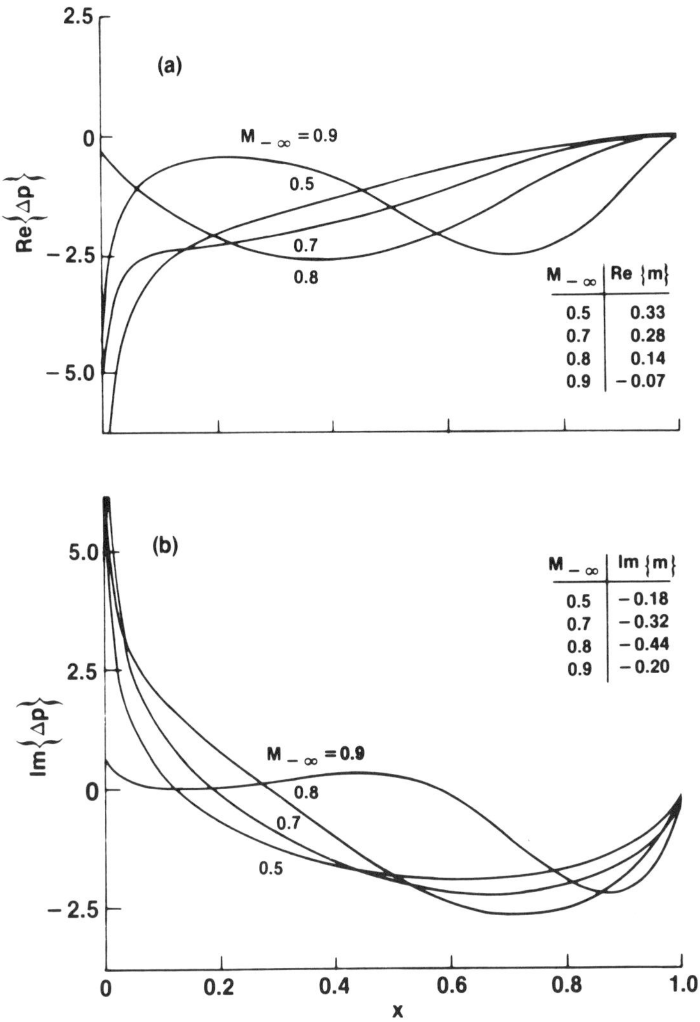

Fig. 23 Effect of Mach number on the unit-frequency response to in-phase torsional blade vibrations of the flat-plate cascade.[42]

Comparisons Between Theory and Experiment

Wind-tunnel experiments have been performed by Carta[58,59] at United Technologies Research Center (UTRC) and by Shaw et al.[60] at NASA Lewis Research Center to provide aerodynamic data in support of the theoretical analyses being developed for blade flutter prediction. The UTRC experiments were performed on a linear compressor-type cascade consisting of 11 NACA 65 series airfoils operating at low subsonic inlet Mach number and at two inlet flow angles. This cascade has a 55-deg stagger angle and a gap-to-chord ratio of 0.75; the blades were driven in

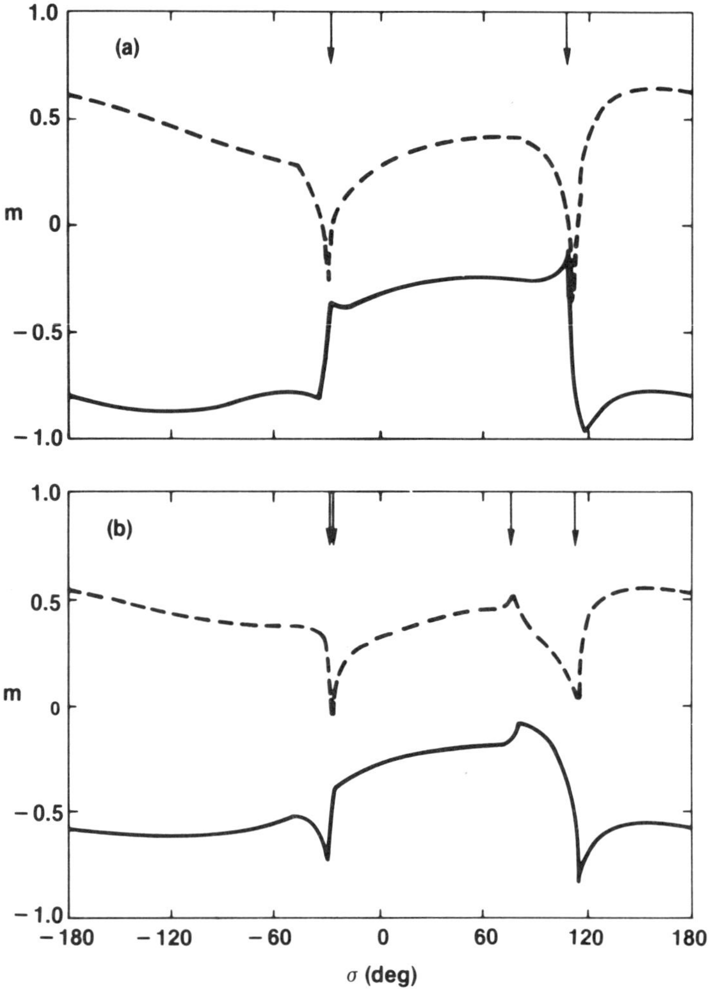

Fig. 24 Unsteady moments due to unit-frequency torsional blade vibrations at $M_{-\infty} = 0.7$: a) flat-plate cascade; b) DCA cascade. Dashed lines, in-phase component; solid lines, out-of-phase component.[10]

pitch (torsion) at amplitudes $|\alpha|$ of 0.5 and 2.0 deg around an axis near midchord, i.e., $X_P, Y_P = 0.5$, 0.115. Response measurements were taken at three frequencies ($\omega = 0.144$, 0.244, and 0.302) and at eight interblade phase angles ($\sigma = 0$, ± 45, ± 90, ± 135, and 180 deg). The resulting pressure data were reduced to Fourier coefficient form for direct comparison with theoretical predictions and processed to yield integrated loads.

Theoretical predictions for mean surface pressures, unsteady pressure differences at $\sigma = -45$ and 135 deg, unsteady surface pressures at $\sigma = -135$ deg, and unsteady moments are shown along with Carta's

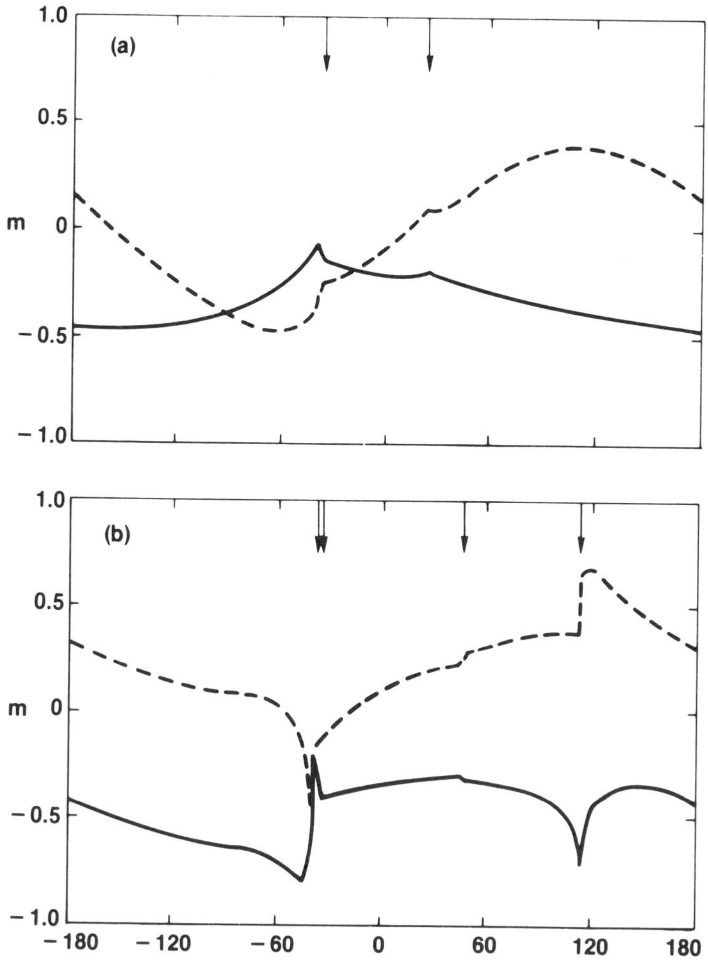

Fig. 25 Unsteady moments due to unit-frequency torsional blade vibrations at $M_{-\infty} = 0.9$: a) flat-plate cascade; b) DCA cascade. Dashed lines, in-phase component; solid lines, out-of-phase component.[10]

measurements in Figs. 26–29. These results were determined for an inlet Mach number of 0.17, the higher of the two tested mean inlet flow angles, and for torsional motions with $\alpha = 2$ deg and $\omega = 0.244$. The theoretical predictions for mean surface pressures have been found to match closely the measured pressures (see Fig. 26), but only if the inlet flow angle prescribed in the calculation differed from that reported in the experiment. Thus, the inlet flow angle used in the analysis is 62.2 deg, whereas that reported in the experiment is 66 deg. The predicted and measured unsteady response information shown in Figs. 27–29 is in very good qualitative

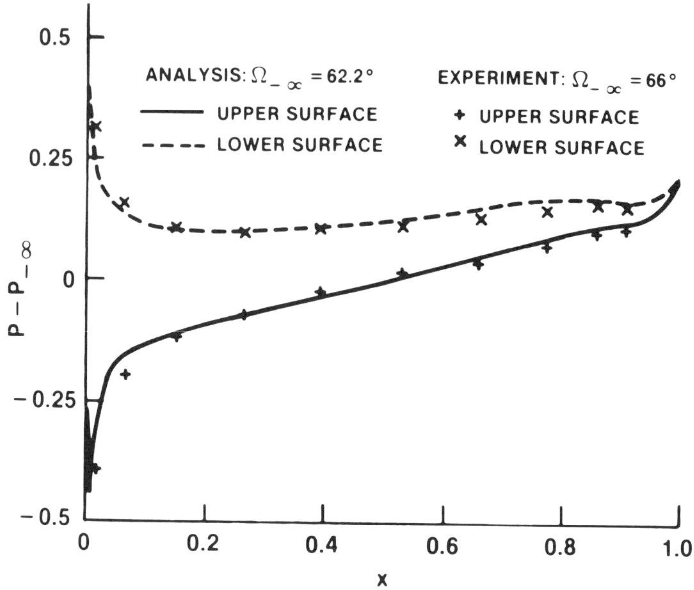

Fig. 26 Mean surface-pressure distributions for the UTRC cascade of NACA 65 series airfoils; $M_{-\infty} = 0.17$.

agreement, but quantitative differences exist. Additional analytic and experimental results for this UTRC cascade can be found in Refs. 58, 59, 61, and 62. It has been generally found that a good quantitative agreement between predictions and measurements can be obtained for the magnitudes of the various response quantities, but the predicted and measured phase angles agree only qualitatively. Also, unsteady response predictions based on a linearization relative to an appropriate nonuniform steady flow are in much better agreement with Carta's measurements than those based on the classical subsonic linearization.

The NASA Lewis experiments were performed on a linear fan-type cascade consisting of nine symmetric biconvex airfoils. These were driven simultaneously to provide a 1.2-deg pitching motion about midchord. This cascade has a stagger angle of 53 deg and a gap-to-chord ratio of 0.767. The airfoils have a maximum thickness-to-chord ratio of 0.076 at midchord and close in leading and trailing edges of small radius. Tests were performed at $M_{-\infty} = 0.65$, $\Omega_{-\infty} = 53$ deg, and at $M_{-\infty} = 0.8$, $\Omega_{-\infty} = 60$ deg. In the latter case the blades operate at a relatively high mean incidence and a shock occurs on the suction surface near the leading edge of each blade. The unsteady response measurements are given in Ref. 60 for torsional vibrations at frequencies of 200 and 500 Hz (i.e., at reduced frequencies of $\omega = 0.442$ and 1.106, respectively, for $M_{-\infty} = 0.65$, and $\omega = 0.366$ and 0.916 for $M_{-\infty} = 0.8$) and interblade phase angles of ± 90 deg.

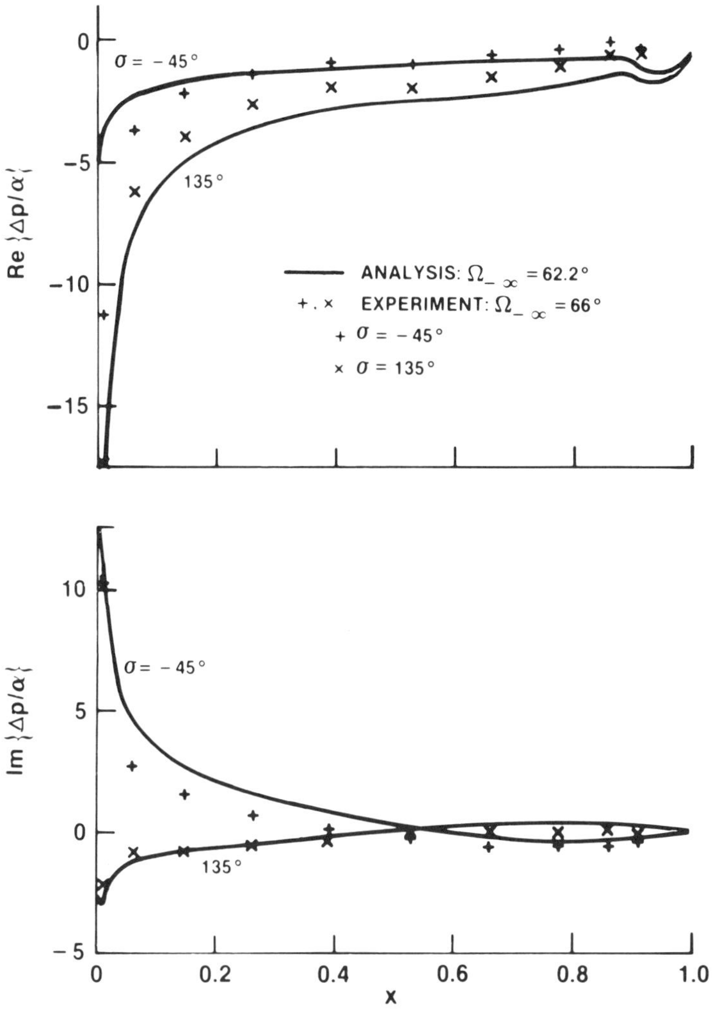

Fig. 27 Unsteady pressure difference distributions due to torsion for the UTRC cascade; $M_{-\infty} = 0.17$, $\alpha = 2$ deg, $\omega = 0.244$, $X_P, Y_P = 0.5$, 0.115.

Theoretical predictions for these cases are reported along with the foregoing measurements in Ref. 62. The blades considered in the theoretical study were extended versions of those used in the experiment. That is, biconvex blades having the same surface radius and the same thickness at midchord as the experimental profiles were studied, but the upper and lower surfaces were extended to close in wedge-shaped leading and trailing edges. As a result, the analytic blade profiles were 4% longer than those used in the experiment. This modification was introduced to simplify

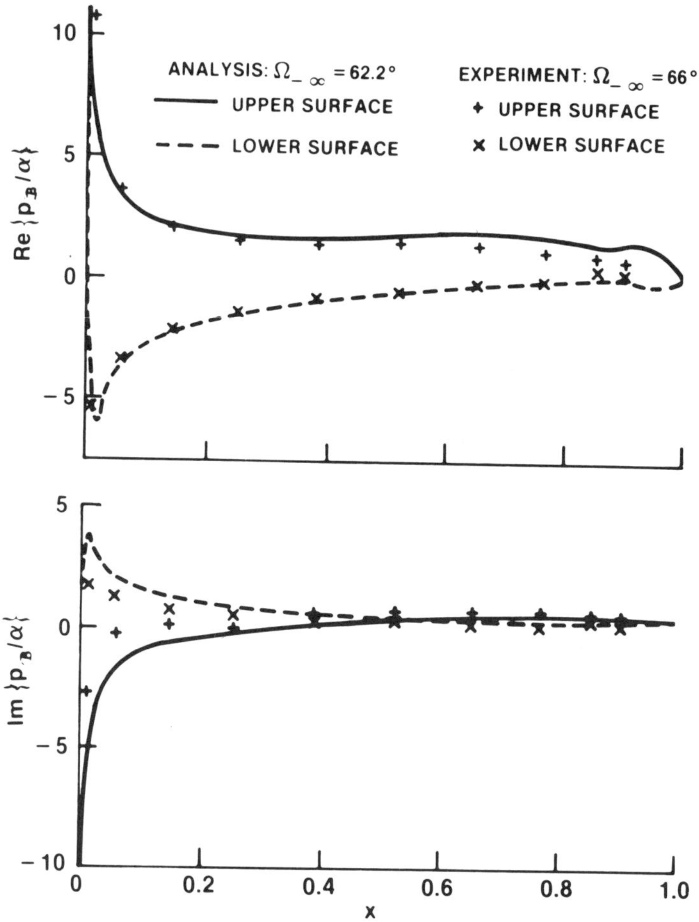

Fig. 28 Unsteady surface-pressure distributions due to torsion for the UTRC cascade; $M_{-\infty} = 0.17$, $\alpha = 2$ deg, $\omega = 0.244$, $X_P, Y_P = 0.5, 0.115$, $\sigma = -135$ deg.

analytic considerations at blade edges. The remaining experimental parameters were retained in the analytic study.

Analytic predictions in the vicinity of a sharp (or small radius) blade leading edge are not accurate because of difficulties in predicting the flow over such blades, especially when they operate at high mean incidence. Indeed, unless a relatively coarse computational grid is employed near the leading edge, iterative solution procedures for the nonlinear steady problem are found to diverge. Therefore at the present time it is not possible to predict effects associated with certain phenomena observed in the NASA experiment,[60] e.g., leading-edge lambda-type shocks and flow separations downstream of these shocks, using current codes.

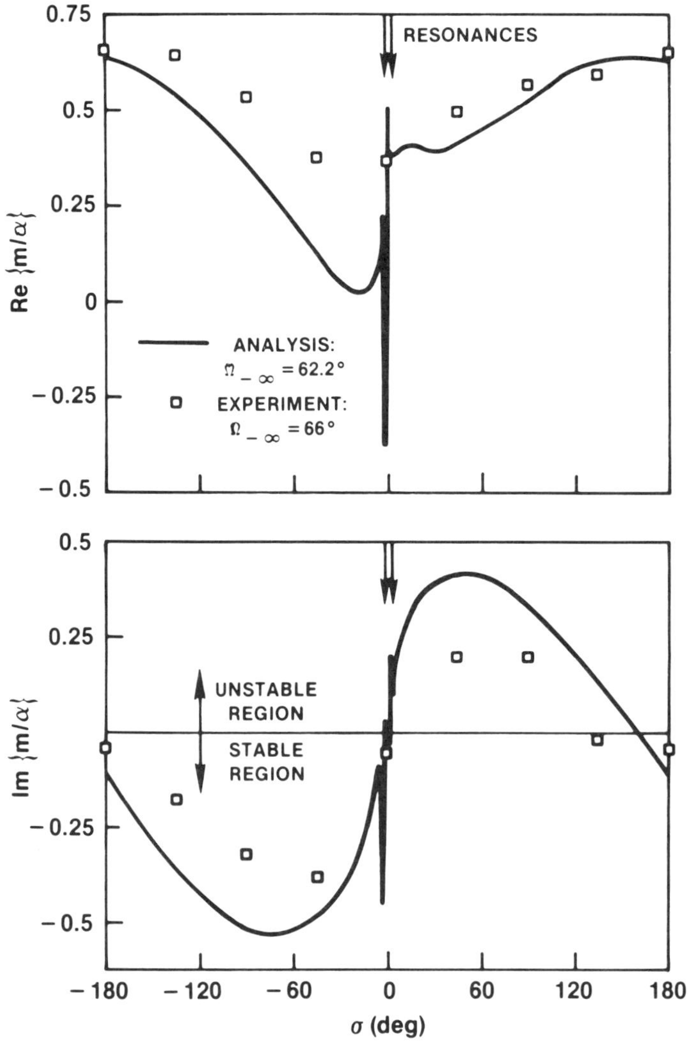

Fig. 29 Unsteady moments due to torsion about midchord for the UTRC cascade; $M_{-\infty} = 0.17$, $\alpha = 2$ deg, $\omega = 0.244$, $X_P, Y_P = 0.5$, 0.115.

In view of these limitations, the correlations between the predicted and measured surface Mach number distributions, unsteady pressure distributions for $\omega = 0.366$ and $\sigma = -90$ deg, and unsteady pressure distributions for $\omega = 0.916$ and $\sigma = -90$ deg indicated in Figs. 30, 31, and 32, respectively, are somewhat encouraging. However, large differences between the predictions and measurements occur over the forward part of each blade. Significant improvements in both steady and unsteady flow calculation procedures will be required to predict the flows through this NASA Lewis

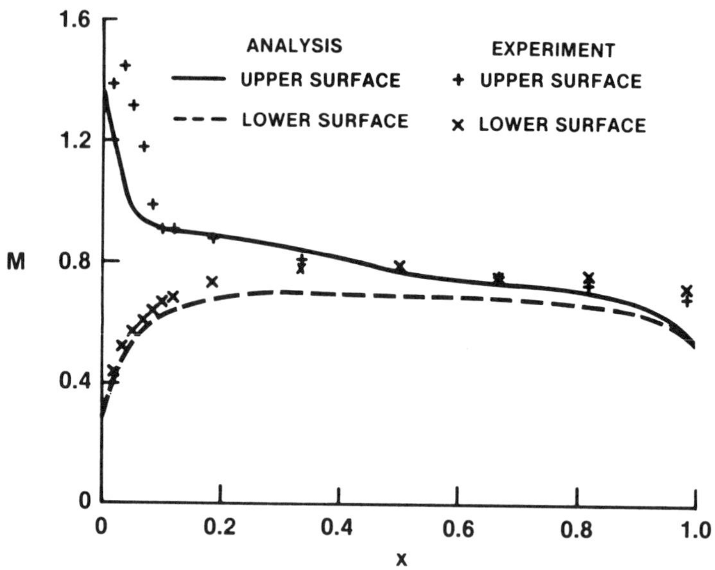

Fig. 30 Steady surface Mach number distributions for the NASA Lewis cascade; $M_{-\infty} = 0.8$, $\Omega_{-\infty} = 60$ deg.[62]

configuration or, in more general terms, the flows through cascades of sharp-edged fan blades operating at high mean incidence.

10. Concluding Remarks

In this chapter the aeroelastic characteristics and aerodynamic environments associated with the flutter and forced vibration of axial flow turbomachinery blading have been described. Also, the requirements placed on unsteady aerodynamic analyses intended for turbomachinery aeroelastic applications and the major assumptions usually invoked in the development of such analyses have been outlined. Because they occupy the central position in the subject area, the discussion herein has been limited to linearized unsteady aerodynamic theories for isolated two-dimensional blade rows and the response predictions based on such theories. In particular, the major emphasis has been placed on the description of an unsteady linearization relative to a fully nonuniform steady background flow. Analyses based on this type of linearization are under active development to meet the special requirements imposed by turbomachinery aeroelastic problems.

In the theory emphasized here the unsteady flow is regarded as a small perturbation from a fully nonuniform isentropic and irrotational mean or steady flow. This perturbation is caused by small-amplitude spatially and temporarily periodic structural (blade motions) and incident external aerodynamic (entropic, vortical, and acoustic) excitations. Thus, the steady flow is determined as a solution of a full potential boundary-value problem and

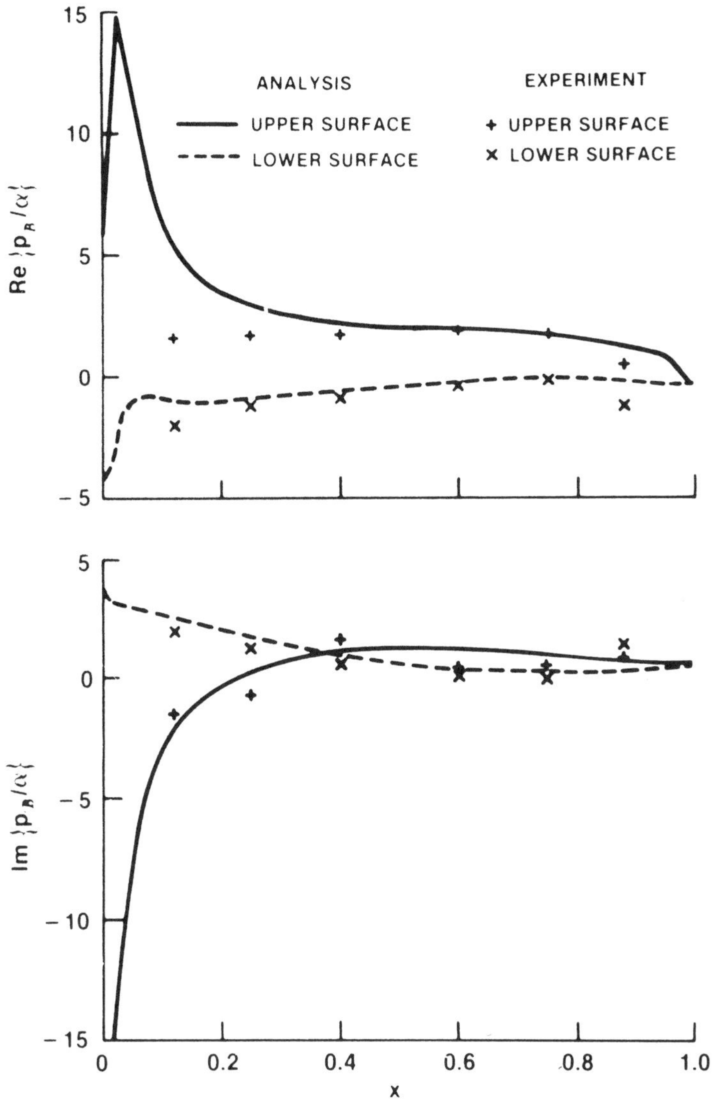

Fig. 31 Unsteady surface-pressure distributions due to torsion for the NASA Lewis cascade; $M_{-\infty} = 0.8$, $\Omega_{-\infty} = 60$ deg, $\alpha = 1.2$ deg, $\omega = 0.366$, $\sigma = -90$ deg.[62]

the linearized unsteady flow as a solution of a time-independent linear variable coefficient problem in which the variable coefficients depend on the underlying steady flow. The latter consists of a system of three field equations along with conditions imposed at the mean positions of blade, wake, and shock surfaces and far upstream and far downstream from the blade row. These equations can be solved sequentially to determine the

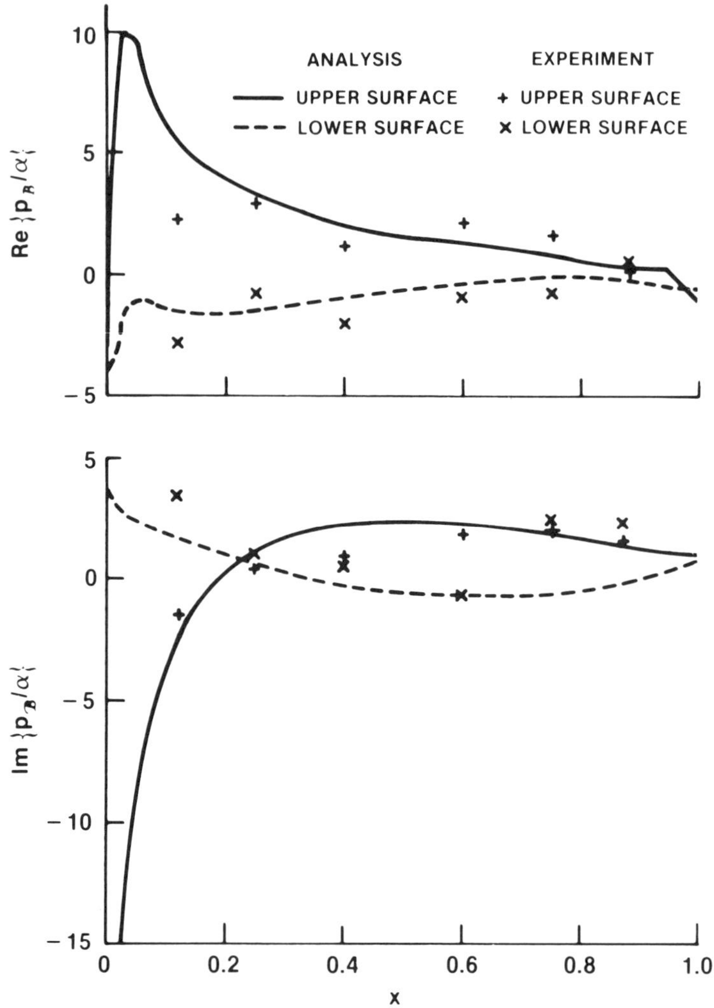

Fig. 32 Unsteady surface-pressure distributions due to torsion for the NASA Lewis cascade; $M_{-\infty} = 0.8$, $\Omega_{-\infty} = 60$ deg, $\alpha = 1.2$ deg, $\omega = 0.916$, $\sigma = -90$ deg.[62]

entropic, rotational velocity, and velocity potential fluctuations (hence, the fluctuations in all fluid dynamic properties) throughout the required solution domain, i.e., a single extended blade-passage region of finite extent. For flutter applications the entropy and rotational velocity fluctuations can be set to zero; therefore, only a single field equation must be solved to determine the unsteady flow. The equations of classical subsonic or supersonic linearized theory and those of high-frequency and time-linearized transonic flow theory can be recovered as special cases of the more general linearization.[10]

Example response predictions have been presented for the blades of fan- and compressor-type cascades undergoing pure torsional motions. We have considered flows in which the coupling between steady and unsteady disturbances could be neglected (i.e., flat-plate cascades with blade mean positions aligned with the inlet flow direction) and flows in which variations in the mean flow have an important impact on unsteady aerodynamic response. Semianalytic solutions, based on the classical linearization, for subsonic and supersonic flat-plate cascades demonstrate the influence of various global physical parameters (Mach number, frequency, etc.) on the aerodynamic response at a moving blade surface and the susceptibility of supersonic blading to torsional flutter. The results for the NACA 0012, DCA, NACA 65 series, and biconvex airfoil cascades operating at subsonic inlet Mach numbers illustrate, to some extent, the status of numerical field methods for solving the steady and the variable coefficient linearized unsteady boundary-value problems. Results for the NACA 0012 and DCA cascades demonstrate effects of blade geometry, mean incidence, and transonic phenomena on the unsteady response to prescribed torsional blade motions, whereas those for the NACA 65 series and biconvex airfoil cascades are indicative of the correlations that have been obtained between theoretical predictions and experimental measurements.

Linearizations relative to nonuniform steady flows offer great potential for meeting the needs of turbomachinery aeroelastic designers for efficient unsteady aerodynamic analyses that contain the essential physics associated with aerodynamically induced blade motions. However, before this potential can be fully realized, significant improvements in numerical solution methods for both nonlinear steady and linearized unsteady flows must be achieved so that reliable unsteady aerodynamic response information can be provided for the wide range of geometric configurations and flow conditions at which blade vibrations are of practical concern. Some needed capabilities include the ability to predict transonic flows (i.e., subsonic flows with embedded supersonic regions) through fan and compressor cascades operating at high positive or negative incidence, supersonic flows with complex moving shock patterns, and the high-frequency unsteady flows associated with incident entropic, vortical, and acoustic disturbances.

Because of the computational requirements associated with time-accurate resolutions of the nonlinear flow equations and the demands for economical aeroelastic predictions, there will be an ongoing need to develop more complete linearized unsteady aerodynamic formulations and solution methods. Some issues that require further study and clarification include: the representation of strong shock phenomena in a potential-based mean flow approximation,[63,64] the prediction of oblique shock motions, the inclusion of closed-form expressions for the entropy and rotational velocity fluctuations[48,49] in the linearized unsteady formulation, and the determination of useful analytic representations for the far-field unsteady potential at transonic ($M \sim 1$) and supersonic freestream Mach numbers. Major advances in our ability to predict turbomachinery aeroelastic behavior should result if future research is directed toward determining the effects of viscid/inviscid interactions and possibly large-scale flow separations on linearized inviscid unsteady aerodynamic behavior. Ultimately, linearized unsteady aerodynamic analyses that account for nonuniform mean flow and viscid/

inviscid interaction phenomena should be extended for application to three-dimensional aeroelastic problems.

Acknowledgments

The preparation of this article was supported under the United Technologies Corporate Research Program. The theoretical analyses and numerical solutions reported herein were determined to a great extent under research contracts with NASA Lewis Research Center and the Naval Air Systems Command. The continued interest in this work and the assistance provided by J. J. Adamczyk, D. R. Boldman, and D. Hoyniak (NASA Lewis) and M. F. Platzer, R. P. Shreeve, and G. Derderian (NAVAIR) are gratefully acknowledged.

References

[1]Mikolajczak, A. A., Arnoldi, R. A., Snyder, L. E., and Stargardter, H., "Advances in Fan and Compressor Blade Flutter Analysis and Predictions," *Journal of Aircraft*, Vol. 12, April 1975, pp. 325–332.

[2]Platzer, M. F., "Unsteady Flow in Turbomachines—A Review of Current Developments," *Unsteady Aerodynamics*, AGARD CP-227, 1977.

[3]Carta, F. O., "Aeroelasticity Research and Unsteady Aerodynamics," *The Aerothermodynamics of Aircraft Gas Turbine Engines*, edited by G. C. Oates, U.S. Air Force Rept. AFAPL-TR-78-52, Air Force Aero Propulsion Lab., Wright-Patterson AFB, OH, July 1978, Chap. 22.

[4]Sisto, F., "Aeroelasticity in Turbomachines," *A Modern Course in Aeroelasticity*, edited by E. H. Dowell, Sijthoff & Noordhoff, Alphen aan der Rijn, The Netherlands, 1978, pp. 383–409.

[5]Fleeter, S., "Aeroelasticity Research for Turbomachine Applications," *Journal of Aircraft*, Vol. 16, May 1979, pp. 320–326.

[6]Fleeter, S. and Jay, R. L., "Unsteady Aerodynamic Measurements in Flutter Research," *AGARD Manual on Aeroelasticity in Axial-Flow Turbomachines, Unsteady Turbomachinery Aerodynamics*, Vol. 1, edited by M. F. Platzer and F. O. Carta, AGARD-AG-298, March 1987, Chap. VIII.

[7]Gostelow, J. P., *Cascade Aerodynamics*, Pergamon, Oxford, UK, 1984.

[8]Snyder, L. E. and Burns, D. W., "Forced Vibration and Flutter Design Methodology," *AGARD Manual on Aeroelasticity in Axial-Flow Turbomachines, Structural Dynamics and Aeroelasticity*, Vol. 2, edited by M. F. Platzer and F. O. Carta, AGARD-AG-298, June 1988, Chap. XXII.

[9]Whitehead, D. S., "Unsteady Aerodynamics in Turbomachinery," *Special Course on Unsteady Aerodynamics*, AGARD Rept. 679, 1980, Chap. 6.

[10]Verdon, J. M., "Linearized Unsteady Aerodynamic Theory," *AGARD Manual on Aeroelasticity in Axial-Flow Turbomachines, Unsteady Turbomachinery Aerodynamics*, Vol. 1, edited by M. F. Platzer and F. O. Carta, AGARD-AG-298, March 1987, Chap. II.

[11]Whitehead, D. S., "Classical Two-Dimensional Methods," *AGARD Manual on Aeroelasticity in Axial-Flow Turbomachines, Unsteady Turbomachinery Aerodynamics*, Vol. 1, edited by M. F. Platzer and F. O. Carta, AGARD-AG-298, March 1987, Chap. III.

[12]Whitehead, D. S., "Vibration and Sound Generation in a Cascade of Flat Plates in Subsonic Flow," Cambridge Univ. Engineering Dept. Rept. CUED/A-Turbo/TR 15, Cambridge, UK, 1970.

[13]Kaji, S. and Okazaki, T., "Propagation of Sound Waves Through a Blade Row. II. Analysis Based on the Acceleration Potential Method," *Journal of Sound and Vibration*, Vol. 11, March 1970, pp. 355–375.

[14]Smith, S. N., "Discrete Frequency Sound Generation in Axial Flow Turbomachines," British Aeronautical Research Council, R&M 3709, London, UK, 1971.

[15]Verdon, J. M., "Further Developments in the Aerodynamic Analysis of Unsteady Supersonic Cascades. I. The Unsteady Pressure Field. II. Aerodynamic Response Predictions," *Transactions of American Society of Mechanical Engineers, A: Journal of Engineering for Power*, Vol. 99, No. 4, Oct. 1977, pp. 509–525.

[16]Nagashima, T. and Whitehead, D. S., "Linearized Supersonic Unsteady Flow in Cascades," British Aeronautical Research Council, R&M 3811, London, UK, 1978.

[17]Adamczyk, J. J. and Goldstein, M. E., "Unsteady Flow in a Supersonic Cascade with Subsonic Leading-Edge Locus," *AIAA Journal*, Vol. 16, Dec. 1978, pp. 1248–1254.

[18]Ni, R. H., "A Rational Analysis of Periodic Flow Perturbation in Supersonic Two-Dimensional Cascade," *Transactions of American Society of Mechanical Engineers, A: Journal of Engineering for Power*, Vol. 101, No. 3, July 1979, pp. 431–439.

[19]Bendiksen, O. and Friedmann, P., "Coupled Bending-Torsion Flutter in a Supersonic Cascade," *AIAA Journal*, Vol. 19, June 1981, pp. 774–781.

[20]Perumal, P. V. K., "Thin Airfoil in Eddy Array and Part-Stalled Oscillating Cascade," Ph.D. Dissertation, Stevens Institute of Technology, Hoboken, NJ, 1976.

[21]Chi, M. R., "Unsteady Aerodynamics in Stalled Cascade and Stall Flutter Prediction," American Society of Mechanical Engineers, ASME Paper 80-C2/Aero-1, Aug. 1980.

[22]Goldstein, M. E., Braun, W., and Adamczyk, J. J., "Unsteady Flow in a Supersonic Cascade with Strong In-Passage Shocks," *Journal of Fluid Mechanics*, Vol. 83, Pt. 3, Dec. 1977, pp. 569–604.

[23]Surampudi, S. P. and Adamczyk, J. J., "Unsteady Transonic Flow over Cascade Blades," *AIAA Journal*, Vol. 24, Feb. 1986, pp. 293–302.

[24]Namba, M., "Three Dimensional Flows," *AGARD Manual on Aeroelasticity in Axial-Flow Turbomachines, Unsteady Turbomachinery Aerodynamics*, Vol. 1, edited by M. F. Platzer and F. O. Carta, AGARD-AG-298, March 1987, Chap. IV.

[25]Salaun, P., "Three Dimensional Flow," *AGARD Manual on Aeroelasticity in Axial-Flow Turbomachines, Unsteady Turbomachinery Aerodynamics*, Vol. 1, edited by M. F. Platzer and F. O. Carta, AGARD-AG-298, March 1987, Chap. V.

[26]Tijdeman, H. and Seebass, R., "Transonic Flow Past Oscillating Airfoils," *Annual Review of Fluid Mechanics*, Vol. 12, 1980, pp. 181–222.

[27]Fung, K.-Y., Yu, N. J., and Seebass, R., "Small Unsteady Perturbations in Transonic Flows," *AIAA Journal*, Vol. 16, Aug. 1978, pp. 815–822.

[28]Ehlers, F. E. and Weatherill, W. H., "A Harmonic Analysis Method for Unsteady Transonic Flow and Its Application to the Flutter of Airfoils," NASA CR-3537, May 1982.

[29]Kerlick, G. D. and Nixon, D., "A High-Frequency Transonic Small Disturbance Code for Unsteady Flows in a Cascade," AIAA Paper 82-0955, June 1982.

[30]Vogeler, K., "The Unsteady Pressure Distribution on Parabolic Profiles in a Finite Cascade Oscillating in Supersonic Flow," *Symposium Proceedings: Unsteady Aerodynamics of Turbomachines and Propellers*, Cambridge Univ. Engineering Dept., Cambridge, UK, Sept. 1984, pp. 171–190.

[31]He, L. and Zhou, S., "Numerical Prediction of Transonic Stall Flutter in Axial Compressors," *Symposium Proceedings: Unsteady Aerodynamics of Turbomachines and Propellers*, Cambridge Univ. Engineering Dept., Cambridge, UK, Sept. 1984, pp. 119–128.

[32]Verdon, J. M., Adamczyk, J. J., and Caspar, J. R., "Subsonic Flow Past an Oscillating Cascade with Steady Blade Loading—Basic Formulation," *Unsteady Aerodynamics*, Vol. 2, edited by R. B. Kinney, 1975, pp. 827–851.

[33]Atassi, H. and Akai, T. J., "Effect of Blade Loading and Thickness on the Aerodynamics of Oscillating Cascades," AIAA Paper 78-227, Jan. 1978.

[34]Atassi, H. and Akai, T. J., "Aerodynamic and Aeroelastic Characteristics of Oscillating Loaded Cascades at Low Mach Number. I. Pressure Distribution, Forces and Moments," *Transactions of American Society of Mechanical Engineers, A: Journal of Engineering for Power*, Vol. 102, No. 2, April 1980, pp. 344–351.

[35]Verdon, J. M. and Caspar, J. R., "Subsonic Flow Past an Oscillating Cascade with Finite Mean Flow Deflection," *AIAA Journal*, Vol. 18, May 1980, pp. 540–548.

[36]Carstens, V., "Unsteady Pressure Distributions in a Cascade Taking into Account Profile Thickness and Steady Circulation," *Proceedings of the 2nd International Symposium on Aeroelasticity in Turbomachines*, edited by P. Suter, Juris-Verlag, Zurich, Switzerland, 1981, pp. 65–83.

[37]Caruthers, J. E., "Aerodynamic Analysis of Cascaded Airfoils in Unsteady Rotational Flow," *Proceedings of the 2nd International Symposium on Aeroelasticity in Turbomachines*, edited by P. Suter, Juris-Verlag, Zurich, Switzerland, 1981, pp. 31–64.

[38]Whitehead, D. S. and Grant, R. J., "Force and Moment Coefficients for High Deflection Cascades," *Proceedings of the 2nd International Symposium on Aeroelasticity in Turbomachines*, edited by P. Suter, Juris-Verlag, Zurich, Switzerland, 1981, pp. 85–127.

[39]Caspar, J. R. and Verdon, J. M., "Numerical Treatment of Unsteady Subsonic Flow Past an Oscillating Cascade," *AIAA Journal*, Vol. 19, Dec. 1981, pp. 1531–1539.

[40]Whitehead, D. S., "The Calculation of Steady and Unsteady Transonic Flow in Cascades," Cambridge Univ. Engineering Dept. Rept. CUED/A-Turbo/TR 118, Cambridge, UK, 1982.

[41]Verdon, J. M. and Caspar, J. R., "Development of a Linear Unsteady Aerodynamic Analysis for Finite-Deflection Subsonic Cascades," *AIAA Journal*, Vol. 20, Sept. 1982, pp. 1259–1267.

[42]Verdon, J. M. and Caspar, J. R., "A Linearized Unsteady Aerodynamic Analysis for Transonic Cascades," *Journal of Fluid Mechanics*, Vol. 149, Dec. 1984, pp. 403–429.

[43]Whitehead, D. S. and Newton, S. G., "A Finite Element Method for the Solution of Two-Dimensional Transonic Flows in Cascades," *International Journal for Numerical Methods in Fluids*, Vol. 5, Feb. 1985, pp. 115–132.

[44]Caspar, J. R., "Unconditionally Stable Calculation of Transonic Potential Flow Through Cascades Using an Adaptive Mesh for Shock Capture," *Transactions of American Society of Mechanical Engineers, A: Journal of Engineering for Power*, Vol. 105, No. 3, July 1983, pp. 504–513.

[45]Habashi, W. G., Hafez, M. M., and Kotiuga, P. L., "Computation of Choked and Supersonic Turbomachinery Flows by a Modified Potential Method," *AIAA Journal*, Vol. 23, Feb. 1985, pp. 214–220.

[46]Acton, E. and Newton, S. G., "Numerical Methods for Unsteady Transonic Flow," *AGARD Manual on Aeroelasticity in Axial-Flow Turbomachines, Unsteady Turbomachinery Aerodynamics*, Vol. 1, edited by M. F. Platzer and F. O. Carta, AGARD-AG-298, March 1987, Chap. VI.

[47]Hunt, J. C. R., "A Theory of Turbulent Flow Round Two-Dimensional Bluff Bodies," *Journal of Fluid Mechanics*, Vol. 61, Dec. 1973, pp. 625–706.

[48]Goldstein, M. E., "Unsteady Vortical and Entropic Distortions of Potential Flows Round Arbitrary Obstacles," *Journal of Fluid Mechanics*, Vol. 89, Pt. 3, Dec. 1978, pp. 433–468.

[49]Goldstein, M. E., "Turbulence Generated by the Interaction of Entropy Fluctuations with Non-uniform Mean Flows," *Journal of Fluid Mechanics*, Vol. 93, Pt. 2, July 1979, pp. 209–224.

[50]Aris, A., *Vectors, Tensors and the Basic Equations of Fluid Mechanics*, Prentice-Hall, Englewood Cliffs, NJ, 1962.

[51]Hounjet, M. H. L., "Transonic Panel Method to Determine Loads on Oscillating Airfoils with Shocks," *AIAA Journal*, Vol. 19, May 1981, pp. 559–566.

[52]Williams, M. H., "Linearization of Unsteady Transonic Flows Containing Shocks," *AIAA Journal*, Vol. 17, April 1979, pp. 394–397.

[53]Tijdeman, H., "Investigations of the Transonic Flow Around Oscillating Airfoils," National Aerospace Laboratory (Amsterdam) Rept. NLR TR-77090-U, Dec. 1977.

[54]Fung, Y. C., *An Introduction to the Theory of Aeroelasticity*, Wiley, New York, 1955, pp. 166–168.

[55]Miles, J. W., *The Potential Theory of Unsteady Supersonic Flow*, Cambridge Univ. Press, Cambridge, UK, 1959.

[56]Snyder, L. E. and Commerford, G. L., "Supersonic Unstalled Flutter in Fan Rotors: Analytical and Experimental Results," *Transactions of American Society of Mechanical Engineers, A: Journal of Engineering for Power*, Vol. 96, No. 4, Oct. 1974, pp. 379–386.

[57]Abbott, I. H. and Von Doenhoff, A. E., *Theory of Wing Sections*, Dover, New York, 1959.

[58]Carta, F. O., "An Experimental Investigation of Gapwise Periodicity and Unsteady Aerodynamic Response in an Oscillating Cascade," NASA CR-3513, prepared under Contract NAS3-22018 for NASA Lewis Research Center, June 1982.

[59]Carta, F. O., "Unsteady Aerodynamics and Gapwise Periodicity of Oscillating Cascaded Airfoils," *Transactions of American Society of Mechanical Engineers, A: Journal of Engineering for Power*, Vol. 105, No. 3, July 1983, pp. 565–574.

[60]Shaw, L. M., Boldman, D. R., Buggele, A. E., and Buffum, D. J., "Unsteady Pressure Measurements on a Biconvex Airfoil in a Transonic Oscillating Cascade," Transactions of American Society of Mechanical Engineers, A: Journal of Engineering for Gas Turbines and Power, Vol. 108, Jan. 1986, pp. 53–59.

[61]Fransson, T. H., "Workshop on Aeroelasticity in Turbomachine-Cascades," *Symposium Proceedings: Unsteady Aerodynamics of Turbomachines and Propellers*, Cambridge Univ. Engineering Dept., Cambridge, UK, Sept. 1984, pp. 361–395.

[62]Verdon, J. M. and Usab, W. J., Jr., "Application of a Linearized Unsteady Aerodynamic Analysis to Standard Cascade Configurations," NASA CR-3940, prepared under Contract NAS3-22257 for NASA Lewis Research Center, Nov. 1985.

[63]Hafez, M. M. and Lovell, D., "Entropy and Vorticity Corrections for Transonic Flows," AIAA Paper 83-1926, July 1983.

[64]Klopfer, G. H. and Nixon, D., "Nonisentropic Potential Formulation for Transonic Flows," *AIAA Journal*, Vol. 22, June 1984, pp. 770–776.

Chapter 8

Alternative Methods for Modeling Unsteady Transonic Flows

David Nixon*
Nielsen Engineering & Research, Inc., Mountain View, California

1. Introduction

The most common methods of predicting unsteady transonic flow are those based on finite-difference techniques. These methods require a reasonably large computer, and before such computers became available considerable effort was put into developing simpler approaches to the problem. Some of the earlier methods are described briefly in this chapter, in case some of the knowledge gained in the development of these techniques can be used profitably in the future. For a more complete description the reader should consult the references.

The transonic small-disturbance (TSD) equation or the full potential equation can model only certain transonic flow phenomena, although it is possible to upgrade this model by empirical or semiempirical additions. Some of these techniques are described in this chapter. Since the TSD equation can be relatively expensive to compute, a final topic addressed in this chapter is the use of the transonic "indicial" method, which can considerably reduce the amount of computation and thus reduce the cost required for practical calculations.

2. Basic Equations

Most of the analysis in this chapter is based on the TSD equation derived in Chapter 2, namely,

$$(1 - M_\infty^2)\phi_{xx} + \phi_{yy} + \phi_{zz} - \frac{2M_\infty^2 c}{U_\infty}\phi_{xt} - \frac{M_\infty^2 c^2}{U_\infty^2}\phi_{tt} = k\phi_x\phi_{xx} \tag{1}$$

where k is a transonic parameter given by Eq. (37) in Chapter 2. The boundary conditions are that the flow is tangent to the body, the wake cannot support a load, and only outgoing waves are permitted in the far field.

*President.

3. Integral Equation Methods for Predicting Unsteady Transonic Flows

In the subsonic and supersonic flow regimes, integral equation methods are commonly used to predict the flow characteristics. This is because linear differential equations can adequately represent such flows, and such equations can be transformed easily into a boundary integral equation. At transonic speeds, the flow cannot be represented by a linear equation, and the classical integral equation now involves the flow in the interior of the flow domain. However, if the time-dependent component of the flow can be regarded as a small perturbation on a mean flow, then it seems possible to use a straightforward perturbation analysis to derive a linear governing equation for the unsteady component with perhaps some simplification of the transonic integral equation. If the velocity potential $\phi(x,y,z,t)$ is split into components such as

$$\phi(x,y,z,t) = \phi_0(x,y,z) + \varepsilon\phi(x,y,z,t) + - - \tag{2}$$

then a perturbation analysis combining Eqs. (1) and (2) gives the following equation for ϕ_1:

$$(1 - M_\infty^2)\phi_{1_{xx}} + \phi_{1_{yy}} + \phi_{1_{zz}} - \frac{2M_\infty^2 c}{U_\infty}\phi_{1_{xt}} - \frac{M_\infty^2 c^2}{U_\infty^2}\phi_{1_{tt}} = k(\phi_{0_x}\phi_{1_x})_x \tag{3}$$

This is a linear equation with variable coefficients that depend on the initial steady flow ϕ_0. A two-dimensional form of this equation has been solved by Nixon[1,2] for shock-free flows. Although solutions of Eq. (3) have been obtained by finite-difference techniques by Traci et al.[3] and Weatherhill et al.[4] for flows with shock waves there are some points that should be considered in such a formulation.

If the initial steady flow solution ϕ_0 has a shock wave at a location x_{s_0}, then a solution of Eq. (3) will also have a shock wave at x_{s_0}, since by definition ϕ_{1_x} is small and cannot annihilate a jump in ϕ_{0_x}. If, in a physical unsteady motion, the shock wave moves, then Eq. (3) cannot represent the shock motion. Also, it is possible that the physical ϕ_{1_x} may have more than one discontinuity; hence, the numerical solution could give one shock wave for a physical flow that has two. In a finite-difference representation of Eq. (3) it is possible to remove this discrepancy at the shock if a "shock-fitting" algorithm is used. Such a technique has been developed by Fung et al.[5] An integral equation technique for two-dimensional harmonically oscillating flows has been developed by Nixon[6] which treats the shock motion by means of a strained coordinate analysis. This method is described briefly below.

As noted earlier, one of the main difficulties associated with unsteady transonic flows when shock waves are present is to treat the shock motion correctly. It is noted in Chapter 2 that this problem is easily overcome by the use of a strained coordinate system in which the shock location is invariant. A more complete description of the method is found in Ref. 7. The main restriction is that no loss or generation of shock waves may occur during the perturbation. The theory is simplified if it is assumed that the shock wave is normal to the freestream, since in this case only the streamwise coordinate need be strained. For simplicity in presentation it is assumed that there is only one shock wave on each surface of the airfoil. Following the basic ideas of Ref. 7, x is expanded as a series in time. Thus,

$$x = x' + x_1(x', \pm 0)\, e^{i\omega t} + \cdots \tag{4}$$

where x' is the strained coordinate and from Ref. 6

$$x_1(x', \pm 0) = \delta x_s(\pm 0) x_1(x', \pm 0)$$

$$x_1(x', \pm 0) = \frac{x'(1 - x')}{x_s'(\pm 0)[1 - x_s'(\pm 0)]}, \qquad 0 \leqslant x' \leqslant 1$$

$$x_1(x', \pm 0) = 0; \qquad x' < 0, \qquad x' > 1 \tag{5}$$

where $x_1(x', \pm 0)$ is a specified function for the upper and lower surfaces, respectively, and $x_s'(\pm 0)$ and $\delta x_s(\pm 0)$ denote the steady-state shock location and the magnitude of the shock oscillation in the upper and lower half planes, respectively.

If $\beta = (1 - M_\infty^2)^{\frac{1}{2}}$ and $\bar{y} = \beta y$, then $\phi(x,y,t)$ is expanded in the series

$$\phi(x,y,t) = \beta^2/k\{\bar{\phi}_0(x',\bar{y}) + \bar{\phi}_1(x',\bar{y})\, e^{i\omega t}\, e^{iM_\infty^2 \Omega x'} + \cdots\} \tag{6}$$

where

$$\Omega = (\omega c)/(U_\infty \beta^2) = \nu/\beta^2 \tag{7}$$

and $\nu = \omega c/U_\infty$ is the reduced frequency. The terms $\bar{\phi}_0$ and $\bar{\phi}_1$ are given by

$$\bar{\phi}_0(x',\bar{y}) = k/\beta^2 \phi_0(x',y) \tag{8a}$$

$$\bar{\phi}_1(x',\bar{y}) = k/\beta^2 \phi_1(x',y) \tag{8b}$$

Substitution of the series Eqs. (4) and (6) into the two-dimensional version of Eq. (1) leads to the linear equation

$$\bar{\phi}_{1_{x'x'}} + \bar{\phi}_{1_{\bar{y}\bar{y}}} + K^2 \bar{\phi}_1 = g_{1_{x'}} + \delta x_s(\pm 0) f(\phi_{0_{x'}}) \tag{9}$$

where

$$g_{1_{x'}} = (\phi^*_{1_{x'}} \phi_{0_{x'}})_{x'} + iM_\infty^2 \Omega (\phi^*_{1_{x'}} \phi_{0_{x'}}) \tag{10a}$$

$$K = M_\infty \Omega \tag{10b}$$

$$f = [x_{1_{x'}}(\phi_{0_{x'}} - \phi_{0_{x'}}^2)]_{x'} + x_{1_{x'}}\left[\phi_{0_{x'}} - \frac{\phi_{0_{x'}}^2}{2}\right]_{x'} \tag{10c}$$

and

$$\phi^*_{1x'} = \phi_{1x'} + iM_\infty^2\Omega\phi_1 \tag{11}$$

Consider the equation

$$L(\phi) = F \tag{12}$$

where $L(\ )$ is a second-order linear differential operator and F and ϕ are functions with certain smoothness restrictions in a domain S bounded by a curve C. Equation (12) can be written in integral form using Green's identity to give

$$\int_C \left[\psi\frac{\partial\phi}{\partial n} - \phi\frac{\partial\psi}{\partial n}\right] \mathrm{d}C = -\iint_S [\psi L(\phi) - \Phi L(\psi)]\ \mathrm{d}S \tag{13}$$

where n is an inward drawn normal to C and ψ is a yet to be specified function, with the same smoothness criteria as ϕ. From Eq. (12) the term $L(\phi)$ in Eq. (13) can be replaced by F.

If the operator $L(\ )$ is given,

$$L(\ ) = \frac{\partial^2}{\partial x'^2} + \frac{\partial^2}{\partial y^2} + K^2 \tag{14}$$

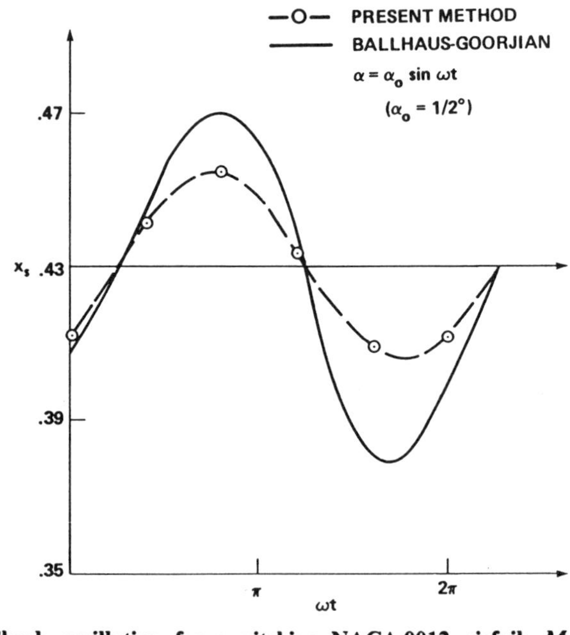

Fig. 1 Shock oscillation for a pitching NACA 0012 airfoil; $M_\infty = 0.8$, $\nu = 0.2$, $\alpha_0 = 0.5$ deg.

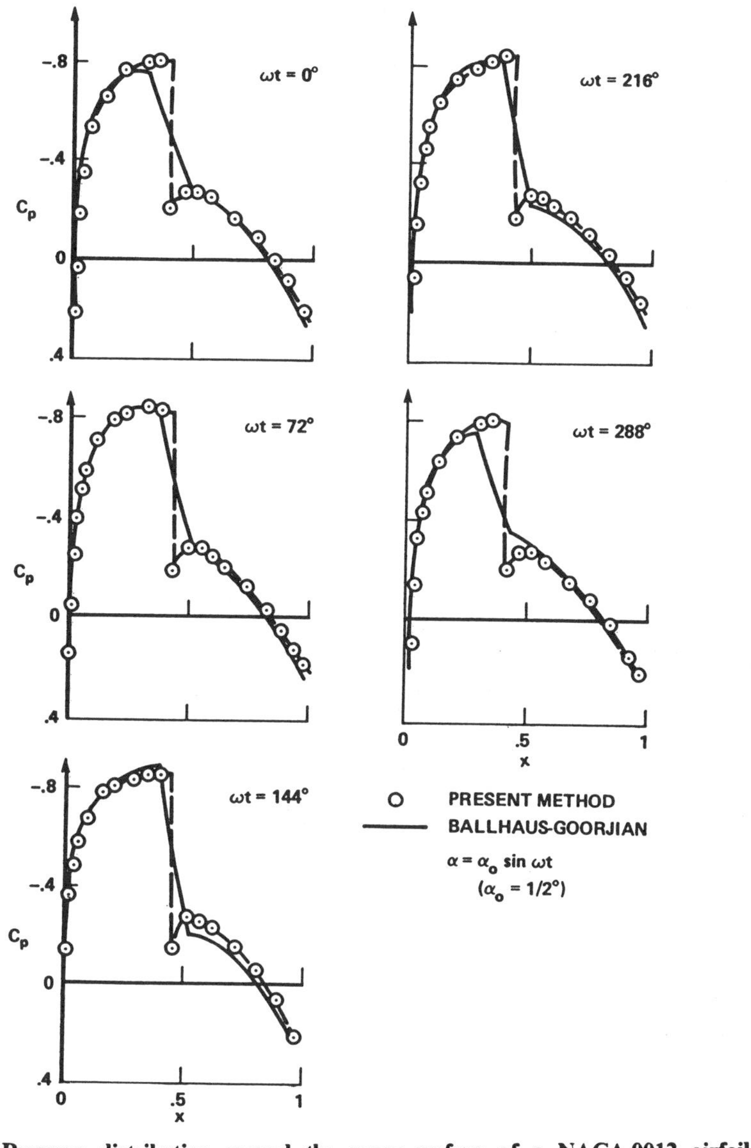

Fig. 2 Pressure distribution around the upper surface of a NACA 0012 airfoil pitching about midchord; $M_\infty = 0.8$, $\nu = 0.2$, $\alpha_0 = 0.5$ deg.

ψ is given by

$$L(\psi) = \delta(r) \tag{15}$$

where $\delta(r)$ is the delta function and F is given by the right-hand side of Eq. (9), then Eqs. (12) and (13) are the basic equations used by Nixon.[6] After much algebraic manipulation the final integral equation is

$$\bar{\phi}_{1_{x'}}(x',\bar{y}) - g(x',\bar{y}) = I_L(x',\bar{y}) + I_S(x',\bar{y}) + \delta x_s(\pm 0) I_f(x',\bar{y}) \tag{16}$$

where I_L is a line integral over the wing and wake and is a function of the boundary conditions, I_S is a field integral over the domain S and is a function of f and thus depends on the unknown $\bar{\phi}_{1_{x'}}$, and I_f is a function of the known $\phi_{0x'}$. The shock motion $\delta x_s(\pm 0)$ is found by enforcing an "entropy" condition (see Chapter 2) in which no expansion shocks exist. Results of a low-frequency version of this theory (labeled "present method") are shown in Figs. 1 and 2 for a NACA 0012 airfoil oscillating in pitch at a freestream Mach number of 0.8 at a reduced frequency of 0.2; the amplitude of oscillation is 0.5 deg. In Fig. 1 the shock motion is compared to results obtained by the method of Ballhaus and Goorjian,[8] and in Fig. 2 a similar comparison is made for the pressure coefficient. The agreement between these results is quite satisfactory. The computation time for these integral equation solutions is considerably faster than a finite-difference solution.

If there is no shock motion in the flow, then the term $\delta x_s(\pm 0)$ in Eq. (16) is zero and the problem is similar to the subsonic theory of Nixon.[1] However, even in such a simple case, the solution requires the evaluation of the field integral I_S, and it is this computation that greatly increases the computer time over a conventional, linear, subsonic solution. It is, in principle, possible to eliminate this integral if the function F in Eq. (12) is zero and if a Green's function can be found for the operator $L(\ )$. An attempt to develop such a procedure has been made by Liu and Pi.[9] In their method a series of kernel functions suitable for subsonic, supersonic, and transonic flows are derived, and these are then used in a similar way to the conventional subsonic, linear, panel methods.

This method does not, at present, treat shock-wave motions and is formulated in terms of an acceleration potential Φ, which in the present notation is defined for harmonic oscillation as

$$\Phi = \nu\phi_1 + (1 + \phi_{0_x})\phi_{1_x}$$

After some algebra a differential equation for Φ is developed, namely,

$$\beta_0^2\Phi_{xx} + \Phi_{yy} + \Phi_{zz} + A\Phi_x + B\Phi = 0 \tag{17}$$

where β_0^2, A, and B are functions of the steady-state solution Φ_{0_x}.

Equation (17) is of the form of Eq. (12), and the Green's identity can be applied. If the domain S is divided into elements, ΔS, in each of which ϕ_{0x} can be considered constant, then a Green's function can be found for each element. The Green's function (ψ) is a solution of

$$\beta_0^2\psi_{xx} + \psi_{yy} + \psi_{zz} + A\psi_x + B\psi = \delta(r) \tag{18}$$

In this piecewise manner the field integral in Eq. (13) can be made to vanish. The movement from one element ΔS to its adjacent element

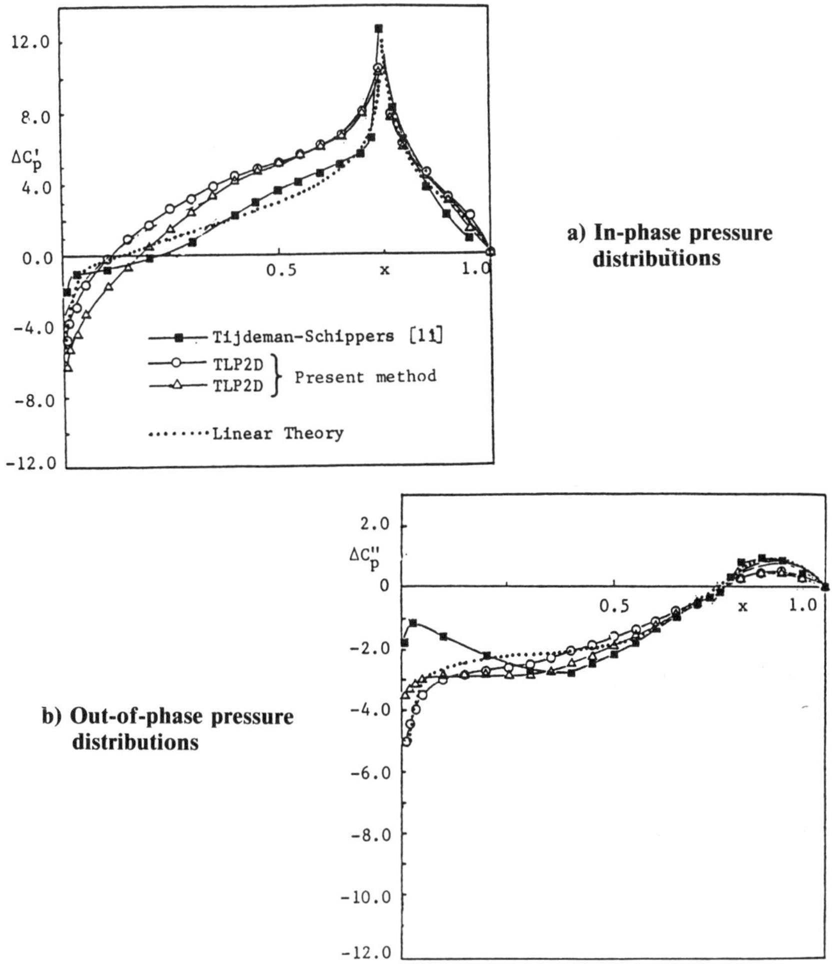

a) In-phase pressure distributions

b) Out-of-phase pressure distributions

Fig. 3 NACA 64A006 airfoil with oscillating flap at $M_\infty = 0.8$ and $k = 0.253$ with hinge point at three-quarter chord.

involves a probable discontinuous change in the coefficients β_0, A, B in Eq. (17), and it is likely that derivatives of ψ and Φ are themselves discontinuous across these interfaces. However, unless ϕ_{0_x} is considered a constant everywhere in S, this technique violates the smoothness criteria for the application of Green's identity, which requires a continuity of second derivatives of Φ and ψ. It has been found in steady[10] flow that the variations of Φ_{0_x} in the $\bar{y}$ direction play a critical role in determining the shock location in an integral equation method. It is, therefore, possible that the method of Liu and Pi[9] is not capable of accurately predicting flows with shock waves. However, for subcritical flows the method may be quite satisfactory. An example of a NACA 65A006 airfoil is shown in Fig. 3. The Mach number is 0.85, and the reduced frequency is 0.06. The two forms of

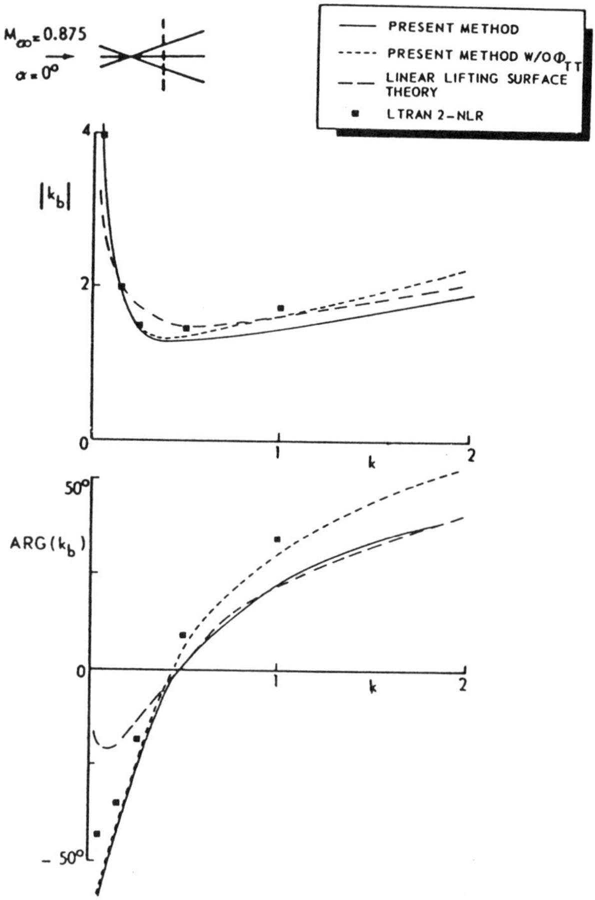

Fig. 4 Unsteady lift coefficient on a pitching NACA 64A006 airfoil.

the mixed kernel function results agree quite well with the experimental results of Tijdeman and Schippers.[11] It should perhaps be noted that linear theory is also quite accurate for this case. Although a computation time is not given for this method, it is to be expected that it would be considerably less than that required for a finite-difference calculation.

An interesting variant of the integral equation method has been proposed by Hounjet.[12] If only the integrated flow quantities such as lift and moment coefficients are required, then the detailed breakdown in a conventional perturbation theory over the region traversed by the shock wave is not important. If the oscillatory loads over this region can be estimated using an equivalent continuous flow model, then a subsonic panel method can be used to compute the required lift and moment coefficients. This panel method does use a "local linearization" of the steady flow velocity and is thus subject to a similar criticism to that given earlier for the "mixed kernel

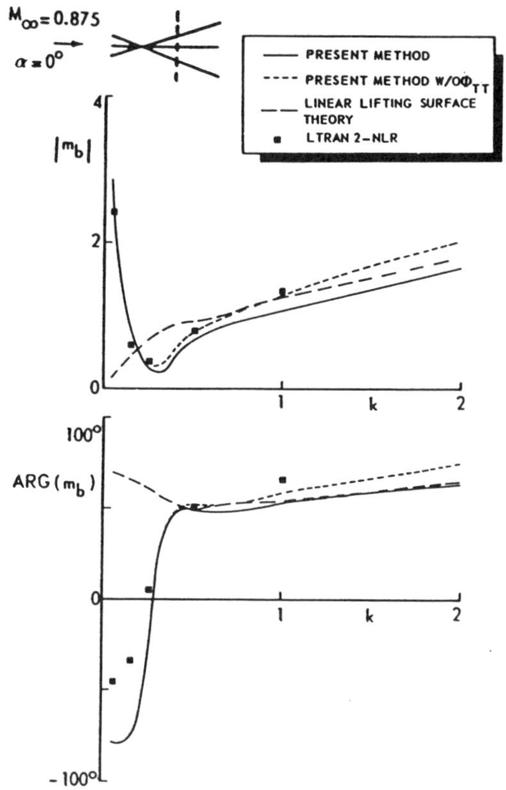

Fig. 5 Unsteady moment coefficient on a pitching NACA 64A006 airfoil.

function" method of Liu and Pi.[9] The computation time is 5% of that required for an equivalent finite-difference calculation. The magnitude and phase lag of the unsteady lift and moment coefficients for a NACA 64A006 airfoil at $M_\infty = 0.875$ and $\alpha_0 = 0$ are shown in Figs. 4 and 5 and are compared to the finite-difference solutions from the code LTRAN2-NLR, a variant of the Ballhaus-Goorjian code.[8] It may be seen that the methods give a considerable improvement over linear theory.

A version of the integral equation method that is close to that of Nixon[6] is that of Hounjet.[13] In this approach no coordinate straining is used, and the field integral in Eq. (16) is evaluated using the mixed-difference scheme of Murman[14] which allows shock capture rather than the shock fitting used by Nixon.[6] Since a shock capture algorithm does not give a discontinuity, no coordinate straining may be necessary, in an engineering sense. However, in the limit of zero grid size a true discontinuous shock will emerge. It is difficult to see how a moving shock wave can be treated with a linear equation if coordinate straining is not used. The real and imaginary components of the pressure coefficient for the flow over a NACA 64A006

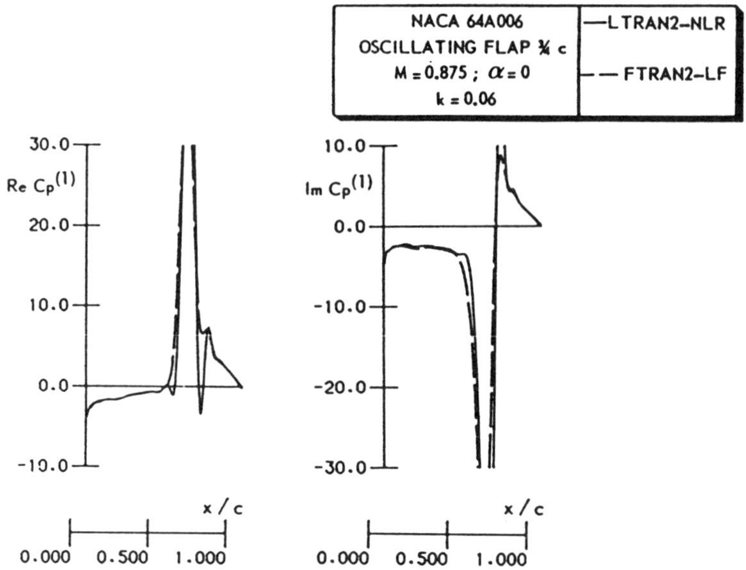

Fig. 6 Unsteady pressure distribution on a NACA 64A006 airfoil with oscillating flap calculated with low-frequency methods.

at $M_\infty = 0.875$ with an oscillating flap at $v = 0.06$ are shown in Fig. 6 and are compared to finite-difference results of LTRAN2-NLR.

A more satisfactory approach than that of Ref. 13 is given by Hounjet,[15] in which a finite-difference method is used for the inner part of the flow and an integral equation solution for the outer part; this hybrid technique allows the proper shock representation near the wing and removes the need for a large finite-difference grid by representing the outer flowfield by an integral equation. The method has been applied to both two- and three-dimensional wings. The real and imaginary components of the section lift and moment for a fighter wing are shown in Fig. 7 and compared with experimental data.

In an aeroelastic analysis these integral equation methods will only supply the aerodynamic transfer functions. The method is only of use when the variations of the generalized forces are linear functions of the deflection q. At transonic speeds this is true[16] for a small-amplitude oscillation, provided shock waves are not generated or destroyed during the motion.

In summary, an integral equation method can be developed to predict unsteady transonic flow characteristics even if there is shock motion. The computation time is less, perhaps considerably less, than an equivalent finite-difference calculation. However, the method is only useful for aeroelastic computations if the amplitude is small and if shock waves are not generated or destroyed during the motion. This last restriction is important for a supercritical wing oscillating about its design point; in such cases, shock waves may be generated or destroyed during the motion, and it is probable that the integral equation methods are invalid.

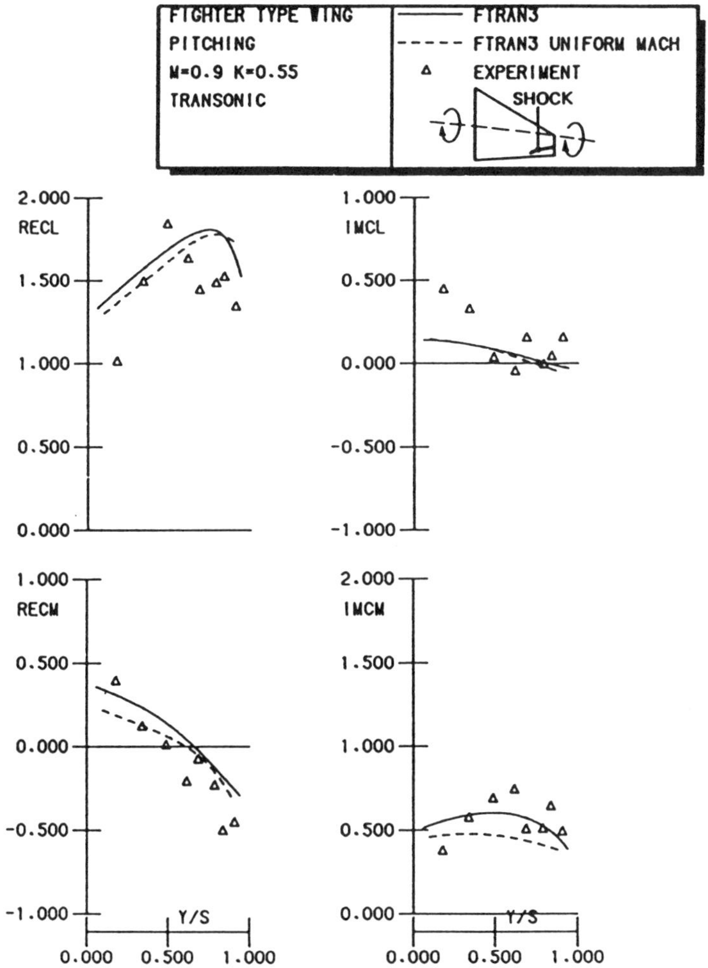

Fig. 7 Unsteady airloads on a pitching NACA 64A010 airfoil (AGARD notation).

4. Local Linearization Techniques

The principal difficulty in solving the TSD equation [Eq. (1)] is the presence of the nonlinear term

$$k\phi_x\phi_{xx}$$

This nonlinearity is essential if shock waves are to be modeled, since it provides the necessary mechanism to give a double-valued solution. However, if there are no shock waves, that is, the flow is subcritical, or if the shock waves are at the leading or trailing edge of the wing, then the flow on the wing is shock-free. In such cases the nonlinear term may be approximated by some linear function with some expectation of adequate results. One such method is termed local linearization and is a development of the earlier techniques proposed by Spreiter and Alksne.[17]

For example, consider the two-dimensional steady flow version of Eq. (1), namely,

$$(1 - M_\infty^2)\phi_{xx} + \phi_{yy} = k\phi_x\phi_{xx} \tag{19}$$

which may be written as

$$\lambda\phi_{xx} + \phi_{yy} = 0 \tag{20}$$

where

$$\lambda = 1 - M_\infty^2 - k\phi_x \approx 1 - M^2 \tag{21}$$

and M is the local Mach number.

The tangency boundary condition of Eq. (20) for a symmetric airfoil is written as

$$\phi_y(x, \pm 0) = \pm y_s'(x) \tag{22}$$

where $y = y_s(x)$ denotes the surface of the airfoil.

The local linearization concept assumes that λ is constant in Eq. (20), which then has the solution

$$\phi_x = \frac{1}{\lambda^{\frac{1}{2}}} \cdot \frac{1}{\pi} \int_0^1 \frac{y_s'(\xi)}{x - \xi}\,\mathrm{d}\xi \tag{23}$$

which is similar to the classic Prandtl-Glauert solution. The quantity λ has still to be determined. Equation (23) can be written as

$$\phi_x = \frac{1}{\lambda^{\frac{1}{2}}} u_i \tag{24}$$

where u_i is the incompressible result given by

$$u_i = \frac{1}{\pi} \int_0^1 \frac{y_s'(\xi)}{x - \xi}\,\mathrm{d}\xi \tag{25}$$

If Eq. (24) is differentiated, then

$$\phi_{xx} = \frac{\mathrm{d}u_i}{\mathrm{d}x} \bigg/ \lambda^{\frac{1}{2}} = \frac{\mathrm{d}u_i}{\mathrm{d}x} \bigg/ [1 - M_\infty^2 - k\phi_x]^{\frac{1}{2}} \tag{26}$$

and if λ is now treated as a variable, Eq. (26) is a simple differential equation for ϕ_x which can be integrated to give

$$u = \frac{1}{k}\left\{1 - M_\infty^2 - \left[(1 - M_\infty^2)^{\frac{3}{2}} - \frac{3}{2}ku_i\right]^{\frac{2}{3}}\right\} \tag{27}$$

In spite of the inconsistencies the technique does give reasonable results for simple cases. There are similar analyses for sonic and supersonic flows. A full description can be found in Ref. 18. If a steady version of Eq. (3) is to be solved rather than Eq. (19), that is, the nonlinear term is linearized and given in terms of ϕ_{0_x}, then the equivalent equation to Eq. (20) is

$$\lambda_1\phi_{xx} + \lambda_2\phi_x + \phi_{yy} = 0 \tag{28}$$

where

$$\lambda_1 = 1 - M_\infty^2 - k\phi_{0_x} \approx 1 - M_0^2 \tag{29}$$

$$\lambda_2 = -k\phi_{0_{xx}} \approx \frac{d}{dx}(1 - M_0^2) \tag{30}$$

and M_0 is the steady local Mach number. Equation (28) is not as simple as Eq. (20), but, if it is assumed that the base Mach number M_0 varies slowly over the airfoil, then λ_2 is negligible and Eq. (28) reduces to Eq. (20) but with λ known in terms of ϕ_{0_x}. The solution is therefore

$$\phi_x = \frac{1}{\lambda_1^{\frac{1}{2}}}u_i = u_i/[1 - M_0^2]^{1/2} \tag{31}$$

which is similar to the Prandtl-Glauert result but uses a local Mach number rather than the freestream value.

The preceding descriptions of local linearization are for steady flow, but the concept has been extended to unsteady transonic flow. The analyses are based on the linearized equation concept given by Eq. (28). One of the earlier theories is that of Stahara and Spreiter,[19] who examined two-dimensional airfoil flows at sonic conditions, and Ruo and Theisen,[20] who examined three-dimensional wings, again for sonic flow. The magnitude and phase lag of the unsteady pressure distribution in a sonic freestream for a 6% parabolic arc airfoil is shown in Fig. 8 for a number of reduced frequencies k. The results of linear theory are shown for comparison. It may be seen that at low frequencies there is a considerable difference between the results of linear theory and that of local linearization whereas at high frequency the results of both theories are nearly identical. This is to be expected, since at high frequencies the ϕ_{xt} term in Eq. (3) dominates the nonlinear term $k(\phi_{0_x}\phi_{1_x})_x$. Cunningham[21] developed the idea further to treat mixed transonic flow. A result of Cunningham's for a trapezoidal wing at $M_\infty = 0.937$ and $\nu = 0.218$ is shown in Fig. 9. It can be seen that the real part of the pressure distribution agrees quite well with

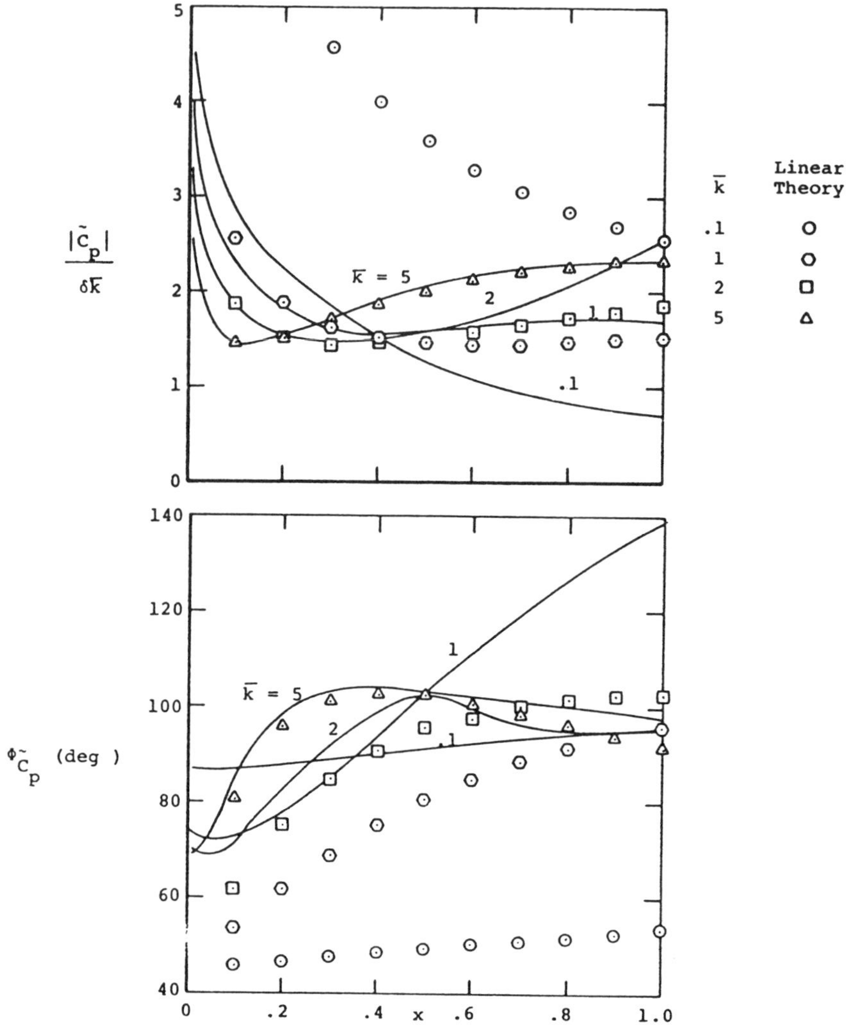

Fig. 8 Comparisons of local linearization unsteady pressure distributions with linear theory.

experimental data, whereas the imaginary component is not in such good agreement. The steady local Mach number used in the calculations is also shown in Fig. 9. It should be pointed out that unsteady flow measurement at transonic speeds is difficult; therefore, the early data may not be as accurate as more recent data.

5. Indicial Method

In linear models of subsonic or supersonic flow an alternative to expanding the governing equation in a Fourier series is to use an indicial function.

An indicial function analysis depends on the governing equation being linear, which is not the case for transonic flow. However, if the amplitude of motion is small and if shock waves are not generated or destroyed during the motion, then a linear equation similar in concept to Eq. (9) can be derived to model the unsteady motion. This is the basis of work in Ref. 22. A brief outline is given in the following paragraphs.

Consider the change in the pressure coefficient $C_p(x,y,t)$ due to an infinitesimal change in some parameter $\varepsilon(\tau)$ at some time τ. If the pressure distribution varies continuously with ε, then a Taylor series expansion gives

$$\Delta C_p(x,y,t) = C_{p_\varepsilon}(x,y,t,\tau)\,\frac{d\varepsilon(\tau)}{d\tau}\,\Delta\tau + \text{higher-order terms} \tag{32}$$

where t is time and $C_{p_\varepsilon}(x,y,t,\tau)$ is the rate of change of $C_p(x,y,t)$ with ε at some time τ; $\Delta C_p(x,y,t)$ is the change in pressure distribution. Neglecting the higher-order terms, the total effect of all such steps up until t is then

$$\Delta C_p(x,y,t) = \varepsilon(0)C_{p_\varepsilon}(x,y,t,0) + \int_0^t C_{p_\varepsilon}(x,y,t,\tau)\,\frac{d\varepsilon(\tau)}{d\tau}\,d\tau \tag{33}$$

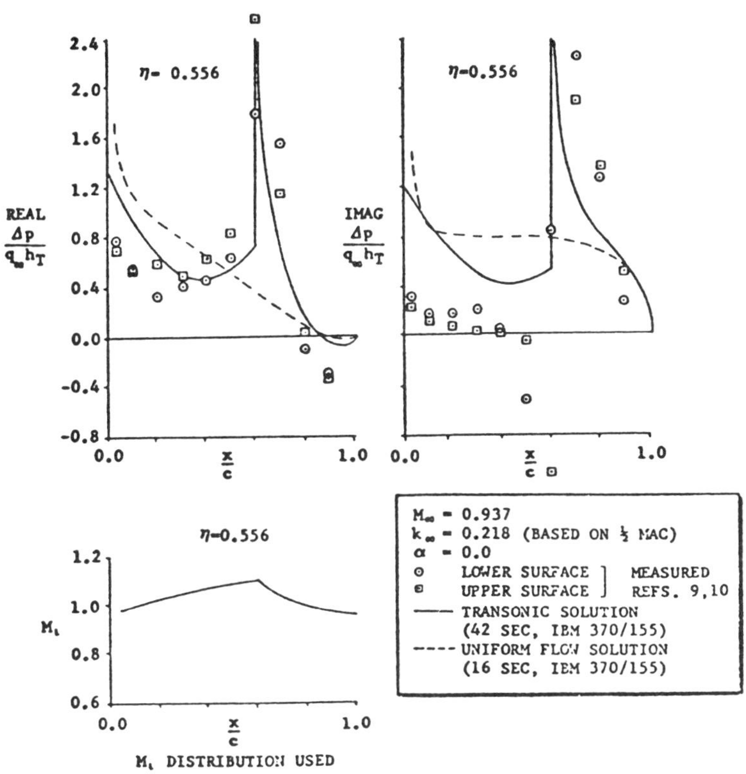

Fig. 9 Results for the oscillating trapezoidal wing at $M_\infty = 0.937$.

If it can be assumed that the behavior of C_p with ε is linear, then $C_{p_\varepsilon}(x,y,t,\tau)$ can be represented by its value at $\tau = 0$, provided that the t is taken relative to τ; thus,

$$C_{p_\varepsilon}(x,y,t,\tau) = C_{p_\varepsilon}(x,y,t-\tau,0) \tag{34}$$

Equation (29) then becomes

$$\Delta C_p(x,y,t) = \varepsilon(0)C_{p_\varepsilon}(x,y,t,0) + \int_0^t C_{p_\varepsilon}(x,y,t-\tau,0)\,\frac{\mathrm{d}\varepsilon(\tau)}{\mathrm{d}\tau} \tag{35}$$

By a simple change of variable, Eq. (36) can then be written as

$$\Delta C_p(x,y,t) = \varepsilon(0)C_{p_\varepsilon}(x,y,t) + \int_0^t C_{p_\varepsilon}(x,y,\tau)\,\frac{\mathrm{d}\varepsilon(t-\tau)}{\mathrm{d}\tau}\,\mathrm{d}\tau \tag{36}$$

where the functional form of $C_{p_\varepsilon}(x,y,t,0)$ has been contracted to $C_{p_\varepsilon}(x,y,t)$ for convenience in presentation. Equation (36) is the form of a typical relation in the indicial method. It is assumed that $C_{p_\varepsilon}(x,y,t)$ can be obtained by some means, either analytically or numerically. In the foregoing analysis it has been assumed that the variation of $C_p(x,y,t)$ with ε is linear. For continuous flows governed by a nonlinear equation, this assumption may well be adequate for small values of ε. However, as pointed out in Ref. 6, if there are shock waves that move during the motion, the assumption of linear variation is totally inadequate in the region bounded by the extremities of the shock motion. This is because the pressures in this region can jump from a preshock value to a postshock value, or vice versa, as the shock traverses this region. This difficulty can be overcome by the use of a strained coordinate system in which the shock remains at the same location.

Let the strained coordinate system be given by $(x',y't)$, which is related to the physical coordinate system (x,y,t) by

$$x = x' + \delta x_s(t)x_1(x') \tag{37}$$

where $\delta x_s(t)$ is the change in shock location at some t, and $x_1(x')$ is a known straining function. This is a slightly different formulation of Eq. (5). In Eq. (37) it is assumed that the shock waves are normal to the freestream.

For low reduced frequencies, the pressure coefficient $C_p(x,y,t)$ is given by

$$C_p(x,y,t) = -\frac{2\beta^2}{k}\,u(x,y,t) \tag{38}$$

where $u(x,y,t)$ is the scaled perturbation velocity in the x direction, which, following the ideas of Ref. 6, is given by

$$u(x,y,t) = u_0(x',y)[1 - \delta_{x_s}(t)x_{1_{x'}}(x')] + u_1(x',y,t) \tag{39}$$

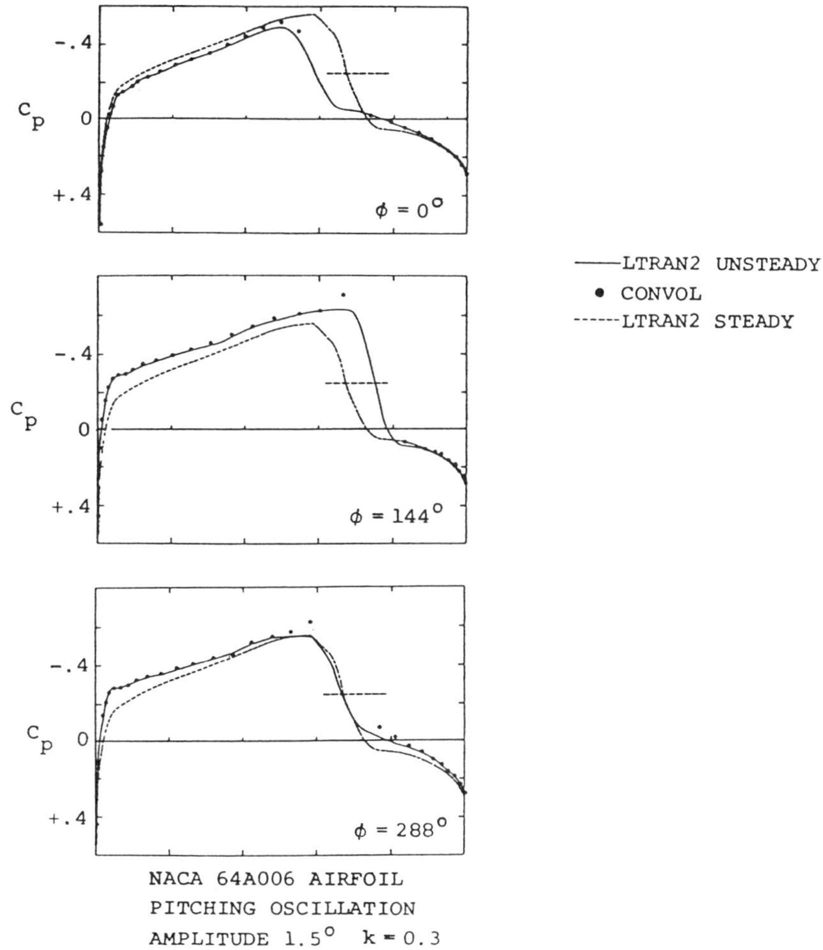

Fig. 10 Unsteady pressures for a pitching NACA 64A006 airfoil; $m_\infty = 0.875$.

where

$$k = (\gamma + 1)M_\infty^2$$

$$\beta = (1 - M_\infty^2)^{\frac{1}{2}}$$

and $u_0(x',y)$ is the value for the mean steady state. The coordinates (x',y) and (x,y) are related by Eq. (37).

In Eq. (39) $\delta x_s(t)$ and $u_1(x',y,t)$ depend linearly on the parameter $\varepsilon(\tau)$ and can therefore be treated by the indicial method. The nonlinear effect appears implicitly through the transformation from the strained coordinates to the physical coordinates. Thus,

$$\delta x_s(t) = \delta x_{s_\varepsilon}(t)\varepsilon(0) + \int_0^t \delta x_{s_\varepsilon}(\tau)\,\frac{\mathrm{d}\varepsilon(t-\tau)}{\mathrm{d}t}\,\mathrm{d}\tau \tag{40}$$

$$u_1(x',y,t) = u_{1_\varepsilon}(x',y,t)\varepsilon(0) + \int_0^t u_{1_\varepsilon}(x',y,\tau)\frac{d\varepsilon(t-\tau)}{d\tau}\,d\tau \tag{41}$$

where $\delta x_{s_\varepsilon}(t)$ and $u_{1_\varepsilon}(x',y,t)$ are the indicial response of $\delta x_s(t)$ and $u_1(x',y,t)$, respectively.

The expression for the pressure coefficient can then be obtained from Eqs. (38–41). Thus,

$$C_p(x,y,t) = \frac{2\beta^2}{k}\left\{u_0(x',y) + u_{1_\varepsilon}(x',y,t)\varepsilon(0) + \int_0^t u_{1_\varepsilon}(x',y,\tau)\frac{d\varepsilon(t-\tau)}{dt}\,d\tau\right.$$
$$\left. - u_0(x',y)x_{1_{x'}}(x')\left[\delta x_{s_\varepsilon}(t)\varepsilon(0) + \int_0^t \delta x_{s_\varepsilon}(\tau)\frac{d\varepsilon(t-\tau)}{d\tau}\,d\tau\right]\right\} \tag{42}$$

where x is given by Eq. (37). Hence, if the indicial responses δx_{s_ε} and u_{1_ε} (or $C_{p_{1_\varepsilon}}$) are known, then the pressure distribution and the shock motion for any motion defined by $\varepsilon(\tau)$ can be found. The indicial responses are obtained by computing the transient behavior of the shock motion and the pressure coefficient due to a step change in ε. This transonic calculation need only be performed once for the range of $\varepsilon(\tau)$.

A considerable advantage of this technique is that the fundamental equations, Eqs. (40) and (41), are linear; hence, the principle of superposition can be used. For example, if an airfoil is oscillating with a flap oscillating at a different frequency, then Eqs. (40) and (41) become

$$\delta x_s(t) = \sum_{i=1}^{2}\left\{\delta x_{s_{\varepsilon_i}}(t)\varepsilon_i(0) + \int_0^t \delta x_{s_{\varepsilon_i}}(\tau)\frac{d\varepsilon_i(t-\tau)}{dt}\,d\tau\right\} \tag{43}$$

$$u_1(x',y,t) = \sum_{i=1}^{2}\left\{u_{1_{\varepsilon_i}}(x',y,t)\varepsilon_i(0) + \int_0^t u_{1_{\varepsilon_i}}(x',y,\tau)\frac{d\varepsilon_i(t-\tau)}{dt}\,d\tau\right\} \tag{44}$$

where the subscript i denotes the motion and indicial response of each of the two oscillating components. An example of indicial calculations (denoted by the label CONVOL) is given by Kerlick and Nixon[22] for a NACA 64A006 airfoil oscillating in pitch with amplitude of 1.5 deg and reduced frequency of 0.3 and is shown in Fig. 10. It can be seen that the indicial method agrees fairly well with the direct calculation (denoted by the label LTRAN2) obtained by the computer code of Ballhaus and Goorjian.[8]

The technique has been applied by Nixon and Tzuoo[23] to two-dimensional transonic flows with gusts and has been coupled with an optimizer to reduce gust response by judiciously moving the control surfaces.

In the calculation of the indicial response the numerical error can degrade the initial transient so that the final result is inaccurate. One way to avoid this difficulty is to use piston theory to model the initial step. This is advantageous in that, apart from the initial step, the indicial responses are relatively smooth functions and can be approximated by a truncated series of exponentials. If this is done, then the convolution integrals in Eqs. (43) and (44) can be integrated analytically. This idea is described briefly in the following paragraphs.

In the early phase of subsonic and supersonic flutter analysis, great use was made of the piston theory limit. Briefly, the supposition was that, if the airfoil oscillated at very high frequencies, the velocity of the airfoil surface was an order of magnitude greater than the streamwise velocity. As a consequence, the linearized two-dimensional version of Eq. (1) becomes

$$\phi_{yy} = M_\infty^2 \frac{c^2}{U_\infty^2} \phi_{tt} \tag{45}$$

The solution is

$$\phi = f\left(y \mp \frac{U_\infty}{cM_\infty} t \right) \tag{46}$$

For $y > 0$, the positive sign is taken; for $y < 0$, the negative sign is used. This is to ensure outgoing waves. The pressure coefficient is given by

$$C_p(x, \pm y, t) = -2 \frac{c}{U_\infty} \phi_t(x, \pm y, t) \tag{47}$$

From Eq. (46),

$$\phi_y(x,y,t) = f'\left(y \mp \frac{U_\infty ct}{M_\infty} \right) \tag{48}$$

For a step change in the normal velocity component of $-\alpha$, Eq. (48) gives

$$f'(0) = -\alpha \tag{49}$$

Combining Eqs. (47) and (48) gives the result

$$C_p(x, \pm 0, 0) = \mp \frac{2\alpha}{M_\infty} \tag{50}$$

This is the initial pressure coefficient experienced by the airfoil due to a step change in angle of attack.

In the transonic equation [Eq. (1)], there is no mechanism to invalidate the basic premise of piston theory, namely, that streamwise velocities are small relative to the normal velocity. Hence, Eq. (50) is valid for transonic flow. Furthermore, since streamwise velocities contain the shock information for Eq. (1), the initial movement of shock does not enter Eq. (45). Therefore, it is assumed that the shock movement does not start with the discontinuity that characterizes the pressure coefficient.

An example from Ref. 23 is shown in Fig. 11, where the variation of the lift coefficient with time is shown for a NACA 64A006 airfoil at

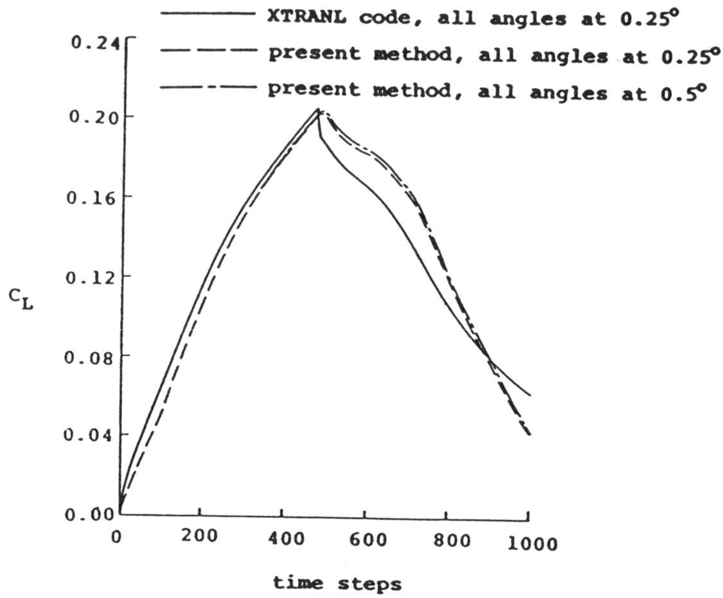

Fig. 11 Test case comparison of C_L for a NACA 64A006 airfoil at $M_\infty = 0.875$ and $\alpha_m = 0$ deg.

$M_\infty = 0.875$ responding to a gust and the reaction of two control surfaces, one at the leading edge and one at the trailing edge. The control surface deflections are taken to be 0.25 deg and for a second case, 0.5 deg. The gust amplitude is taken to be equivalent to an angle of attack equal to these deflections. It can be seen that for C_L the agreement with the results of the XTRAN2L[24] code is quite reasonable. For comparison the values at an amplitude of 0.5 are halved. Figure 11 shows that C_L is a linear function of amplitude if shock waves are not generated or destroyed during the motion. In Fig. 12 the variation of the hinge moment C_H for the aft control surface is shown for amplitude of 0.25 deg and 0.5 deg, and the results differ considerably. The difference between the two calculations is that at the higher amplitude the shock is not always present on the control surface. This indicates the nonlinear effect of the shock motion on the hinge moment.

6. Indicial Theory for Cascade Flows

For cascade flutter the dominant parameter is the interblade phase angle and a theory, similar to the indicial theory for wings, can be constructed for cascades.[25,26] For a given transonic indicial response calculation for an infinite cascade with one blade moving and the others held stationary, the complete transonic solution for any interblade phase angle can be constructed by superposition. A brief outline of the theory and the method of reducing a computation for an infinite cascade to a more tractable computation is given in the following paragraphs.

Although the indicial theory can give pressure distributions, only lift and pitching moments are considered here for reasons of clarity in presentation. The gap-to-chord ratio is denoted by h and the freestream Mach number by M_∞.

The lift is given as[25]

$$C_L(t) = \sum_{n=-\infty}^{\infty} C_{L_n}(t - n\sigma) \tag{51}$$

where C_{L_n} is the lift induced at blade zero by the motion of the nth blade with the other blades stationary and σ is the interblade phase angle. If the flow is harmonic in time, with frequency ω,

$$C_L(t) = \tilde{C}_L\, e^{i\omega t}, \qquad C_{L_n}(t - n\sigma) = \tilde{C}_{L_n} \exp[i\omega(t - n\sigma)] \tag{52}$$

In a cascade, the unsteady flow induced on blade zero by the nth blade can be characterized by the asymptotic form of the unsteady flow, provided that the nth blade is sufficiently far from blade zero (for example, several chord lengths). The asymptotic behavior for the induced normal velocity is

$$v(x,y) \approx A\, \frac{e^{-i\bar{K}|y|}}{|\bar{y}|^{\frac{1}{2}}} \tag{53}$$

where A is a complex constant, $\bar{y} = (1 - M_\infty^2)^{\frac{1}{2}} y$, and K is defined by Eq. (10). This result is obtained from classical subsonic airfoil theory.[1] The lift induced on a blade is proportional to the induced upwash parameter, which is given by Eq. (53). Hence, it follows that

$$C_{L_n} = C_{L_m} \left|\frac{m}{n}\right|^{\frac{1}{2}} \exp[-iK(|n| - m)\bar{h}] \tag{54}$$

where the relation $\bar{y} = n\bar{h}$ is used, and m is a given blade beyond which Eq. (53) is a good approximation. It is assumed that C_{L_m} is known. Note that the actual value of $v(x,y)$ is not given by Eq. (53), since this result only holds if there are no other blades. However, the actual upwash will vary with y in a form similar to Eq. (53). Using Eqs. (51–53) gives

$$\begin{aligned} C_L = \sum_{n=-m+1}^{m-1} \tilde{C}_{L_n} e^{-in\sigma} + \sum_{n=m}^{\infty} \tilde{C}_{L_m} \left(\frac{m}{n}\right)^{\frac{1}{2}} \\ \times \exp\{-iK(n - m)\bar{h} - in\sigma\} + \sum_{n=-\infty}^{-m} \tilde{C}_{L_m} \left(\frac{m}{n}\right)^{\frac{1}{2}} \\ \times \exp\{-iK(n - m)\bar{h} - in\sigma\} \end{aligned} \tag{55}$$

It can be seen that the effect of the cascade blades diminishes as $n^{-\frac{1}{2}}$; hence, a large number of blades is necessary in the initial computation of

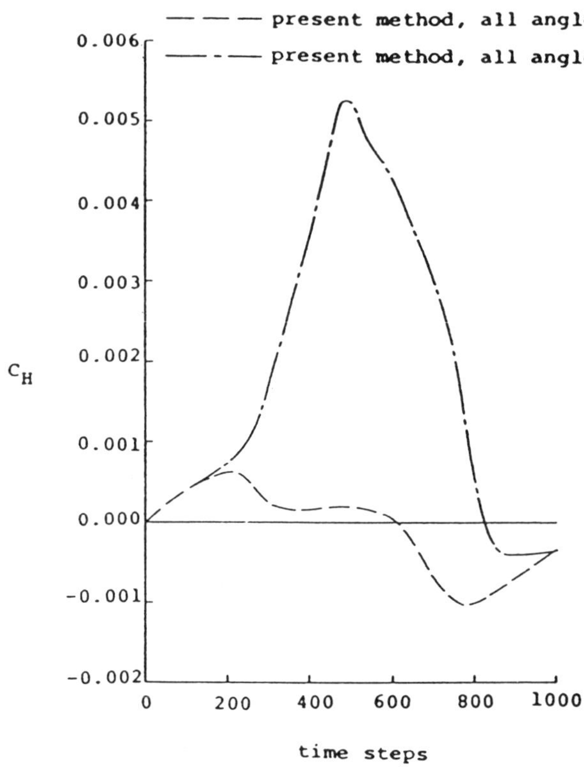

Fig. 12 Test case comparison of CH for a NACA 64A006 airfoil at $M = 0.875$ and $= 0$ deg.

C_{L_n} unless the approximation of Eq. (54) is used. It is also of interest to further examine Eq. (55). The two infinite sums are bounded unless

$$\pm K\bar{h} - \sigma = 2p\pi \tag{56}$$

where p is an integer. Thus, for an interblade phase angle given by

$$\sigma = \pm K\bar{h} - 2p\pi \tag{57}$$

The expression $\tilde{C}_L$ will be infinite. Results for a flat-plate cascade at $M_\infty = 0.5$, $v = 1.0$, and $h = 1.0$ computed using Eq. (55) are shown in Fig. 13. In this calculation, the total number of blades in the numerical calculation is 13 and $m = 3$. The number of blades used in the sum in Eq. (55) is 100. Figure 14 shows a result for a transonic case, namely, a cascade with a NACA 0012 section, $M_\infty = 0.75$, $h = 2.0$, and $v = 1.0$. It can be seen that there is a considerable difference between a subsonic case and a transonic case.

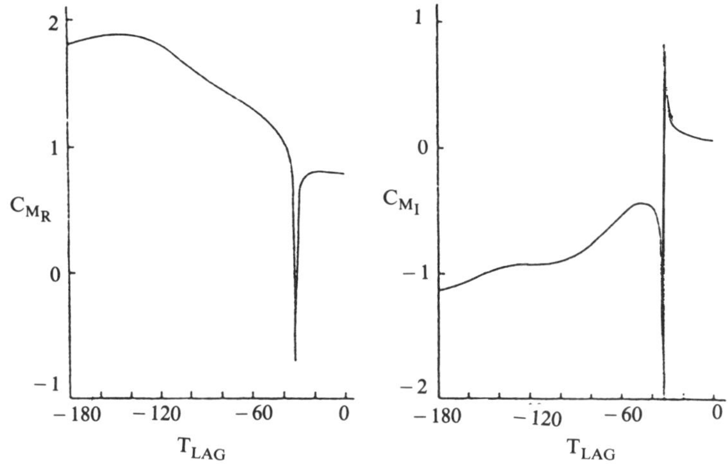

Fig. 13 Moment due to torsion for a flat plate; $M_\infty = 0.5$, $v = 1.0$, $h = 1.0$.

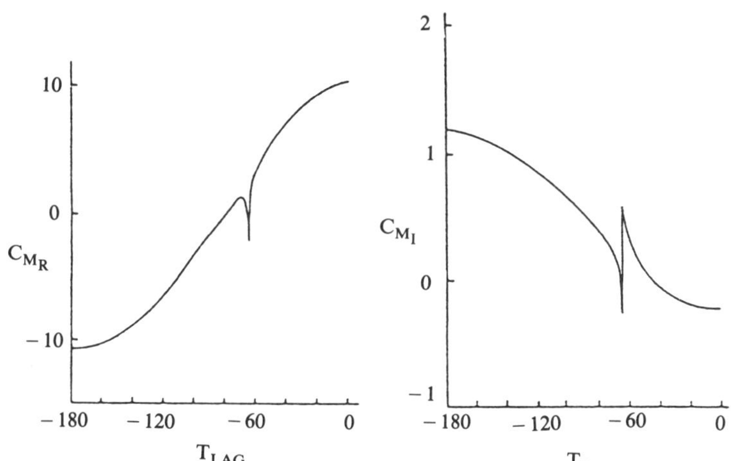

Fig. 14 Moment due to torsion for a NACA 0012 section; $M_\infty = 0.75$, $v = 1.0$, $h = 2$.

7. Strong Shock Potential Theories

The basic assumptions of isentropy and irrotationality in potential theory are only valid when there are no shock waves[27] in the flow. Although the potential theory is valid strictly only for shock-free flows, the results obtained by such a theory for flows with shock waves are sufficiently close to experimental data for practical use, provided the shock waves are weak. Thus, when the TSD solutions are compared to the more realistic Euler equation solutions, it is found that the agreement is satisfactory, provided that the basic restriction to weak shock waves is not violated.

If the flow has strong shock waves, however, then there is considerable disagreement between both TSD solutions and Euler equation solutions. Generally the predicted shock location for the potential theories is much further aft than that for the Euler equation solutions. The causes of the error in the shock location in the steady TSD theory for two-dimensional flow have been examined in Ref. 28, where the basic justification for a correction procedure has been derived. The basic hypothesis of the theory is that the error in shock location is primarily due to the stronger shock predicted by TSD theory compared to the shock strength of the Euler equations. It is also assumed that, if the shock strength is suitably corrected, then the shock location should be approximately correct. These correction theories have been applied to both steady[28] and unsteady[29] solutions of the TSD equation and to the steady full potential equation.[27] A simple description of the method is given below.

The basic differential equation used in the present analysis is

$$(a + b\phi_x + c\phi_x^2)\phi_{xx} + \phi_{yy} + A\phi_{xt} + B\phi_{tt} = 0 \tag{58}$$

where a, b, and c are parameters to be chosen:

$$A = -2M_\infty^2 \nu, \qquad B = -M_\infty^2 \nu^2 \tag{59}$$

where ν is the reduced frequency, c is the airfoil chord, and U_∞ the freestream velocity.

Equation (58) is similar to Eq. (1), but with the addition of the formally negligible term

$$c\phi_x^2\phi_{xx}$$

The idea is to pick the constants a, b, c such that Eq. (58) gives results closer to those of the Euler equations than the classical potential equation. Two of the conditions are that

$$a = 1 - M_\infty^2$$

and that the normal shock strength should be equal to the Rankine-Hugoniot value. The first condition lets the equation approach the classical linear equation when the nonlinear effects are negligible, and the second condition gives the correct shock strength. This equation has been examined by Nixon,[30] but the method is relatively cumbersome. The most rational of the various strong shock formulations is described by Fuglsang and Williams.[31] In an alternative description of this method, Eq. (58) is written in the form

$$(1 - M_\infty^2)(V + CV^2/2)_x + \phi_{yy} + A\phi_{xt} + B\phi_{tt} = 0 \tag{60}$$

where

$$V = \frac{\phi_x}{1 + D\phi_x} \tag{61}$$

and C and D are constants. If V is expanded as a series in ϕ_x, it resembles Eq. (58). The conditions enforced by Fugslang and Williams[31] are that

$$(1+\phi_x^+)(1+\phi_x^-) = a^{*2} = [2+(\gamma-1)M_\infty^2]/[(\gamma-1)M_\infty^2] \tag{62}$$

which is the Prandtl relation, and that Eq. (60) becomes parabolic at sonic conditions; thus,

$$1 + C\phi_x^*/(1+D\phi_x^*) = 0 \tag{63}$$

where

$$\phi_x^* = a^* = \{[2+(\gamma-1)M_\infty^2]/[(\gamma-1)M_\infty^2]\}^{\frac{1}{2}}$$

and nondimensionalized by the freestream velocity, the plus and minus signs denote values ahead of and behind a shock wave, respectively, and the asterisk denotes a value at sonic conditions. Equations (62) and (63) are sufficient to determine C and D.

A result of Ref. (31) is shown in Fig. 15, where the real and imaginary components of the pressure difference are compared to results of the unmodified TSD equation and experimental data; it may be seen that the modified TSD equation does provide some improvement on the TSD equation. It should be noted that a significant advantage of the strong shock theories is that they apparently reduce the occurrence of the multiple solutions described in Chapter 2.

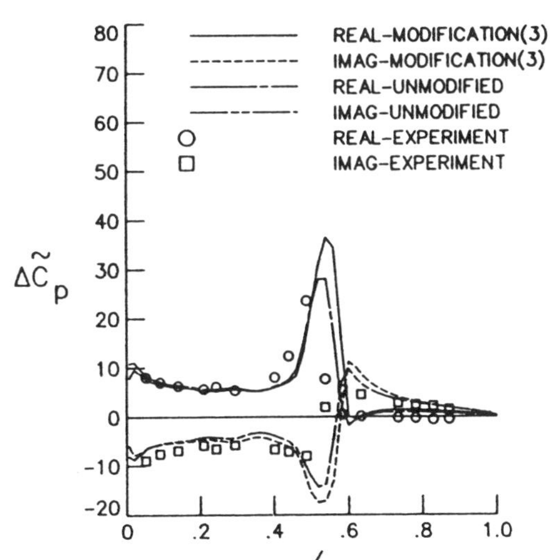

Fig. 15 Unsteady pressure distribution for a NACA 64A010A airfoil at $M_\infty = 0.796$, $\alpha_m - 0$ deg, $\alpha_0 = 0.102$ deg.

8. High-Frequency Solutions

It has been noted in Chapter 2 that at high frequencies the shock does not move; hence, the equation can be linearized. Since the frequency is high, the ϕ_{xt} term in Eq. (1) dominates the ϕ_{xx} term, and the resulting equation, due to Landahl,[32] is

$$\phi_{yy} + \phi_{zz} - 2M_\infty^2 c/U_\infty \phi_{xt} - \frac{M_\infty^2 c^2}{U_\infty^2} \phi_{tt} = 0 \tag{64}$$

If the motion is harmonic in time with ω, then Eq. (64) becomes

$$\bar{\phi}_{yy} + \bar{\phi}_{zz} - 2iM_\infty^2 \nu \bar{\phi}_x + M_\infty^2 \nu^2 \bar{\phi} = 0 \tag{65}$$

where the ν is given by

$$\nu = \omega c/U_\infty \tag{66}$$

and

$$\phi(x,y,z,t) = \bar{\phi}(x,y,z)\, e^{i\omega t} \tag{67}$$

If the Fourier transform with respect to x of Eq. (65) is taken, that is,

$$\tilde{\phi} = \frac{1}{(2\pi)^{\frac{1}{2}}} \int_{-\infty}^{\infty} \bar{\phi}\, e^{-iux}\, dx \tag{68}$$

then Eq. (65) can be written as

$$\tilde{\phi}_{yy} + \tilde{\phi}_{zz} + K^2 \tilde{\phi} = 0 \tag{69}$$

where

$$K^2 = (2ku + k^2) M_\infty^2 \tag{70}$$

Thus, the mixed-type three-dimensional partial differential equation [Eq. (1)] has been reduced to a linear two-dimensional equation [Eq. (69)]. Equation (69) has been solved by Landahl[32] in closed form for a number of two- and three-dimensional flows.

References

[1]Nixon, D., "A Two-Dimensional Aerofoil Oscillating at Low Frequencies in High Subsonic Flow," British Aeronautical Research Council, London, Current Paper 1285, 1974.

[2]Nixon, D., "A Two-Dimensional Aerofoil with a Control Surface Oscillating at Low Frequency in High Subsonic Flow," *Aeronautical Quarterly*, Vol. XXV, 1974, pp. 186–198.

[3]Traci, R. M., Albano, E. D., and Farr, T. L., "Small Disturbance Transonic Flows About Oscillating Airfoils and Planar Wings," AFFDL TR-75-10, Air Force Flight Dynamics Lab., Wright-Patterson AFB, OH, 1975.

[4]Weatherhill, W. H., Ehlers, F. E., and Sebastian, J. D., "Computation of the Transonic Perturbation Flow Fields Around Two- and Three-Dimensional Oscillating Wings," NASA CR-2599, 1975.

[5]Fung, K.-Y., Yu, N. J., and Seebess, R., "Small Unsteady Perturbations in Transonic Flows," *AIAA Journal*, Vol. 16, Aug. 1978, pp. 815–822.

[6]Nixon, D., "Calculation of Unsteady Transonic Flows Using the Integral Equation Method," *AIAA Journal*, Vol. 16, Sept. 1978, pp. 976–983.

[7]Nixon, D., "Perturbation of a Discontinuous Transonic Flow," *AIAA Journal*, Vol. 16, Jan. 1978, pp. 47–52.

[8]Ballhaus, W. F. and Goorjian, P. M., "Implicit Finite-Difference Computations of Unsteady Transonic Flows about Airfoils" *AIAA Journal*, Vol. 15, Dec. 1977, pp. 1728–1735.

[9]Liu, D. D. and Pi, W. S., "Transonic Kernel Function Method for Unsteady Flow Calculations Using a Unified Linear Pressure Panel Procedure," AIAA Paper 80-0737, May 1980.

[10]Nixon, D. and Hancock, G. J., "Integral Equation Methods—A Reappraisal," *Symposium Transonicum II, Göttingen*, edited by K. Oswatitsch and D. Rues, Springer-Verlag, Berlin, FRG, 1976.

[11]Tijdeman, H. and Schippers, P., "Results of Pressure Measurements on a Lifting Airfoil and Oscillating Flap in Two-Dimensional High Subsonic and Transonic Flow," NLR TR-73018L, National Aerospace Lab., Amsterdam, The Netherlands, 1973.

[12]Hounjet, M. H. L., "Transonic Panel Method to Determine Loads on Oscillating Airfoils in Transonic Flow with Embedded Shocks," *AIAA Journal*, Vol. 9, May 1981, pp. 559–566.

[13]Hounjet, M. H. L., "Calculation of Unsteady Transonic Flows with Shocks by Field Panel Methods," *AIAA Journal*, Vol. 20, June 1982, pp. 857–859.

[14]Murman, E. M., "Analysis of Embedded Shock Waves Calculated by Relaxation Methods," *AIAA Journal*, Vol. 12, May 1974, pp. 626–633.

[15]Hounjet, M. H. L., "A Field Panel/Finite Difference Method for Potential Unsteady Transonic Flow," *AIAA Journal*, Vol. 23, April 1985, pp. 537–545.

[16]Nixon, D., "Notes on the Transonic Indicial Method," *AIAA Journal*, Vol. 16, June 1978, pp. 613–616.

[17]Spreiter, J. R. and Alksne, A. Y., "Theoretical Pressure Distributions on Wings of Finite Span at Zero Incidence for Mach Numbers Near 1," NACA TR-R-88.

[18]Spreiter, J. R. and Alksne, A. Y., "Thin Airfoil Theory Based on Approximate Solution of the Transonic Flow Equation," NACA Rept. 1359, 1958.

[19]Stahara, S. S. and Spreiter, J. R., "Development of a Nonlinear Unsteady Transonic Flow Theory," NASA CR-2258, 1973.

[20]Ruo, S. Y. and Theisen, J. G., "Calculation of Unsteady Transonic Aerodynamics for Oscillating Wings with Thickness," NASA CR-2259, 1974.

[21]Cunningham, A. M., "An Oscillating Kernel Function Method for Lifting Surfaces in Mixed Transonic Flow," AIAA Paper 74-359, April 1974.

[22]Kerlick, G. D. and Nixon, D., "Calculation of Unsteady Transonic Pressure Distributions by the Indicial Method," *Journal of Applied Mechanics*, Vol. 49, June 1982, pp. 273–278.

[23]Nixon, D. and Tzuoo, K.-L., "Prediction of Gust Loadings and Alleviation at Transonic Speeds," *Journal of Aircraft*, Vol. 24, Oct. 1987, pp. 703–709.

[24]Whitlow, W., "XTRAN2L: A Program for Solving the General Frequency Unsteady Transonic Small Disturbance Equation," NASA TM-85723, 1983.

[25]Nixon, D., "Computation of Unsteady Transonic Cascade Flows," *AIAA Journal*, Vol. 21, May 1983, pp. 790–792.

[26]Nixon, D., Tzuoo, K.-L., and Ayoub, A., "Rapid Computation of Unsteady Transonic Cascade Flows," *AIAA Journal*, Vol. 25, May 1987, pp. 760–762.

[27]Klopfer, G. H. and Nixon, D., "Nonisentropic Potential Formulation for Transonic Flows," *AIAA Journal*, Vol. 22, June 1984, pp. 770–776.

[28]Nixon, D., "Transonic Small Disturbance Theory with Strong Shock Waves," *AIAA Journal*, Vol. 18, June 1980, pp. 717–718.

[29]Kerlick, G. D., Nixon, D., and Ballhaus, W. R., "Unsteady Transonic Small Disturbance Approximation with Strong Shock Waves," *AIAA Journal*, Vol. 21, June 1983, pp. 918–920.

[30]Nixon, D., "Validation of a Computer Code to Treat Strong Shock Waves in a Transonic Flow," Nielsen Engineering & Research, Inc., Mountain View, CA, TR-316, Dec. 1983.

[31]Fuglsang, D. and Williams, M. H., "Nonisentropic Unsteady Transonic Small Disturbance Theory," AIAA Paper 85-0600, April 1985.

[32]Landahl, M. T., *Unsteady Transonic Flow*, Pergamon, Oxford, UK, 1961.

Author Index

PROGRESS IN ASTRONAUTICS AND AERONAUTICS SERIES VOLUMES

VOLUME TITLE/EDITORS

*1. **Solid Propellant Rocket Research** (1960)
Martin Summerfield
Princeton University

*2. **Liquid Rockets and Propellants** (1960)
Loren E. Bollinger
The Ohio State University
Martin Goldsmith
The Rand Corporation
Alexis W. Lemmon Jr.
Battelle Memorial Institute

*3. **Energy Conversion for Space Power** (1961)
Nathan W. Snyder
Institute for Defense Analyses

*4. **Space Power Systems** (1961)
Nathan W. Snyder
Institute for Defense Analyses

*5. **Electrostatic Propulsion** (1961)
David B. Langmuir
Space Technology Laboratories, Inc.
Ernst Stuhlinger
NASA George C. Marshall Space Flight Center
J.M. Sellen Jr.
Space Technology Laboratories, Inc.

*6. **Detonation and Two-Phase Flow** (1962)
S.S. Penner
California Institute of Technology
F.A. Williams
Harvard University

*7. **Hypersonic Flow Research** (1962)
Frederick R. Riddell
AVCO Corporation

*8. **Guidance and Control** (1962)
Robert E. Roberson
Consultant
James S. Farrior
Lockheed Missiles and Space Company

*9. **Electric Propulsion Development** (1963)
Ernst Stuhlinger
NASA George C. Marshall Space Flight Center

*10. **Technology of Lunar Exploration** (1963)
Clifford I. Cummings
Harold R. Lawrence
Jet Propulsion Laboratory

*11. **Power Systems for Space Flight** (1963)
Morris A. Zipkin
Russell N. Edwards
General Electric Company

12. **Ionization in High-Temperature Gases** (1963)
Kurt E. Shuler, Editor
National Bureau of Standards
John B. Fenn, Associate Editor
Princeton University

*13. **Guidance and Control—II** (1964)
Robert C. Langford
General Precision Inc.
Charles J. Mundo
Institute of Naval Studies

*14. **Celestial Mechanics and Astrodynamics** (1964)
Victor G. Szebehely
Yale University Observatory

*15. **Heterogeneous Combustion** (1964)
Hans G. Wolfhard
Institute for Defense Analyses
Irvin Glassman
Princeton University
Leon Green Jr.
Air Force Systems Command

16. **Space Power Systems Engineering** (1966)
George C. Szego
Institute for Defense Analyses
J. Edward Taylor
TRW Inc.

17. **Methods in Astrodynamics and Celestial Mechanics** (1966)
Raynor L. Duncombe
U.S. Naval Observatory
Victor G. Szebehely
Yale University Observatory

18. **Thermophysics and Temperature Control of Spacecraft and Entry Vehicles** (1966)
Gerhard B. Heller
NASA George C. Marshall Space Flight Center

*19. **Communication Satellite Systems Technology** (1966)
Richard B. Marsten
Radio Corporation of America

*Out of print.

20. Thermophysics of Spacecraft and Planetary Bodies: Radiation Properties of Solids and the Electromagnetic Radiation Environment in Space (1967)
Gerhard B. Heller
NASA George C. Marshall Space Flight Center

21. Thermal Design Principles of Spacecraft and Entry Bodies (1969)
Jerry T. Bevans
TRW Systems

22. Stratospheric Circulation (1969)
Willis L. Webb
Atmospheric Sciences Laboratory, White Sands, and University of Texas at El Paso

23. Thermophysics: Applications to Thermal Design of Spacecraft (1970)
Jerry T. Bevans
TRW Systems

24. Heat Transfer and Spacecraft Thermal Control (1971)
John W. Lucas
Jet Propulsion Laboratory

25. Communication Satellites for the 70's: Technology (1971)
Nathaniel E. Feldman
The Rand Corporation
Charles M. Kelly
The Aerospace Corporation

26. Communication Satellites for the 70's: Systems (1971)
Nathaniel E. Feldman
The Rand Corporation
Charles M. Kelly
The Aerospace Corporation

27. Thermospheric Circulation (1972)
Willis L. Webb
Atmospheric Sciences Laboratory, White Sands, and University of Texas at El Paso

28. Thermal Characteristics of the Moon (1972)
John W. Lucas
Jet Propulsion Laboratory

29. Fundamentals of Spacecraft Thermal Design (1972)
John W. Lucas
Jet Propulsion Laboratory

30. Solar Activity Observations and Predictions (1972)
Patrick S. McIntosh
Murray Dryer
Environmental Research Laboratories, National Oceanic and Atmospheric Administration

31. Thermal Control and Radiation (1973)
Chang-Lin Tien
University of California at Berkeley

32. Communications Satellite Systems (1974)
P.L. Bargellini
COMSAT Laboratories

33. Communications Satellite Technology (1974)
P.L. Bargellini
COMSAT Laboratories

34. Instrumentation for Airbreathing Propulsion (1974)
Allen E. Fuhs
Naval Postgraduate School
Marshall Kingery
Arnold Engineering Development Center

35. Thermophysics and Spacecraft Thermal Control (1974)
Robert G. Hering
University of Iowa

36. Thermal Pollution Analysis (1975)
Joseph A. Schetz
Virginia Polytechnic Institute

37. Aeroacoustics: Jet and Combustion Noise; Duct Acoustics (1975)
Henry T. Nagamatsu, Editor
General Electric Research and Development Center
Jack V. O'Keefe, Associate Editor
The Boeing Company
Ira R. Schwartz, Associate Editor
NASA Ames Research Center

38. Aeroacoustics: Fan, STOL, and Boundary Layer Noise; Sonic Boom; Aeroacoustic Instrumentation (1975)
Henry T. Nagamatsu, Editor
General Electric Research and Development Center
Jack V. O'Keefe, Associate Editor
The Boeing Company
Ira R. Schwartz, Associate Editor
NASA Ames Research Center

39. Heat Transfer with Thermal Control Applications (1975)
M. Michael Yovanovich
University of Waterloo

40. Aerodynamics of Base Combustion (1976)
S.N.B. Murthy, Editor
Purdue University
J.R. Osborn, Associate Editor
Purdue University
A.W. Barrows
J.R. Ward, Associate Editors
Ballistics Research Laboratories

41. Communications Satellite Developments: Systems (1976)
Gilbert E. LaVean
Defense Communications Agency
William G. Schmidt
CML Satellite Corporation

42. Communications Satellite Developments: Technology (1976)
William G. Schmidt
CML Satellite Corporation
Gilbert E. LaVean
Defense Communications Agency

43. Aeroacoustics: Jet Noise, Combustion and Core Engine Noise (1976)
Ira R. Schwartz, Editor
NASA Ames Research Center
Henry T. Nagamatsu, Associate Editor
General Electric Research and Development Center
Warren C. Strahle, Associate Editor
Georgia Institute of Technology

44. Aeroacoustics: Fan Noise and Control; Duct Acoustics; Rotor Noise (1976)
Ira R. Schwartz, Editor
NASA Ames Research Center
Henry T. Nagamatsu, Associate Editor
General Electric Research and Development Center
Warren C. Strahle, Associate Editor
Georgia Institute of Technology

45. Aeroacoustics: STOL Noise; Airframe and Airfoil Noise (1976)
Ira R. Schwartz, Editor
NASA Ames Research Center
Henry T. Nagamatsu, Associate Editor
General Electric Research and Development Center
Warren C. Strahle, Associate Editor
Georgia Institute of Technology

46. Aeroacoustics: Acoustic Wave Propagation; Aircraft Noise Prediction; Aeroacoustic Instrumentation (1976)
Ira R. Schwartz, Editor
NASA Ames Research Center
Henry T. Nagamatsu, Associate Editor
General Electric Research and Development Center
Warren C. Strahle, Associate Editor
Georgia Institute of Technology

47. Spacecraft Charging by Magnetospheric Plasmas (1976)
Alan Rosen
TRW Inc.

48. Scientific Investigations on the Skylab Satellite (1976)
Marion I. Kent
Ernst Stuhlinger
NASA George C. Marshall Space Flight Center
Shi-Tsan Wu
The University of Alabama

49. Radiative Transfer and Thermal Control (1976)
Allie M. Smith
ARO Inc.

50. Exploration of the Outer Solar System (1976)
Eugene W. Greenstadt
TRW Inc.
Murray Dryer
National Oceanic and Atmospheric Administration
Devrie S. Intriligator
University of Southern California

51. Rarefied Gas Dynamics, Parts I and II (two volumes) (1977)
J. Leith Potter
ARO Inc.

52. Materials Sciences in Space with Application to Space Processing (1977)
Leo Steg
General Electric Company

53. Experimental Diagnostics in Gas Phase Combustion Systems (1977)
Ben T. Zinn, Editor
Georgia Institute of Technology
Craig T. Bowman, Associate Editor
Stanford University
Daniel L. Hartley, Associate Editor
Sandia Laboratories
Edward W. Price, Associate Editor
Georgia Institute of Technology
James G. Skifstad, Associate Editor
Purdue University

54. Satellite Communications: Future Systems (1977)
David Jarett
TRW Inc.

55. Satellite Communications: Advanced Technologies (1977)
David Jarett
TRW Inc.

56. Thermophysics of Spacecraft and Outer Planet Entry Probes (1977)
Allie M. Smith
ARO Inc.

57. Space-Based Manufacturing from Nonterrestrial Materials (1977)
Gerard K. O'Neill, Editor
Princeton University
Brian O'Leary, Assistant Editor
Princeton University

58. Turbulent Combustion (1978)
Lawrence A. Kennedy
State University of New York at Buffalo

59. Aerodynamic Heating and Thermal Protection Systems (1978)
Leroy S. Fletcher
University of Virginia

60. Heat Transfer and Thermal Control Systems (1978)
Leroy S. Fletcher
University of Virginia

61. Radiation Energy Conversion in Space (1978)
Kenneth W. Billman
NASA Ames Research Center

62. Alternative Hydrocarbon Fuels: Combustion and Chemical Kinetics (1978)
Craig T. Bowman
Stanford University
Jorgen Birkeland
Department of Energy

63. Experimental Diagnostics in Combustion of Solids (1978)
Thomas L. Boggs
Naval Weapons Center
Ben T. Zinn
Georgia Institute of Technology

64. Outer Planet Entry Heating and Thermal Protection (1979)
Raymond Viskanta
Purdue University

65. Thermophysics and Thermal Control (1979)
Raymond Viskanta
Purdue University

66. Interior Ballistics of Guns (1979)
Herman Krier
University of Illinois at Urbana-Champaign
Martin Summerfield
New York University

***67. Remote Sensing of Earth from Space: Role of "Smart Sensors"** (1979)
Roger A. Breckenridge
NASA Langley Research Center

68. Injection and Mixing in Turbulent Flow (1980)
Joseph A. Schetz
Virginia Polytechnic Institute and State University

69. Entry Heating and Thermal Protection (1980)
Walter B. Olstad
NASA Headquarters

70. Heat Transfer, Thermal Control, and Heat Pipes (1980)
Walter B. Olstad
NASA Headquarters

71. Space Systems and Their Interactions with Earth's Space Environment (1980)
Henry B. Garrett
Charles P. Pike
Hanscom Air Force Base

72. Viscous Flow Drag Reduction (1980)
Gary R. Hough
Vought Advanced Technology Center

73. Combustion Experiments in a Zero-Gravity Laboratory (1981)
Thomas H. Cochran
NASA Lewis Research Center

74. Rarefied Gas Dynamics, Parts I and II (two volumes) (1981)
Sam S. Fisher
University of Virginia at Charlottesville

75. Gasdynamics of Detonations and Explosions (1981)
J.R. Bowen
University of Wisconsin at Madison
N. Manson
Université de Poitiers
A.K. Oppenheim
University of California at Berkeley
R.I. Soloukhin
Institute of Heat and Mass Transfer, BSSR Academy of Sciences

76. Combustion in Reactive Systems (1981)
J.R. Bowen
University of Wisconsin at Madison
N. Manson
Université de Poitiers
A.K. Oppenheim
University of California at Berkeley
R.I. Soloukhin
Institute of Heat and Mass Transfer, BSSR Academy of Sciences

77. Aerothermodynamics and Planetary Entry (1981)
A.L. Crosbie
University of Missouri-Rolla

78. Heat Transfer and Thermal Control (1981)
A.L. Crosbie
University of Missouri-Rolla

79. Electric Propulsion and Its Applications to Space Missions (1981)
Robert C. Finke
NASA Lewis Research Center

80. Aero-Optical Phenomena (1982)
Keith G. Gilbert
Leonard J. Otten
Air Force Weapons Laboratory

81. Transonic Aerodynamics (1982)
David Nixon
Nielsen Engineering & Research, Inc.

82. Thermophysics of Atmospheric Entry (1982)
T.E. Horton
The University of Mississippi

83. Spacecraft Radiative Transfer and Temperature Control (1982)
T.E. Horton
The University of Mississippi

84. Liquid-Metal Flows and Magnetohydrodynamics (1983)
H. Branover
Ben-Gurion University of the Negev
P.S. Lykoudis
Purdue University
A. Yakhot
Ben-Gurion University of the Negev

85. Entry Vehicle Heating and Thermal Protection Systems: Space Shuttle, Solar Starprobe, Jupiter Galileo Probe (1983)
Paul E. Bauer
McDonnell Douglas Astronautics Company
Howard E. Collicott
The Boeing Company

86. Spacecraft Thermal Control, Design, and Operation (1983)
Howard E. Collicott
The Boeing Company
Paul E. Bauer
McDonnell Douglas Astronautics Company

87. Shock Waves, Explosions, and Detonations (1983)
J.R. Bowen
University of Washington
N. Manson
Université de Poitiers
A.K. Oppenheim
University of California at Berkeley
R.I. Soloukhin
Institute of Heat and Mass Transfer, BSSR Academy of Sciences

88. Flames, Lasers, and Reactive Systems (1983)
J.R. Bowen
University of Washington
N. Manson
Université de Poitiers
A.K. Oppenheim
University of California at Berkeley
R.I. Soloukhin
Institute of Heat and Mass Transfer, BSSR Academy of Sciences

89. Orbit-Raising and Maneuvering Propulsion: Research Status and Needs (1984)
Leonard H. Caveny
Air Force Office of Scientific Research

90. Fundamentals of Solid-Propellant Combustion (1984)
Kenneth K. Kuo
The Pennsylvania State University
Martin Summerfield
Princeton Combustion Research Laboratories, Inc.

91. Spacecraft Contamination: Sources and Prevention (1984)
J.A. Roux
The University of Mississippi
T.D. McCay
NASA Marshall Space Flight Center

92. Combustion Diagnostics by Nonintrusive Methods (1984)
T.D. McCay
NASA Marshall Space Flight Center
J.A. Roux
The University of Mississippi

93. The INTELSAT Global Satellite System (1984)
Joel Alper
COMSAT Corporation
Joseph Pelton
INTELSAT

94. Dynamics of Shock Waves, Explosions, and Detonations (1984)
J.R. Bowen
University of Washington
N. Manson
Université de Poitiers
A.K. Oppenheim
University of California
R.I. Soloukhin
Institute of Heat and Mass Transfer, BSSR Academy of Sciences

95. Dynamics of Flames and Reactive Systems (1984)
J.R. Bowen
University of Washington
N. Manson
Université de Poitiers
A.K. Oppenheim
University of California
R.I. Soloukhin
Institute of Heat and Mass Transfer, BSSR Academy of Sciences

96. Thermal Design of Aeroassisted Orbital Transfer Vehicles (1985)
H.F. Nelson
University of Missouri-Rolla

97. Monitoring Earth's Ocean, Land, and Atmosphere from Space – Sensors, Systems, and Applications (1985)
Abraham Schnapf
Aerospace Systems Engineering

98. Thrust and Drag: Its Prediction and Verification (1985)
Eugene E. Covert
Massachusetts Institute of Technology
C.R. James
Vought Corporation
William F. Kimzey
Sverdrup Technology AEDC Group
George K. Richey
U.S. Air Force
Eugene C. Rooney
U.S. Navy Department of Defense

99. Space Stations and Space Platforms — Concepts, Design, Infrastructure, and Uses (1985)
Ivan Bekey
Daniel Herman
NASA Headquarters

100. Single- and Multi-Phase Flows in an Electromagnetic Field Energy, Metallurgical, and Solar Applications (1985)
Herman Branover
Ben-Gurion University of the Negev
Paul S. Lykoudis
Purdue University
Michael Mond
Ben-Gurion University of the Negev

101. MHD Energy Conversion: Physiotechnical Problems (1986)
V.A. Kirillin
A.E. Sheyndlin
Soviet Academy of Sciences

102. Numerical Methods for Engine-Airframe Integration (1986)
S.N.B. Murthy
Purdue University
Gerald C. Paynter
Boeing Airplane Company

103. Thermophysical Aspects of Re-Entry Flows (1986)
James N. Moss
NASA Langley Research Center
Carl D. Scott
NASA Johnson Space Center

104. Tactical Missile Aerodynamics (1986)
M.J. Hemsch
PRC Kentron, Inc.
J.N. Nielsen
NASA Ames Research Center

105. Dynamics of Reactive Systems Part I: Flames and Configurations; Part II: Modeling and Heterogeneous Combustion (1988)
J.R. Bowen
University of Washington
J.-C. Leyer
Université de Poitiers
R.I. Soloukhin
Institute of Heat and Mass Transfer, BSSR Academy of Sciences

106. Dynamics of Explosions (1986)
J.R. Bowen
University of Washington
J.-C. Leyer
Université de Poitiers
R.I. Soloukhin
Institute of Heat and Mass Transfer, BSSR Academy of Sciences

107. Spacecraft Dielectric Material Properties and Spacecraft Charging (1986)
A.R. Frederickson
U.S. Air Force Rome Air Development Center
D.B. Cotts
SRI International
J.A. Wall
U.S. Air Force Rome Air Development Center
F.L. Bouquet
Jet Propulsion Laboratory, California Institute of Technology

108. Opportunities for Academic Research in a Low-Gravity Environment (1986)
George A. Hazelrigg
National Science Foundation
Joseph M. Reynolds
Louisiana State University

109. Gun Propulsion Technology (1988)
Ludwig Stiefel
U.S. Army Armament Research, Development and Engineering Center

110. Commercial Opportunities in Space (1988)
F. Shahrokhi
K.E. Harwell
University of Tennessee Space Institute
C.C. Chao
National Cheng Kung University

111. Liquid-Metal Flows: Magnetohydrodynamics and Applications (1988)
Herman Branover, Michael Mond, and Yeshajahu Unger
Ben-Gurion University of the Negev

112. Current Trends in Turbulence Research (1988)
Herman Branover, Michael Mond, and Yeshajahu Unger
Ben-Gurion University of the Negev

113. Dynamics of Reactive Systems Part I: Flames; Part II: Heterogeneous Combustion and Applications (1988)
A.L. Kuhl
R & D Associates
J.R. Bowen
University of Washington
J.-C. Leyer
Université de Poitiers
A. Borisov
USSR Academy of Sciences

114. Dynamics of Explosions (1988)
A.L. Kuhl
R & D Associates
J.R. Bowen
University of Washington
J.-C. Leyer
Université de Poitiers
A. Borisov
USSR Academy of Sciences

115. Machine Intelligence and Autonomy for Aerospace (1988)
E. Heer
Heer Associates, Inc.
H. Lum
NASA Ames Research Center

116. Rarefied Gas Dynamics: Space-Related Studies (1989)
E.P. Muntz
University of Southern California
D.P. Weaver
U.S. Air Force Astronautics Laboratory (AFSC)
D.H. Campbell
The University of Dayton Research Institute

117. Rarefied Gas Dynamics: Physical Phenomena (1989)
E.P. Muntz
University of Southern California
D.P. Weaver
U.S. Air Force Astronautics Laboratory (AFSC)
D.H. Campbell
The University of Dayton Research Institute

118. Rarefied Gas Dynamics: Theoretical and Computational Techniques (1989)
E.P. Muntz
University of Southern California
D.P. Weaver
U.S. Air Force Astronautics Laboratory (AFSC)
D.H. Campbell
The University of Dayton Research Institute

119. Test and Evaluation of the Tactical Missile (1989)
Emil J. Eichblatt Jr.
Pacific Missile Test Center

120. Unsteady Transonic Aerodynamics (1989)
David Nixon
Nielsen Engineering & Research, Inc.

(Other Volumes are planned.)